D1609603

PRAISE FOR *COMMUNITY-SCALE COMPOSTING SYSTEMS*

"With *Community-Scale Composting Systems*, James McSweeney makes a powerful contribution to the growth of our organics recycling movement. Easy to read, the book is chock full of practical information folks can use to get started or to grow their operation. As a card-carrying member of what James calls the Cult of Compost, I highly recommend it."

—CARY OSHINS, associate director, US Composting Council

"*Community-Scale Composting Systems* is far and away the best reference on composting at the community scale. Indeed, it is one of the best references about composting altogether. The book's range is impressive, offering current and practical advice to practitioners of many applications of composting, from small, on-site systems to larger systems that border on commercial scale. Community-scale and on-site composting are often the most logical and most sustainable options for managing organic residuals, especially food waste. Mr. McSweeney comprehensively covers this important recycling niche within this useful and enjoyable book."

—ROBERT RYNK, professor, SUNY Cobleskill

"If agriculture is to transition away from a dependence on petroleum-based fertilizers to local fertility sources in a meaningful way, then we will need to relearn the art of using our neighbors' leaves, grasses, manures, and other 'wastes' to fuel our farms. This book shows how to do just that on a community scale. It is the most urgent and important work I know of, and McSweeney explains it better than anyone."

—BEN HARTMAN, author of *The Lean Farm* and *The Lean Farm Guide to Growing Vegetables*

"Too often we hear that food waste cannot be diverted from landfills and incinerators due to lack of composting facilities. *Community-Scale Composting Systems* debunks that myth by providing a much-needed roadmap for how to compost successfully at a community scale. Micro-composting is now the fastest growing segment of the composting industry and will undoubtedly grow even faster with this resource."

—BRENDA PLATT, director, Composting for Community Initiative, Institute for Local Self-Reliance

"James has written an almost encyclopedic handbook to address the needs of small- to medium-sized community compost efforts. This book provides a good and detailed consideration of the broad range of possible approaches and challenges to community aspirations to feed the soil while mitigating wastefulness. If you are thinking of rescuing wasted food in your community, you would be well advised to start by reading this book."

—KARL HAMMER, founder and president, Vermont Compost Company

"*Community-Scale Composting Systems* is a well-written and organized practical guide to the diverse systems used for food waste composting. It presents community collection program examples; methods for the design of small- to large-scale composting facilities; and the properties used to characterize feedstocks, the composting process, and finished composts. Detailed case studies are used to describe creative and successful circular food systems. It is essential reading for 'rotstars' committed to understanding and mastering the complexities of organics recycling."

—FREDERICK C. MICHEL, JR., professor of biosystems engineering, The Ohio State University

"As composting is increasingly being adopted by neighborhoods, farms, community gardens, schools, businesses, and entrepreneurs, James McSweeney's timely book provides comprehensive information to help them effectively recycle organic waste into soil amendments that can be used to grow more food. McSweeney's in-depth guide provides the answers needed to design and manage successful composting operations."

—RHONDA SHERMAN, author of *The Worm Farmer's Handbook*

"*Community-Scale Composting Systems* taps into James McSweeney's experience as both a community composting practitioner and educator. His guidance and how-to tips are invaluable to newcomers and old-timers alike."

—NORA GOLDSTEIN, editor, *BioCycle*

COMMUNITY-SCALE COMPOSTING SYSTEMS

A Comprehensive Practical Guide for Closing the Food System Loop and Solving Our Waste Crisis

James McSweeney

Foreword by Marguerite Manela

Chelsea Green Publishing
White River Junction, Vermont
London, UK

Unless otherwise noted, all photographs and illustrations by James McSweeney.

Portions of this book were adapted in part from the work of Highfields Center for Composting courtesy of the Vermont Sustainable Jobs Fund.

Portions of this book were adapted in part from the work of the Institute for Local Self-Reliance.

Lyrics from the song "Ashes of American Flags" on page 400 written by Jeff Tweedy and Jay Bennett. Published by Words Ampersand Music and You Want a Piece of This Music, administered by BMG Rights Management US. Reprinted with permission.

Cover photographs, *clockwise from top left*, used courtesy of Compost Pedallers, taken by and used courtesy of Scot Nelson, and taken by James McSweeney and used with permission from the Brooklyn Grange Rooftop Farm.

Acquiring Editor: Makenna Goodman
Developmental Editor: Fern Bradley
Editor: Michael Metivier
Editorial Assistant: Nick Kaye
Copy Editor: Laura Jorstad
Proofreader: Katherine R. Kiger
Indexer: Linda Hallinger
Designer: Melissa Jacobson

Printed in the United States of America.
First printing January 2019.
10 9 8 7 6 5 4 3 2 1   19 20 21 22 23

Our Commitment to Green Publishing

Chelsea Green sees publishing as a tool for cultural change and ecological stewardship. We strive to align our book manufacturing practices with our editorial mission and to reduce the impact of our business enterprise in the environment. We print our books and catalogs on chlorine-free recycled paper, using vegetable-based inks whenever possible. This book may cost slightly more because it was printed on paper that contains recycled fiber, and we hope you'll agree that it's worth it. The text paper used for *Community-Scale Composting Systems* was supplied by LSC Communications and contains at least 10% postconsumer recycled fiber.

Library of Congress Cataloging-in-Publication Data
Names: McSweeney, James, 1980- author.
Title: Community-scale composting systems : a comprehensive practical guide for closing the food system loop and solving our waste crisis / James McSweeney.
Description: White River Junction, Vermont : Chelsea Green Publishing, [2019] | Includes bibliographical references and index.
Identifiers: LCCN 2018035586| ISBN 9781603586542 (hardcover) | ISBN 9781603586559 (ebook)
Subjects: LCSH: Compost.
Classification: LCC TD796.5 .M38 2019 | DDC 631.8/75—dc23
LC record available at https://lccn.loc.gov/2018035586

Chelsea Green Publishing
85 North Main Street, Suite 120
White River Junction, VT 05001
(802) 295-6300
www.chelseagreen.com

MIX
Paper from responsible sources
FSC® C132124

Dedicated to Miles, Mary, and Rowan
May your roots grow deep
To fill the sky

And to Amanda
For healthy dirt makes
healthy mamas and healthy babies!

When to Make the Compost

Begin your compost heap now; do not delay; let every spare moment be put in in the woods raking up leaves or in the swamps pulling up muck.

—George Washington Carver
"What Shall We Do for Fertilizers Next Year?"
November 1916

CONTENTS

FOREWORD

In the summer of 2011, I moved to New York City, convinced I wouldn't stay more than a year or two. I couldn't survive without trees and mountains, and obviously there were none of those in the city! After my first long winter, a sense of emptiness hit me hard. When the following spring finally emerged, I knew I needed to become more involved in my community and began seeking volunteer opportunities. When I learned about the Western Queens Compost Initiative (WQCI), a small group of community composters who were collecting food scraps from the public and composting them in several locations throughout Western Queens, I signed up to volunteer. The next weekend I joined them in a park in my neighborhood. They handed me a crowbar, and I spent the day liberating street trees from the bricks that restricted their roots and mixing finished compost into their small allotments of starved soil. Dirty, sweaty, tired, and grinning from ear to ear, I was hooked. My initial assessment proved wrong; New York City had plenty of trees, and there were in fact mountains—mountains of food scraps ready to be transformed into compost. And over the past seven years, I have witnessed the growth of an incredible movement doing just that.

For the next year and a half after my first experience with WQCI, I volunteered almost every weekend. I spoke with hundreds of people who dropped off their food scraps with us at the farmers market and helped engage new people in our work. I joined the staff and other volunteers to build windrows by hand under the Queensboro Bridge, processing several tons of food scraps at a time. These came from the growing network of food scrap drop-off sites that were emerging throughout the city, started by various community groups and organizations and propelled with funding from the NYC Department of Sanitation (DSNY).

The current decade has seen a marked increase in the number of community compost sites in New York City, and in the technical knowledge and skill level of the composters who are establishing and managing the sites that make up this vibrant network. The NYC Compost Project (NYCCP), a program created by DSNY in 1993 to promote composting through outreach and education, has played a key role in laying the groundwork for and supporting this growth over the past 25 years. I have worked with the NYCCP in several different positions since 2012, initially as a volunteer, then managing the NYCCP team at Queens Botanical Garden, one of seven teams throughout the city. In 2015 I joined DSNY as Senior Manager of Community Composting and Compost Distribution, where I've been tasked with shaping the current and future role and direction of community composting within the rapidly changing landscape of waste management in New York City.

I first met James McSweeney in 2015 at a BioCycle Conference in Boston, Massachusetts. We both attended a workshop called Food for People, Food for Soil. The day-long workshop brought together a group of individuals working to transform our food system into one that produces abundant food and little to no waste while also improving our soil, one of our most precious resources. James presented on his work in Vermont helping farms build composting operations that incorporate animals and animal wastes into the process, describing the methods he used to help clarify his clients' goals and to design operations tailored to each individual farm and organization.

From the start it was clear to me that James balanced a strong passion for composting and community building with a practical, intelligent, and organized approach to project management. I was impressed by James's level of industry knowledge, professionalism, and curiosity. At a moment when my own career in compost project management was shifting its focus from one county to an entire city of over 8 million residents, it was helpful to meet someone who understands that while anyone can make compost, there is not a one-size-fits-all model that works for everyone. Furthermore, he recognizes that community-scale composting presents an opportunity to expand the reach and positive impact of composting. In *Community-Scale Composting Systems*, James translates this understanding into a serious, technical guide that will support the existing community-scale composting industry while inspiring more to begin learning and composting. His years of researching, building, and operating different sizes and models of compost systems have led to a book that carefully and thoroughly describes the many considerations and choices that a compost operator needs to make when designing and developing a composting operation at any scale. The book guides the reader through the steps involved in each decision-making process, and it brings together a wide array of examples from around the country, showcasing the variety of models in the community composting movement.

Over the past three decades, the NYC Compost Project has evolved and grown in size and scope while maintaining a consistent mission: to reduce waste by providing New Yorkers with the knowledge, skills, and opportunities they need to produce and use compost locally. In the nineties the program focused on backyard-scale composting and yard waste reduction. In 2000 the NYCCP developed the NYC Master Composter Certificate Course, and has since trained and supported hundreds of volunteers to conduct composting outreach and education in their own neighborhoods and communities. These volunteers help us reach entirely new audiences, and their work exponentially increases the number of people receiving information about composting and waste reduction in New York City. Building on the work of several standout Master Composters, including those who started WQCI, the NYCCP expanded its programming to include public food scrap drop-off sites and a network of community compost sites that have scaled up and collectively diverted over 14,000,000 pounds of food scraps from landfills in the past seven years.

Community-scale composting provides opportunities for residents to get involved in ways that have implications far beyond reducing waste. These sites are social hubs that spark curiosity and build community. People who live and work in these spaces benefit from exercise, inspired conversation, and friendships, and develop skills in construction, problem solving, community organizing, and civic engagement.

Community composters are constantly experimenting and building off of one another's ideas. As a result, the scale of the network, the variety of potential solutions, the number of people involved, and the skills of those involved are constantly growing. Each new model adds brilliance to the movement. Operating well-managed, carefully monitored compost sites at any and all scales is our best tool for promoting the growth of this key industry. Community-scale compost sites are distinguished in their ability to welcome people in to visit, see, and learn about the process of turning food scraps and yard waste into a product that helps maintain the life we need in our soils. These sites have an opportunity to engage more people in the conversation about composting, soil health, and reducing waste. With *Community-Scale Composting Systems*, James has elevated this group of sites to their proper place in this industry. This book will serve as a tool for a wide range and growing number of composters for years to come.

—MARGUERITE MANELA
Queens, New York

ACKNOWLEDGMENTS

First, I feel an enormous gratitude to the mentors and teachers who schooled me in the science and practice of sustainable agriculture early on. Your knowledge, ideas, and passion are a continuous resource and inspiration. Thanks: Andrew Millison for giving me my permaculture glasses; Tim Crews—without you, I doubt I would have made a career out of gathering nutrients in space and time; Bill Litzinger, for challenging me to question my assumptions, always; Gary Seals, Hendrikus Schraven, and Elaine Ingham, for teaching me when to go in with a knife and the awesome power of good compost; Chris Jaeb, for moving me to Hawaii, surrounding me with extremely knowledgeable people, and giving me a tractor, really denuded soils, and no choice but to learn how to make better compost.

As a composting technical mentor and food systems thought leader, no one has been more influential to me than Tom Gilbert (Black Dirt Farm). You have shared so much practical knowledge, and you gave me the opportunity to learn by doing. So much of the subject matter in this book comes from the work that was done under your leadership at the Highfields Center for Composting, and for that I am greatly indebted to you and the entire Highfields clan: Chris Duff, June Van Houten, Josh Kelly, Tyler Buswell, Karen Wiseman, Alex Utevsky, Noah Fishman, Maia Hanson, Jason Bernier, Jenn Davis, Kim Mercer; to Highfields's founder Tod Delaricheliere; and to the many dedicated board members who served over the years. Highfields provided visionary and practical leadership in developing regenerative community food systems in Vermont and nationally.

A huge thanks is also owed to the forward-thinking people and organizations that funded our work at Highfields, some of which was adapted for use in this book: USDA SARE and Rural Development, Block Family Foundation, High Meadows Fund, John Merck Fund, Ben & Jerry's Foundation, Vermont Agency of Natural Resources, Vermont Agency of Agriculture Food & Markets, Holborn Foundation, Claneil Foundation, Grassroots Environmental Fund, Canaday Family Foundation, and the Vermont Sustainable Jobs Fund. Enormous thanks to the Vermont Sustainable Jobs Fund for granting me permission to adapt from the curated work of Highfields, as well as to the Vermont Food Cycle Coalition and the Compost Association of Vermont. Their collective work continues to drive community food scrap recycling in my home state of Vermont, and is an aspirational model for sustainable food systems thinking.

Two people have been at the forefront of the national community composting movement, providing the foundational glue (and humus-like compounds) through their sponsorship and coordination of the national gatherings, and through their thought leadership and seemingly tireless work on so many fronts. These are Brenda Platt, co-director of the Institute for Local Self-Reliance (ILSR), and Nora Goldstein, editor of *BioCycle* magazine. In my admiration of and gratitude to both, I know that I am far from alone. I haven't counted, but *BioCycle* will no doubt be the most cited source in my endnotes. Brenda has been an enormous collaborator in two recent resources—*Growing Local Fertility: A Guide to Community Composting* and *Micro Composting: A Guide to On-Site and Small Scale Food Scrap Composting Systems*—parts of which were adapted for this book with the institute's generous permission. ILSR's contribution to these reports was made possible by the 11th Hour Project and other donors.

There are several organizations that have provided me with gainful employment as well as learning opportunities and partnership through the duration of this book's authorship. These include the Vermont Agency of Natural Resources Department of Environmental Conservation, Massachusetts

RecyclingWorks and the Center for Ecological Technology, the Institute for Local Self-Reliance, and Recover Green Roofs.

The concept for this book was initiated by Chelsea Green Publishing. An editor there, Makenna Goodman, reached out around the time that myself and others of the Highfields crew were transitioning on to other things. It was clearly too good an idea to let go of, because here it is. Makenna left for a prolonged maternity leave, and I worked with Fern Bradley for the duration of the manuscript phase. She was my stalwart guide through the matrix that is the development of a book like this. Michael Metivier took up where Fern left off and never missed a beat. He and his editorial assistant, Nick Kaye, have shaped this book into the readable and user-friendly form that it is today. It's been an honor and a privilege working with them and the rest of the visionary team at Chelsea Green Publishing.

There was an enormous volume of research involved in creating this book, and I could not have done it without my intrepid research team: Madeline Seibert, Deanna Nappi, Janice McPhillips, and Elspeth Pennell. Thank you for your contributions, ideas, and camaraderie.

Several names came up again and again in our research: people who have contributed greatly to the modern body of work on composting, and to my own understanding of the science and technical aspects of composting. They include Bob Rynk (and the *On-Farm Composting Handbook*), Will Brinton (and Woods End Laboratories), Roger Haug, Eliot Epstein, and Nora Goldstein (and *BioCycle*).

To the many composters whom I've had the honor to work with and/or learn from: This book is really dedicated to you. There are several folks I need to mention: Eric Paris, Richard Hudak, Karl Hammer, Chris Duff, Dan Goossen, Trevor Mance, Matt Proft, Marissa DeDominicis, Charlie Bayrer, Leah Reathorford, Shakara Petteway, Erik Martig, Lynn Fang, Brian Jerose, and Gaelan Brown. Thank you all for your tremendous contributions to the book. Your unique stories and expertise added enormous depth.

For those of you who knew David Buckel, it may come as a surprise to see his contribution here so soon after his death. I did not know David as well as some, but I know that his dedication to justice, to our planet, and to community will long be remembered both within and far beyond the composting community. I am grateful to have known David and to be able to include some small part of his story in the pages of this book.

Finally, endless thanks to my family and friends for your support and encouragement over the last four years. Thanks to Steph Miller for lovingly and wisely raising up our young 'uns; to Chani and Seth for being all-around inspiring human beings, and for sharing your passions for food, travel, nature, and living a life of integrity and quality; to Dad, for passing on your creative mind, for showing me the power of ideas, and for giving me the tools (literally) to put ideas into action; to Mom, for being an all-around incredible model of human integrity and spirit; to my wife, Amanda, for your tremendous support and companionship, both spoken and unspoken; and to you, my children, and Mary Margaret, thanks for making room for organic matter in various states of degradation in our small urban dwelling for all these years and for all of the ways, both large and small, that you contributed to making this book possible.

INTRODUCTION

Recycling Organics at the Community Scale

The book in your hands is designed for practitioners of community-scale composting. It was created as a technical resource for farmers, designers, service providers, and organics recycling entrepreneurs and advocates of all types, with a focus on developing small- and even micro-scale infrastructure that can enable communities to return sustenance to soils in their local food system. The main scope of the book is dedicated to compost system options and design, from basic sizing and layout to advanced techniques such as aerated static pile composting. Management techniques and operational considerations are also covered, including testing, developing analytically based compost recipes, and system-specific best management practices. As a consultant working with composters in the development and start-up phase, I have found that end uses and market planning—even when applied to small-scale composting scenarios—is extremely useful, and so the book concludes with that subject.

As you will see, community-scale composting is coming of age, marked by a profusion of new interest and agri-preneurial composting ventures. Great respect is owed to its early practitioners and thought leaders, those who have remained unshaken in their vision of an ecologically integral closed-loop food system. Thanks to them, there are an abundance of models and experiences to build from, and build we have. Throughout this book we'll invite several of these practitioners to give us a tour of their operations, so you can better understand what makes them tick.

But before we start, what is this movement we're calling community composting? *Community composting* has been labeled a buzzword,[1] and I think it has truly earned this status. Take two things, composting and community—each held in high esteem by the eco-conscious of today, and two things that have never been more needed on our planet—and put them together. This is why we have a movement on our hands.

Although the main focus of this book, and of community composters, is on growing the movement through practical implementation, there is an ongoing discussion about how we as a movement identify ourselves. Much of this discussion centers on the word *community* and how it relates to composters—or rather, how composters relate to their communities. Early on in writing, I chose to use the term *community-scale* as a way to define the book's scope, rather than *community-based* composting, which is open to much broader interpretation. The ideas behind community-based development are quite comprehensive with a whole history, philosophy, and methodology of their own. For example, Asset Based Community Development (ABCD) is a methodology for engaging with and building from the existing assets of communities.[2] In many ways, truly community-*based* composting is the gold standard of community composting, ideals I myself believe in and strive for, but on which I am by no means an expert. For this reason, while the book is designed to be useful to those who identify as community-based composters, I deliberately chose to not use the term for the book.

Scale, on the other hand, is technical and quantifiable in nature, areas where I have significant expertise to share as a consultant and designer. Since this is a technical manual, *community-scale composting* seemed to appropriately fit the underlying theme of the book.

So what is a community-scale composting system? Inherent in the concept is the idea that the system has the capacity to capture and recycle a community's organic waste stream. In reality most community-scale composting systems target a specific portion of the waste stream, usually the most local, and usually the food scraps and yard waste of the community, as opposed to human or animal wastes. There is a whole methodology to identifying, estimating, and targeting organics streams that is outlined in chapters 4, 5, and 12. Community-scale composting may involve one main composter who grows to meet the needs of the community as program participation develops or may involve a series of partner composters who collectively service the community. Either way, inherent in this scale of composting is infrastructure that is simple, diverse, and dispersed—principles that unite this movement with other modalities of ecologically modeled thinking.

Community-Scale Composting

The fundamental building blocks of composting systems are the same regardless of scale.* In every system organic refuse is *generated*, *collected*, and then *composted*. These three basic components are necessary whether the source comprises a single residence composting food scraps in the backyard or all of the organics generated by a city. Everyone on the planet generates organic "wastes," just as we all eat and breathe. A 2014 estimate by the US Environmental Protection Agency reports that 38.4 million tons of food scraps are generated in the United States each year, with only 5.1 percent recovered and recycled.[3] In that wasted material is embodied tremendous resource potential. Naturally, an enormous diversity of systems to capture, transport, and recycle organics has developed from this resource niche. But composting wouldn't exist without the fourth fundamental component of these systems, the *end uses*. These include a wide range of compost applications and markets, as well as other forms of soil fertility, animal feed, and energy.

The subject of this book is organics recycling systems composed of these four basic components, handling organics at volumes above the scale of a single home and below the scale of a whole large city—in other words, *community-scale*. When planned appropriately, community-scale composting systems can handle the organics generated by a specific population, neighborhood, or region. Across the United States composters operating at this scale are emerging at a growing pace, as is an awareness of the benefits and services provided by composting. In my experience this sector of composters is typically focused on producing a very high-grade end product. Small-batch, horticultural, boutique, organic, bicycle-powered, artisan, community-based: There is no end to the ways in which these producers can be classified. What is increasingly apparent from this boom is that (1) there remains an abundance of waste; (2) awareness of the problem is growing; and (3) people are actively intervening to build the systems to collect, recycle, and reuse available organic resources in their communities.

The rapid growth and maturation of composting has led to a healthy discussion about what a functional and efficient organics recycling infrastructure can and should look like. One result of this discussion has been the solidification of a self-identifying segment of composters referring to themselves as "community

* Throughout this book, I use the word *composting* to refer generally to the recycling of organic materials. Although at times it is used to refer specifically to composting systems and processes, when speaking generally it can also refer to other methods of organics recycling.

composters." Appropriately, the concept of community composting seems to have developed organically and with no single particular origin. Rather, organizations, projects, and even news articles appear to have taken on the term for its succinctness, though with their own slightly varying definitions. I first started using the term to describe the services we provided while working at the Highfields Center for Composting. At first I found it hard to define exactly what we were doing, but clearly it involved composting, community, education, and infrastructure. Over time I learned to describe what we did as "supporting the development of community composting programs," and eventually our motto became "your partner in community composting." It wasn't until the 2012 US Composting Council Conference in Orlando, Florida, that the presence of a plethora of similarly minded practitioners around the United States helped the concept truly come into focus.

I wasn't the only one whose eyes were opened at that conference. Within the year conference calls and planning for the first national convergence of community composters were ongoing. In the fall of 2013, the composting industry's leading publication, *BioCycle*, and the Institute for Local Self Reliance (ILSR) hosted the first Cultivating Community Composting Forum in Columbus, Ohio, convergence that has continued annually.

In 2014 Highfields Center for Composting (HCC) and ILSR wrote the first comprehensive guide on the subject, *Growing Local Fertility: A Guide to Community Composting* (which Brenda Platt of ILSR, Jenn Davis, and I co-authored, and from which portions of this book are adapted). We faced the question of defining community composting for both ourselves and our audience, which took a great deal of time and thought. We decided that it was critical to involve other community composting practitioners and advocates in the process and reached out to several of the players that were involved in the development of the national community composting forum for their input. The team also agreed that the development of *defining principles*, rather than a set of exclusive criteria, would have the greatest long-term impact on the culture of composting as a whole. We wanted the vision for community composting to feel inclusive to the wide array of practitioners who identify with the term, and ideally to act as a standard of social and environmental ethics toward which anyone in the organics recycling sector could strive.

The team arrived at the following set of Core Community Composting Principles:

1. **Resources recovered.** Waste is reduced; food scraps and other organic materials are diverted from disposal and composted.
2. **Locally based and closed loop.** Organic materials are a community asset and are generated and recycled into compost within the same neighborhood or community.
3. **Organic materials returned to soils.** Compost is used to enhance local soils, support local food production, and conserve natural ecology by improving soil structure and maintaining nutrients, carbon, and soil microorganisms.

Defining *Organic* and *Organics*

One definition of the word *organic* is "from life"—something that is, or was once, alive. *Organics* refers to once living materials: the raw materials of composting. *Organic* has of course come to represent a biologically based method of agriculture that has a list of certified practices associated with it. By this definition, organic compost would require entirely certified organic inputs to be labeled organic. However, standards of compost management, if met, produce compost that is suitable for use on organic farms. For more on this topic, see the *Producing Compost Suitable for Use on Organic Farms* sidebar on page 128.

4. **Community-scaled and diverse.** Composting infrastructure is diverse, distributed, and sustainable; systems are scaled to meet the needs of a self-defined community.
5. **Community-engaged, -empowered, and -educated.** Compost programming engages and educates the community in food systems thinking, resource stewardship, or community sustainability while providing solutions that empower individuals, businesses, and institutions to capture organic waste and retain it as a community resource.
6. **Community-supported.** Programming aligns with community goals (such as healthy soils and healthy people) and is supported by the community it serves. The reverse is true, too. A community composting program supports community social, economic, and environmental well-being.[4]

What Is Compost?

No book on the subject of composting can escape this inevitable question. The definition of *composting* that I have been using for a number of years, which is a combination of several definitions I've come across, is:

> *The return of organic materials to a rich, stable, humus-like material through a managed oxidative decomposition process that is mediated by microbe metabolism.*

But *compost* is much trickier to define. If you've ever purchased compost, you may have ended up with any number of things: rotted manure rife with weed seed; mulch that someone tried to pass off as compost; leaf mold; dehydrated digestate solids; or beautiful, mature, organic-matter-rich, earthy, crumbly, black, life-giving . . . well, what else? . . . compost.

It's hard to say with any authority what is or isn't compost, because compost has both a history and a future of transformation. It was once something entirely different; now it's compost; and soon it will yet again be something else. The challenge with defining compost perhaps stems from the definition of composting. The criteria for each of composting's defining traits—*organic waste, stability, managed, oxidative*—can be met in any number of ways, which are often not the least bit equivalent. *Metabolized by microbes* might actually be the most straightforward characteristic of compost, and that's saying something.

I believe in an inclusive definition, though looking at the end product of the composting process, I view some composts as being more compost than others. At the "very compost" end of the spectrum, the material is mature and stable, meaning there is little food energy (carbon) available to microbes from the original parent material. Virtually all of that potential energy has been consumed, and what remains is an enormous population of living, breathing organisms, the most complex forms of carbon—well on their way to becoming those mysterious "humic substances"—and the non-organic minerals endogenous to the original raw materials. The organisms and stable organic matter in compost contain the building blocks of healthy soil. Embodied within are energy, nutrients, and a structure that mimics a fully developed soil ecosystem capable of sustaining plant life. Compost is both the hardware and the operating system needed for creating healthy soil, ready to support the life of the land.

At the "sort-of compost" end of the spectrum are products created in a hurry, or with a high percentage of materials that are non-compostable, or based on misinformation about the composting process. Often one or two minor management adaptations can make a huge difference in quality. There may be high levels of non-organic matter, including minerals or plastics, or of the original raw non-decomposed organic matter. Many scenarios can lead to composts on the low-quality end of the spectrum; once you know what you're looking for, you'll see examples everywhere you look. While these compost-like materials may offer some value depending on the circumstances, it's important to understand the benefits or drawbacks to using the specific product in a gardening or farming application. Don't waste your time (or someone else's) utilizing or creating a product that will not produce the desired effects. With some basic knowledge, anybody can reliably produce great material. Throughout this book, I'll cover some of the desirable and undesirable qualities of different composts and the scenarios that lead to these qualities.

The Cult of Compost

For many people the need to compost, once awakened, is insatiable. Composting calls, it speaks from the beyond, drawing in believers. Seriously, farmers and others who make compost often describe the positive effects composting has on their minds and spirits almost like a meditation or yoga practice. A large number maintain a deep belief in composting as part of a holistic way of life.

I share this for a few reasons. First, if you are an inadvertent member of the Cult of Compost, know you are not alone. We are legions strong and spreading from root to sky. Second, you should know that a species of soil bacterium (*Mycobacterium vaccae*) releases serotonin. Chris Lowry, a neuroscientist studying the role of bacteria like *M. vaccae* in serotonin release, describes the effects of the bacteria as "basically no different from all the SSRIs" (such as Prozac).[5] There are no studies to date about dependency concerns with *Mycobacterium*. And there are no doubt other microbes with similar effects. Third, to define compost strictly as a substance that has undergone *biologically mediated oxidation* misses the fact that there is a huge social aspect to composting. Similar intersections between technology, social order, and microbial communities have a long and beautiful history (I'm thinking of the fermentation traditions of the Tarahumara and Lacondon peoples of Mexico). Those traditions are also a path to both community and the divine.

This is all to say, composting may play a more important role than we recognize in creating greater wellness in our communities. Have no fear in taking on a holistic view (but try to leave the dogma out).

Defining This Book's Focus

This book focuses on recycling systems that include food scraps—the fastest-growing area of community-scale composting—rather than systems designed for composting yard wastes, animal manures, or biosolids (sewage). Across the country 58 percent of yard wastes are already composted.[6] Animal manures and bedding are usually handled at the home or farm scale, although they certainly play a big role in many food scrap recycling operations as well, and with the growth of backyard animal husbandry, there may be new opportunities for community composters to capture these manures. Human wastes are usually handled at the municipal

The Term *Food Scraps* as a Best Management Practice

Food is not waste unless food is wasted. Therefore the term *food waste* is a misnomer, despite its widespread use. In transitioning from a waste paradigm to a resource paradigm, the language that we use matters. We are not managing wastes; we are managing resources. When food can be diverted to feed people, *food surplus* or similar terminology is much better than *food waste*. A growing number of people use the term *food scraps* to describe food that is no longer edible by humans but can still feed animals or microbes (that is, be composted or digested anaerobically). This includes New York's Department of Sanitation, the State of Vermont, and my birthplace of Cambridge, Massachusetts, to name just a few. I advocate that this term (for lack of a better one) be considered a compost best management practice (BMP), alongside pile monitoring and meeting temperature treatment standards. If composters don't lead the way on this, then how can we expect the general public's ideas about waste to change?

Figure 0.1. America's largest waste hauler promoting the idea of reuse.

level or at the home (humanure systems). However, although the focus of this guide is on food scraps, much of the information presented here is informed by on-farm and biosolids composting and there is a great deal of overlap among composting system types and the raw materials they process.

Where Banana Peels Go and Never Come Back

The alternative to organics recycling is landfilling and incineration, options that carry a host of festering issues where organics are concerned, not the least of which is that the nutrient value in organics, once wasted, can never again be reclaimed in a useful form. Landfills are a one-way ticket to banana-peel hell. Consider that this earth is literally made of stardust, and that that stardust somehow developed into the only known life in the universe, and that that life is composed of a limited quantity of nutrients that only happen to be in surplus abundance at this time due to a bonanza of fossil fuel energy, and that all of this is happening in the blink of an eye in geological terms.* Putting all of the rigorous science of greenhouse gas emissions and toxic leachate polluting groundwater from landfilling organics aside, the story of the broken link between all the life that ever was on this planet and all of the life that ever will be should be enough to convince anyone to compost. It's a really hard message to get across to a wide audience (although I strongly encourage you to try).

The ~~Waste~~ Resource Paradigm Shift

The recognition that resources are being wasted and that that waste is costly to our environment and society is, without a doubt, spreading. Just within the past few years, laws have been enacted banning food scraps from landfills and incinerators in cities such as Seattle, San Francisco, and Portland. At the time of writing, five states are in the process of rolling out bans and composting mandates: Vermont, Massachusetts,

* My friend Timothy Crews, an agroecology professor, coined the phrase *fossil fuel bonanza*. It's one that has always stuck with me.

Connecticut, Rhode Island, and California.[7] There is even a ban that applies to large generators that went into effect in 2016 in New York City.[8]

Composting on such a massive scale is nothing new. It often surprises people that one of the largest sectors of the composting industry is based on processing organic materials known as biosolids. Biosolids are, yes, human excrement . . . and everything else that goes into the sewer. This is not your backyard humanure composting system, it's a veritable bio-geo-pharma-chemo cocktail of everything that is modern humanity. In 2010 an estimated 562,000 dry tons of biosolids were composted annually,[9] which was on par with food scrap composting in the United States in 2014 at 1.94 million wet tons, according to the EPA (dry tons would be one-third to one-fifth of the wet weight of biosolids).[10] Early efforts to develop methods of biosolids composting go back to research conducted by the US EPA in the 1970s, and this industry created the foundation for much of what the composting industry is today.

I first encountered a biosolids composting operation in college, when my agroecology professor, Tim Crews, took our class to the Albuquerque Soil Amendments Facility. They composted the solid portion of the city's sewage or sludge after it had gone through an anaerobic digester at the wastewater treatment plant. I was in awe of the 8-foot-tall windrows of composting shit (*shit* is not a dirty word in the composting world) and the massive self-propelled windrow turner named after the infamous Egyptian symbol of regeneration, the Scarab (figure 0.2), and excited by the realization that this was happening—that and the pungent smell of ammonia (NH_3) that permeated everything. I don't think that many people knew then or are aware now that massive volumes of anthropogenic nutrients are captured, processed, and then reapplied to the land in this way.

Figure 0.2. A compost windrow turner at the Albuquerque Soil Amendments Facility, 2003.

Making Waste Management Compatible with Our Food System

Ten-plus years after touring the Albuquerque Soil Amendments Facility, I still perceive a disconnection between similar massively scaled organics recycling operations and the goal of producing high-quality local food. To many readers it may be a given that the composts you produce are intended for food production, but I can assure you, this is commonly not the case. Much as you see with foods in organic agriculture, the scale, localness, distribution, processing, and quality of the composts that are available to the end consumer can vary hugely from producer to producer and product to product. In fact, today's organics recycling is as much a waste management business as it is an agricultural practice. There is sometimes tension between these two sectors, despite the common goal of minimizing waste and conserving resources. In unique cases, operations from the waste industry display an awareness of the needs of agricultural producers, but this is by no means a given, which many in the waste management sector openly admit.

While managing food scraps as a high-quality agricultural input holds enormous value, that potential frequently goes unrealized. Instead composts go on the market that are immature, contain significant plastic nanoparticles,* or are commingled

* Sometimes the particles are not nano; they are downright macro! Reports range from parts of Dunkin' Donuts cups in compost from a commercial source to a baby doll head in a California farm field. Needless to say that farmer was pretty turned off.

with sewage, reducing the options for end use. The emergence of these new products brings up many questions, and their uses must be better understood and targeted appropriately. One 2009 study found a correlation between facility scale and the presence of fecal indicators (bad bugs in human and animal feces), with levels of *E. coli* and salmonella increasing with facility scale.[11] Another undesirable trend is utilizing water resource recovery facilities or WRRFs (formerly known as wastewater treatment plants or WTPs). Combining food scraps with sewage reduces the food scraps' potential end uses, creating an end product that is less marketable and contains traces of everything that went into the sewage system including heavy metals and industrial wastes.[12] While these approaches keep organics resources out of landfills, the yields of these processes can be highly variable when it comes to their quality as an agricultural input, offsets of greenhouse gas (GHG) emissions and renewable energy status, and overall net benefit to the environment and society. In cases where the solids fraction coming out of the WRRF are not separated and composted, many people are rightly calling this practice out as "composting" in name only.

There are some key differences and benefits to small-scale composting that I hope will become clearer by the end of this book. Yet until quite recently, depending upon where you live, small-scale producers of high-quality composts have been either the norm or nonexistent. The tremendous growth in composting has led to an explosion of composting at all scales, and hence a stratification of distinct sectors of organics recyclers. In many places small-scale high-quality producers are nonexistent or just now starting to appear. In other places community-scale composting forged the path, and now laws and other factors have initiated larger operations that potentially threaten smaller producers.

All of that said, neither this book nor the community-scale composting movement is about condemning any one sector of organics recyclers. Organics recycling is nuanced and there is a need for infrastructure that meets the scale of our population, as well as a need for diverse grades of composts for a variety of applications, including non-food-producing lands. Instead, community-scale composting's role is to make great, local-food-system-grade composts and to be a growing model of local, closed-loop ecological principles.

How to Use This Book

As a technical manual, this book is an information-dense resource for use by community-scale composter practitioners when developing composting infrastructure and managing their operations. If you're using it as a reference, you can jump in anywhere to extract a single bit of information. It's also organized in a progression, with steps guiding specific assessment and design processes.

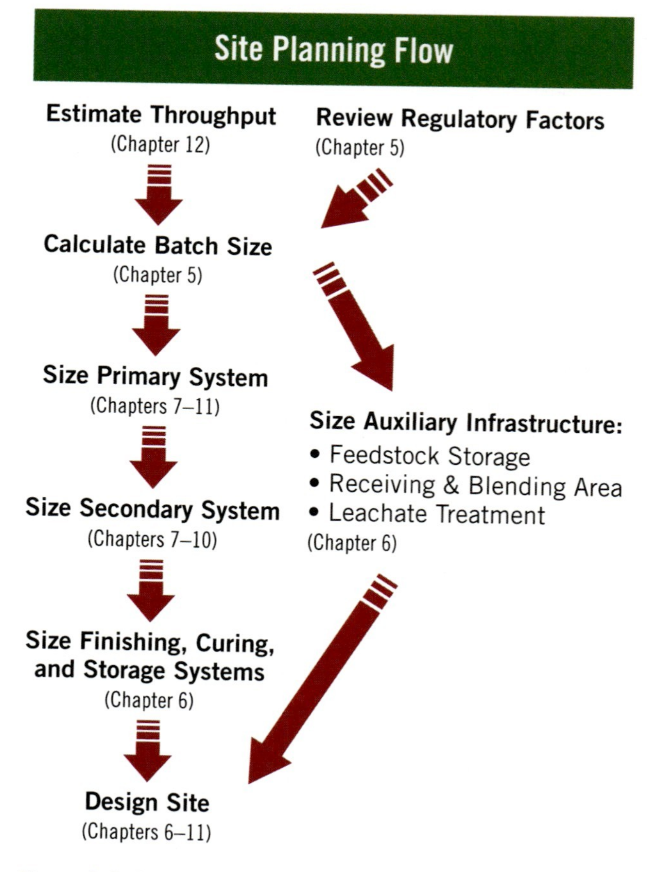

Figure 0.3. The planning flow as laid out in this book in illustrated form.

Some readers might find the depth of information and complexity intimidating. I'm just as intimidated—more so even—knowing firsthand how much information has been compiled here. But it was my intention to bridge the gap between the highly technical realm of engineers and scientists and that of small-scale practitioners, and to do so in a complete and rigorous manner. I hope that's what my research team and I have done.

To make the book as user-friendly as possible, I have made efforts to guide the reader through the process. When you're using this book, there are two tools that might help you organize your thinking. First, there are a series of checklists throughout the design and planning chapters. These can be adapted for your own purposes and used to cross-check the completeness of your planning. Second, the book is organized based on a logical progression that I use for phased project management called the Five D's:

1. Discover
2. Define
3. Design
4. Develop
5. Deploy

As a consultant managing as many as 15 projects at a time, I have found it is absolutely critical to have a method for thinking about the phases of taking a new project from concept to implementation. As an introduction to help you use this book effectively, I'll talk about the first three D's briefly.

Discover

The first phase of a project, *Discover*, is designed to challenge the original assumptions of a given project, in essence asking and answering questions to the point where rough concepts become vetted concepts. One of the fundamental uses of this book is to assist the *Discovery* process of your project. Only if you challenge your assumptions, discover the realities, and then adapt your concept appropriately will your project result in functional and sustainable outcomes.

Another way that Phase 1 *Discovery* has been very helpful to me as a consultant is in weeding out unpromising projects or concepts before they take up too much of anyone's time. I will greatly limit my involvement in projects that have not undergone a thorough *Discovery* phase and have been known to spend an hour on the phone convincing people that they are not ready to start a composting project. Those same composters may come back to me a year later, with their ducks in a row, ready to move forward.

Define

Define is where you take all of the information generated in *Discovery* and compile it in a way that is useful to you and others working on the project. Before moving forward with detailed designs or purchasing new equipment, I highly recommend that you create a simple written description of the system scale and method/s, a list of the required permits where applicable, as well as a budget and, ideally, some goals and expected outcomes. There are multiple reasons for this, but the most important is that if you have done a good job in the *Define* phase, you can also be assured that the *Discovery* was adequate, and that you are ready for *Design*.

Define should force you to identify unanswered questions and make solving them part of your game plan.

Design

The majority of this book is devoted to the *Design* of composting operations. However, you may need to reference design sections of the book in other phases and vice versa. Such is the nature of the five D's. Use them when they are useful to mapping your project's progress; ignore them when they are a hindrance.

The five D's are designed to be somewhat overlapping, although the outputs of each phase are distinct. For example, to develop a budget, you may need to begin a rough design concept in the *Discovery* phase.

If any one phase is incomplete, you will likely have to go back to it at some point later in the project (this is usually okay, but it can also be frustrating and costly). Chapters 1, 2, and 5 in particular should be helpful in the *Discovery* and *Define* phases.

Whose Book Is This?

Although the title of this book speaks most directly to those who already identify themselves as community composters, many readers may just be becoming familiar with this concept. You may be thinking, *I just want to make compost. What does my composting operation have to do with community?* The reality is that this is a reference on Composting with a capital *C*, with tons of useful information for any composting application. Ultimately, how we categorize composters is not the subject of this book; it will be useful for farmers, dirt lovers and agro-eco-geeks, policy makers, home composters, teachers of composting, compost technical service providers, extension agents, Master Composters and Gardeners, municipal and urban planners, municipalities and public-sector waste managers, the waste industry, permaculturalists, ecological designers, activists, community organizers, architects, engineers, international aid workers—the list goes on.

Right now we're just seeing the tip of the iceberg of what the community composting movement will ultimately offer our communities, food systems, local and regional economies, and planet. Nora Goldstein, editor of *BioCycle* magazine and thought leader in the movement, had this to say about the diversity of perspectives in how community composters view their work:

> *One of the community composting movement's greatest strengths is that it doesn't fit into one mold. Yet there is a collective identity that's been coalescing for some time now. Community composters embody the spirit of innovation and creativity and this helps build the foundation for positive change and civic engagement. Community composters make it possible for households and neighborhood businesses to recycle their food scraps locally to make compost for community gardens, urban farms and planters. Community composters provide jobs with living wages for youth and adults, and enterprise opportunities that come in all shapes and sizes, from worker-owned co-ops to for-profit businesses offering collection and composting services.*[13]

It is this very diversity that she has documented within the movement that defines community composting and makes it so strong. And isn't that the nature of community?

Another theme present throughout most of the book is the composting of food scraps. Even if you are not a food scrap composter, many parts of this book will still apply to you, because the fundamentals and technical function of composting are the same regardless of the raw materials.

My hope is that this book becomes a repository and reference of the highest quality. I have had the honor of learning from some of the very best in both composting and sustainable agriculture. And so it is humbly, and with great admiration, that I acknowledge and thank them for their hard work, dedication, and thought leadership. Similarly, this book is for our most important teachers, the composting practitioners whose models inspire and guide us.

Composters, this book is yours. There are no boundaries to confine where this information can be applied. Take it, test it, adapt it, make it your own, and share what you've learned.

CHAPTER ONE

Common Models in Community-Scale Composting

Whether a composting system comprises a single residence's food scraps in a backyard or all of the organics generated by a city, the basic components are the same: organic refuse is *generated*, *collected*, and then *composted*. Although not technically part of the composting system, the fourth fundamental component of any model are its products and *end uses*, which can include soil building, animal feed, and energy.

At the community scale some common composting models have emerged. They have both unique and overlapping elements in terms of where the composting is taking place in the community and who is operating the system. These models include:

- Composter networks
- On-farm composters
- Commercial composters
- Schools and other institutions
- Community gardens and farms
- Worker cooperatives
- Demonstration and training sites
- Collection services
- Drop-off programs
- Home composting initiatives

In this first chapter we'll first look at the common components across composting models, and then delve into the core concepts, composting methodologies, and unique considerations associated with each of 10 models.

rotstar

(rät-stär)

DEFINITION: Ecoslang for a person who is a leader in the implementation of organics recycling programming and infrastructure on a community scale.

It's worth noting that in 2014, while working on *Growing Local Fertility: A Guide to Community Composting*[1] in collaboration with the Institute for Local Self-Reliance and others at the Highfields Center for Composting, we first worked on categorizing these model "species," if you will. I have adapted and built on some of those ideas here, not merely as an academic exercise, but in an attempt to illustrate the higher-level workings of these systems and their unique niche in the sustainable community ecosystem. When we consider the technical aspects of composting operations, we risk reducing the models to trucks, tractors, blowers, and bacteria. While it is integral to consider the function and design of each component within the system, it is important also to step back and consider the relationships among components. Only together do they create the system as a whole.

Model Components

The most basic formula of composting models is *generator, collection, recycler, end user*. Let's look at the individual roles and basic economics of these elements.

The Generator

Compostable materials are being generated literally everywhere we look, from leaves senescing in fall to leftovers passing their prime right now in my refrigerator. Living involves the creation and expiration of unutilized organic matter. Enter decay.

These points of generation are opportunities for composters in our communities to intervene. Food scraps in particular are a unifying material for the most recent generation of community-scale composters. The use of food scraps as a feedstock is rapidly growing; the US Environmental Protection Agency estimates that food scrap composting increased from 2.2 percent of generation in 2000 to 5.1 percent in 2014.[2]

Sources of food scraps are what we at the former Highfields Center for Composting called food scrap generators (FSGs). Many people just call them generators. FSGs = everybody. FSGs can be classified as *residential, commercial, institutional,* and *manufacturing* (see chapter 4). The most likely sources of food scraps will depend on factors within your individual community, such as regional food scrap densities and the FSG sectors that local food scrap collection providers serve (for instance, many haulers serve either residential FSGs or commercial, institutional, and manufacturing FSGs).

Composting food scraps requires the addition of other raw materials to create a balanced recipe. Depending upon the community, any number of *secondary feedstocks* might be generated and available for composting. In general, food scrap composters seek dry, high-carbon feedstocks such as leaves, wood chips and shavings, animal bedding, and clean shredded paper (to name just a few) to balance out the wet, nitrogenous character of food scraps.

Figure 1.1. Food scraps are source-separated at a festival. Note Tyler Buswell (*right*) educating festival-goers at the Close the Loop recovery station. Courtesy of Highfields Center for Composting and Vermont Sustainable Jobs Fund.

Where there is a need for disposal in the community, sourcing additional feedstocks may be simple and free. In many places, however, obtaining adequate feedstocks may take considerable effort and money. Composting operations' demand for feedstocks increases as food scrap processing increases, so keeping the two in balance requires planning and thought. In general, one part food scraps by volume requires three to five parts other materials in order to create a balanced recipe and mitigate problems. Feedstock characteristics, estimation, and recipe development are covered in depth in chapter 4.

Basic Generator Economics

At the time of writing, diverting compostable materials is essentially an economic wash for most organics generators. Sometimes generators save money, sometimes they break even, and sometimes it costs them a little bit more for composting versus disposal. It all depends on the constantly shifting local economics of trash and organics recycling. In many cases composting through a collection service is a separate expense from trash, but renegotiating trash collection fees once the organics component has been removed can make up for it.

Collection and Transport

Diverting local food scraps and other feedstock to composting requires some form of collection and transport. Organic material is separated from other discards at the point of generation, and then transported to the compost site or to another transfer point. If the composting takes place at the point of generation, it's typically called on-site composting and transport is minimal. If the composting takes place elsewhere, more transport is involved.

Some models involve self-hauling, such as residential food scrap drop-offs, where individual households take their scraps to a local drop point. There are also commercial or municipally operated food scrap collection services. Collection services are the literal go-betweens

Figure 1.2. Bicycle collection is a thriving model in many cities around the country. Here Madeleine Froncek of Compost Pedallers in Austin, Texas, moves food scraps and other residential organics using a bicycle and cargo trailer. Courtesy of Compost Pedallers.

for FSGs and compost sites, and are sometimes operated as side businesses by composters themselves.

Composters also sometimes self-haul their secondary feedstocks or contract haulers to do it. The dramatic increase in food scrap recycling has spurred a diversity of innovative new collection apparatuses at the community scale, from bicycle-powered systems to trucks with capacities from 1 to 10 tons of food scraps per load (see chapter 12).

In order to collect clean organic material, it is vital to provide education on *source separation* to FSGs and other feedstock sources. While in some cases third parties can provide this education to FSGs, it is often the responsibility of the collection service, and those who can't get clean material risk paying more for tipping fees or losing a compost processor altogether. For more on source separation, see *Generator Training and Education* in chapter 12, page 347.

Basic Collection Economics

Typically haulers charge the generators for collection of food scraps. They then pay a composter a tipping fee to receive and process the material. Transportation distance and route density are crucial factors in the profitability of collection services. The faster they can fill their hauling unit, the better the economic return. Many food scrap haulers report that they can make money in the current economic climate.

The economics of transporting secondary feedstocks varies considerably depending upon supply and demand for the given material (see *Feedstock Economics* in chapter 13, page 352). In general, though, this cost is paid by either the composter or the generator depending upon the material.

Composter (Organics Recycler)

Once material has been captured and collected, it is then transported to the compost site. There are a wide variety of options and methodologies for composting food scraps and other organics. All of the common composting methods are viable options for community-scale composters. The system needs to have the capacity to manage the intended volume of organics, and some methods are better suited for certain applications than others. Beyond the standard types of composting, other options that can add value for composters include integrating animals into your composting system, composting with worms, and adding heat recovery infrastructure to your composting system. All of the standard composting methods, as well as composting with animals, are covered in great depth throughout this book. (I ran out of time and room to cover vermicomposting and compost heat recovery in as much depth, but hopefully there will be more to come on those topics.)

On-Site and Off-Site Composting

Two subsets of composter types are *on-site composters* and *off-site composters*. Where the composting takes place at the point of generation, it's typically called on-site composting. All other composters are considered off-site, although we just refer to them as composters and leave the *off-site* part out. When we look at specific models, the location where the composting takes place is distinguished in this way.

In the ILSR guide *Micro Composting: A Guide to Small Scale & On-Site Food Scrap Composting Systems*,[3] we wrote about the choice to compost off-site versus on-site and what it takes to make composting on-site work. For many businesses, schools, and other food scrap generators, the simplest option is letting a well-managed composting operation handle the actual composting process off-site, assuming there is an acceptable food scrap collection option available. When we look at the on-site and other non-commercial composting models in this guide, we focus a lot on the motivations behind each, which have a large impact on the sustainability of these models.

Basic Composter Economics

Composter revenues typically include compost sales and *tipping fees*. It is standard practice for composters to receive a tipping fee from a hauler when they accept and process food scraps. The fee is a justifiable cost to haulers, because it replaces the tipping fee that they would be charged at the landfill. Ultimately that cost is passed down to the generator.

Composter tipping fees currently range from $30 to $65 per ton, and they appear to be increasing with the changing economics of trash.

Many food scrap recycling models are also based on saved costs or other values yielded as a result of the compost and composting process. These might include:

- Improved soil quality and crop productivity on a farm or in a garden
- Heat captured generating or offsetting energy (methane in the case of anaerobic digestion)
- Food scraps and compost offsetting animal feed costs

Costs associated with composting typically fall under the basic categories of site/infrastructure, feedstocks, marketing and sales, equipment, labor, and permitting. These costs are often not insignificant. Indeed, composting can be an economically challenging proposition given the current economic valuation of organic matter and waste, which is highly disproportionate to composting's environmental and social value in my humble opinion.

The economics are different in almost every case; some composters make money, many break even, some lose money. Any veteran composter will tell you that the economics of the composting business are challenging, and for many this will be a conversation ender. The reality is that, as for so many businesses, economic conditions change every day, and composting mandates, rising trash prices, transportation, and food prices are all working to (hopefully) make localized composting and compost use more attractive economically. But the economics can also be deceiving, and I've seen many would-be-composters with dollar signs in their eyes. Their projects usually don't amount to much. For this reason, I *strongly* encourage composters to consider the economics of composting for their specific operation.

End User

Composters can create a wide variety of compost products from their community's organic refuse. Some systems also deliver secondary yields such as eggs, thermal energy, or methane in the case of anaerobic digestion. Composts can be used to build soil and support fertility along with the use of complementary practices such as cover cropping, crop and grazing rotations, and manure application for sustainable nutrient and soil management.

Compost can be sold through a wide range of markets including as bagged products such as potting mixes, in bulk retail to home gardeners (for instance, by the individual cubic yard), or wholesale to other farms, landscapers, distributors, and the like. Evaluating potential markets for compost products is critical if the sale of compost is a part of your operation's business model. Even in many rural regions, there is a demand for high quality local composts and sales in the 300–500 cubic yard/year range can be met by primarily bulk retail sales (e.g. 1–20 cubic yards at a time), assuming that the local market isn't

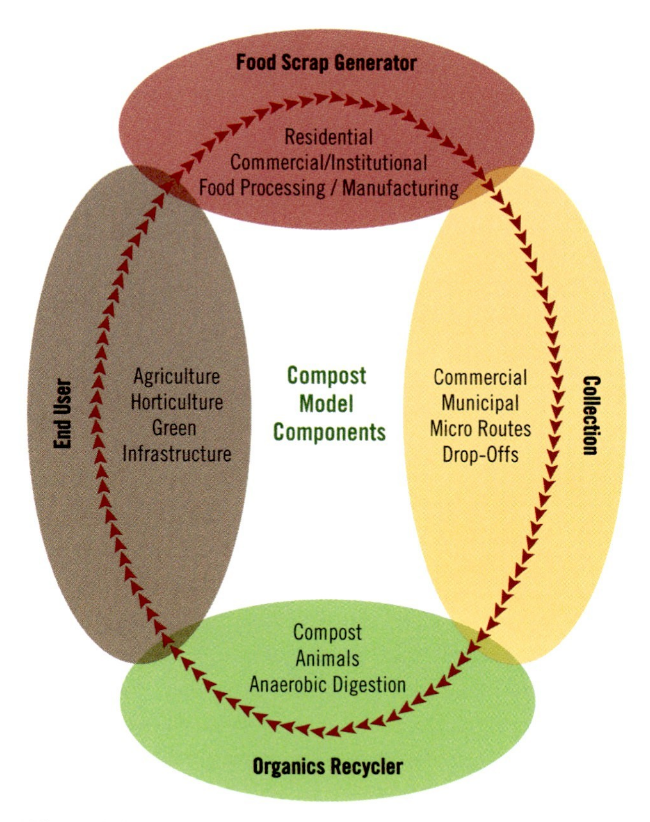

Figure 1.3. The basic components of the community compost model.

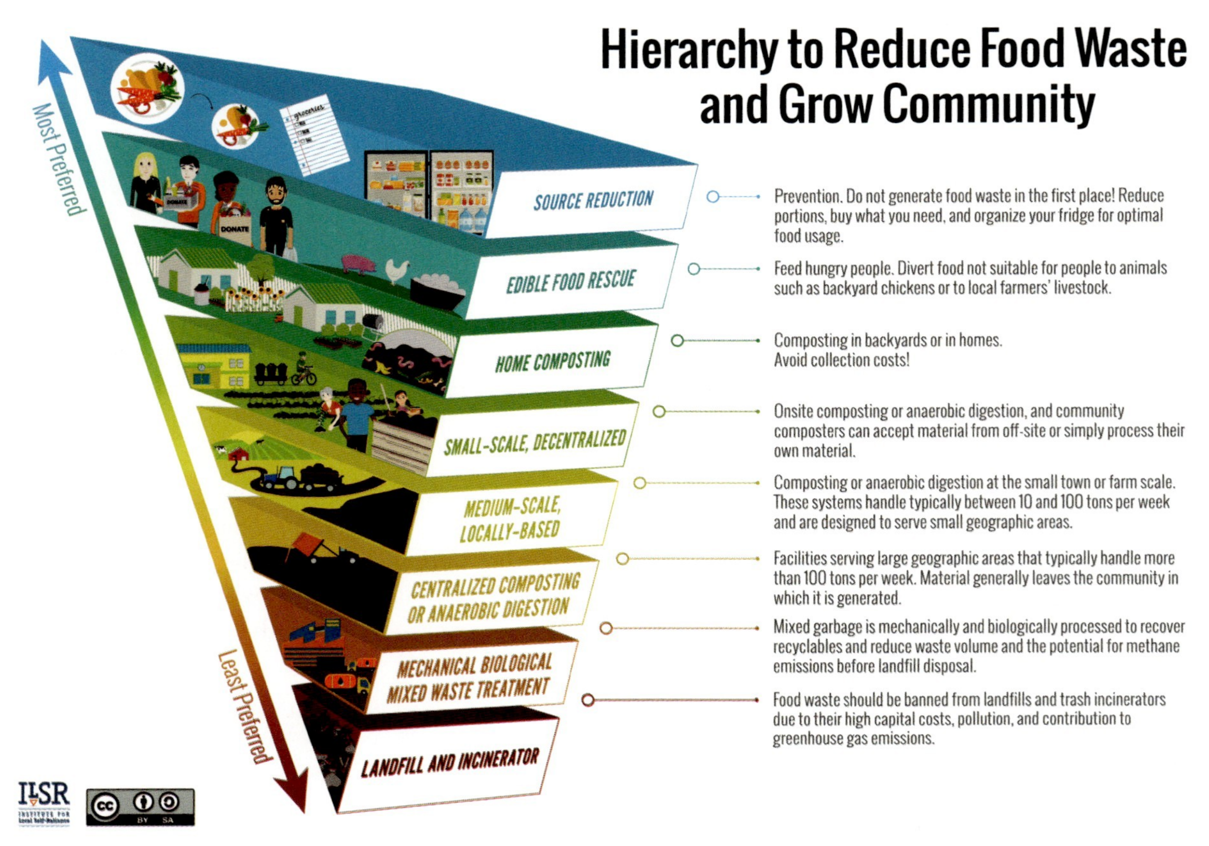

Figure 1.4. A food scrap recovery hierarchy that prioritizes local- and community-oriented solutions. Courtesy of Institute for Local Self-Reliance.

already saturated. Chapter 14 discusses this topic in much greater detail.

Basic End User Economics

The economic return on compost end use varies widely from situation to situation. Consider that wine growers in California often pay over $500 a yard for vermicompost. I couldn't even begin to explain the economics of that or the literally thousands of other cost benefits associated with one end use or another. What I can say is that in most cases, the use of compost is an investment, and that investment's returns increase over time. As soil improves, costs decrease, pollution decreases, biological diversity increases, and so on. Exactly how that is valued is extremely scenario-specific, but it is worth understanding to the greatest extent possible.

10 Common Models

In the following section we'll look at the common components across the 10 community-scale composting models, and then delve into the core concepts, composting methodologies, and unique considerations associated with each.

Composter Networks

Managing organic resources locally at the community scale in theory involves a diverse and dispersed network of composters. If you took the nine other community-scale models and applied them across any region of our country, what you'd end up with would be a patchwork of composting activity: collection services, residential drop-offs, on-farm composters,

COMPOSTER NETWORKS

Common Components

Income/Value

- Compost sales
- Food scrap tipping fees
- Compost use
- Food scraps as animal feed
- Renewable energy

Expenses

- Permitting
- Infrastructure
- Feedstocks
- Marketing and sales
- Equipment
- Labor

Food Scrap Sources

- Self-haul
- Commercial and/or municipal collection
- Residential food scrap drop-offs

Secondary Feedstock Sources

- Manures
- Leaves (urban)
- Sawdust/shavings
- Wood chips
- Bark

Composting Methodology (On-Site and Off-Site)

- Turned windrow
- Aerated static pile
- Composting with animals
- Bays (urban, demonstration site)
- Vermicomposting
- In-vessel (urban)

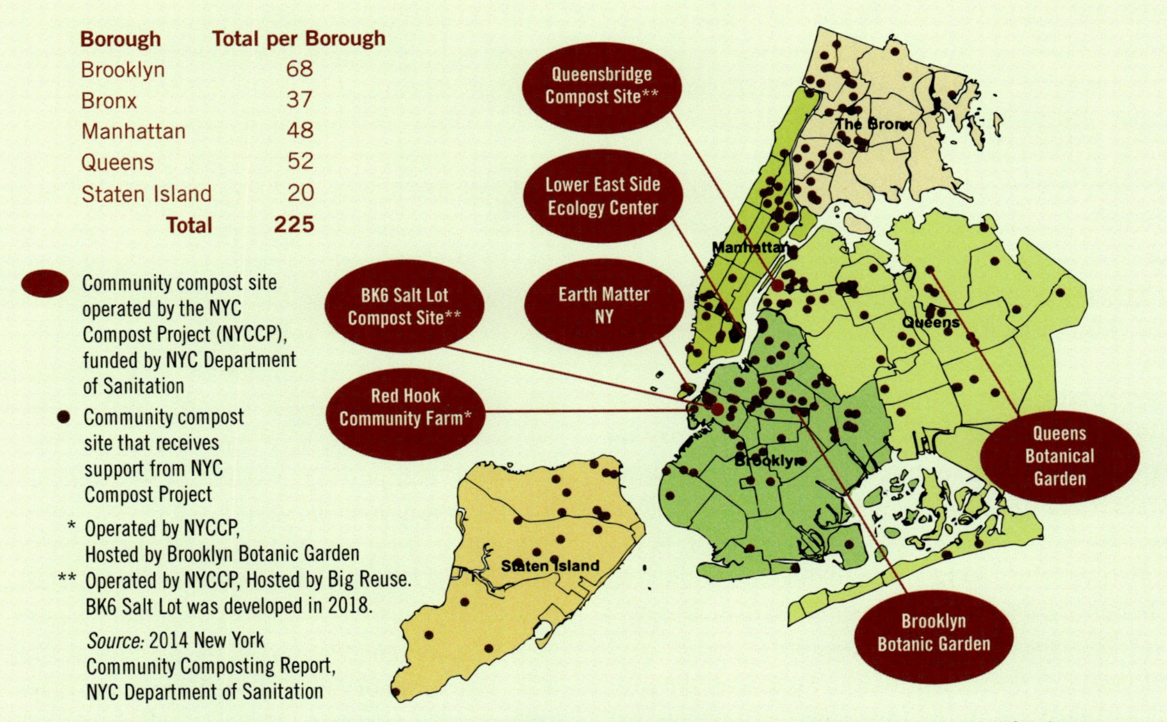

Borough	Total per Borough
Brooklyn	68
Bronx	37
Manhattan	48
Queens	52
Staten Island	20
Total	**225**

Figure 1.5. New York City is home to a continuously expanding network of community compost and food scrap drop-off sites. This 2014 map shows community-compost sites affiliated with the NYC Compost Project. Courtesy of the NYC Compost Project.

small commercial enterprises, on-site systems at schools, community garden micro sites, demonstration sites, home composting, and so on. There will be some inevitable service overlap and competition between entities, but there will also be strength in numbers and lots of opportunity for partnership.

Over time a culture of and regional pride in composting can develop. These regional composter networks can form a solid identity, as in the case of the NYC Compost Project, or they can remain informal. It's important to recognize that there is a model for building significant regional composting capacity that is both decentralized and stable. In essence, this is *the model* that the community composting movement is working to develop (and that this book is designed to support).

Unique Considerations

Composting systems and networks don't just happen organically; it takes boots on the ground making shit happen, as well as vision and community leadership. There are roles for policy makers, NGOs, and trade associations in facilitating collaboration, creating funding opportunities, and providing educational programs. Without grassroots initiatives and robust community involvement, regional composting networks cannot develop and be sustained.

The success of the Close the Loop programs at the Highfields Center for Composting leads me to believe that there is a huge amount of interest in many places for developing networked composting infrastructure that's locally owned and operated. At its core that program included (1) compost pilots, demonstrations, and research; (2) composter and FSG technical assistance; and (3) education focused on both the vision and the action-oriented how-to of implementing community-scale composting models. (As an aside, the Highfields Center for Composting closed in 2014 due to financial shortfalls, but the Close the Loop program was thriving at the time. Had the organization been able to pivot to these three core competency areas, Highfields would probably still exist today.)

Anticipate cynicism on the part of many that taking a community-oriented approach can be effective. There is significant effort involved without a doubt, but there is arguably a higher cost and less benefit to the community in taking a more centralized and less inclusive approach.

On-Farm Compost Operations

Traditionally, farms have acted as the roots of our communities. What better way to feed those roots (quite literally) than to return local nutrients and organic matter directly to our local food sources? The simple act of closing this loop, if done sustainably, can have a ripple effect throughout a community, giving local residents, businesses, schools, waste haulers, gardeners, and community leaders all a functional role in their local food and soil systems.

Farms can be converted to compost sites fairly efficiently when they have underutilized infrastructure and equipment that can reduce start-up and operating costs. Old concrete manure staging areas and feed bunks, for example, litter my home state of Vermont, similar to abandoned lots in some cities. Throw in an aerated static pile (ASP) system, some stormwater diversion, and leachate treatment and the site is practically ready to go. In terms of equipment needs, most farms already have a tractor with a bucket loader. As long as the farm can afford to share the tractor with the site when it's needed, the composting can help spread out the cost of the tractor for the farm. By comparison, a commercial site would have to cover the whole cost of the loader on its own.

Historically, farmers let little go to waste, and this is still true for many farmers today. In my experience they rarely miss the logic of composting. Although some farms have adopted practices that degrade soil and pollute our waterways at astonishing rates, the ability to use resources efficiently remains in farming's DNA. Farms' direct link back to the local food system, farmers' knowledge of efficient resource management and practical know-how, and the connections that farmers have with their local communities all make farms a potentially logical fit for locating community-scale composting operations.

ON-FARM COMPOST OPERATIONS

Common Components

Income/Value

- Compost sales
- Food scrap tipping fees
- Compost use
- Food scraps as animal feed
- Renewable energy

Expenses

- Permitting
- Infrastructure/site
- Feedstocks
- Marketing and sales
- Equipment
- Labor

Food Scrap Sources

- Self-haul
- Commercial and/or municipal collection
- Residential food scrap drop-offs

Secondary Feedstock Sources

- Manures
- Leaves (urban)
- Sawdust/shavings
- Wood chips
- Bark

Composting Methodology (Off-Site)

- Turned windrow
- Aerated static pile
- Composting with animals
- Bays (urban, demonstration site)
- Vermicomposting
- In-vessel (urban)

Figure 1.6. Hudak Farm's compost site in Swanton, Vermont, processes community food scraps, manures, and yard debris. The farm is diversified, producing vegetables and greenhouse stock among other things. They sell some of their compost, but the majority of it goes directly on their land. Courtesy of Highfields Center for Composting and Vermont Sustainable Jobs Fund.

Unique Considerations

I urge readers, farmers or otherwise, not to assume that food scrap composting on farms is a solution that fits all or even most scenarios. Communities generate food scraps nonstop, so managing that material is at minimum a weekly and more often a daily commitment. Adding the management of community food scraps on top of farming has to align well with the farm's interests, operational goals, and capacities. But beyond that, I can say that without exception, every on-farm composter I know who composts community food scraps has an underlying passion for composting itself. I wish that I could say the economic benefits alone are the driving factor, but after working with farm after farm, I know better. It's the love of compost, the love of soil, the love of community, and the aversion to waste.

When considering importing food scraps onto a farm for fertility, it's also important to consider the fact that some farms have an inherent nutrient deficit, while others have a surplus. Vegetables and other field crops equate to an export of nutrients off the farm. There is no doubt that the fertility in the food scraps will be needed. Most large dairy farms on the other hand import grain and fertilizer, and the nutrients in their manure alone are often already more than their fields can handle. Accepting food scraps would mean exporting even more nutrients. Not necessarily a deal breaker, but an important factor to understand in relation to any particular farm.

Lastly, regulation of food scrap composting and animal feeding on farms differs from state to state, and the specifics can change periodically. I talk generally about permitting and applications on farms in chapter 5. Assessing applicable permits and knowing how your agricultural operation would fall within them is a critical step.

Commercial Composters

The term *commercial composter* refers to any operation where making and selling compost is the primary business. These are traditionally entrepreneurial ventures, but they could also be a commercial element of a nonprofit, a worker-owned cooperative, or even a public-private partnership (say, a private composter leases public lands for $1 and in return provides composting services to the community). What distinguishes commercial enterprises from other community-scale ventures is that (1) the entity is not primarily in the business of farming (depending on whether or not you define composting as farming); (2) composting is happening off-site from the point of generation; and (3) compost sales and tipping fees are a source of revenue for the operation.

Unique Considerations

Since the viability of commercial composting operations largely depends on compost sales and tipping fees, their likely markets require significant consideration. Most small composting operations need to get top dollar for their product, which means that the product needs to be of a high quality. Compost marketing and sales take thought and effort, especially when producing a value-added product.

Tipping fees are competitive in some areas and can make a huge difference in the economics of an operation. Since regional tipping fees are not necessarily public knowledge, researching the local market can take some digging. Some commercial operations focus on tipping fees as their main revenue source, but there are obvious dangers when the quality of the processing and end product aren't adequately valued.

Schools and Other Institutions

When growing a composting program in your community, schools, colleges, and other institutions are an obvious place to start. School composting programs are an opportunity to foster an ethic of environmental stewardship and participatory ecology in young minds. Young composters grow into old composters, and so investment in the training and education of today's youth will have a long-term payback for community composting efforts. Students, staff, and school volunteers find composting to be a rewarding experience that provides the community with multiple opportunities to practice

COMMERCIAL COMPOSTERS

Common Components

Income/Value

- Compost sales
- Food scrap tipping fees
- Compost use
- Food scraps as animal feed
- Renewable energy

Expenses

- Permitting
- Infrastructure
- Feedstocks
- Marketing and sales
- Equipment
- Labor

Food Scrap Sources

- Self-haul
- Commercial and/or municipal collection
- Residential food scrap drop-offs

Secondary Feedstock Sources

- Manures
- Leaves (urban)
- Sawdust/shavings
- Wood chips
- Bark

Composting Methodology (Off-Site)

- Turned windrow
- Aerated static pile
- Composting with animals
- Bays (urban, demonstration site)
- Vermicomposting
- In-vessel (urban)

Figure 1.7. Snow-covered windrows at the former Highfields Center for Composting site in Wolcott, Vermont. Highfields was a nonprofit, but operated its composting site commercially while also providing educational and research programs. Courtesy of Highfields Center for Composting and Vermont Sustainable Jobs Fund.

SCHOOLS AND OTHER INSTITUTIONS

Common Components

Income/Value
- Compost use
- Educational value
- Renewable energy

Expenses
- Infrastructure
- Feedstocks
- Equipment
- Labor

Food Scrap Sources
- Entity's kitchen

Secondary Feedstock Sources
- Manures
- Leaves (urban)
- Sawdust/shavings
- Wood chips
- Bark

Composting Methodology (On-Site and Off-Site)
- Turned windrow
- Aerated static pile
- Composting with animals
- Bins/bays (urban, demonstration site)
- Vermicomposting
- In-vessel (urban)

Figure 1.8. The on-site composting shed at the Charlotte Central School has a highly educational focus. The system processes less than half of the school's total food scraps, because the total volume would be massive. The other half therefore is composted off-site. Splitting the material up in this way provides compost for the school's gardens, meets the educational mission, and allows all of the material to be recycled without overwhelming the small human-powered system.

integrative thinking and take responsibility for the conservation of resources.

Several options exist for composting at schools. Where collection services are available, many schools simply separate their food scraps from recyclables and trash and then pay to transport the compostables to the nearest off-site composting operation, often a local farmer. Programs such as this already exist in many communities. A composting program at the Bellows Free Academy (a public high school in St. Albans, Vermont), for example, grew into a regional composting program involving multiple schools and businesses, a waste district hauler, and a local on-farm composter. That's the power of grassroots community activism!

Where collection programs don't already exist, the off-site model is easily adaptable and scalable with the right planning. For instance, a teacher or parent might have a small flock of laying hens that they wouldn't mind feeding with the school's scraps.

Some schools choose to make compost on-site, particularly when there are existing gardens, providing a full soil-to-soil loop that few children would ever experience otherwise. Collection programs are particularly well suited for more dense urban settings where other options—such as locating a system on-site or keeping chickens—might face logistical challenges. On-site composting or sending the scraps to a local chicken flock are both options more likely to apply to rural settings or sites that have ample space to manage the materials.

Unique Considerations

Every successful school composting program requires at least one person to serve as leader. This is especially critical for on-site programs. Since "composter" isn't in most existing school job descriptions, the importance of this role is often overlooked. The compost coordinator (or similar title) gets the ball rolling and facilitates system implementation, student and staff training, and program management. This person could be staff, an administrator, a parent, or a community volunteer, and may change over time, but someone will need to facilitate the program as long as composting is happening on-site. It is essential that school leadership validates this role and its purpose, to ensure that the school is committed to filling and supporting this function over time. Ideally the role is specified in a job description. The principal, and ideally the school board, should make clearly defined written commitments to the program to ensure success.

The size and design of on-site composting systems at schools should meet the school's requirements for handling all food scraps generated over the long term, as well as for cleanliness and food safety, pollution mitigation, and overall manageability. Again, this is a given requirement of any composting system, but with entities whose primary mission is not making compost, the point requires frequent reiteration. The quality and efficacy of the system influences the value of the educational experience for the students. To help clarify how a school will successfully implement and manage its composting system, a clearly defined strategy for implementation and a plan for managing the system should be developed. The sustainability of active, dynamic systems like composting will require similarly active management and the capacity to respond to challenges.

For sites that can't or choose not to handle all of the food scraps a school generates, there are other options. For example, a school might just make a single batch of compost in the fall, then manage and study that pile throughout the rest of the school year. In this case, or in a case like that described in figure 1.8, a school might set up school-wide composting through a collection program, then siphon off a portion to compost on-site for their own uses.

Community Gardens

Composting at community gardens is very common, but most often the feedstocks are just what is produced on-site—garden debris, stems, leaves, veggies, weeds, and the like. As a community-scale model, a garden in theory composts off-site material as well, typically food scraps from the gardeners' homes, but also compost from the surrounding community

COMMUNITY GARDENS

Common Components

Income/Value
- Compost use
- Food scraps as animal feed

Expenses
- Infrastructure
- Feedstocks
- Equipment

Food Scrap Sources
- Self-haul
- Residential food scrap drop-offs

Secondary Feedstock Sources
- Garden debris (chopped)
- Manures
- Leaves (urban)
- Sawdust/shavings
- Wood chips
- Bark

Composting Methodology (On-Site and Off-Site)
- Turned windrow
- Aerated static pile
- Bins/bays
- Vermicomposting
- In-vessel (urban)

Figure 1.9. You can't go far in the New York metropolitan area without stumbling upon a community garden compost site, like this solar-powered ASP system in Williamsburg, Brooklyn.

in some cases. There are examples of this all across New York City, where a large number of micro compost sites are located at community gardens. A 2014 report listed 225 sites in total.[4] Community garden sites that compost material from the surrounding community are typically handling food scraps from local residential drop-off programs and are primarily volunteer-managed.

Unique Considerations

Even at what is typically a micro scale, managing food scraps at community garden sites takes significant thought and effort. Because these systems are a model for composting in the community and they tend to replicate, it's incumbent on garden composting operations to create something worthy of replication.

There tends to be turnover in both leadership and volunteer roles over time, so those who initiate the compost project should consider ways to create continuity in management practices. I probably sound like a broken record, but the human elements of these systems are often where things fall apart. The systems themselves are not actually that hard to manage properly with training.

The best ways to prevent management failures are to (1) document protocols in a management plan, keeping it simple and being specific; (2) create redundancy in roles (say, by ensuring that at least three people are trained and experienced in monitoring piles; and (3) make sure that composting is a written part of the garden's mission and mandate (adopted by garden board or whatever formal organization the garden has in place). If the composting systems are part of a network that's replicating and improving on the community garden model, connect with that network. Work with whatever organizations exist to support this work (or create one if it doesn't already exist) and build a *community of practice* around it.

Worker Cooperatives

One unique business model that several community composters have adopted is that of the worker cooperative. The worker co-op is a distinct form of cooperative enterprise where workers own and manage the business. It's a business structure that creates equality across its worker-owners. Most of the worker-owned composting enterprises that I'm aware of are collection services, but the model could apply just as easily to processors.

Unique Considerations

I have very little experience with co-ops, but understand that they operate very differently than more traditional business or nonprofit structures. Those who are considering adopting the model should research what's involved and learn from other co-ops. There are numerous local, regional, national, and international organizations, associations, and networks that support co-op development.

Demonstration and Training Sites

Most compost sites provide some amount of basic public education about what they do. Composting can seem mysterious, and people are interested in it. Whether school buses of students on field trips or compost agro-eco-geeks like myself, most sites attract people who want to experience composting for themselves. So for most composters, creating an environment conducive to learning is a benefit, if not a requirement.

Beyond this norm, however, are compost sites that include community and composter education as part of their core mission. These sites can be commercial, on-farm, or community garden sites—any model really—but a dedication to education beyond its commercial necessity is what defines this group of composters as a distinct model.

Demonstration and training sites typically house a variety of composting methods and systems, have educational signage, and offer hands-on opportunities for people to be involved in the composting process. Some sites might showcase a dozen or more different styles of home composters; others may focus on larger systems, or model one type of system in particular.

Many demonstration and education sites often also offer a valuable research component, partnering

WORKER COOPERATIVES

Common Components

Income/Value

- Compost sales
- Food scrap tipping fees
- Compost use
- Food scraps as animal feed
- Renewable energy

Expenses

- Permitting
- Infrastructure
- Feedstocks
- Marketing and sales
- Equipment
- Labor

Food Scrap Sources

- Self-haul
- Commercial and/or municipal collection
- Residential food scrap drop-offs

Secondary Feedstock Sources

- Manures
- Leaves (urban)
- Sawdust/shavings
- Wood chips
- Bark

Composting Methodology (Off-Site)

- Turned windrow
- Aerated static pile
- Composting with animals
- Bays (urban, demonstration site)
- Vermicomposting
- In-vessel (urban)

Figure 1.10. CERO, a worker co-op founded in Dorchester, Massachusetts, collects a combination of residential food scraps consolidated by Metro Pedal Power, weeds from Recover Green Roofs, and cut flowers from Forêt Design Studio for composting.

DEMONSTRATION AND TRAINING SITES

Common Components

Income/Value

- Compost sales (maybe)
- Food scrap tipping fees (maybe)
- Compost use
- Food scraps as animal feed
- Renewable energy

Expenses

- Permitting (maybe)
- Infrastructure
- Feedstocks
- Marketing and sales
- Equipment
- Labor

Food Scrap Sources

- Self-haul
- Commercial and/or municipal collection
- Residential food scrap drop-offs

Secondary Feedstock Sources

- Manures
- Leaves (urban)
- Sawdust/shavings
- Wood chips
- Bark

Composting Methodology (On-Site and Off-Site)

- Turned windrow
- Aerated static pile
- Composting with animals
- Bins/bays (urban, demonstration site)
- Vermicomposting
- In-vessel (urban)

Figure 1.11. Field exercises at the Rot Star Boot Camp, an all-day training that was targeted to micro- and small-scale composters at the Highfields Center for Composting Research and Education Site. Courtesy of Highfields Center for Composting and Vermont Sustainable Jobs Fund.

with academic institutions and running full-scale research projects, as Highfields often did with the University of Vermont's Plant and Soil Science Department. Other sites might adopt more of a citizen science model or conduct research and development of their own composting systems and management techniques with the goal of simply learning and replicating what works.

Unique Considerations

The embedded educational value created by demonstration and education sites goes well beyond the specifics of composting. For example, some community compost/farm sites are designed to create leadership opportunities for youth or to build diverse skill sets that can expand someone's employment opportunities, including operating and maintaining machinery, using compost on landscapes or in farming, and environmental science and design skills. Education sites can also help build management skills that are useful for running a business, leading a summer youth program, and much more. The auxiliary learning opportunities are limitless; it's just a matter of where the site wants to take it.

It probably goes without saying that when a site's primary focus is on education or research, there are associated costs. Many such sites are largely volunteer-operated and dependent on grants and nontraditional funding sources. Even if a demonstration and educational site contains a commercial component, given the slim margins in commercial composting, creating an economic model where that commercial element funds significant educational or research activities could be challenging. It might take a lot of time and investment up front before such a facility could become economically self-sufficient.

Food Scrap Collection Services

There are a growing number of services around the country that collect food scraps from businesses, institutions (such as schools), and residences. Some pick up other organics as well, including yard debris, paper, and compostable service ware. Food scrap collection services can take many forms and scales. In the urban community composting movement, there is a growing contingent of bicycle collection services. These are often food scrap consolidators, moving organics from dispersed residential generators to a drop-off point where a collection truck can then transport the material to a composter located farther away than would be practical for bicycle transportation. A few bicycle services, like Compost Pedallers in Austin, Texas, deliver directly to a distributed network of small compost sites. There are also a growing number of small commercial haulers who work with small trucks, vans, dump trailers, box trucks, and customized trash trucks of various models.

Several factors affect how collection services fit into community-scale models. The largest food scrap collection trucks are relatively small-scale by design, typically 10 tons per load or less. Larger trucks might be possible, but it would require a high density of very large generators to justify, and access to collecting the material would be challenging. At 10 or fewer tons per load, the largest collection route can easily fit into a community-scale model. I would argue that where haulers send their organics is a more critical factor. For example, a large hauler could take a community-scale approach by delivering to a small compost site or to a network of small compost sites. On the other hand, a bicycle hauler could deliver to a drop-off location whose contents are eventually delivered to a water resource recovery facility that may or may not compost and is typically not well integrated into the local food system.

Since collection services are the middleperson between the generator and the organics recycler, they have a large influence on the ultimate destiny of a community's organic resources.

Unique Considerations

The success of commercial food scrap collection services at the community scale is a very promising trend, but it's not a given. One of the most important factors in making the model work economically is food scrap density. Profitability depends on creating

FOOD SCRAP COLLECTION SERVICES

Common Components

Income/Value

- Collection fees

Expenses

- Marketing, sales, and generator education
- Tipping fees
- Collection equipment
- Supplies (carts, buckets)
- Labor

Food Scrap Sources

- Food scrap generators
- Residential food scrap drop-offs

Secondary Feedstock Sources

- N/A

Composting Methodology (Off-Site)

- Collection apparatus
  - Trucks
  - Trailers
  - Bikes

Figure 1.12. Marketing the growth of residential curbside collection in Cambridge, Massachusetts. Unfortunately, at the time of writing, this material was being slurried and trucked to a water resource recovery facility, rather than composted.[5]

a collecting route with enough volume to pay for the route plus profit. Larger, more consolidated pickups are more efficient than smaller, more distributed ones, so to remain profitable, a hauler would need to charge more for the latter. The converse of this equation is that dispersed compost sites create shorter distances from route to recycler, which increases collection efficiency.

Conventional wisdom is that food scrap bans, which are currently being enacted every year in states and cities, increase participation in composting services, which increases collection density. The market research involved in planning food scrap collection routes is pretty intuitive, but it requires due diligence as well as some guesswork. For some strategies and metrics in this area, see chapter 12.

If you are considering starting a collection service that consolidates material for pickup by another service (such as bike collection from residences), you will need to research the legality of such operations in your state. This type of activity might fall under transfer station or other solid waste management laws, in which case the location where the drop-off/consolidation takes place would need to be permitted for that specific activity. My impression is that many small collection services operate under the radar in this regard, but if your collection model depends on consolidation for pickup, I recommend finding ways

Micro Routes

Micro routes—or the one-to-three generator to one-composter model—are another style of community collection partnership that is important to acknowledge. They are typically small composters or farms, often incorporating animals, that self-haul food scraps from a small number of collection points.

This type and scale of composting are often small enough not to require permitting, although regulations vary considerably by state. As a consultant who works in a lot of rural areas, I have grown to love these models, because they can be super simple and fast to get up and running. Once started, micro routes manage themselves. They operate on such a small scale that any problems are minimal and good relationships between the generator and the composter are almost inherent to the model itself.

These types of community composting relationships are happening all around us and we don't even realize it. Often they happen organically; other times they take a little bit of work to initiate. The most common example that I have worked with occurs when one or several schools send food scraps to a local chicken farm for feed. Usually the composter does the collection, but there are all kinds of variations. Maybe the food scraps travel home with the kid on the bus to the farm, or maybe the bus driver is the composter.

As diverting residential food scraps becomes more of a priority in communities, another model that I think is going to really take off is residential food scrap drop-offs being handled at this sort of micro scale. This is in fact happening throughout New York City as part of the NYC Compost Project (NYCCP). All over the city material is collected at drop-off points, then processed at nearby community garden sites as well as larger community sites.[6] (See the *NYC Compost Project Hosted by the Brooklyn Botanical Garden at Red Hook Community Farm*, page 191, and *Space and Place: Adapting to the Constraints of Urban Composting*, page 210.

to do this legally. (Ideally, states are working to create regulations that support this model.)

Chapter 12 delves into models, methods, and equipment in much greater depth.

Drop-Off Programs

There are typically three pathways for residential food scraps to be composted: home composting, curbside collection, and drop-offs. The name *residential food scrap drop-off* is pretty self-explanatory. Organics generated by households are transported or self-hauled to these consolidation points, where a collection service can pick them up in one efficient swoop. In some cases, the food scraps are composted at the drop-off site, say at a community garden site.

Drop-offs are a no-brainer option for communities to offer composting to those who aren't able or inclined to compost at home, because it requires minimal investment compared with running a residential collection program. Typically the location is nothing more than a compost collection cart, and ideally some educational signage about what is and is not allowed.

Unique Considerations

There are very few red flags or contingencies with food scrap drop-offs. They are a low-cost, simple, and logical option. There are a few requirements,

DROP-OFF PROGRAMS

Common Components

Income/Value

- Collection fees (free is better)

Expenses

- Generator education
- Collection or tipping fees
- Supplies (carts, buckets)
- Labor

Food Scrap Sources

- Local residents

Secondary Feedstock Sources

- Yard debris
- Backyard livestock manures (not dog/cat wastes)

Composting Methodology (Off-Site)

- N/A

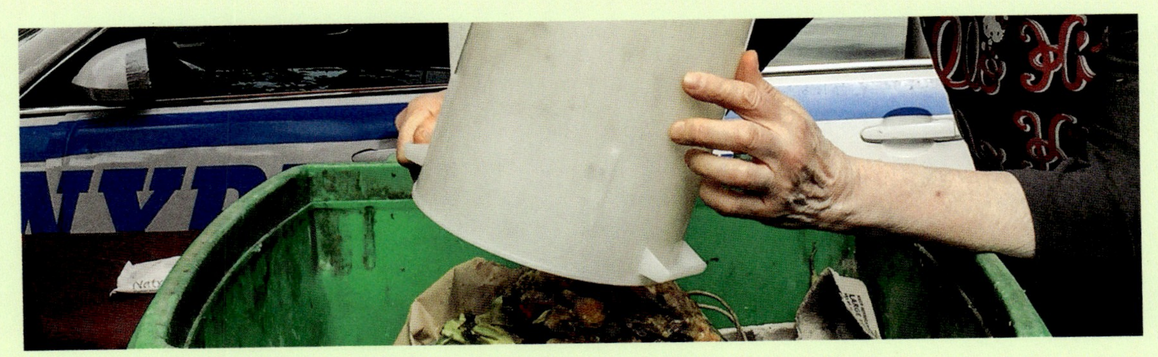

Figure 1.13. A member of the community drops off the food scraps she's been collecting over the past week in a bucket in her freezer. Courtesy of Jimena Gracia.

however. First, someone needs to collect from the drop-off, which can be an issue in areas that don't have composting collection services already available. In these cases, or in cases where the available collection route doesn't align well with the program's goals, it is possible to set up a micro route or one-to-one generator to micro composter partnership. These options might involve a local farm collecting from the drop-off for chicken feed in a rural area or collecting the material for processing at a community farm in a more urban area. These options are slightly more involved, but they take advantage of existing community assets and can be very sustainable once the right partnership is in place.

Another requirement is for drop-offs to be located in a spot that's both convenient and legal. Possible locations in each community depend upon state and local regulations. Common drop-off sites to explore include waste transfer stations, grocery stores, subway stations, and compost sites. Farmers markets are also common drop-off sites, but keep in mind that these and other seasonally available options such as schools will only serve (or conveniently serve) participants for part of the year.

There are a number of other questions and decisions involved in making drop-off programs successful. One key question is: Who will pay for the composting service? These and other considerations, including generator outreach and education and providing source separation materials such as buckets and compostable bags, are discussed further in chapter 12.

Home Composting Initiatives

Home composting is the epitome of small-scale distributed composting, and any comprehensive regional strategy for composting embraces and supports residents in managing as much of their own organics as possible. If you consider that about a third of the food scraps generated in the United States come from homes, even a small increase in the rate of home composting would make a measurable impact. And compared with the cost of collection and off-site infrastructure, the cost of home-scale composting is typically a small, onetime system investment mostly covered by the resident. (Sometimes the cost is slightly subsidized by local government, but home composting systems can also be easily made from free repurposed materials

HOME COMPOSTING INITIATIVES

Common Components

Income/Value
- Workshop fees (free is better)
- Compost everywhere!

Expenses
- Presenters
- Supplies
- Location
- Labor (coordination, management)

Food Scrap Sources
- Residents

Secondary Feedstock Sources
- Residents
- Wood chips (check with municipality)
- Leaves (check with municipality)
- Scavenged

Composting Methodology (On-Site)
- Backyard bins, tumblers, and open piles
- Vermicomposting
- Composting with animals
- Residential food scrap drop-offs

such as pallets.) The real costs are public education and promotion. In most cases, however, education and promotion are going to be a cost regardless of the composting model if the strategy is to be effective. There is a strong argument to be made for dedicated efforts at initiating effective home composting or what we'll call *home composting initiatives.*

There is a lot more that could be done to promote and support home composting in most places. Master Composters, a composting initiative focused on the home and micro scale that was developed in the Seattle area in the late 1980s, has been replicated, adapted, and is now well established in many parts of the country.[7] It represents what is often referred to as a *train-the-trainer* model, where interested citizens learn home composting best management practices, then become composting ambassadors in their neighborhoods and communities. Master Composter–inspired programs such as NYC Master Composter and ILSR's Neighborhood Soil Rebuilders have upped the ante somewhat by creating multi-tiered certification levels. The higher levels require significant hands-on learning and volunteer commitments.

Effectively leveraging train-the-trainer programs involves not only training community members, but also supporting them as they are let loose in their communities. Here are few examples of what these compost instigators might do:

- Plan and give home composting workshops
- Create home composting social media and video campaigns
- Engage schools to find creative means to take the message of composting to the next level
- Create simple three-bin or other backyard compost system demonstrations in public locations, frequently staffed with volunteer educators
- Build a mobile home composting educational setup (say, an old trailer) and take it on the road to schools and the like.

Unique Considerations

Similar to drop-offs, home composting is a low-risk endeavor. The most challenging thing is that inexperienced composters often focus on unnecessary details like the compostability of orange peels. As educators and others on the inside of the community composting movement, we've got to make composting a lot simpler. To me, that means expanding the idea of home composting initiatives to include not just backyard composting but also drop-offs, collection, and compost end use.

Not everyone is going to compost in his or her backyard. I think a lot more households can and will, but every community should at least have a drop-off option, and communities with sufficient density should make collection options available. Awareness of those options and creating them if they don't exist are some of the first steps in home composting initiatives. And while we're doing the work of presenting folks with their composting options, why not also point out that they can also support composting simply by using compost! This is crucial if we want to keep the movement growing.

I tell people who want to compost in their backyard two things: (1) Keep the food scraps covered and/or contained;* and (2) give the compost a lot of time (a year or more). If they do these two things, they can pretty much compost everything they generate, with the one exception of weeds and diseased plant parts. Coverage/containment deals with odors and pests, and time deals with potential human pathogens through microbial competition and antagonism. Without temperature treatment, I wouldn't risk adding weed seeds or plant pathogens, though. If people can't manage these three C's of home composting—cover, contain, and complete—then I think they should consider composting through a drop-off or collection program.

The fourth C would of course be C (carbon), but having adequate carbon is sometimes a roadblock for home composters. I tell people that adding carbon will improve their process and end product, but it's not a necessity; lots of people compost in their

* Containment in areas where animals are a problem means securing bins with hardware cloth or using sealed systems, like tumblers.

backyard without adding much C, and telling them they need C can actually be a deterrent. That said, one part of a home composting initiative could be to source carbon materials that residents could pick up for use at home. For example, you could have a reverse drop-off point: a residential food scrap drop-off site where there is also a pile of wood chips and shredded leaves available for residents to pick up and add to their compost piles at home. It's also good to remind people that they can always use compost or *overs* (screened-out particles) as a cover material, whereas their carbon materials are most valuable blended into their food scraps, although they are certainly good cover as well.

CHAPTER TWO

Composting Methods and Technologies

This chapter introduces a number of different styles of organics recycling, five of which are given full-chapter treatment later in the book. In each system-specific chapter, the details of design and function are covered in much greater depth. Here, the methods are described and discussed in simple terms, with an eye on context and common applications. Some of these methods, including *turned windrows* and *bin/bay systems*, are quite common. Others, such as *composting with animals* and *compost heat recovery*, are less common, or at least less prominently discussed.

Choosing a composting method can be very straightforward, but in most cases there is value in considering multiple options before pursuing any specific design in great detail. There is no one perfect or best method; each system has both strengths and weaknesses, which can make it hard to accurately compare them. Identifying the value that you aim to derive from the system early on is key. For example, an institutional on-site composter (such as a university cafeteria) may value low labor costs above all else, so making the up-front investment into an in-vessel system might make a lot of sense. A community farm that views the input of human labor as an engagement tool might choose a lower-tech solution, such as turned windrows. The more knowledgeable you are about the different options, the more straightforward the assessment and decision process will be.

aeration
(a(-ə)r-ˈā-shən)
DEFINITION: The act or process of dispersing, circulating, or mixing air into a material.

Many sites employ more than one method to manage different compost phases, or they evolve from one method to another as they grow. Piloting new methods is almost always a good idea when possible. Typically, the more advanced methods are utilized for the primary and secondary phases, and windrows or open piles are used for finishing and curing. Table 2.1 organizes the methods in terms of the composting phases to which they are best suited (see *The Phases of Composting* on page 58).

Composting also allows plenty of room and opportunities for pioneering new and adapting old methods. Community-scale composting is uniquely positioned to innovate, collaborate, and potentially revolutionize the way that we recycle organics in the coming decades. We are at a scale where there is little risk in trying new things and big incentive to be frugal. (Most small operations operate on a shoestring.) Also, composting offers great research and education potential if we engage public, private, and nonprofit sectors, as well as research institutions.

Chapters 7 through 11 capture the key planning considerations for their respective styles of composting. In them I use the example of a 10-ton-per-week

Table 2.1. Composting and Food Scrap Recycling Methods by Phase

Composting Phase	Composting and Food Scrap Recycling Method
Pre-processing	Feeding/composting with animals Anaerobic digestion (AD) Pulping and slurrying Dehydration
Primary composting	Turned windrows Bins/bays Aerated static pile (ASP) Passively aerated static pile In-vessel Compost heat recovery
Secondary composting	Turned windrows Bins/bays Aerated static pile (ASP) Passively aerated static pile In-vessel (systems with long retention time) Vermicomposting Compost heat recovery (systems with long retention time)
Finishing	Turned windrows / open piles Bins/bays
Curing	Windrows / open piles Bins/bays

facility (food scraps processed) consistently, as I do throughout the book, so that each method can be easily comparable. I hope that this chapter, along with a more detailed look at the method you're considering, will help your decision-making process.

Turned Windrows

The turned windrow method is by far the most common technique for composting beyond the home scale.[1] A windrow is an elongated pile that is generally turned or "rolled" from the side with a bucket loader, tractor, or specially engineered machine called a windrow turner. Windrows can also be turned by hand. The long shape of a windrow makes it easy to turn the piles and provides surface area for passive airflow into the compost. Windrows also provide a simple means to organize a compost site, by combining and tracking materials of a similar age in a scalable volume or batch.

Several of New York Compost Project's community composting sites started with manual turning of windrows. Teams of volunteers with shovels would get together to turn a windrow and take part in the action of managed composting. Some sites now use skid-steer loaders for turning and

Figure 2.1. Curing windrows on the right, secondary windrows on the left, following ASP. Courtesy of Highfields Center for Composting and Vermont Sustainable Jobs Fund.

are permanently staffed, with five-day workweeks. Red Hook Community Farm continues to hand-turn piles (see the operational profile by David Buckel in chapter 8, page 191).

A drawback with turned windrows is that the piles and work space around them have a large footprint compared with other methods. Windrow turners can speed up the process and take up less space when working between the piles than is needed for loader-turned windrows. However, small turners only turn small windrows, reducing much of the spatial benefit, and large turners are quite expensive. These site capacity dynamics are one of the main subjects covered in the later chapters.

Figure 2.2. An insulated bin primed with a base of about one foot of horse manure hits 140°F before any food scraps have even been added.

Bin and Bay Systems

Composting in *bins* is probably the most common style for backyard home-scale systems, but the concept applies to larger-volume systems as well. As the scale of the system grows, I often refer to a bin as a *bay*, just because it conveys a larger capacity than I associate with a bin. That said, the two words are interchangeable for our purposes; some readers may call a small bin a bay, or vice versa.

Bins are commonly used for demonstration sites, community gardens, neighborhood drop-off networks, and K–12 schools. Bays are utilized on farms, at commercial sites, and in larger institutional settings. Forced aeration is sometimes integrated into the design at both a small and a large scale.

The concept of these systems is quite straightforward: Materials are contained in a bay with sides—a wire bin, straw bales, or any number of other designs—that support the compost pile's vertical edge. Walls allow more materials to fit into a smaller space than would open piles or windrows with angled sides. The material in the bin is turned for aeration and can be accessed from one or more sides for loading and unloading.

There are commercially available bins of many different sizes and shapes designed for backyard use, but most are too small to meet the needs of on-site and small-scale composters beyond a very small school or office. Systems designed for home use would quickly be overwhelmed.

In climates where winter temperatures fall below freezing, bins with insulation and at least several cubic yards of material are helpful to bring the compost into the optimal temperature range. Larger bins or bays are typically custom-built, and while they can be constructed from repurposed materials such as pallets or concrete blocks, in most cases bins that are fully contained in hardware cloth are required to prevent animal intrusions.

One of the potential downsides of bin systems is that they encourage some sites to fill the bin, then leave it unturned and unmanaged until they need to empty it to make room for new material. While this type of neglect can occur with other methods (windrows, for example), with the more advanced composting methods (ASP and in-vessel) regular turnings and aeration are built into the system's management.

Material in well-managed bins flows from bay to bay, getting mixed and aerated along the way. Typically once the first bin is full, it is considered a *batch*, and no new material is added to it unless the recipe needs adjusting. Each batch is then rotated to the next bin and then on to the next bin in a constant succession based on the rate of input. Three-cubic-yard

Figure 2.3. Moving large high-density polyethylene (HDPE) pipes with a pipe wrench. Green Mountain Compost in Williston, Vermont, uses below-grade aeration pipe for their primary ASP and aboveground pipes for their secondary ASP.

Figure 2.4. At a much smaller scale, this O2 Compost Micro-Bin System utilizes ASP on a rooftop at the Brooklyn Grange Rooftop Farm.

or pallet-size bins are common for backyard systems, but at larger scales the number and dimensions of the bins must be specifically sized to meet the needs of the operation.

Aerated Static Pile Composting

The aerated static pile (ASP) method involves running perforated pipes or ductwork underneath compost piles, then actively aerating the material by pushing (positive aeration) or pulling (negative aeration) fresh air through the material with blowers. If designed properly, ductwork distributes the airflow relatively uniformly throughout the material; controls such as timers and temperature sensors are used to operate the fans, which supply fresh oxygen to microbes and also cool the material if need be. ASP systems can be small- or large-scale, and can be custom-built or purchased as fully pre-engineered systems.

ASP systems can come with significant benefits, as well as a learning curve, as is the case with more complex composting systems. Operators typically find temperature and uniform moisture control to be some of the most challenging factors to manage. As a result of these and other factors, some pile turning is still advisable.

ASP systems require monitoring, as well as access to electricity and, ideally, water, but once the pile is built, there is less need to use a tractor, windrow turner, or shovel to provide aeration. ASP systems can save labor and equipment costs over time, and they are space-efficient. Active aeration helps the microbes do their job quickly, thus shortening the composting process by months. Some designs even cut the processing time by as much as half or two-thirds. Shorter processing time means finished compost can leave the site faster or be consolidated into less space, leaving more room for new material, thereby increasing the throughput capacity.

The material in ASP systems can also be piled very closely, because there is less need for work space, which is where people and equipment move in managing the material. Material can be stacked up against other batches of material. The internally supplied aeration means that it can be stacked into taller piles as well. Ultimately this adds up to a great deal of efficiency in space and time.

Another major advantage to ASP systems is that food scraps stay contained, giving them time to break down before they are exposed through turning or moving. Physical containment of food scraps by "capping" freshly blended material with a layer of organic matter or compost reduces the likelihood of attracting birds or other critters, which can become nuisances or vectors. (This is also a best management practice for all open piles.)

With *negative aeration* an external biofilter can be created to treat the odorous exhaust that has been pulled out of the pile by the blower. The exhaust is forced through a pile of high-carbon, high-surface-area, and yet porous organic media. With *positive aeration* a biofilter layer is used to cap the surface of the pile. Either way, by filtering air from the active composting process through the biofilter, the chance for nuisance smells is reduced.

Well-designed and -managed systems can control pile temperatures in the ideal range of 120 to 150°F, and they easily hit the 131°F threshold for the three days required by organic standards and other regulations, including the new FDA Food Safety Modernization Act (FSMA).[2]

ASPs can have their downsides, too. Controlling temperatures, managing moisture levels, and homogenization all take time to get right. With aboveground (pipe-on-grade or POG), aeration pipes have to be worked around and moved, get taken apart, and be set up again. Belowground channels require more cleaning. Still, ASPs are a great solution in many cases, and I've developed a strong preference for this method.

Passively Aerated Static Pile

Passive aeration of compost is the process that all composting methods rely on to some degree when not being actively aerated through turning or forced aeration with blowers. The process relies on the interstitial space (pore space) within the matrix of organic particles, and the processes of convection and diffusion of gases (namely oxygen) through the pile. Large particles (½ to 3 inches) in the pile create a porous architecture that allows passive aeration to take place, an important factor in any composting recipe that is often overlooked. As compost heats, it creates a convective "chimney" effect, with the heat rising out of the top of the pile, pulling fresh cool air in through its base (figure 3.3, page 55).

A compost pile whose base is designed to assist this natural process is often called a passively aerated static pile. Designs are highly variable and can be as simple as piling the material on a thick base of coarse wood chips. Others build air channels beneath the pile as for an ASP system and allow convection to do the work of the blower. Small bin systems are often built on wooden pallets or flooring, which is itself an effective form of passive aeration, because the gap increases the flow of oxygen at the pile's base.

Composters using this method should make sure that the aeration system does not provide an access point for critters, particularly with larger systems that serve the public. Covering the ends of plastic pipes with hardware cloth, for example, allows air to flow in and is rodent-proof (rodents like tunnels). As with every system, turning periodically will decrease the likelihood of critters taking up residence. Turning will also speed up the process and create a more uniform and finished-feeling product. While static piles are commonly used in community gardens, urban farms, and other community-based projects, I do not recommend passively aerated compost that does not involve some turning and heat treatment.

In-Vessel Composting

The term *in-vessel composters* refers to a wide array of compost system styles whose unifying feature is that they are enclosed, giving them a wide range of advantages over open methods in many situations. The terms *in-vessel*, *reactor*, and *digester* are often used interchangeably,[3] but we'll stick to *in-vessel* for simplicity's sake.

These systems come in sizes aimed at a variety of scales, from backyard to large industrial facilities.

Plastic tumblers and enclosed bins are widely used at the home scale, and although several larger tumblers, including insulated tumblers, can reach thermophilic temperatures, the largest of these options would only be capable of meeting the needs of the smallest institutions. In Montreal a tumbler called the Joracan is used in a city program, where multiple families share a single unit, but this only works during warmer months. There are numerous options that fall between the home and the large industrial scale that are ideal for on-site composting and still quite small in scale.

The most common styles of small-scale in-vessel systems include rotating drums, containerized ASP systems, and auger-turned systems. In the more high-tech systems, oxygen, moisture, and temperature can be automatically monitored and controlled. Non-food scrap composting applications for in-vessel systems include biosolids at the industrial scale, as well as animal mortalities (dead livestock) on farms.

In-vessel systems are popular for applications where space is limited. They can take up little space relative to other composting systems and process material efficiently. Nuisances such as odors and pests are mitigated through containment, aeration, and active biofiltration. Most in-vessel systems are designed for a hot and fast primary composting stage, but will require secondary composting, finishing, and curing. Typically what comes out of the vessel initially will not be mature enough for most uses. However, if the system is sized effectively, visible food scraps and the most offensively odorous compounds should be broken down.

The companies that produce in-vessel systems promote their efficiency and control. There are a huge range of designs, costs, capacities, and product

Figure 2.5. The auger in the Green Mountain Technologies Earth Flow mixes and moves material from one end of a shallow containerized bed to the other over about a three-week processing period.

features across the market. There are also technologies that sometimes call themselves composters or market themselves in the composter genre, but actually produce a very different type of end product (namely dehydrators and flushing systems).

Vermicomposting

Composting with worms, or *vermicomposting*, involves introducing specific species of worms into the composting process to decompose raw or partially composted organic materials into a rich finished compost. People of all ages are drawn to worms, which, alongside springtails, pseudoscorpions, and other creatures, are excellent visual aids for public engagement and education. *Eisenia fetida* and *E. andrei*, often called red wigglers, are the most popular types of worms for vermicomposting, although a wide variety of other species are used throughout the world. The ideal temperatures for these worms is between 59 and 77°F, although they remain active even in 50°F basements.[4] Thus, vermicomposting is a mesophilic process, but not hot enough to inactivate weeds and other type of seeds. Worms have been shown to be very effective at reducing many human pathogens (*Salmonella* spp., fecal coliforms, enteric viruses) in biosolids (humanure compost).[5] However, they do not appear quite as effective in terms of parasites (*Ascaris suum*).[6] Many composters who want extra assurance will first use a thermophilic composting process, followed by vermicomposting as a secondary phase.

Vermicomposting produces *vermicompost*, which worm aficionados recognize as distinct and often preferable to purer worm *castings* (worm excrement). Vermicompost typically contains a diverse matrix of organic components compared with castings, which is believed to be a benefit in terms of crop performance. Concentrated castings are usually marketed as such, whereas products with a higher diversity of other compounds are considered worm compost or vermicompost. Since many laypeople don't know what vermicompost is, from a marketing perspective using the term *worm castings* or just plain *worm compost* may be more attractive.

Figure 2.6. Inoculating a continuous flow vermicomposting bed with *Eisenia fetida*. Courtesy of Highfields Center for Composting and Vermont Sustainable Jobs Fund.

Worm composts typically contain higher levels of plant-available nitrogen than straight compost and contain natural hormones (produced in the worms' bodies) that promote desirable traits in plants.[7]

In terms of design, worm composters are distinct from other compost systems. Good vermicomposter designs facilitate the harvesting of finished compost while minimizing the need to manually move or separate the worms after harvesting. The volume and depth of the material needs to be limited to prevent heat buildup, although even at the 2-foot depth standard for most commercial worm beds, temperatures can rise well above the worms' inhabitable range. To prevent overheating, the age, flow rate, and moisture content of the feedstock need to be closely managed.

While there are batch methods for composting with worms, continuous flow designs offer advantages. These designs work on a simple principle: Feed from the top, harvest from the bottom. This is harmonious with the worms' natural tendency to migrate to the top layers of the soil, where they can feed on fresh organic matter and abundant microbes. Continuous flow vermicomposting beds (often called reactors) are open boxes, usually 2 feet deep and various widths and lengths. Fresh material is fed at the top of the bed, and finished material is cut from the bottom with a blade. The worms tend to stay at the top, free and clear of the blade, although occasionally worms will emerge at the bottom. Many popular worm bins for the home and classroom work similarly to the continuous flow bins and have interchangeable layers so older material can be harvested off the bottom trays, which, once emptied, are then moved to the top where they can be filled with fresh material. Some larger operations will combine multiple bins or create slightly larger versions for their purposes.

Mechanized vermicomposting feeding and harvesting systems have been developed for the large commercial scale, but small-scale systems are usually manually operated. Where labor and space allow, vermicompost systems are excellent for transforming a small amount of material into high-value compost. For example, a 40 × 5 × 2-foot (L × W × H) continuous flow bed can process about 1 cubic yard per week of pretreated compost (four to six weeks old). That would equate to 300 to 400 pounds per week of food scrap throughput.

A vermicomposting setup like this also requires work space and area for the pre-composting of materials, with overall space requirements of about 800 square feet, ideally covered and enclosed, and possibly heated. If you plan to sell vermicompost, you probably also need space to dry, screen, and possibly bag it. If you compare this option with other composting methods in terms of start-up costs, labor, and space, most other systems are more efficient pound for pound in terms of food scraps processed. The economic justification of vermicomposting would depend upon a number of variables, including access to high-end horticultural markets.

During the start-up period of a vermicomposting system, the worm population grows and the system gets charged with material. It takes 6 to 12 months for a system to reach full capacity, depending upon how many worms you start with, and plans should be made to increase throughput gradually in the interim. Once the system is up and running, the processing time should be three to four months to produce mature vermicompost from raw material. The system described above would produce 15 to 25 cubic yards per year of finished vermicompost. Generally, vermicompost requires cover and may require heat depending upon your location (there are worm composters in in the northeastern United States that supply no additional heat to their buildings).

Vermicomposting is a topic worthy of in-depth discussion that is beyond the scope of this book. For more, see *The Worm Farmer's Handbook*, by vermicomposting expert Rhonda Sherman.

Static Pile

A static pile is simply a compost pile that is formed and then left primarily unturned. With an adequate recipe, the pile may still achieve high temperatures and maintain some level of aerobic activity; however, static piles often take a year or two to fully break down, and the finished material will be less then homogeneous. Static pile composting is not a best management practice for food scrap composters. In general, some level of active management is greatly encouraged to achieve a hot pile that will inactivate pathogen and weed seeds, deter pests, speed up the process, and educate and engage the public about the art and science of managed composting. I strongly advise against an unturned approach for systems handling community food scraps.

Food Scraps as Animal Feed, and Composting with Animals

With the increasing focus on food scraps as resource rather than waste comes greater recognition and interest in using community-collected food scraps to feed animals. Feeding scraps to animals was in fact the norm (and still is in many places) up until very recently in our history. From an environmental and food production perspective, there is no better use for foods that are no longer consumable by humans than to feed them to a local animal.

The majority of food residuals in the United States is fed to pigs. However, composting with chickens is the main area focus in this book due to a growing interest in chickens by community composters. Some conscientious consumers prefer buying eggs from chickens fed on human food where they are available, recognizing both the environmental and health benefits of avoiding entirely grain-based diets. Vermont Compost Company in East Montpelier, Vermont, is well known for having fed hundreds of chickens without grain for years, and there is a voracious market for their high-quality eggs. They have integrated laying hens into their unique composting process, which starts by making a compost recipe with the food scraps, then letting their birds forage on the compost blend as it becomes active. They then segregate the compost once it's been heat-treated to ensure the chickens do not introduce any untreated excrement.

Other sites, such as Earth Matter on Governors Island in New York City (see figure 2.7), have a food scrap chicken feeding area bedded with high-carbon amendments. They frequently remove the bedding and any uneaten food scraps for composting at a nearby location.

If the number of birds is adequate to consume the volume of food scraps you're feeding, the total volume of scraps can be reduced by 90 percent or more. There is a lack of detailed research on this practice, but one rubric that is used and appears to be relatively reliable is to feed approximately 2 pounds of scraps per bird per day.

Figure 2.7. The hen feeding and pre-composting area at Earth Matter on Governors Island, New York City (Charlie Bayrer, *left*; Marisa DeDominicis, *right*).

Figure 2.8. Following a hydronic heat exchange, ASP exhaust at Jasper Hill Farm in Greensboro, Vermont, is dispersed into a growing bed, which acts as both biofilter and radiant heater while supplying CO_2 to greenhouse plants. Courtesy of Highfields Center for Composting and Vermont Sustainable Jobs Fund.

The report *Leftovers for Livestock: A Legal Guide for Using Excess Food as Animal Feed* provides excellent summaries of state-by-state regulations along with applicable federal laws.[8] Check with local authorities about your state's requirements before feeding community scraps to animals.

There is much more on nutrition, regulations, and other subjects in the *Composting with Animals* section in chapter 11, page 287.

Compost Heat Recovery

As aerobic microbes consume organic materials in the composting process, their metabolism generates heat, which is released into the surrounding environment. What if that heat could be captured and put to work? Several composters around the country have developed new approaches to do just that.

Vermont-based Agrilab Technologies Inc. has developed a number of compost heat recovery (CHR) systems that involve negative aeration pulling heat and vapor from hot compost, then using a heat exchanger to transfer that heat into water. These systems were based on earlier work pioneered by Bruce Fulfurd (City Soil and Greenhouse) and others at the New Alchemy Institute. The prototype pulled the heat and vapor into a biofilter or planter bed to supply CO_2 and heat to a greenhouse. This type of compost heat recovery is both scalable and economical, although the design needs to balance inputs, heat, and CO_2 while filtering nitrogen gases in order to avoid damage to plant and human health.

Other heat recovery models include the "Jean Pain" mound and a containerized ASP system developed by dairy farmer Conan Eaton (figure 10.10, page 285), to name just two. CHR could also theoretically be combined with numerous existing composters on the market. Agrilab's Compost Drum Dragon, for example, is designed to easily connect with rotating in-vessel drums, creating a continuous supply of hot water for heating and other uses.

Anaerobic Digestion

Without oxygen, microbial degradation of organic matter produces different by-products than aerobic (with oxygen) composting. One of those by-products is methane (CH_4), a naturally occurring flammable gas. Anthropogenic (human-caused) forms of methane that escape into the atmosphere (such as from landfills, agriculture, and energy production) are responsible for 10.6 percent of global greenhouse gas emissions.[9] Cutting greenhouse gas emissions is one critical reason to keep organics out of landfills, where they produce and emit methane. Of all the greenhouse gases generated by the waste sector in the United States, 86.3 percent were emitted by landfills, which is 20.4 percent of total methane emissions according to the US EPA.[10]

A controlled process of anaerobic decomposition, called anaerobic digestion (AD), captures methane for use as a fuel. Technically and biologically, AD is more complex than composting. Microbes called methanogens generate methane, and the process involves handling and combusting a flammable and potentially corrosive gas.

Nevertheless, large-scale AD is a widespread practice in northern climates, and on a smaller and even residential scale in many warmer parts of the world, where proper temperatures for methanogens can be easily maintained (around 100°F). There are also a number of small-scale AD models in the United States, although very few correspond to the scale and scope of the composting models covered elsewhere in this guide. I lack deep familiarity with AD systems, but several potential systems to research include Impact Bioenergy, Sistema BioBolsa, and Hestia Home Biogas.

While the AD and composting sectors of organics recycling can be at odds with each other to some degree, there is also broad recognition that neither is the panacea that it might appear on the surface. Functionally, the two can actually be complementary when AD is used as a preliminary treatment, followed by a secondary hot composting phase. With liquid AD systems, this is possible by separating out the solids following the anaerobic phase. Solids extracted from the digester's out-coming material, *digestate*, are then composted, which adds value to the outputs. Some AD systems known as *high-solids* or *dry fermentation* digesters are designed to work at a lower moisture content range. These systems do not require a solids separator, because the digestate is already a relatively solid feedstock. There are even high-solids AD systems that can go from anaerobic to fully aerobic composting in the same unit.

Food Scrap Dehydrators

A number of products on the market process food scraps into a temporarily stable form by grinding and removing moisture from the material. I refer to these as *dehydrators*, but they are called by other names and are even confused sometimes with composters. Some systems reportedly reduce the total volume and/or weight of the food scraps by as much as 85 to 95 percent.[11] While these machines are marketed as producing a soil amendment, the end product would never be confused with compost by anyone who knew what they were looking for. The material that I've seen had an orange-ish hue, a slightly sour smell, and very uniform particles, almost like cornmeal. Users of the dehydrated end product as a soil amendment have reported issues with the product promoting fungus and attracting vectors.[12]

Large generators typically install these systems on-site, rather than sending materials to an off-site processor. Some vendors say that the end product makes a great animal feed. I am aware of at least one system in California that uses a similar process to create a heat-treated commercial animal feed.

While most composters won't have any use for dehydrator technology, they are part of the food scrap recycling technology industry, so at the very least it's good to be aware that they exist. Also, I haven't looked into the economics of these systems in much depth, but there may be times when it would be cost-effective from a collection perspective to install a unit on-site at a large generator, in order to reduce the volume and hence reduce hauling costs for collecting the material for further processing off-site. Often the systems are leased, so that also makes them attractive from a cost perspective.

I would be remiss if I did not mention that my first encounters with dehydrator systems involved several horror stories, including one about a large waste hauler who distributed these as a composting solution to a few very large businesses, including one ski resort that generated a few tons of food scraps a week. The system would randomly spew out unfinished material that was described as vomit-like. Not a pretty picture, and certainly not an experience that we want associated with composting. This happened several years ago, and there are success stories,[13] so I think that the technologies are getting more reliable, but there is always a slightly higher risk when you are (1) processing food scrap materials on-site (composting or otherwise) and (2) relying on technologies that have a high level of complexity. The bottom line is that "food scrap processor" is rarely a role generators want to play; it's just not core to their business, and when problems arise, it can be a real distraction. Needless to say, if you want to use a system like this, do your research and make sure that the vendor you work with has ironed out the kinks.

For on-site composters, some vendors sell de-waterers, which are a different class of pretreatment than dehydrators. By removing water, the total volume and, assumedly, the dry matter needed to compost effectively are reduced as well. Essentially this reduces the size of the composter (or dehydrator) required or the volume that would need to be hauled, whatever the case may be.

Pulping and Flushing Equipment

Another class of equipment that overlaps frequently with composting systems basically turns food scraps into flushable or pumpable slurry. For our purposes these systems can be grouped together broadly as

Figure 2.9. High-capacity slurrying equipment under construction in Charlestown, Boston. The system decontaminates organics and transforms them into a pumpable form that's acceptable to AD facilities. The slurry is currently transported to a wastewater treatment plant in North Andover, Massachusetts.

pulping and *flushing systems*. Some flushing systems are also known as biodigesters, because they are designed to reduce biological oxygen demand (BOD; for more on this subject see *Biological Oxygen Demand*, page 53) before flushing. Since *biodigester* is a term that is commonly used for AD, I find it confusing in this application and try to avoid it.

On a home or individual generator scale, the slurry usually goes down the drain where it can be managed as sewage. Most people are familiar with in-sink food disposals in homes. Larger food scrap generators take a whole range of forms, but the concept is similar and the end destination is usually the sewage treatment plant. At large businesses, like grocery stores, picture something that looks like a small in-vessel composter attached to the dishwasher. It grinds and slurries food scraps while removing large contamination that would clog the plumbing.

On a *really* large scale, material is typically trucked to the slurrying operation, processed, then transported again to a digester, which may or may not also be a water resource recovery facility (WRRF), formerly known as a wastewater treatment plant (see figure 2.9).

As with dehydrators, most composters won't have any use for pulping systems—or rather, wouldn't find them economical to implement. I wanted to mention them in this section because it's good for composters to be aware of them as one option for a food scrap generator.

On the surface these systems make a lot of sense, particularly for operations where on-site composting would be challenging. Using existing sewer pipes for transporting, and existing WRRF for processing, reduces the need for transportation and organics recycling infrastructure (which may or may not exist). Often some of the energy in the material is captured, because many wastewater treatment plants include anaerobic digestion as a pre-processing step to reduce biological oxygen demand prior to treatment. Also, solids are sometimes captured and composted or turned into fertilizer on the back end of the treatment plant, so often some of the nutrients have been retained and recycled.

Where many people take issue with the sewage treatment plant approach is when it is considered as an equivalent or replacement for creating high-quality composts for communities. As previously mentioned, the solids fractions coming out of the WRRFs are not always separated and composted, making this practice "composting" in name only. Even where the by-products of the WRRF are composted, it's important to acknowledge that among food producers, compost made from food scraps is typically considered a desirable feedstock whereas compost made from biosolids (municipal sewage) is not (see *Making Waste Management Compatible with Our Food System* on page 7). Also, there are still big unanswered questions about the carbon footprint and larger environmental/social/community benefits of this model compared with composting and other food scrap recycling options.

Community-Scale Food Scrap Collection

Although food scrap collection is not technically a composting method, it is an essential component of the closed community compost loop. Collection is a large subject and is covered in much greater depth in chapter 12. On the surface, the equipment and methods look very similar to recycling or trash collection, but don't be fooled; to collect organics well is a nuanced endeavor. A lot can be learned from what others in the trade are doing well or would like to be doing better.

As an overview, I will briefly mention the current common collection models. At the smallest scale are the micro routes, described in chapter 1, where there is a one-to-three-generator to one-composter relationship, run in personal cars and pickup trucks. Bicycle collection systems can move 200 to 300 pounds per load (more with electric assist). Box trucks, non-dump trailers (such as horse trailers), and pickup trucks that are retrofitted with a dump body and tote tipping equipment can typically haul 0.5 to 2 tons per load. Dump trailers like the one in

Table 2.2. Applications and Challenges of Composting and Food Scrap Recycling Methods

	Methodology	Common Applications	Common Challenges
Common methods	**Turned windrows**	On-farm composting systems Sites where space is not limiting Sites with few or no nearby neighbors Early-stage sites	Undersizing a site for the volume of material it will handle Inadequate space for loader operation, leading to inefficient/suboptimal pile management Insufficient vector controls (rodents, birds)
	Bin systems	Schools Community gardens Combined with small-scale forced aeration (aerated static pile)	Under-turning Piles not heating in winter (insulation can help) Insufficient rodent-proofing Drying and uneven moisture
	Aerated static pile (ASP) composting	On-farm, commercial, and community composting sites Sites where space is limiting Sites with nearby neighbors Used for the primary (active) composting phase followed by secondary turned windrows Expansion strategy for later-stage sites	Overheating Uneven air distribution due to poor design Drying Preferential air channeling Non-uniform end product Need for mixer Higher upfront cost Lack of access to electricity and water in remote locations Insufficient vector controls (rodents, birds)
	In-vessel composting	Small to large institutions and community composting sites Sites where space is limiting Sites with nearby neighbors Used for primary phase followed by secondary turned windrows/piles or other	Upfront cost User error (such as improper recipe, adding large materials that causes clogs)
	Vermicomposting (worm composting)	Micro-scale or large commercial operations, demonstration sites, on-farm composting, K–12 schools, and universities Sites where space is not limiting (e.g. low throughput) Used for secondary phase following thermophilic primary composting phase Used where high-end crops justify high compost production costs	Often labor-intensive Worms have unique habitat requirements, which require careful management Not space-efficient Flies, mites, and other insects in indoor systems

figures 11.6 and 12.24 (pages 303 and 342 respectively) can potentially move and tip in the 4-to-5-ton range. Then there are collection trucks that have been retrofitted for food, often roll-off dumpster trucks and sometimes packer trucks and other traditional trash trucks. Some trucks are capable of multi-stream collection, simultaneously picking up food scraps with recycling and/or trash in one trip, with separate compartments for storing each.

One important distinction among collection systems is that at the point of generation, food scraps are typically contained in a plastic cart (picture one

	Method	Applications	Challenges
Less common methods	**Composting with animals**	On-farm composters Used where daily labor is available for animal husbandry Compost pretreatment through early active phase	Undersizing system Insufficient vector controls (rodents, birds)
	Compost heat recovery	On-farm composters Applications with demand for thermal energy (greenhouses, farm process water, drying compost)	Relatively new technology, requires significant learning May require year-round thermal demand to justify investment Low-cost systems are labor-intensive Lack of tested and affordable systems at the small scale
	Passively aerated static pile	Static piles that are turned semi-frequently are used in similar applications to turned windrow and bin composting methods Unturned piles are common, but this is not recommended as a strategy	Insufficient vector controls (rodents, birds) Aeration pipes can provide access to rodents (cover ends) Static piles can create habitat for critters Non-uniform end product
	Static piles	Home composting On-farm composting of manures NOT SUITED FOR FOOD SCRAP COMPOSTING AT COMMUNITY SCALE	Does not meet temperature treatment best management practices (BMP) Insufficient vector controls (rodents, birds)
Non-composting methods	**Anaerobic digestion**	Pre-processing with composting of solid portion of digestate (solids separator following AD) Residential, small-farm, and community applications in warm climates Large-farm, municipal, and industrial applications in colder climates Community-scale AD may become more common in near future Good for food processing and liquid food residuals	Technically complex Not well suited for physical contamination High up-front investment Status as renewable energy dependent on handling and end uses of digestate
	Dehydration	Used on-site to reduce volume and moisture content (MC) Potential to create animal feed	Finished product not a stable soil amendment, and has limited applications compared with compost Reports of problems with material being released prematurely
	Slurrying Systems	Used as a pretreatment for AD and WRRCs May do a good job of removing contamination	Typically involves off-site pre-processing (not always), then transportation to processor Systems that combine scraps with sewage limit options for end use

of those trash cans with two wheels) or in a bucket for small generators or residences. The material is picked up and transported to a composting facility. Some collection equipment is designed to tip the food scraps into the truck or trailer, wash the tote or bucket (ideally) on-site, and return the same tote or bucket to the generator. Other equipment collects the full cart and swaps it with a different clean, empty one. The hauler then transports the full carts to the compost facility, tips the cart or buckets, and washes them. I often refer to this as cart *tipping* versus *swapping*. There are also collection services that tip but

Figure 2.10. A Compost Now van loaded full of 5-gallon buckets on their residential food scrap collection route. Courtesy of Compost Now.

don't wash and leave a dirty tote. While this reduces the cost of collection, it is not ideal for obvious reasons, not the least of which is that if a competitor cleans totes they will have a big advantage in terms of client satisfaction.

System Compatibility and Comparison

I hope it will become abundantly apparent as you read this book that different composting methods often make sense at different stages of the composting process. Therefore, operations often use multiple methods. Learning about and weighing the possible system options is a critical part of the assessment work described in chapter 5.

The inherent challenge in choosing a system you have never used is simply that it's hard to know exactly what will work for you. This will feel especially risky if you're looking at making a significant investment, which is why due diligence is so important and also why I like systems that are scalable and can be piloted before investing in a full-scale system.

Tables 2.1 and 2.2 summarize the composting methods in terms of their common utility during the four phases of composting, as well as some common applications and challenges.

CHAPTER THREE

The Composting Process

In order to design and manage the systems that are the focus of the book, a solid foundation in composting fundamentals (Compost 101) is critical. While chapter 13 looks at the management techniques composters use to produce quality, consistent results, this chapter covers the fundamentals from a more theoretical perspective—the *what* and *why*, rather than the *who*, *where*, and *how*.

In this book I discuss a "managed" style of composting, which is critical when handling food scraps at scale. Without going too heavily into the science, I'll give an overview of what is involved in creating conditions conducive to effectively composting food scraps: the use of recipes, aeration, agitation, and containment. We'll look at *thermophilic* (hot) composting and the chemical and biological factors that underlie it, as well as how to meet temperature treatment targets, which is a goal for most operators.

Later in the chapter we'll look at four phases of composting: the primary, secondary, finishing, and curing phases. The way that I classify these phases may be slightly differently than in other literature on the subject. I explain why I find it useful to think of the phases in this way, in terms of systems design and process management. The chapter starts with a discussion of the intensity of management practiced by different styles of operators and the reasons for implementing compost *best management practices* or *BMPs*.

putrescible
(pyo͞o'tresəbəl)
DEFINITION: Subject to decay and decomposition.

Managed Versus Non-Managed Systems

First, a recap of the definition of *composting* that I use:

> *The return of organic wastes to a rich, stable, humus-like material through a managed oxidative decomposition process that is mediated by microbe metabolism.*

The key word here is *managed*. While the bumper sticker COMPOST HAPPENS is technically correct, it's a misconception that what compost operations managers do equates to the dispersed decomposition taking place around us all the time.

Nature does not combine moisture, protein, and other dead putrescible compounds in the volumes that food scrap composters handle and then leave them to rot unattended. This would be the equivalent of dropping a dead elephant once a week at your operation and expecting only microbes to show up (and hoping the neighbors won't mind the smell of the decomposition that ensues). The defining line between composting and, for lack of a better word, "natural" decomposition is that composting is managed.

Composters often go to great efforts to control what is otherwise an entirely natural set of processes. We operate somewhere on the boundary between farming, biology, manufacturing, and existentialism (and yet we are often regulated as solid waste management operations, go figure).

The degree to which an operator intercedes in the process can vary greatly by composter and method. Some have argued for the value of less intensively managed methods, while others prefer more intensive management. For example, in his paper "Sustainability of Modern Composting," Dr. William Brinton found that very little research had been conducted to verify the need for intensified compost methodologies such as windrow turners and aerated static pile (ASP) composting. His research on the topic looked at the costs and effects of turning frequency in manure-based composts and found the benefits of more frequent turning to be minimal in terms of stabilization and pathogen reduction, while the costs per wet ton went up over 1300 percent. With this rise in cost came increased loss in nitrogen and organic matter.[1]

The sustainable concept of "less is more" is an extremely valuable perspective to keep in mind, and on the whole I tend to agree with the assessment that the economic and ecological benefits of technological interventions need to be thoroughly vetted. There are reasons why operations may take a more aggressive approach based on the materials they are handling and their intended end use.

With food scraps in particular, there is no getting away from the need to adopt and religiously implement BMPs, even if it sometimes feels like we are verging on compost dogma. In my experience trouble ensues when food scrap composters overlook management norms. My perspective as a consultant is that the compost operations that take pride in implementing BMPs are the ones that succeed, both in terms of growth and in terms of their overarching mission of engaging communities in recycling organic wastes. Ultimately, every composter will need to identify and adopt the management practices required to produce a product that is the right quality for their intended end uses, while keeping potential nuisance and pollution issues at bay and ensuring regulatory compliance.

Decomposers' Basic Needs: Food, Air, Water, and Warmth

The first time I visited a compost site with my mentor, Tom Gilbert, he explained to our clients how to create the conditions conducive to hot composting in a way that stuck with me. He said that the most active microbes in the pile need the same things that we do. They need oxygen (not as much as we do, but some). They need water, or pile moisture content, as we call it. They need food: essentially, digestible carbon for energy and proteins to build cells and multiply. And for the microbes to be really active consumers, they need warmth.

If the right balance of food, air, and moisture is struck, the microbes will generate their own heat through metabolic activities, creating a positive feedback loop where rising temperatures increase microbial activity, which in turn increases temperature further, and so on. When the process is working well, enough metabolic energy is released that temperatures will easily rise well above the 131°F required to inactivate human pathogens, plant pathogens, or weed seeds that may be present in the parent material. In their wake the microbes leave behind the rich, black, humus-like soil amendment that we call compost.

That is the simple version.

The art of composting is how you understand, combine, and manage your different raw *feedstocks* (raw materials) to create optimal conditions throughout the process. Learning to compost involves managing and eventually mastering the interplay between these four needs of microbes. The basic tools for doing this include *recipe*, *pile monitoring*, *aeration*, and *agitation*.

In the first part of this chapter, we're going to talk about the roles of aeration, agitation, temperature, and temperature treatment. Recipe and feedstocks, which are the food and moisture part of the composting equation, are covered in more depth in the next chapter.

Air

When consuming organic matter, microbes use oxygen at a measurable rate. Without a consistent supply, oxygen becomes depleted, leading to anaerobic conditions and a reduced rate of decay. With an adequate supply of oxygen, the rate of decomposition is maximized. Think of the provision of oxygen to compost as a breath of air. Over time the pile exhales. Through various aeration methods, we act as the pile's lungs, giving it a fresh breath at as optimal a frequency as possible.

Early in the process, the pile's exhale is fast, whereas later it grows much slower until eventually the pile can basically breathe on its own. A more scientific way of understanding the oxygen demands of decomposers is BOD and COD.

Biological Oxygen Demand

Biological oxygen demand, or *BOD*, refers to the measurable rate at which oxygen is consumed by aerobic microbes over a specific time period.[2] (BOD can also stand for *biochemical oxygen demand*, which means the same thing.) *Chemical oxygen demand*, or *COD*, refers to the total oxygen required to oxidize organic matter into CO_2.[3] I find BOD and COD to be helpful concepts toward understanding the relationship between oxygen supply/demand and compost stabilization.

The science on compost and BOD/COD shows that an optimal level of oxygen exists that could be supplied at any given moment, and that over time that level decreases as carbon is metabolized (although initially BOD actually increases as composting initiates). When oxygen is limited, composting slows. When an operation meets the oxygen demand, the microbes are extremely efficient consumers.

A study by Singh et al. points to the dynamic between recipe and BOD and COD rates. Too much carbon can slow the process, as can too little. The research compared the BOD, COD, and BOD:COD ratio over 20 days in composts with carbon to nitrogen (C:N) ratios of 16, 22, and 30, processed in a large rotating drum. In the compost that had a C:N of 22, BOD jumped and dropped again by over 100 percent between days 3 and 8, meaning the microbial demand for oxygen at day 5 was more than twice that of day 3. Over the 20-day period this compost had a BOD and COD reduction of 82 and 70 percent, respectively, showing rapid stabilization that was greater than in either of the other mixes.[4]

Measuring BOD and COD is common in compost research, but for most compost operations, knowing the precise BOD or COD of a compost is not necessary. What may be more helpful is to do a cost-benefit assessment as to whether providing more oxygen would be beneficial, assuming it appears limiting. Some operations have all the time and space in the world and can afford a slower process. Most operations reach a point where space becomes limiting, and therefore more closely aligning the site's aeration rates with microbial demand speeds up the process, which is highly beneficial.

Aeration and Agitation

Consistent provision of fresh air and frequent agitation (or mixing) of compost are two widely accepted best management practices. Each of these is a distinct topic, but because in practice they often take place simultaneously, I'll discuss both here.

Active aeration refers to the processes of physically introducing fresh air, while *passive aeration* refers to the movement of air and oxygen without intervention. Both are important aspects of good system and recipe design. All of the composting methods that are promoted in the industry address aeration and agitation, though there is certainly debate about the degree of effectiveness of different methods.

Aeration and agitation take various forms depending upon the composting method. With turned windrows, active aeration takes place during agitation. In the aerated static pile method, aeration takes place through forced aeration (pushing air through

Figure 3.1. Turning aerates the compost, but it also repositions material from the edges of the piles into the active core where it can heat and break down. Turning speeds up the process and leads to a more homogeneous end product.

the pile with fans). With bins/bays and in-vessel systems, aeration can be supplied through agitation, forced aeration, or a combination of the two.

When designing or assessing a composting system, efficient and low-cost active aeration and agitation is the gold standard. Agitation provides fresh air and also redistributes food sources and moisture in the pile. Systems that provide aeration alone, such as ASP, are cost-effective, but the end results are less uniform than those that incorporate more frequent mixing. Materials in different areas of a pile tend to break down at different rates for a number of reasons, including non-uniform oxygen distribution, drying out, or uneven moisture saturation. The areas at the edge of the pile are also cooler, serving as insulation for the pile's active core, but this also hinders that material's ability to break down.

These stagnant portions of the pile represent a fresh and available food source, but that material

Figure 3.2. A blower acts as the lungs of the compost pile, providing a fresh breath of air at an adjustable interval.

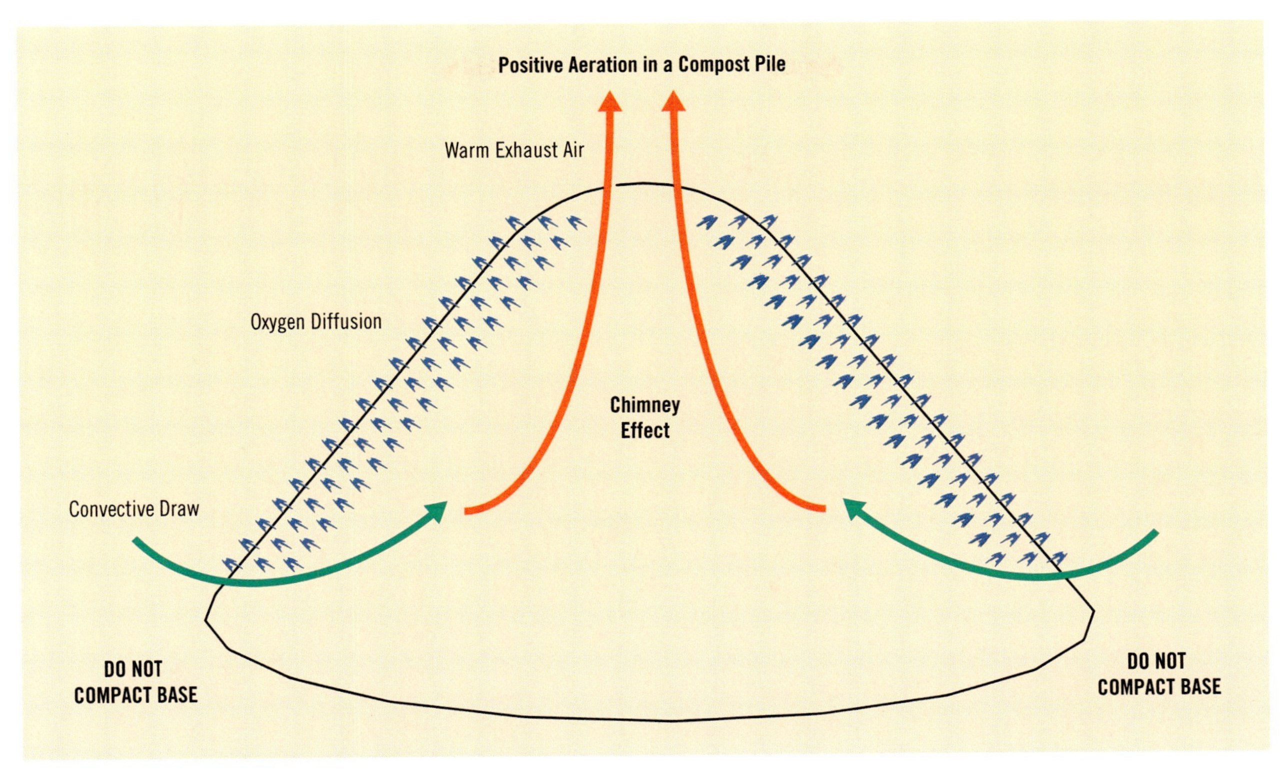

Figure 3.3. Convection and diffusion in a compost pile.

needs to be remixed. The goal is for all of the material in the pile to reach temperature treatment targets (see the following section) to ensure pathogen and weed seed inactivation. When everything else is working properly, the oxygen and fresh food introduced through turning tend to generate a burst of heat (which in turn maintains the composter's endorphin and dopamine levels). The cycle of agitation, oxygen, and fresh food drives the composting and heat treatment process.

Passive Aeration

The passive movement of air plays a critical role in supplying oxygen in between aeration and agitation cycles. Passive oxygen movement in the pile is typically classified in terms of convection or diffusion. *Convection* is the movement of air caused by the heat differential in the pile from top to bottom. Figure 3.3 is a re-creation of the classic compost pile convection graphic, which illustrates how the warm air rising out of the top of the pile pulls cool air in through its base, a process called the chimney effect.

Diffusion is the tendency of substances to intermingle on a chemical gradient. In terms of aeration, this means that high-oxygen-content areas (outside the pile) will essentially bleed into low-oxygen-content areas.

With turned windrows, passive aeration plays an especially vital role because piles are turned daily at a maximum, and more commonly turned every three to seven days during the active phase. Therefore, oxygen supplied when a human is actually doing the turning plays a more minor role in comparison with systems that provide oxygen automatically (ASP or in-vessels that provide virtually constant aeration).

Effectively leveraging passive aeration depends largely upon the porosity of the compost mix (see *Porosity*, chapter 4, page 77). It's a valuable tool and one that it would be a mistake to overlook. However, to create a uniformly composted and stable end product, remixing is required (yes, composters are also DJs) in most systems (with the exception of vermicompost).

Temperature

Microbial activity is much greater under warm conditions. It's why we use refrigerators to preserve our food, why you are more likely to get a staph infection in the tropics (well, the heat and the humidity), and why the composting process works faster as the pile temperature rises. Different microbes are adapted to different temperature ranges, and as luck would have it, the microbes that do well at temperatures that dramatically reduce pathogens also happen to be really efficient at metabolizing organic matter. Research has shown that ideal temperatures for rapid decomposition are in the 122 to 140°F range.[5]

Microbes are classified by the temperature ranges in which they live. When composting takes place in those temperature ranges, it is classified in the same way. Composting at temperatures above 113°F is known as *thermophilic* composting, named after the *thermophiles*. Between 77 and 104°F we call it *mesophilic* composting, after the *mesophiles* that are active in that range (don't ask me what happened to the microbes at 105 to 112°F). Between 32 and 77°F, there is a class of microbes known as *psychrophiles*.[6] Psychrophiles perform some metabolic activities,[7] but during the psychrophilic phase, physical decomposers such as worms, nematodes, and centipedes are also extremely active.

Figure 3.4. Three feet into this frosty pile, it's too hot to touch. The temperature outside is right around 0°F, but compost is a good insulator and can maintain temps above the target 131°F for prolonged periods even during extreme cold events. Turning and over-aeration during extreme cold periods should be avoided.

Hot composting and *cold composting* are common vernacular in home- and small-scale scenarios. While there is no official classification of where cold composting ends and hot composting starts, *hot composting* and *thermophilic composting* are often used interchangeably. It is in fact the case that compost piles can and almost always do encompass both simultaneously. There is a hot zone, surrounded by a warm zone, surrounded by a cool and eventually an ambient zone at the surface of the pile. In large piles the hottest zone is usually only a foot or so into the pile, a zone that has access to oxygen by passive means. I've frequently observed the hot zone moving inward over a several-day time span. My theory is that as oxygen demands lessen on the pile's edges, more oxygen becomes available deeper in the pile, although I've never seen any research on the subject. From a temperature treatment perspective, the goal is to mix and aerate the pile, so that all of the material spends time in this hot zone, as described in the following section.

Temperature Treatment Standards and the Process to Further Reduce Pathogens

A significant body of research has shown that through the high temperatures endemic to hot composting, as well as through biological competition and antagonism and ammonia disinfection, effective composting is able to reduce pathogens to the point that they no longer pose a significant threat to the public.[8] This is not the same as sterilization, which would be counterproductive to the process. We need the beneficial microbes to continue to do their job. It is, in fact, the opposite of sterilization: The material is so abundant with microbial activity that the heat generated reaches temperatures no longer conducive to the undesirable microbes.

Higher heat yields faster heat treatment results. For example, Burge found a 10-fold reduction in *Salmonella* spp. (one of the hardier pathogens) after 89 minutes at 131°F. The same reduction took just 7.5 minutes at 140°F.[9] While high temperature is not the only mechanism for pathogen reduction in compost, it's the factor that is most easily measurable in the field, and therefore it's the primary standard for compost safety in the industry. In addition to pathogen reduction, high-temperature composting inactivates weed seeds in the majority of cases. Several studies have found that temperature-treated compost safely eliminates viable garlic mustard and European buckthorn seeds.[10] When I was at the Highfields Center for Composting, we partnered with the UVM Plant and Soil Science Department as part of a research project to look at weed seed and plant pathogen reductions during thermophilic composting. We composted weed seeds and *Alternaria brassicae* (early blight inoculum) in several different compost mixes, meeting the organic temperature treatment standards. Interestingly, weed seeds that were exposed to both the compost and heat were reduced at a higher rate than heat alone, possibly due to the activity of hydrolytic enzymes in the compost on the weed seed coats.[11]

Certainly anyone who distributes compost for public consumption should be managing the composting process to ensure that all of the material processed has been temperature-treated. The risks of not doing so are just too high as an operation and for the industry as a whole. As far as I know, there has never been a case where a disease outbreak was associated with an untreated compost product. That is a very strong track record, but it would only take one case of salmonella to really hurt the industry and the movement. For this reason, we need to take responsibility for educating both ourselves and our colleagues in the composting community with regard to implementing best practices. When composters take care in following BMPs, we all thrive.

Meeting temperature treatment standards means monitoring temperature in multiple locations throughout each batch and keeping detailed monitoring and pile-turning records, along with logs of other management activities. Pile monitoring and record keeping are discussed in *Compost Pile Monitoring and Management*, page 360.

Depending upon the regulatory framework in which you operate, there may or may not be a specific standard that is required of your operation. If you do not already know what specific temperature treatment standards apply to your operation, you will need to be certain of this as part of your overall regulatory assessment. If no regulations apply to you (for example if you are composting on-site at a school), then I suggest adopting a standard and instituting it as part of your overall management plan.

The most widely accepted temperature treatment standard in the United States is for all material to reach at least 131°F or greater for at least three days, although the exact language changes from regulation to regulation. These standards all stem from regulations set for biosolids (sewage) by the EPA (40 CFR Part 503). This is commonly referred to as the Process to Further Reduce Pathogens, or PFRP. The precise language in part 503 for PFRP is as follows:

> *Using either the within-vessel composting method or the static aerated pile composting method, the temperature of the sewage sludge is maintained at 55 degrees Celsius or higher for three days. A windrow composting system requires the temperature to be maintained at 55 degrees or higher for 15 days or longer with a minimum of five turnings.*[12]

For in-vessels and aerated static pile, most states have duplicated this language for food scrap composters almost to the letter. For turned windrows, you will see some version of this in state regulations, some more prescribed, some less. For example, California's—which I like—reads as follows:

> *If the operation or facility uses a windrow composting process, active compost shall be maintained under aerobic conditions at a temperature of 55 degrees Celsius (131 degrees*

> *Fahrenheit) or higher for a pathogen reduction period of 15 days or longer. During the period when the compost is maintained at 55 degrees Celsius or higher, there shall be a minimum of five (5) turnings of the windrow.*

There are states whose current temp treatment standards are more restrictive, requiring temps to be maintained for 13 of 16 *consecutive* days, for example, with five turnings, which makes consistently meeting the precise letter of the rule difficult at best. Other states provide few or no temp treatment requirements, which creates ambiguity. Ultimately, I tell composters to shoot for meeting whatever standards they fall under, but to follow best composting practices no matter what. For sites that don't have rules that apply to them or are having trouble meeting restrictive rules and simply want to ensure safety, my go-to standard is the language recommended by the National Organic Standards Board (NOSB):

> *(ii) the compost pile is mixed or managed to ensure that all of the feedstock heats to the minimum of 131°F (55°C) for the minimum time (3 days).*[13]

Their language meets the intent of rule 503 and clearly specifies the three-day heating target between turnings. As someone who has managed turned windrows to meet temperature treatment standards in extreme cold, I can tell you that the NOSB standard can be met consistently. As long as a site can show that they've met the NOSB recommendations or EPA 503, they can be assured that they've created a safe product that regulators should recognize.

For those writing regulations (or rewriting rules in your state), I strongly urge the creation of standards that are both *prescriptive* and *achievable*, or there is the risk that people will write them off completely. It certainly makes my life easier as a consultant when I can tell a composter "do these things and you will easily be in compliance." I hate telling them that "the rules are flawed" or "ambiguous" and that they may or may not be in compliance. People shouldn't have to run their business with that kind of uncertainty.

The Phases of Composting

While the composting process is contiguous from start to finish, distinguishing between process phases is extremely valuable. System choice, site planning, and everyday operational decisions should all be based upon an understanding of the physical, biological, and chemical transformation that the compost undergoes.

The phases of composting as I have come to understand them from a functional standpoint include *primary*, *secondary*, *finishing*, and *curing*. While there are a variety of ways to interpret the phases of composting, this is the framing that I have found to be most useful as an operator, planner, and designer.

Regardless of the scale or method, most composting operations are going to have some version of the following steps:

1. Receiving
2. Pre-processing and handling
3. Pile blending
4. Pile formation
5. Primary phase
6. Secondary phase
7. Finishing
8. Curing
9. Post-processing
10. End use

The primary, secondary, finishing, and curing phases are the essential phases of composting, whereas the others are auxiliary. We'll discuss these essential phases here, while the other steps are covered in chapters 13 and 14.

Primary Phase

TYPICAL PRIMARY PHASE TIME LINE

- The first 2–6 weeks
- Longer if the process is stagnant

Once an adequate volume of material is blended and formed into a composting pile, the *primary phase* commences. The term *primary* is commonly used to describe this stage of the process, but it is also called the *active phase* or, as I sometimes call it, *phase one*. For continuity I will use the word *primary*.

At this stage, fresh, raw compost has lots of readily available food for microbes. When these materials are combined following a good recipe, optimal conditions exist for microbes to consume and replicate. Bacteria proliferate as simple food sources such as sugars, carbohydrates, proteins, and fats are consumed.

As microbial activity increases, the massive release of metabolic energy raises temperatures above the mesophilic and into the thermophilic range, where resources including oxygen are rapidly depleted, necessitating intensive management including monitoring, aeration, and agitation. The heat generated allows operators to achieve temperature treatment targets, which is one of the defining characteristics of the primary phase. The composting methods described in chapters 7 through 10 are designed to assist with this period of intensive management.

All of this early pile activity is also essential to rapidly break down odorous compounds as they form. The better the starting recipe and the more active the management, the more rapid the breakdown of all compounds, including odors, will be. During the primary phase, containment of food scraps is essential. This is achieved through capping of piles with a compost biofilter or completely enclosing the compost, which is the concept behind in-vessel composters.

The primary phase has ended once the following goals have been reached:

1. Temperature treatment targets are met.
2. Strongly offensive odors are absent.
3. Potential food sources for vectors are mostly indistinguishable and no longer of interest.
4. Oxygen demand is reduced, and pile turning and aeration rates are less frequent.

Biological Phase

As the compost becomes active, there is a shift from mesophilic to thermophilic microbes.[14] The mesophilic microbes that are present initially begin the process by breaking down readily degradable compounds and causing the compost temperature to begin rising. As the temperature grows, microbial populations multiply, reaching as high as 10^9 to 10^{10} per gram of compost (1 to 10 billion).[15] Thermophilic bacteria remain dominant through the primary stage, as relatively simple food sources continue to be available.

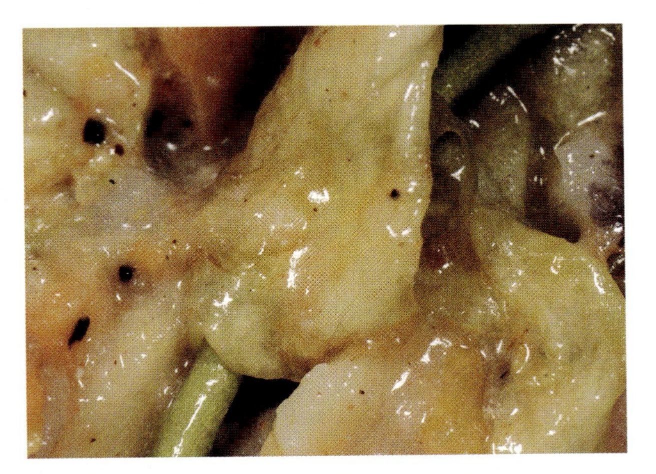

Figure 3.5. An up-close look at food scraps that have undergone the first stages of becoming compost. Photograph by Julian Post, used courtesy of Highfields Center for Composting and the Vermont Sustainable Jobs Fund.

Nitrogen Cycle

Nitrogen is present in its rawest and most volatile forms. Initially we expect much of the N to be in the form of lots of dead but temporarily stable proteins (in food, manure, hay). As these proteins break down, the by-products of protein degradation such as organic acids and ammonia (NH_3) build.[16] There may be volatile forms of N already present in the material, but this depends on the feedstocks. In manures, for example, a great deal of the conversion to ammonia and urea already took place in the animal.

A significant amount of N loss can take place in the primary phase. About 50 to 90 percent of all NH_3 losses typically occur during the first weeks of thermophilic composting. Losses via NH_3 volatilization can range from as low as 13 percent to as high as 70 percent of nitrogen in manure. Nitrogen losses from piles with a C:N ratio of 15:1 have been found to be as much as four times greater than from piles with a C:N ratio of 20:1.[17]

Carbon Degradation

The simplest forms of carbon are quickly accessible to bacteria, which dominate in the primary phase. Sugars, starches, and fats are the most readily available forms of high C, the low-hanging fruit that drive the fast and hot energy cycles in the early stages of the process.[18]

pH

As proteins and fatty acids degrade, acidic compounds volatilize and the pH typically drops significantly, sometimes to as low as 4.5.[19]

Secondary Phase

TYPICAL SECONDARY PHASE TIME LINE

- 4–10 weeks starting in months 0–2
- 2–4 months combined primary and secondary phases (or active phase if you prefer). Only in really unusual or unmanaged cases would active phases last longer.

Following the primary phase of composting, the overall intensity of pile management decreases. I refer to this as the *secondary phase*. Temperatures are still thermophilic (113°F or higher) and may even remain well above 131°F for some time, but your temperature treatment targets have already been met, which means at this point you are managing for product quality and operational efficiency. While oxygen demand is decreasing, some continued aeration is required to maintain aerobicity. At this stage, hungry microbes are eager to consume whatever residual odorous compounds are generated, so the risk of releasing offensive odors from the pile also decreases.

By my definition, the pile is often very "active" during secondary composting. I think it's useful to distinguish between the primary and secondary phases, because there is a tendency when defining the composting process to jump straight from the active to the curing phases, as if immediately following heat treatment, curing commences.

In addition, it's useful to differentiate the primary and secondary phases when considering composting methodology. At the majority of operations, the secondary phase of composting is managed via turned windrows or turned piles. The main reasons for this are economic; other methods require a higher up-front investment. There are fewer associated risks in terms of odors and vectors in the secondary phase compared with the primary, and thus there are also fewer overall benefits from investing in more intensive methods compared with those required in the primary phase. Strategically, people target a disproportionate (yet appropriate) amount of infrastructure investment and management into the primary phase.

That said, more intensive methods can continue to speed up the process in the secondary phase. Two common examples are (1) vermicomposting succeeding a primary phase that meets temp treatment standards; and (2) secondary ASP following either a primary ASP or an in-vessel treatment phase.

The secondary phase has ended once the following goals have been reached:

1. Temperatures have dropped below 115 to 120°F.
2. Acrid smells are gone; some pleasant musky (such as pipe tobacco), moderate ammonia, or other benign smells may remain.

Figure 3.6. Following a primary ASP phase and prior to the secondary ASP phase, compost is temporarily windrowed so that moisture can be added using a water truck (at Green Mountain Compost in Williston, Vermont).

3. No visible food scraps/raw feedstocks remain other than wood and other highly recalcitrant materials.

Biological Phase

During the secondary phase, bacteria are still predominant, although their populations have likely peaked. Even at higher thermophilic temperatures, actinomycetes and some fungi begin the important role of breaking down woody materials.[20]

Nitrogen Cycle

The raw proteins have mostly been broken down and transformed into ammonia (NH_3), which is very volatile. While N loss continues in this phase, it should be slowing and by the end of the secondary phase ammonia should be minimally detectable to the nose.

Carbon Degradation

The simplest forms of C have been consumed, and microbes have moved on to the more complex forms: cellulose, some waxes, and chitin.[21] Lignin degradation is in its early stages.

pH

The drop in pH should be rebounding, and it will slowly continue to rise as organic acids are converted to CO_2. Ammonia interaction may also play a role in raising pH.[22] In the case of high-pH composts, the trend is also toward neutral.

Finishing

TYPICAL FINISHING PHASE TIME LINE

- 1–6 months starting in months 2–6

I call the period following secondary composting, where active material becomes stable material, the *finishing phase*. At the end of this phase, composters should feel completely fine using or selling their compost as a finished product. Many continue to mature or "cure" their compost after it could otherwise be considered finished, but this is an operational choice, not a critical necessity. What is critical is for compost to continue maturing following active composting to the point where it is finished, and this requires every composter to define for themselves what *finished* compost means. I talk more about maturity standards in chapter 14, but for now let's assume that the finishing phase is defined as when material meets your minimum maturity standard.

Very little management is required during the finishing stage. Oxygen demand is low, so it can remain fairly aerobic without active aeration as long as the material is not too wet, too dense, or stored in too large a pile. A couple of turns over a few months will go a long way toward keeping the process moving. Some turning will also help remediate wet and dense conditions (this is common prior to full completion) and ensure that the outside of the pile doesn't fall behind the material in the core. There is very little that can go wrong at this stage, unless the material is sitting in a puddle or gets contaminated with weed seeds or leachate. If you've planned your operation well, these will not be issues, so your job is to monitor it occasionally and make sure that nothing gets in the way of its natural progression.

The *finishing phase* has ended once the following goals have been reached:

1. Your temperature maturity standard has been met—for example, a temperature no greater than 90°F following a turning.
2. The pile smells like rich forest soil.
3. Particles are reduced to crumb-size and smaller (other than large woody and other highly recalcitrant particles).
4. The total volume of material is significantly reduced, typically by greater than 50 percent.

Biological Phase

As the pile cools, the microbe population shifts back from thermophiles to mesophiles, corresponding to the drop in pile temperature. Total bacteria populations decrease dramatically as simple food sources dwindle. Meanwhile the role of fungi increases, breaking down the more complex forms of carbon such as woody particles. Optimal temperatures for lignin decomposition in composting are between

104 and 122°F, right on the boundary between thermophilic and mesophilic.[23] It is not uncommon to see worms inhabit the cooler regions of the piles at this stage, sometimes in large quantities. As the pile continues to mature, there is a shift to higher-trophic-level organisms. Compost's disease-suppressive qualities have been attributed to this shift.[24]

Nitrogen Cycle

In compost, as in soils, there are specialized bacteria that convert ammonia N (NH_3) into nitrite (NO_2-), then nitrate (NO_3-) in a process called *nitrification*.[25] Nitrite and nitrate are more chemically stable forms of plant-available N than ammonia. The role of these bacteria in mediating N stabilization increases fairly late in the process, as temperature drops and maturation commences.[26] As the compost continues to mature, there is very little ammonia left, and nitrite and nitrate increase. Eventually the majority of the nitrogen becomes "organic" once again.[27] The common understanding is that the N in mature compost is contained primarily in the proteins of living organisms. Bacteria are reportedly 7 to 11 percent N and fungi 4 to 6 percent on a dry-weight basis.[28] As long as the compost is not mishandled (say, saturated with water), these living nitrogen sources are considered very stable.

Carbon Degradation

Any carbon that remains at this stage of the process was not readily consumable: mainly lignin (woody cell walls), some waxes, and other complex secondary plant compounds. While these compounds continue to slowly degrade, the rate of C consumption has greatly decreased.

pH

The compost continues to balance itself, acidic conditions becoming more alkaline and vice versa in a gradual progression toward neutral.

Curing

TYPICAL FINISHING PHASE TIME LINE

- 1+ months starting at 4–9 months

Though the final phase of composting is entirely optional, it is intended to impart desirable qualities in the material by further aging it, which allows the ecology of the pile to progress. I use the term *curing* to refer to this biologically distinct phase.

The term is also used in the industry and in regulations to refer generally to maturation beyond the active or heat treatment phases, and so its definition varies significantly based on who you are talking to, which can be a source of confusion. You may hear it used interchangeably with words like *aging*, *maturation*, and *finishing*.

In many cases people refer to curing as everything that happens following intensive management. While this isn't necessarily wrong, it doesn't really tell you anything about what is happening at this specific stage, because the point at which curing started is not defined. *The On-Farm Composting Handbook* describes it this way:

> *There is no specific point at which curing should begin or end. When the windrow temperature no longer reheats after turning, the curing stage begins.*[29]

The first sentence describes how the term *curing* is typically used, with no commonly defined beginning or end. The second sentence more closely aligns with how the term curing is used in this book: Curing is a phase whose defined starting point is the meeting of a temperature-based maturity standard.

Curing has no defined end point because the compost is already "finished" from a stability perspective. Curing therefore relates to qualities beyond stability, such as increased humification and higher-level soil biota including beneficial nematodes and fungi. Curing is the phase when the compost really gains its mythical status, imparting qualities such as disease suppression, levitation, and psychokinesis. Think of curing compost like aging cheese or wine; it's not absolutely necessary, but it creates a better product. That's not to say that uncured compost is bad compost; the two are just not directly equivalent.

Figure 3.7. NYC Compost Project hosted by Big Reuse's project manager Leah Retherford, with some beautifully cured compost delivered to a local community partner. Courtesy of Ron Cauhen.

Curing does come at a cost, however: As the volume of the material continues to reduce, you might lose a few percentage points in total yield. Curing material also takes up space on the site, and although it can be consolidated, real estate can be valuable. As a business you need to consider how to value that additional curing step.

For farmers making compost for themselves, I think taking the time for curing is a no-brainer. Adding quality with no additional effort makes sense when the end volume matters less than the end quality. If you are selling compost, you might set a standard curing time, say a month past the point when compost is finished; then anything beyond a month is gravy. In reality, the seasonality of compost use forces composters to cure certain batches for longer than they might otherwise. Nevertheless, there is always a batch that is just barely approaching salability in the springtime. You want to know ahead of time what the standard is for what you sell, because there is always a temptation to sell as much material as possible.

If the four phases of composting as I've described them here feel overly complex, understand that the curing stage is your insurance policy. If some other stage of the process gets cut short, extra curing can make up for it. Curing can even act as a buffer for inadequate temperature treatment, because over time pathogens are outcompeted by other organisms or simply expire.[30]

The curing phase has ended once the following goals have been reached:

1. Your temperature maturity and curing time standards have been met.
2. The compost has a fluffy, relaxed texture.

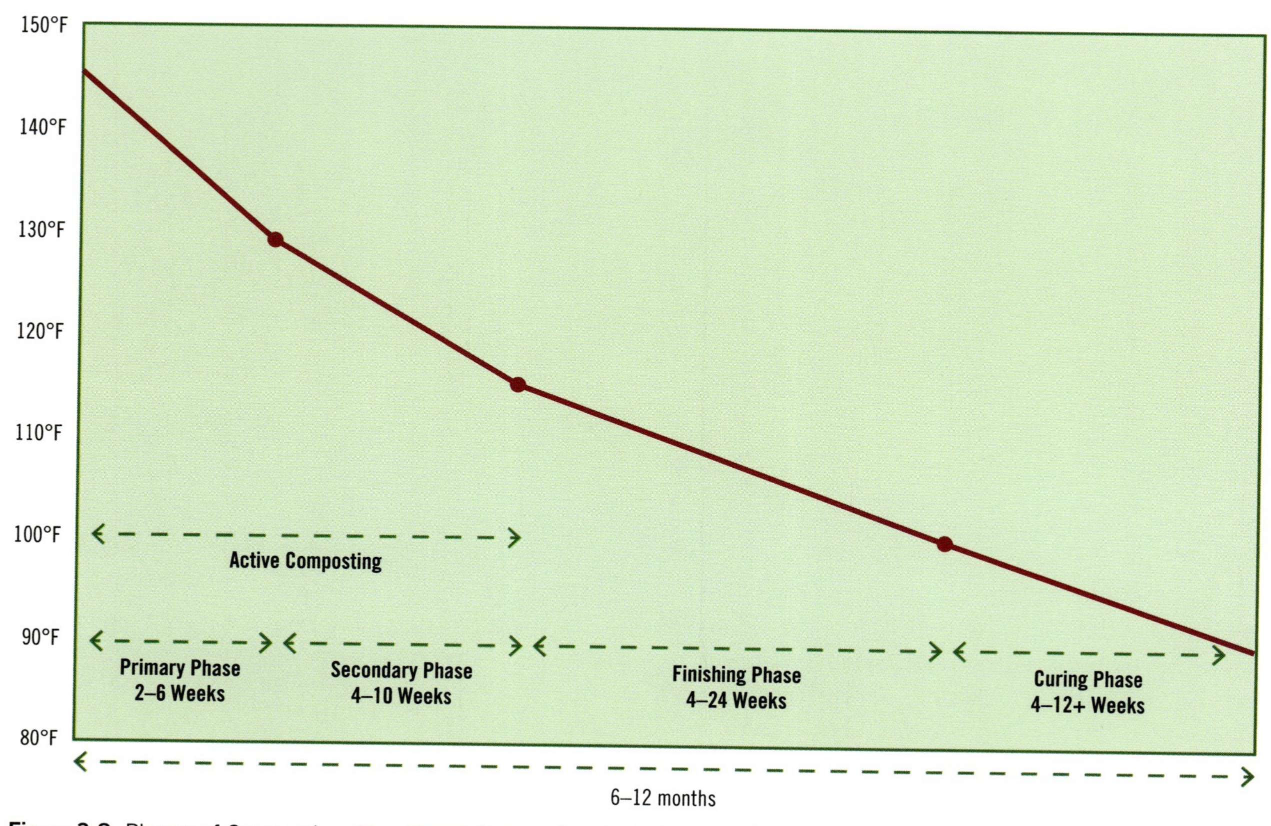

Figure 3.8. Phases of Composting: Time/Target Temperature Relationship. The time range for typical compost phases in managed systems.

Biological Phase

Biologically, the curing phase is marked by an increase in higher-trophic-level biota. While bacterial populations decrease as the compost finishes, their populations are stabilizing during curing. As the compost ecosystem is given time to progress, microorganisms that feed on bacteria and other lower-level biota continue to increase. Well-cured compost can become a rich soil food web, with predators such as nematodes consuming lower in the food chain. Their excretions contain plant-available nutrients, providing a slow but consistent source of fertility.

In their book *Teaming with Microbes,* Lowenfels and Lewis write that one teaspoon of good compost can contain "up to one billion bacteria, 400–900 feet of fungal hyphae, 10,000–50,000 protozoa, and 30–300 nematodes."[31]

Nitrogen Cycle

The soil nitrogen cycle continues, with ammonia being converted to nitrite, then nitrate, then into organic (living protein) forms. Organic N forms now dominate.

Carbon Degradation

Only the most stable and recalcitrant forms of carbon remain. Humification increases over time, producing highly stable and valued compounds like humic and fulvic acids.[32]

pH

The compost may continue to balance itself, acidic conditions becoming more alkaline and vice versa, although this progression should be nearing or at a stable state.

CHAPTER FOUR

Compost Recipe and Feedstocks

Now that we've covered the air and temperature parts of the compost equation, it's time to fill in two major gaps: food and moisture. There is an art and a science to making a mix of compost in which microbes thrive and which does not create major problems. The mix is often referred to as a *recipe*. In the recipe a combination of unique *feedstocks* is formulated based on a set of target parameters.

I've chosen to incorporate recipe and feedstocks into one chapter because there is overlap in the parameters we use to assess them. The combined characteristics of the feedstocks are what make the end result of the recipe. Therefore, feedstocks must be understood in the context of their effect on recipe, and recipe must be understood in terms of the interplay among feedstock qualities.

Most people will probably have heard about carbon, nitrogen, and moisture, the three most talked-about parameters; few actually know how to use them to develop a recipe, however. There are also a variety of other considerations that can affect the composting process, including bulk density, porosity, organic matter, pH, salts, and more. We'll start by reviewing the characteristics we target when creating a compost recipe, which will provide a foundation for understanding feedstock qualities.

organic

(ôr'ganik)

DEFINITION: Involving or derived from living organisms or the products of their life processes.

Compost Recipe and Feedstock Characteristics

A compost recipe is exactly what it sounds like: a guide to combining various materials to create a predictable outcome. It is similar to how a bread recipe takes the guesswork out of baking, but unlike bread, compost's raw materials are less uniform (not bought off a shelf), so each site's recipes are somewhat variable in terms of ingredient proportions. Rather than a compost cookbook, recipe development guides recommend target parameters. Each site must then assess the characteristics of its unique feedstock and use that information to develop a recipe. The targets that are used most widely, shown in table 4.1, are *carbon to nitrogen ratio* (*C:N ratio*), *moisture content* (*MC*), and *bulk density* (*BD*).

While most composters are aware of these targets, putting them into practice is completely foreign for many. The goal of this chapter is to remove the mystery behind achieving these targets.

Table 4.1. Recipe Targets

Parameter	Acceptable Range	Ideal Range
Carbon : nitrogen ratio	20–40:1	25–30:1
Moisture content	50–65%	55–60%
Bulk density	<1,200 lbs. per cu. yd.	700–1,000 lbs. per cu. yd.

Source: Robert Rynk, ed., *On-Farm Composting Handbook*, NRAES 54 (Ithaca, NY: Natural Resource, Agriculture, and Engineering Service, 1992), except for ideal range for bulk density.

Carbon : Nitrogen Ratio

Developing an in-depth understanding of the dynamics between C and N and the roles that each plays individually is absolutely key to mastering the composting process. Carbon is the essential unit of energy in living things, and compost is no exception. Microbes metabolize the carbon in its various forms (carbohydrates, cellulose, hemicellulose, waxes, lignin), and the energy in the carbon drives the process. Under aerobic conditions carbon dioxide (CO_2) is released through respiration; without oxygen (anaerobic), methane (CH_4) can be generated. Anaerobic decomposition is the basis of methane digesters (see *Anaerobic Digestion* in chapter 2, page 45), but composting is predominantly aerobic. Carbon feedstocks are typically drier than nitrogen sources (although there are outliers), and so they can often balance N feedstocks both in terms of MC and C:N.

Nitrogen is a critical building block of proteins. Microbes utilize it to build cells and replicate. This makes N a limiting factor in total microbial population, and since microbes consume the raw material, having more microbes means faster degradation. All proteins are high in N, so higher-protein feedstocks are also higher in nitrogen. Blood, for example, has a C:N of around 3:1, which is very low.[1] Like blood (although blood represents the far extreme), most sources of N are wet and have high odor potential. Typical mixed food scraps have a C:N in the 15 to 16:1 range and can have an MC of almost 90 percent.

The *carbon to nitrogen ratio* (or *C:N* as we usually refer to it) of both feedstocks and compost is the proportionate relationship between the mass of carbon and nitrogen. On a simplified level, C:N should be thought of as the "diet" part of the recipe.

Assuming the target C:N for your compost recipe is within the recommended range of 25 to 30:1, then feedstocks within that range would be considered neutral, feedstocks above 30:1 would be considered a C source, and feedstocks below 25:1 would be an N source. Materials at different ends of the C:N ratio spectrum can be combined to create a balanced recipe. The process of developing a recipe should involve solving for the combined C:N ratio of materials in your raw blend, as well as for moisture content and bulk density (see *Compost Recipe Development*, page 83).

We know that when the C:N ratio is roughly between 25:1 and 30:1, enough N is present to grow adequate microbial populations; these will in turn consume the C in a reasonable amount of time. I think of it like wind power, where the C is the wind and the N is the wind farm. You can have all the wind in the world, but the energy that you can generate is limited by the capacity of the wind farm. Add turbines (N) and you can harness more power (break down C faster), to a point. Add too many turbines (N) and the wind becomes limiting, meaning you are wasting turbines (losing N). It's an energy/consumer dynamic whereby C:N ratio targets represent the optimal equilibrium.

As C is consumed, the C:N ratio decreases. Carbon is released to the atmosphere as CO_2 (or supplied to a greenhouse, as shown in figure 2.8, page 44) through microbial respiration. N is also lost, but the concentration of N increases due to faster rates of C loss.[2] Over the course of the composting process, the C:N ratio decreases by roughly half. Agricultural soils typically have a C:N in the 10:1 to 12:1 range, so with a starting C:N in the target range for compost, the finished compost is relatively similar to ag soils.[3] By this time the residual carbon is stable, often described as humus-like. Respiration and oxygen demand are now minimal. The nitrogen has cycled back to a biological (living) and hence a stable form, through the synthesis of proteins by bacteria and other biota.

Too Much Carbon

When the C:N ratio is too high, meaning there is too much carbon in proportion to the nitrogen available, composting slows primarily due to a lack of microbial populations. The energy is there, but the consumers are limited in how fast they can generate new proteins and replicate. As C is slowly consumed, the N becomes more concentrated, but it will never reach the same richness that it would if there was a greater source of N. The problem is exacerbated by the fact that high C often correlates with low moisture, which also stunts the process, but even if the MC is perfect, the process will still be slower.

Large yard waste or wood chip piles are common examples of high C and often low MC composting. Interestingly, it is thought that free-living nitrogen-fixing bacteria are able to inhabit high C piles and pull nitrogen directly out of the atmosphere, much like on the forest floor. When nitrogen is limited in compost, they may play an increased role.[4] Also, free-living N fixers are most successful in anaerobic environments with lots of C and organic matter present—such as relatively moist, N limited compost piles.[5]

Ultimately, when you are applying finished composts, keep in mind that composts with a C:N of 20:1 or more can tie up N that comes into contact with the compost, reducing plant growth.[6] Lower C:N ratios (below 15:1) correlate with an increase in plant-available nitrogen.[7] This corresponds with the C:N of stable organic matter in soil.[8] High C composts can be applied to soils in combination with an organic nitrogen source to ensure adequate N is available or they can be used as a surface mulch, where there is minimal potential that microbes will compete with plant roots for the available N.

Too Much Nitrogen

As with high C:N ratio composts, when the C:N ratio is too low, the composting process is often stunted, but this depends largely on moisture levels, density, and porosity. Since high N is typically associated with wet and dense materials, oxygen is much more often a limiting factor than carbon. (Keep in mind that a lack of both oxygen *and* carbon is more problematic than just a lack of oxygen.) Picture a mastodon preserved in a peat bog. Anaerobic compost piles are not that different; things take a lot longer to break down when conditions are too wet or too dry, which we'll discuss in much more depth later in this chapter.

What happens in a high N pile that has great MC and porosity (is relatively aerobic)? The answer is everybody's favorite: It depends. Here are some of the downsides and risks associated low C:N ratio composting scenarios:

- All of the available N is not needed biologically, so a larger portion of the N is lost in the forms described in the next bullet. As far back as 1956, Gotaas reported that at C:N ratios of 30 to 35:1, only 0.5 percent is lost, whereas at 22:1 and 20:1, he found N losses of 14.8 and 38.8, respectively.[9]
- Nitrogen and proteins correlate, so high-nitrogen recipes correlate with protein- and nitrogen-related odors (volatile organic compounds, ammonia). Odors are a big concern for small and large composters alike. Odorous compounds are great food for microbes, and that N could be great food for the soil if it were captured. The way that it can be captured is by increasing the carbon content of the mix. Carbon feedstocks create a physical barrier that traps odors, acting as a biochemical sponge that ties up the nitrogenous odors until they can be stabilized through microbial metabolism.
- Vectors (birds, rodents, insects, scavengers) are an issue when odors become a problem or when nitrogenous pest-attracting food sources (such as raw food scraps) sit on the surface of the pile. These conditions don't necessarily correspond with high N, but they often do. The low-tech solution to this issue is for the N sources to be thoroughly encased in your C sources, which often requires incorporating a higher proportion of C feedstocks through blending and capping.

There can actually be upsides to composting at the lower edge of the target C:N range, namely a faster process. When composters can find a good C:N balance point, minimizing C (in combination with good oxygen and MC) initiates a fast process, because there is very little fuel to burn off. Just keep in mind that there can be a much greater downside with high N recipes than with high C recipes.

Measuring C:N Ratio

The percentage of carbon divided by the percentage of nitrogen equals the C:N ratio. You can use either wet or dry weights to calculate this as long as you are consistent with one or the other. Only use dry-weight measurements for making recipe calculations. Commonly available lab testing will provide percentage of carbon and percentage of total nitrogen by weight, and most compost analyses also provide the C:N ratio as part of their standard package. Labs that don't specialize in compost analysis may require you to do the quick math to get the C:N (make sure they give you both C and N percentages in their analysis).

For example, to calculate C:N for a bedded horse manure:

32.2% Total Carbon (Dry Weight)
÷ 0.9% Total Nitrogen (Dry Weight) = 35.7:1 Carbon : Nitrogen Ratio

Since the N portion of the ratio is always 1 with organics (I suppose that some chemical fertilizers might have a ratio where the N is higher than the C), people often leave the 1 out, so I might say that the above horse manure has a C:N of 36 (rounded up).

Carbon to Nitrogen Ratio Is *Not Volumetric*

People often get confused by the concept of C:N ratio because they don't understand that it refers to the *mass* of these two elements, rather than the ratio of either the volume or mass of carbon feedstocks to nitrogen feedstocks. Without analysis and recipe calculations to back it up, C:N ratios are relatively useless to composters, unless you're dealing only with materials that are already in the target range. I'll go into what does and doesn't work when it comes to developing an accurate composting recipe in *Developing Analytically Based Recipes*, page 84, but for now the key is to not confuse the C:N ratio with the actual volumetric recipe.

Moisture Content

The *moisture content* (*MC*) in compost refers to the percentage of the material's weight that is water. Like you and I, microbes require moisture to be active. Aerobic decomposers are most active on the moist surface of the individual particles within the pile. At the edge between particle and void, there is food, water, and air (ideally there is some oxygen in that air)—all the things they require to thrive.

Microbes are adapted to different moisture levels, but the ones most productive in the composting process do best in a target range between 55 and 60 percent. The tolerable MC range for composting is 50 to 65 percent, although what's tolerable depends upon a lot of other variables.

A targeted moisture content is the second variable that must be solved for when developing a recipe, and it can often be more challenging to get the moisture within an optimal range than the C:N ratio, although the two are typically closely correlated.

Assuming the MC for your compost recipe is targeted within the recommended range of 55 to 60 percent, then feedstocks within that range would be considered neutral, feedstocks above 60 percent would be considered a moisture source, and feedstocks below 55 percent would be a dry matter source. Materials at different ends of the MC spectrum can be combined to create a balanced recipe. The process of developing a recipe should involve solving for the combined MC of materials in your raw blend, as well as for C:N ratio and bulk density.

Low Moisture Content

Low moisture content is associated with:

- Slow decomposition
- Higher-than-optimal temperatures

Beyond Browns and Greens

The terms *browns* and *greens* are widely used training tools to teach people about the differences between nitrogen-rich feedstocks (greens), which are typically also wet, and carbonaceous feedstocks (browns), which are typically more dry. These two classifications have served a valuable purpose for educating backyard composters, but this binary simplification doesn't translate well to a larger scale. Separating materials into just two classifications of nitrogenous and carbonaceous has even permeated into some regulations, where states are limiting feedstocks proportions based on these two categories (and miscategorizing one as the other, based on non-representative analysis). Unfortunately, this can tie composters' hands and force them to work with a suboptimal recipe. It is critical for composters of any scale to know whether the materials they are working with fall above or below the target C:N ratio range of 25 to 30:1, and it's natural to sort materials into different categories. In my work I am now advocating that composters classify materials into four categories, in order to create more balanced and diverse recipes.

Greens. Nitrogen-rich materials with C:N ratios below 25:1

Typically

- High MC
- High BD
- High protein
- 20 to 33 percent of recipe by volume

Examples

- Food scraps
- Dairy/beef manure
- Grass
- Coffee grounds/chaff

Browns. Carbon-rich materials with C:N ratios above 40:1

Typically

- Low MC
- Low BD
- High lignin/cellulose
- 20 to 50 percent of recipe by volume

Examples

- Sawdust
- Straw
- Paper
- Well-bedded horse manures

Neutrals. Balanced materials with C:N ratios between 25:1 and 40:1

Typically

- Medium MC
- Medium BD
- High cellulose
- 20 to 50 percent of recipe by volume—can increase proportion without large effect on C:N

Examples

- Horse manures
- Calf/heifer manures
- Hay
- Mixed leaves/grass

Porous. Low-bulk-density materials with mixed and large stable particles that add porosity and structure to the pile

Typically

- High C:N
- Low to medium MC
- High lignin
- 5 to 20 percent of recipe by volume (recommended)

Examples

- Wood chips
- Ground bark
- Ground yard debris

- Nitrogen loss
- False maturity indications
- Labor/cost adding moisture
- Combustion (in very large hot piles)

Composts with MC below 50 percent stagnate significantly. Essentially, the microbes lack optimal habitat and are either very slow at metabolizing the material or go dormant. Low MC scenarios often correlate with high C:N, because feedstocks such as yard debris or wood chips are typically both dry and high in C.

Dry conditions are a common challenge for composters who operate in arid climates. Seasonal variation and drought conditions are also a factor. Dry summers can be a problem for many of the composters that I work with. Composting methodology can also be a cause of drying. For example, composters who work under cover, turn with high frequency (say, a windrow turned operation), or use aerated static pile (ASP) systems tend to experience dry conditions partway into the process.

Although it may seem to contradict my statement about stagnant decomposition, dry piles are more likely to overheat than optimal or high MC piles, which demonstrates that high temperature is not always an indication of optimal pile activity. One reason that drier piles get hotter than wetter piles is because water has a high specific heat capacity, which means that it takes a large amount of energy to raise the temperature of water compared with air or organic matter. For example, it takes 1 BTU (British Thermal Unit) to heat 1 pound of water 1°F, while organic matter would rise 4°F.[10] Water acts as a heat sink and keeps piles cooler as a result. This means that a moist pile that's 150°F might be generating more thermal energy than a dry pile that's 160°F.

From a decomposition perspective, pile temperatures above 160 to 165°F are typically viewed as a cause for concern. By that point, most microbes have died off and only heat-resistant spores formed by certain bacteria and actinomycetes remain.[11] As the temperature decreases, microbial populations recover, but this type of die-off is not considered beneficial to the composting process. The high temperatures exacerbate the drying, creating a positive feedback loop in the wrong direction as the piles become continuously hotter and drier. This is especially a problem with ASP systems.

Another drawback is that dry and hot conditions increase nitrogen volatilization, which removes nitrogen from the compost system. Ammonia (NH_3) stays in solution, but without moisture much of it goes airborne.[12] With dry piles that have a decent supply of nitrogen, the ammonia smell can be staggering, not unlike what happens when manure or digestate is field-applied. Cool and moist field conditions retain more nitrogen,[13] and subsoil injection systems that mitigate contact with the air can reduce N losses by 90 percent.[14]

Normal Versus Premature Drying Over Time

Time is also a factor in drying. Compost tends to dry out as the process progresses. With the exception of some enclosed systems where moisture gets trapped, the moisture is lost at a faster rate than the total volume reduces, and the pile slowly dries out. Although the concentration and distribution of moisture can fluctuate, especially early in the process, this natural drying trend is an overall benefit, as long as it progresses along the same time line as the compost. By the end of the process, most composters want the pile to reach 50 percent MC or lower, to facilitate easy screening, storage, and transportation.

At this stage, lower MC correlates well with reduced microbial activity due to exhausted food supplies. This allows mature compost to be consolidated with less risk of creating severe anaerobic conditions.

Compost that dries out prematurely goes dormant whether it is mature or not, and can give the false impression that it is finished. There are stories of whole pallets of dry-yet-immature compost going out for distribution, then getting rained on and reheating in garden store yards across the country. There are also composters who dry out their compost on purpose to cut the process short for various reasons (but that's a whole other topic).

Although there are prominent composting experts who advocate for an initial target moisture content of approximately 50 percent to ensure a more fully aerobic pile from the start, my experience is that in practice this is difficult to achieve for a number of reasons, especially for food scrap composters. First, unless you start with a very dry feedstock to begin with, it is really expensive to get this much dry matter into the pile, and some composters simply can't source enough dry material for this to work. More important, the natural drying that happens is hard to correct for once it's gone too far, so starting too dry can become a real problem further along in the process. We'll talk a lot more about adding moisture to the pile in other chapters, but for now I'll say that it's challenging and expensive; it's a lot easier to keep your pile in the right MC range than to correct for low MC.

To address any concerns about combustion risk, large piles with a MC between 30 and 40 percent can and do combust, but the conditions have to be just right. Robert Rynk of SUNI Cobleskill, author of *On-Farm Composting Handbook*, for example, advises operations that manage large volumes of compost to keep them under a height of 3 Calvins. The Calvin is a unit any operator can remember: It's the height of Calvin, the proverbial site operator. One Calvin equals approximately 2 meters (Calvin is a big guy). Keeping piles less than 20 feet high will significantly reduce the risk of combustion,[15] which should be no problem at the scale that most readers of this book are operating.

High Moisture Content

High moisture content is associated with:

- Low-oxygen / anaerobic conditions
- High density
- Slow decomposition
- Stagnant temperatures
- Odors and volatile organic compound (VOC) accumulation and release
- Leachate loss
- Methane and nitrous oxide generation

At moisture contents above 65 percent, anaerobic conditions tend to dominate. (I really try to keep recipes at 63 percent max.) Wet piles are typically dense piles, and dense piles can't breathe. The combination of wet and dense is probably the single most challenging aspect of food scraps. Food scraps can contain moisture as high as 90 percent, and once they start to deteriorate, they have the consistency of slop. Without the addition of dry matter, the moisture will remain high, which is literally a recipe for all of the problems you should be most concerned about as a composter: odors, leachate, failing to meet temperature treatment targets, attracting animals and insects, and aggravating neighbors and regulators. (Not to mention making low-grade compost.)

While dry conditions can be inefficient, the troubles associated with wet conditions are several orders of magnitude greater in terms of their inherent risk. Odors, birds, rats, angry neighbors: All of these nuisances can shut down compost sites. The good news is that nuisances are not inherent to composting. It's quite the opposite; most compost sites keep moisture in check and raw material covered, all stuff that we'll cover in much more depth in chapter 13. But it behooves us to understand on a slightly deeper level what happens under wet, dense, and often low-carbon situations.

In his paper "Volatile Organic Acids in Compost: Production and Odorant Aspects," Dr. William Brinton clarifies the cycle of volatile organic acid (VOA) generation and breakdown.[16] The paper describes what many food scrap composters experience, which is that the generation of volatile organic acids, which are associated with "garbagey," putrid smells, is not a result of bad composting. Rather, we should expect a wide range of aerobicity within a compost pile, with some VOA generation in the first 40 days, and in the first 20 days in particular.[17] The key is to have an active composting process, where VOA generation is limited to a volume that can be effectively broken down by microbial activity before escaping to the surrounding environment. Essentially, there is equilibrium in a well-managed pile between material that is generating odor and material that is absorbing and metabolizing it. Dry matter, carbon, and oxygen (porosity and active aeration) are the antidote.

Excess moisture can also lead to the loss of *free moisture* from the pile, which is typically called leachate. Free moisture is essentially pile liquid that doesn't have a solid to absorb it. Therefore it does what water does, which is flow down and out. When it exits the pile we call it *leachate*, the technical term for liquid compost discharge. Leachate is typically high-odor and high-nutrient, could potentially carry pathogens, and should therefore be treated with the same precautions as the raw feedstocks from which it was derived. Leachate should be prevented to the greatest extent possible by controlling MC, and managed by absorbing it with high C dry matter where and when it occurs. At the Highfields Center for Composting, we collected and measured the leachate from beneath an 80-cubic-yard ASP pile in a concrete bay and found that it generated anywhere between 1 and 4 gallons of leachate per cubic yard of compost over a one- to three-week period. That's a lot of stinky liquid!

Lastly, anaerobic conditions have the potential to generate and release the same polluting greenhouse gases that are generated in landfills: methane (NH_4) and nitrous oxide (N_2O).[18] Given that we've worked so hard to mitigate these pollutants by diverting them from disposal, it would be counterproductive to ignore their potential generation at our facilities.

Measuring Moisture Content

MC can be tested using field squeeze tests, in labs, or by using DIY lab methods. For the purposes of developing a compost recipe, commonly available lab testing will provide percent moisture content of both feedstock and composts for you as part of their standard package. Labs that don't specialize in compost analysis often provide percent *dry matter*, which is the inverse of MC (a dry matter reading of 27 percent would be an MC of 73 percent). Some compost sites buy specialty ovens and scales to do MC testing themselves, and though I've never tried it, some people even use microwaves to dry the material. Another tool that I've seen people use is a Koster Moisture Tester, which is designed to measure the moisture content of animal feed (silage, hay).

The concept is pretty simple:

1. Weigh the wet material, being sure to subtract the weight of the container.
2. Dry the material by baking it in an oven (150 to 225°F) until it stops losing weight, then weigh it again.
3. Subtract the dry weight from the wet weight to get the water weight.
4. (Water weight ÷ original wet weight) × 100 = % moisture content.[19]

These tests are particularly important for getting an accurate MC reading on feedstocks that are being used for recipe development, because it is very hard to get an accurate measurement with the field squeeze test, especially with high and low MC

Figure 4.1. An oven in a lab at the University of Vermont's Plant and Soil Science Department can be used for measuring the moisture, organic matter, and carbon content of compost and compost feedstocks.

feedstocks. The accuracy of the MC and bulk density analysis also affects the calculation of C:N, so inaccurate MC readings could skew the recipe altogether.

On the other hand, the field squeeze test for MC can be used for assessing compost once the process is initiated. The goal of the squeeze test is to gauge moisture with relative precision in relation to your target MC. Directions for performing the squeeze test are described in *Monitoring Pile Moisture*, page 364. In most cases, checking MC using the squeeze test weekly during the active phase of composting is adequate unless there is an urgent issue, such as an odor that requires attention.

Inevitably, people ask about using probe-type moisture meters for testing MC in the field. I have steered clear of these on the advice of multiple mentors, who tell me that these are not accurate when it comes to compost. It would certainly be nice to have a quick and accurate probe to give a precise MC reading, but on the other hand, people would lose the physical contact of the squeeze test, which tells you so much more than just the MC.

Bulk Density

The term *bulk density* (BD) refers to the weight of a specific volume of a given material. It is used for weight/volume conversions and it's a metric of whether a material or recipe is overly dense. Dense compost leads to anaerobic conditions. The unit of measurement composters in the United States tend to use is pounds per cubic yard (lbs. per cu. yd.).

Composters can use bulk density in several ways, which are described in the following sections, including:

- Recipe development
- Volume to weight and weight to volume conversions
- Feedstocks assessment

Bulk Density for Recipe Development

When developing a compost recipe, BD is the third characteristic that we solve for mathematically, along with C:N ratio and moisture content. High BD is an indicator of dense conditions, which typically correlate with low porosity. The target for a compost recipe is below 1,100 pounds per cubic yard,[20] with the ideal range being 700 to 1,000 pounds per cubic yard in most cases.

Assuming the target BD for your compost recipe is within the ideal range, feedstocks within that range would be considered neutral, feedstocks above 1,000 pounds per cubic yard would be considered dense, and feedstocks below 700 pounds per cubic yard would be considered light. Low BD materials, which are often porous, are referred to as *bulking agents*, and are used to reduce density and create porosity. Materials at different ends of the BD spectrum can be combined to create a balanced recipe. The process for developing a recipe should involve solving for the combined BD of materials in your raw blend, as well as for C:N and MC.

Analysis of the BDs of each individual material used in the recipe are also necessary for calculating C:N and MC, and for converting the recipe from weight to volume. The recipe formulas are weight-based, but for practical operation, volume is usually involved, both when entering weights into the calculator and for making recipes usable in the field.

Converting from Weight to Volume

Knowing the BD of a specific material allows you to convert from volume to weight and vice versa, which is critical in planning capacity, to estimate throughput, and when making or using recipes. Loader buckets are generally sized by the cubic yard (for instance, ½, 1, 2 cubic yards, and so on), so everything that happens on the site is typically tracked or converted into yards, including truckloads and recipe inputs. With the help of a spreadsheet or calculator and the BD of the materials you're handling, yards can easily be converted to pounds or tons for record-keeping or reporting purposes.

An example of a useful BD is that of food scraps. Although their BD tends to vary depending on what form is being handled, a good rule of thumb for material coming onto the site is roughly 1,000 pounds per

Figure 4.2. Weighing this 5-gallon bucket of residential food scraps (my own) with a luggage scale confirms a bulk density of right around 5 pounds per gallon.

cubic yard (see *Converting Food Scrap to Tons and Cubic Yards*, page 88). This is really nice information to have in your back pocket, because lots of folks in the waste and composting industry speak in tons, and 1 ton = 2 cubic yards.*

Bulk Density for Assessing Raw Feedstocks

Not surprisingly, compost feedstocks have a wide spectrum of densities. A material's BD can tell you a lot in relation to other characteristics.

Bulk density corresponds with:

- High or low moisture content
- High or low mineral content
- Abundance or lack of pore space
- Some combination of the above

* On a much smaller scale, composters may work in gallons, using 5-gallon buckets for example. There are 202 gallons per cubic yard, so 1,000 pounds per cubic yard equals roughly 5 pounds per gallon.

When looking at BD, it's important to determine *why* the material has the density that it does in order to understand its value and/or handling requirements in the recipe. The reality is that BD alone can't really tell you which combination of these six characteristics you're dealing with. Luckily, analysis should give you both MC and organic matter (or volatile solids, which are equivalent). For many materials, the culprit is obvious; for others not so much. As you look at more and more material qualities, you'll be able to better anticipate a material's characteristics in advance.

Bulk Density and Moisture Content

Moisture is often the obvious reason that a material is very heavy, water having a bulk density of approximately 1,680 pounds per cubic yard. But high MC material is not always dense. For example, the dairy manure press cake (separated manure solids) that I've tested has had MCs in the 70 percent range, but BDs of only around 400 pounds per cubic yard. On the other hand, dairy manure from a tie stall barn can be close to 90 percent MC and 1,600 pounds per

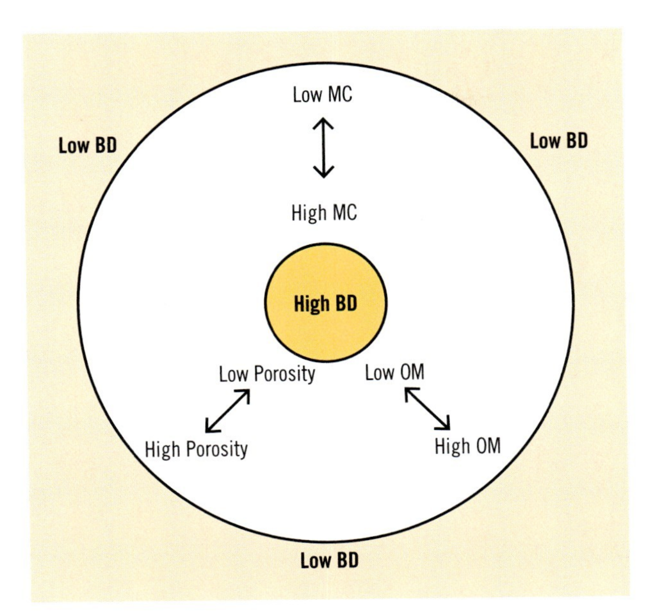

Figure 4.3. While there are outliers, in general bulk density can be understood on a spectrum in terms of three important characteristics: moisture content, porosity, and mineral versus organic matter content.

cubic yard. Both would be considered high MC from a recipe standpoint, but they have incredibly different densities. I know from experience that the press cake heats up within hours and stays hot. The tie stall manure might heat slightly, but that activity probably has a very limited life span unless you reduce its MC and density with dry matter and bulking agents. The takeaway is that high-moisture, high-density material is more challenging than high-moisture low-density. If the press cake had 70 percent MC and a 1,200 BD, I would be just as concerned that it would stagnate.

Bulk Density and Mineral Content / Organic Matter

High mineral content can also be a source of density. Organic matter in itself is typically not very heavy, but organic materials contain minerals, which are generally quite heavy (mineral content, or non-volatile solids, is the inverse of organic matter or volatile solids).

High-mineral feedstocks might be high or low MC, but incredibly dense in either case. For example, a cubic yard of dry beach sand can weigh over 2,500 pounds. The two manures described in the previous section had a high organic matter content (80 to 90 percent range dry weight, or 10 to 20 percent mineral content), which explains why they were so light when their water was removed.

Bulk Density and Porosity

Porosity is an absolutely critical factor in understanding the character of raw materials and how they will perform in a compost blend. Logically, the density of a feedstock correlates with how porous it is, because pore space is essentially air. Porosity or the amount of free air space in a material can be measured, but it is not standard in most compost analysis (although maybe it should be), which is why BD is often used in its place as an indicator.

Bulk Density Outliers

Composters frequently encounter examples of outlier feedstocks, where the analysis of the material just doesn't line up with the normal trends or characteristics you'd expect. These examples are why analysis is so critical. One such material that I've encountered was a batch of sugar sand, or niter, which is the residual diatomaceous earth (silica) often used in filtering maple syrup and other products, including beer (see figure 4.4). As expected, it had a high BD, not unlike sand at 2,395 pounds per cubic yard. But unlike sand, an analysis of the materials showed relatively low mineral content, with an organic matter (OM dry weight) of 69.2 percent. This might point to a high MC, but this was also low at just 24 percent. At first I thought that the analysis was just wrong, but then I looked at the OM closer. On a weight per volume basis, the OM alone accounted for 1,248 pounds per cubic yard, more than half of its total weight. Now the BD started to make sense: The organic matter is really dense, it's a gummy sugary clump, like maple candy. As a feedstock, it's actually more valuable than I originally thought, because it's not just mineral. It will, however, require a lot of blending, moisture, bulking agents, and nitrogen to balance out its unique combination of characteristics. Luckily, most food scrap recipes are going to have all of those things already.

Measuring Bulk Density

Bulk density is easy to test by yourself, and also available through most lab tests, but it may be an add-on (it currently costs an additional $12 from Penn State Ag Labs). Labs give you the option of providing your own bulk density, though I think it's useful to get their measure of BD, even if I were to do my own (see *Measuring Bulk Density On-Site*).

PENNSTATE
1 8 5 5

(814) 863-0841 Fax (814) 863-4540
Agricultural Analytical Services Laboratory
The Pennsylvania State University
University Park PA 16802
www.aasl.psu.edu

Analysis Report For:	Copy To:
Highfields Institute P.O. Box 503 Hardwick VT 05843	

LAB ID:	SAMPLE ID:	REPORT DATE:	SAMPLE TYPE:	FEEDSTOCKS	COMPOSTING METHOD	COUNTY
C05051	Butternut Mtn. NITRE	6/3/2011	Feedstock			

COMPOST ANALYSIS REPORT

Compost Test 1A

Analyte	Results (As is basis) (Weight basis)	(Volume Basis*)	Results (Dry weight basis)
pH	6.3	—	—
Soluble Salts (1:5 w:w)	3.26 mmhos/cm	—	—
Bulk Density*	—	2395 lb/yd³	—
Solids	75.3 %	1804 lb/yd³	—
Moisture	24.7 %	590 lb/yd³	—
Organic Matter	52.1 %	1248 lb/yd³	69.2 %
Total Nitrogen	0.13 %	3 lb/yd³	0.2 %
Carbon	21.1 %	506 lb/yd³	28.1 %
Carbon:Nitrogen Ratio	166.40	—	166.40

*Volume results are calculated on the basis of laboratory-determined compost bulk density

Figure 4.4. Niter analysis. Niter is not a material that most composters will encounter, but it's an interesting example of an outlier in terms of the density of the organic matter present. Courtesy of Highfields Center for Composting and the Vermont Sustainable Jobs Fund.

Measuring Bulk Density On-Site

Measuring BD of a material is really quite simple, using the following method.[21]

Materials

- 5-gallon bucket
- Shovel
- Scale (bathroom or luggage scale)

Methodology

1. Weigh the empty 5-gallon bucket (usually around 2 pounds).
2. Collect samples from your pile at various depths until the 5-gallon bucket is half full.
3. Drop the bucket 10 times from a 6-inch height. The goal is to gently pack the material (heavy packing will give an inaccurate density).
4. Continuing to collect representative samples, fill the empty portion of the bucket halfway again, then repeat the dropping process.
5. Fill the bucket to the top. Repeat the dropping process.
6. Fill the bucket to the top.
7. Weigh the full bucket.

Bulk Density Calculation

(Filled Bucket Weight – Empty Bucket Weight) × 40 = Pounds per Cubic Yard

Field Assessment of Pile Density

Composters need to have the capacity to quickly assess pile density in the field. For more on this, see the *Assessing Pile Structure* section in chapter 13, page 366.

Porosity

Porosity refers to the amount of *free air space, voids,* or *interstitial* space between particles. The porosity of a pile, or lack thereof, impacts how well gases can flow, allowing fresh air to enter and old air to escape. While we don't typically calculate porosity during recipe development, it is really the fourth critical characteristic of any recipe and needs to be considered along with C:N ratio, MC, and BD. Optimal porosity is between 30 and 60 percent free air space, but since in most cases we don't actually measure it, the porosity of your recipe is based on what you know about your feedstocks, how your mix feels to work with, and ultimately how it performs.[22]

While it's tempting to think of porosity as the counterpoint to dense and often wet pile conditions, porosity should not be confused with bulk density. Porous materials tend to be lighter, but many lightweight materials lack pore space. Dry sawdust, for example, can be very light, but when super fine it acts like glue when wet.

The primary means of creating porosity is by adding feedstocks that composters refer to as bulking agents. The most common bulking agents are wood chips and wood grindings. Wood chips have been found to have as much as 43 to 88 percent free air space depending upon the method of measurement.[23] Adding 5 to 10 percent wood chips to a recipe by volume can create adequate porosity in most food scrap or manure recipes, assuming there are also plenty of other carbon and dry matter sources. The two key factors to consider when thinking about the value of a material as a bulking agent are particle size and structural integrity.

Measuring Porosity

In most cases, composters design for porosity during development of a compost recipe, then assess it in practice through diligent observation and monitoring. But some cases call for more accuracy. Researchers might be interested in quantifying free air space (FAS) as a reference point or process variable. Aerated static pile composting performs better with higher porosity, and testing might assist in optimizing a recipe without overbulking.

Unfortunately, the commonly used field test for measuring FAS, known as the water saturation and draining procedure, has been found to be inaccurate, consistently underestimating FAS in comparison with lab methods. Under the uncommon circumstance that precise porosity is useful to you, it is best to send a sample to a lab and make sure that they use either the air or water pycnometer methods or the Haug equation.[24] The water saturation and draining procedure may still be useful as a method for comparing mixes, but the results should only be compared in terms of that method, not others.

Particle Size

When consolidated, larger particles create more free air space than smaller particles. Consider a pile of gravel or sand, compared with a clump of clay. Possessing the smallest soil particles, clay acts like one solid particle and can even be impermeable to water. Gravel and sand on the other hand remain pervious to water, even if they are compacted with heavy machinery.

When lots of small particles combine, they become dense and act like one massive particle. No air can move. Water gets trapped. Many compost feedstocks act almost like clay, and until they start to break down, they might as well be. Incorporating some larger particles into a dense mix helps to break up those uniform masses.

There are benefits to having a mix with some smaller particles, however, if porosity can be maintained. Small particles have more surface area (per volume), where microbes can easily interface with them as a food source, along with oxygen and moisture. The surface of particles is where all the action is, and as long as small particles aren't stacked too densely, they will have higher levels of microbial activity and break down faster.

Conversely, large woody particles don't typically break down completely in one composting cycle due to their low surface area to volume ratio, which is one of the reasons why a common recommendation is for

Figure 4.5. Feedstocks with a natural diversity of particle sizes are advantageous in creating a blend that is porous and also has a lot of surface area for microbes to inhabit.

composters to only use 5 to 10 percent in the recipe by volume. Particles 1 inch long or smaller seem to be ideal, but particles up to 3 inches long (such as some wood grindings) will work fine; they'll just be slower to break down.

The ideal compost blend has a diversity of particle sizes, balancing density, porosity, and surface area.

Measuring Particle Size

Particle size is an uncommon test in most cases, but might be useful when analyzing the value of a particular feedstock. It is typically assessed in terms of the percentage of particles that fall within a certain size range. Labs use different-size screens, screening down from large to small. For example, if I'm buying a lot of bark from two different sources, I might analyze them both for particle size, then weigh their cost against the product with the most diversity of particle sizes, or the one with the greatest proportion of fine particles, or large particles (depending upon what I'm looking for).

Figure 4.6. At the NYCCP hosted by Earth Matter (Governors Island, New York City), a cross section of a compost pile shows the vertical stability of a pile that has good structural integrity.

Structural Integrity

As a material breaks down, its particle size reduces. As a general rule, light materials become more dense, and heavier materials become less dense. Therefore, we cannot think about the density and porosity of a pile as static. There are a lot of factors involved, including release and rebalancing of moisture, particle size reduction, and gravity.

There can be periods where the density in a pile increases, typically in the primary phase, as decomposition first initiates and cell walls release moisture. At those times, particles that maintain their structural integrity are invaluable, because they keep the interstitial pathways open. Compare wood chips versus hay. Lignin, the structural cell wall dominant in woody plants, holds up in a compost pile quite well compared with cellulose, which is more dominant in hay. This is why hay breaks down faster, but it's also why it clumps more.

Piles that lack structurally integral particles also tend to slump. I refer to this as the *stackability* of a blend. If the mix of particles can't maintain a vertical structure, it's not conducive to passive aeration.

Other Key Factors in Compost Recipes and Feedstocks

C:N ratio, moisture content, and bulk density are the three key characteristics that we calculate in recipe development. Porosity is considered the fourth fundamental recipe characteristic for compost performance, although it is estimated, not calculated.

There can be no doubt, however, that these four variables, while essential, are just the beginning when it comes to understanding recipe and feedstocks. In this section we'll discuss several additional factors that are relevant and sometimes overlooked, including:

- Available carbon
- pH
- Salts (conductivity)
- Organic matter
- Feedstock age

Available Carbon

Given that carbon is the primary building block of life on earth, you can imagine the diversity of its forms. As composters, we can make sense of these seemingly infinite C forms by training ourselves to assess their usefulness in our process. Carbon to nitrogen ratio is one way to think about the C in a material. Another important way is to assess the qualities of a material's carbon as a food source for microbes. We refer to this as *available carbon* or *biologically available carbon*. Carbon availability affects the rate of composting, so more available carbon is preferable in terms of a fast processing time.

Think of biological C availability as a spectrum, much like C:N ratio, with simple forms of carbon like sugars, carbohydrates, and fat being the most readily consumable by microorganisms. The more complex carbon forms are less biologically available, with woody lignified carbon at the low end of C availability.

There are also organic compounds that are high C, but the C is basically not consumable by microbes. Charcoal and biochar are examples of this, where the available C has already been combusted away, and only biologically inert forms remain. The availability of C in a feedstock affects the composting process in a number of ways and therefore should be considered when sourcing feedstocks and developing recipes.

To illustrate carbon availability in practice, compare low-availability carbon sources like wood chips versus something like hay, which is more readily available. Even in the best possible conditions, wood chips will typically not break down completely in a single compost cycle, whereas hay can be no longer identifiable in weeks or even days.

Carbon availability also impacts the accuracy of the recipe development process. For example, wood chips in recipe calculations appear to add a lot of C, increasing the C:N significantly, but because the C is mostly lignin and because they have a low surface area to volume ratio, little C is immediately available to the microbes—in other words, the functional C:N ratio is actually much lower. This could lead to stagnant conditions, odors, leaching, and other issues associated with a low C:N (and high moisture). For this reason, the common wisdom is to leave wood chips out of the recipe when calculating the C:N ratio, then add them in at 5 to 10 percent by volume to bulk up the pile.

There are correlations between available carbon and C:N ratio. Most neutral and low C:N ratio feedstocks have very available forms of carbon (hay and food scraps, respectively). However, at a high C:N ratio there are both very available and very unavailable high-carbon feedstocks. Shredded paper is an example of a high C:N feedstock that can be unidentifiable in a matter of days. With high C woody feedstocks, carbon availability can vary depending on particle size and surface area. Larger woody particles such as wood chips have a lower surface area to volume ratio compared with smaller particles such as sawdust, making the carbon at the particle's core less immediately available.

Table 4.2. Biological Carbon Availability of Different Organic Carbon Forms

Availability	Organic Carbon Form
High availability	Sugars
	Carbohydrates
↕	Starches
	Cellulose
	Hemicellulose
	Chitin
Low availability	Lignin
	High C wood ash/biochar (unavailable C)

pH

pH is a scale of acidity and alkalinity. Pure water, being absolute neutral, is a 7 on the pH scale. Substances with pH lower than 7 are more acidic, while those with higher pH are more alkaline. Extreme pH in compost can have a number of implications for the composting process, from stunting biological activity to increasing nitrogen loss and odor generation.

The acidity of compost fluctuates throughout the process as materials mix and degrade. The pH trend with food scrap composts, which are typically more acidic, is first downward as acids degrade and CO_2 is released. Regardless of whether compost started out low or high in pH, the long-term trend is toward neutral.[25] This is because organic acids lower pH, while ammonia raises it. Both act as neutralizing agents, moderating pH.[26]

Despite compost's neutralizing activity, extreme pH in the initial recipe can cause major issues, and pH should therefore be targeted as follows:[27]

Target pH Range: 6.5–8.0
Acceptable pH Range: 5.5–9.0

Problems Associated with Low pH

The primary concerns with low pH (less than 5.5) composts appear early in the process, when materials are in their most raw and volatile forms. During early active composting, the generation of organic acids causes the pH to drop. If the compost already has a low pH, this creates even more acidic conditions and could cause an unmitigated buildup and release of organic acids (associated with odors). Low pH has been found to correspond with higher levels of anaerobic and facultative anaerobic microbial species and fewer thermophilic species.[28] Low pH also severely limits microbial function and is therefore not only risky but counterproductive.[29] Once stable, however, low pH composts shouldn't cause major problems as long as they are used for an application where slightly more acidic composts are suitable. Common acidic feedstocks include mixed food scraps, coffee grounds, apples, and citrus, the last of which has a pH of 2 to 4.5.[30]

Problems Associated with High pH

High pH is more forgiving than low pH in terms of nuisance concerns. The primary concern is that higher pH correlates strongly with higher nitrogen loss, due to the transformation of ammonium (NH_4) to ammonia (NH_3). Ammonia losses increase dramatically at a pH somewhere between 8 and 9.[31] In addition, high alkalinity is not as conducive to microbial diversity as are more neutral conditions. That said, one of the reasons that the composting process is so forgiving is the adaptation of specific microbes to a wide range of pH conditions.[32] High pH composts shouldn't cause major issues as long as they are used for the appropriate application. Common alkaline feedstocks include dairy manure, horse manure, and wood ash (used only at very low volumes).

pH in Your Feedstocks and Recipe

Although I have been advised that it is nearly impossible to accurately calculate the pH of a recipe based on a blend of multiple feedstocks, this really shouldn't be necessary. Before accepting a large volume of a feedstock that you know or suspect will bring the recipe out of the target range, the best approach is to do a small-scale trial.

Blend the recipe, including a neutralizing feedstock as needed (alkaline for acidic feedstocks or vice versa), then either observe the process or test the pH of the blend. If you're considering handling a feedstock that has a high odor risk, this approach could prevent major problems. Play with the recipe and test the pH on a small scale, even first at a kitchen-blender scale. In theory, you could find the right proportions of the problem feedstock and neutralizing agent without the more neutral elements of the recipe, because these should only act as an additional buffer.

Measuring pH

pH can be self-tested or lab-tested. It comes standard on compost and feedstock analysis from labs. It can also be tested in-house using a pH meter. A common self-testing method is the saturated paste method, which is described in the *On-Farm Composting Handbook*.[33] Since the pH of your compost is determined by the pH of your feedstocks, most composters find that testing feedstocks and finished compost occasionally is enough of an insurance policy.

Salts (Electrical Conductivity)

It is common knowledge that above certain concentrations, salt kills plants by inhibiting their ability to absorb water. The salinization of soil reduces germination and creates veritable drought conditions, even when there is plenty of water available. Salt levels in compost must therefore be monitored, so that its presence does not inadvertently poison plants.

Salts in compost and soils are most commonly measured on a scale of *electrical conductivity (EC)*. You may encounter various units of EC, but the most common unit of is millimhos per centimeter (mmhos/cm). Although the literature is very vague about what constitutes safe concentrations of salts in compost, the general recommendation is to keep compost under EC 5 mmhos/cm and to use compost above that level with caution. All compost contains some salt, and at low levels these minerals are potentially beneficial sources of soil nutrients. But since ECs in the 5 mmhos/cm range are relatively common, testing finished compost periodically is advisable.

In addition, testing the salt content in raw feedstocks can be used as a preventive measure. Typically, operations should be on the lookout for changes in salt content in feedstocks or for high salts in a new feedstock. If finished compost shows acceptable ranges, then you simply need to maintain those levels by testing inputs and outputs. Sites that are at risk of high EC may want to purchase their own testing equipment, which is relatively inexpensive.

It's important to be aware of two things that happen with regard to the salt content of feedstocks during composting. First, by combining higher-salt with lower-salt feedstocks, the salt gets diluted, which is helpful. Second, the process of composting and the loss of organic matter concentrates salts.[34] This natural concentration is logical, because like other minerals, salts don't volatilize; they stay solid, so to speak. With composts that have high salts, it's important for the composter to be informed about the risks and to inform their end users. It is, of course, up to end users to monitor their own soils, if they are comfortable managing those risks.

Measuring Salts (Conductivity)

Salts, or electrical conductivity, can be self-tested or analyzed in a lab. EC comes standard on compost and feedstock analysis from labs. A commonly accepted procedure is the saturated paste method (TMECC 04.10).[35]

Organic Matter (Volatile Solids)

We tend to think of organic materials as synonymous with organic matter, but composters quickly realize that the percentage of our feedstocks that is actually degradable can vary dramatically and is often much less than we expect. *Organic matter (OM)*, also called *volatile solids*, describes the percentage of a material that is combustible and therefore partially compostable. *Non-volatile solids* or *mineral content* describes the percentage of the material that is non-organic or mineral in nature and therefore will not combust. One is the inverse of the other, so 10 percent non-volatile solids content indicates 90 percent organic matter content.

From both a composting and an end user perspective, organic matter matters. (I'm sure I'm not the first one to say that and definitely won't be the last.) On the processing side, carbon-rich organic matter provides the energy source for the composting process. Low-organic-matter material actually suppresses the composting process. Examples of this are manures that contain a lot of sand bedding. There's some energy there, but not as much as if the animals were bedded with hay or sawdust.

The organic matter in the compost feedstocks ultimately impacts the organic matter in the end product. As the carbon is metabolized, the percentage of organic matter decreases dramatically, which means an increase in the inorganic portion. Although most buyers are not savvy enough to look at the organic matter content, sometimes low organic matter is actually visible to the naked eye (say, as visible sand or minerals). Producing high-organic-matter-content compost is a good target for any operation.

Target Organic Matter Content

Organic matter content is one of the critical quality indicators that I recommend including in any quality control standard for finished product. To manage the OM of the finished product, start by considering OM of your raw materials.

Typical Feedstock: > 60% Organic Matter or Volatile Solids Content (Dry Weight)
Finished Compost Targets: 30–60% Organic Matter or Volatile Solids Content (Dry Weight)[36]

Measuring Organic Matter

Organic matter, volatile solids, non-volatile solids, or mineral content will be standard in analysis of compost and compost feedstock.

Feedstock Age

When sourcing organic materials or using them in a recipe, the age of those materials is important to consider. The age of a feedstock relates strongly to both organic matter content and the C:N ratio of that material. With the exception of extremely dry feedstocks, as most materials are stored, they begin to decompose, releasing carbon and losing energy.

As a general rule, fresh material will have a higher C:N ratio, will activate the pile, and will have a longer life cycle within the pile. An older material that is further along in terms of decomposition will have a lower C:N ratio, will have less biologically available carbon, and therefore will not activate the pile and may even suppress the process, which could potentially be a problem. Let's say you go to a farm and collect their manure every six months. The old manure at the back of the pile will have lost a lot of its energy, whereas the fresh material is still at the beginning of its composting life cycle.

Older material does make a good cover or capping agent for fresh material, and in small volumes it can be blended into the pile without overly suppressing the process (20 percent or less by volume for older material is my usual recommendation).

With materials that naturally take a long time to break down, there is the potential to speed up the composting process by using aged material as a feedstock. With materials like bark or woodchips, using aged material could be a good strategy as long as there is enough energy in your other fresh materials to drive a thermophilic process. Using a small percentage of any aged material, for that matter, could reduce operating costs; if the manure has already lost 50 percent of its volume prior to composting, that's 50 percent less handling for roughly the same volume of finished product.

Compost Recipe Development

I recently spent a significant amount of time interviewing small compost system operators, many of whom were using in-vessel composters that cost upward of $100,000. My number one takeaway was that no matter how advanced the composting method is, without a good recipe, the process quite literally goes to shit. At the backyard scale, lots of people get away with just piling food scraps and forgetting about them (also known as feeding the neighbor's dog and local wildlife). Those who "compost" straight animal manures without adding any carbon aren't really bothering anyone; I just feel sorry for anyone who buys their rotted manure chock-full of weed seeds thinking that they are getting good compost. I will reserve my judgment for those who take on composting food scraps from their community without first considering the recipe they will use. Big mistake. Just don't do it.

The composting recipe is the one thing that stands between a pile of cooking goodness and a mass of putrefying mess. This isn't to say that your recipe needs to be perfect, however. One of the nice things about composting is that as long as you follow a few rules, it's actually a fairly forgiving process. If all you do is wrap your mind around what your "browns," or carbonaceous materials, are, add 2 to 4 parts of those to 1 part food scraps, and make sure that any food is contained either in a vessel or in a thick (6- to

Step 1. Solve for Moisture Content

Where the target range is between 55 and 60 percent and:

M1 = first material
M2 = second material
M3 = third material
MC = moisture content (expressed as a decimal; for instance, 80 percent is 0.80)

Moisture Content of Recipe:

(Weight of M1 × MC of M1) + (Weight of M2 × MC of M2) + (Weight of M3 × MC of M3) ÷ Weight of M1 + Weight of M2 + Weight of M3 = MC of recipe

Note: This formula works for any number of combined materials.

Repeat the calculations as necessary, adjusting the weight of materials as you go until the target MC is satisfied.

Tip: These calculations are weight-based, but compost recipes are typically volume-based. To make the last step of converting from weight to volume easier, use weight increments that correspond to the material's bulk density. For example, if a horse manure has a bulk density of 880 pounds per cubic yard, use weights of 440, 880, 1,320, 1,760, and so on, which would correspond to ½, 1, 1½, and 2 cubic yards. Some recipe calculators do this for you, allowing you to input volume and then converting to weight in their calculations.

Next, calculate the C:N ratio of your recipe based on the proportions you found for a desirable MC. If your first calculations provided an unsatisfactory C:N, you must look at the results and assess how feasible more acceptable results will be. Think about which feedstock or feedstocks, if adjusted, will move the MC and C:N in the desired directions. Remember that obtaining a desirable mix with two ingredients may not be possible. For instance, a two-feedstock recipe that generates a perfect 60 percent MC with a C:N of 150:1 is unlikely to be salvaged by revisiting the calculations alone. In this case additional feedstocks should be considered to keep the recipe largely moisture-neutral and drop the C:N if possible. (A desirable material in this case would be dry but have a low-neutral C:N ratio—hay, for example).

Step 2. Solve for Carbon : Nitrogen Ratio

Where the target range is between 25:1 and 30:1 and:

M1 = first material
M2 = second material
M3 = third material
WW = wet weight
% C = percent carbon dry weight
% N = percent nitrogen dry weight
MC = moisture content (expressed as a decimal)

[% C in M1 × WW of M1 × (1 – MC of M1)] + [% C in M2 × WW of M2 × (1 – MC of M2)] + [% C in M3 × WW of M3 × (1 – MC of M3)] ÷ [% N in M1× WW of M1 × (1 – MC of M1)] + [% N in M2 × WW of M2 × (1 – MC of M2)] + [% N in M3 × WW of M3 × (1 – MC of M3)] = Carbon : Nitrogen Ratio of Recipe

Note: This works for any number of combined materials.

Repeat steps 1 and 2 as necessary, adjusting the weight of materials until both the target MC and C:N ratio are satisfied.

Balancing the MC to targeted levels will prevent a saturation of pore space, but this does not always ensure good pile density. A final step to ensuring the recipe will perform well is solving for bulk density. In most cases BD will be fine if the MC has been adjusted properly, but it is best to double-check.

In order to calculate the BD, you must first convert the feedstocks from weight to volume. If you followed my tip in step 1 about keeping the weights proportional with feedstock BD, this should be really simple.

Step 3. Translate Feedstock Weight to Volume

Where:

M = material
BD = bulk density of material

Volume of M = Weight of M ÷ BD of M

Repeat step 3 for each material in recipe, then solve for bulk density.

Step 4. Solve for Bulk Density

Where the target range is between 700 and 1,000 pounds per cubic yard and:

M1 = first material
M2 = second material
M3 = third material
BD = bulk density of material

Bulk Density of Recipe:

(Volume of M1 × BD of M1) + (Volume of M2 × BD of M2) + (Volume of M3 × BD of M3) ÷ Volume of M1 + Volume of M2 + Volume of M3 = BD of Recipe

Note: The volume and BD units must be consistent across the recipe. This works for any number of combined materials.

If the BD is over 1,000 pounds per cubic yard, you should consider adjusting your recipe to reduce its density. This will require you to perform steps 1 through 4 again, although in this case you may want to start with step 4 and work backward. Wood chips are the most common bulking agent. If you haven't added 5 to 10 percent wood chips by volume yet, this would be advisable. In this case you could repeat step 4 but skip repeating steps 1 and 2, because in practical terms a low percentage of wood chips will have little impact on either the MC or C:N ratio in terms of biologically available carbon.

Source: Adapted from the Highfields Center for Composting, courtesy of the Vermont Sustainable Jobs Fund.[37]

What the Recipe Won't Tell You

In addition to these parameters, you want to consider other factors such as *porosity*, *stackability*, *pH*, *conductivity* (salinity), and *volatile solids* (organic matter), which are not calculated. Factors such as feedstock age, biologically available carbon, and the structural integrity of the mix will also need to be accounted for. Most important, you need to be able to adjust the "as-is recipe" as it's built, to account for variability in feedstocks.

Recipes in the Field

For a compost recipe to be useful, it needs to be understandable and efficient for use by the site operator. First, that means that it needs to use volume units, not weight units. Second, the other feedstocks should be proportional by volume to a single unit of the primary feedstock (food scraps).

It should look something like this:

1 part food scraps
1 part horse manure
2 parts leaves
½ part cow manure
½ part wood chips

With small volumes of food scraps, you simply multiply each feedstock proportional to the volume of food scraps. Let's use four 5-gallon buckets as an example. To follow the above recipe, use the buckets as your volume unit and multiply everything by 4, which would give you:

4 parts food scraps
4 parts horse manure
8 parts leaves
2 parts cow manure
2 parts wood chips

User-friendly recipe calculators will work in volume units rather than weight or do the conversions to volume for you. In the event that you need to convert from weight to volume and you are unclear about how to do this, simply divide the weight of the material by its bulk density to get the total volume units (see "Step 3. Translate Feedstock Weight to Volume" in the *Compost Recipe Formulas* sidebar).

Converting Food Scraps to Tons and Cubic Yards

Dealing with large amounts of loose food scraps is probably the only instance when you will need to convert from weight to volume in the field with any frequency. Actually in some cases, the operation has to convert from volume (carts) to weight (tons), then back to volume (cubic yards). Food scraps come to the compost site in lots of different units or vessels, but almost never do they come by the yard or loader bucket. Tons, totes, or 5-gallon buckets—these are the numbers composters have to work with.

To make your recipe accurate (and probably to bill the hauler for receiving the materials), measurements need to be converted to a common unit. This seems obvious, but the number of experienced composters who seem to lack confidence in this step often surprises me. Luckily a BD of 1,000 pounds per cubic yard for food scraps and 2 yards per ton are both really easy to remember. This is why pretty much everything that I do with my work is calculated in tons per week. Tons are the most practical common unit between haulers and composters.

Some sites have the ability to weigh what comes onto the site and some haulers have cart tippers that can weigh their carts, but most small sites have neither option. Rather, small sites count the totes or have weights verified by a third party (such as a scale station), or they just trust the hauler to report how much is coming in. In this case, to get to tons without actually weighing the material in the truck every time, the hauler needs to track the units of food coming in: the number of 5-gallon buckets, or 32-, 48-, or 64-gallon carts, and so on. If this is the case for you, either you or the hauler will need to add up the total material, then convert to tons for tipping fee billing purposes. Then you can convert to yards for use in your recipe by dividing by 2, because food scraps translate to roughly 2 cubic yards per ton.

Table 4.3. Conversion Factors for a BD of ~1,000 Pounds per Cubic Yard

Volume Unit	Weight per Unit	# per Yard	# per Ton	Assumption
5-gallon bucket	25	41	82	Weighing full 5-gallon buckets repeatedly
12-gallon cart	55	18	36	Based on 48 gallons
32-gallon cart	147	7	14	Based on 48 gallons
48-gallon cart	220	4.5	9	Reported by hauler who weighed totes on numerous occasions
64-gallon cart	293	3.5	7	Based on 48 gallons

For converting to tons and cubic yards, you could use the conversion factors in table 4.3, which all equate to a bulk density of around 1,000 pounds per cubic yard.

To ensure accuracy of tipping fees, it's always a good idea to check the weight of your material occasionally, because small differences can add up to a lot over time. For example, one hauler reports that 48-gallon totes in weight tests averaged 220 pounds, but the weight in the truck load divided by the number of totes collected average only 200 pounds. It could be that a lot of the totes aren't quite full or that there is a high percentage of lighter-weight materials being collected. If the food scraps also contain a lot of compostable service ware, paper, or yard debris, the bulk density will be lower.

In my research for this book, several composters shared analyses of food scraps that included bulk density. These ranged from 1,060 to 1,820 pounds per cubic yard, pointing to the variability of food scrap densities generally, as well as to the differences in sampling methodology and changes in density over time (you can imagine that the material going into the mail was not the same exact density as the material arriving at the lab). The same is probably true at times with materials coming onto the site. While the volume-to-weight conversions in table 4.4 are useful

Figure 4.8. Round bales of hay often go to waste and can be sourced very inexpensively. They are very labor-intensive to process without the right equipment, however, and need to be covered or used before they rot and become extremely soggy and dense. Courtesy of Highfields Center for Composting and Vermont Sustainable Jobs Fund.

when it comes to estimating weight coming onto the site, for recipe purposes we want our recipe to accurately reflect the weight of the material going into the mix. If the total units of secondary feedstock going into the mix are based on a good estimate of the weight of food scraps that you have received, then you are probably good. But some composters mix secondary feedstocks in proportion to food scrap volume (the number of tractor bucket loads), in which case the BD will very likely be higher than 1,000 pounds per cubic yard, due to degradation, settling, and sometimes added wash water. In this case a BD of 1,200 to 1,500 pounds per cubic yard is probably more representative.

Feedstocks

Organic materials with uncertain destinies are to composters as running water is to beavers; their mere existence calls to us incessantly, "Capture us, we are a resource." Tree service trucks full of fresh wood chips, the Christmas tree on the neighbor's curb, the stacks of leaf bags in the fall. Instead of fading into the background as we go about our daily lives, those materials are in the forefront of our minds. Potential feedstocks are everywhere in our communities, but their ends are not certain. Which brings us to the feedstock-specific portion of this chapter (though again, this section can only be somewhat

Compost Feedstock Sampling Protocol

When sampling feedstocks, you are attempting to obtain a representative sample of a given material. You will eventually secure a 1-quart to 1-gallon sample (check with your lab for their preference); however, the pile you are sampling from could contain 10, 100, or 1,000 (or more) cubic yards of material. As a result, a sample from one location in the pile will not represent all of the differences in the material throughout the pile. For this reason, you need to take multiple subsamples that, combined, will represent the entire pile in a quart- or gallon-size sample. You will want to take at least 16 subsamples following this protocol, and combine them in order to obtain the representative sample that will be sent to the lab. For large volumes of material (100 cubic yards or more) or where the material is highly variable, it is best to take more than one sample from a given pile in order to fully represent the variability of the material.

Additionally, it is important to prevent cross-contamination of the feedstock you are sampling with other feedstocks or materials that you may have sampled recently. Therefore, you must ensure that all tools and equipment used for sampling are clean.

Tools

- 5-gallon bucket (clean)
- Spaded shovel (clean)
- Trowel (clean)
- Sample container or ziplock bag (new)
- Permanent marker

Sample Collection

1. Take at least 16 subsamples (one shovelful) from around the pile, including all sides and at various depths. Approximately 75 percent of subsamples should be obtained at a depth of at least 12 inches.
2. Remove any distinctly unrepresentative matter (trash, branches, and the like).
3. Collect the subsamples in the 5-gallon bucket and thoroughly mix the materials together.
4. With your trowel, remove roughly 1 quart to 1 gallon (check with your lab) and place in

feedstock-specific, because feedstocks and recipe can never really be entirely disentangled). The craft in composting is in knowing (1) a specific material to the point where you can predict and manage how it will behave when composted; (2) how to physically and economically capture a material; and (3) what risks, downsides, or added handling might be associated with managing a particular material.

Feedstock Analysis

The qualities of organic materials can vary significantly, even those that on the surface appear similar. Composters who are doing it for a living, who are investing a lot of time and resources, or who face significant risks if things go wrong almost always benefit from using a more targeted composting recipe. The only way get an accurate recipe is by testing your feedstocks. Even if you aren't doing a lot of recipe calculations, a feedstock analysis will allow you to more accurately gauge how a specific material impacts a composting recipe.

To illustrate why feedstock character changes, consider horse manure as an example. Horse manure and bedding can vary significantly based on (1) how frequently stalls are cleaned out; (2) how much bedding is used; (3) what types of bedding are used; and (4) what season it is. These are just a few of the potential

a sample container (ziplock bags will work, but with liquid or semi-solid dairy or hog samples, plastic screw-capped jars are best).

5. Label and date each sample clearly in permanent marker.

Sample Shipping and Handling

1. Store the sample on ice while in transit and freeze as soon as possible (to preserve nitrogen).
2. Complete the necessary lab forms, making sure that the sample is identified the same way on both the container and the form. For example: "Tall Grass Farm Heifer Manure—01/08/18."
3. When the sample is completely frozen, send it to the lab via Priority Mail to ensure rapid delivery.
4. If freezing the sample is not possible, ship it with ice packs.

Submitting Your Sample for Analysis

While there is a wide variety of feedstock characteristics you may be interested in testing for, if you are intending to use the lab analysis for compost recipe development, you will, at the very least, want to request the following information from your lab:

- % moisture or % solids
- Bulk density
- % carbon
- % nitrogen
- Carbon: nitrogen ratio

Not critical, but informative:

- Conductivity (salts)
- pH
- % organic matter or volatile solids

The US Compost Council (USCC) provides a list of Seal of Testing Assurance (STA)–qualified labs on their website. These labs all follow Test Method for Examining Composting and Compost (TMECC) protocols, which are the industry standard for compost analysis.

Source: Adapted from the Highfields Center for Composting, courtesy of the Vermont Sustainable Jobs Fund.[38]

PENNSTATE
1855

(814) 863-0841 Fax: (814) 863-4540
Agricultural Analytical Services Laboratory
The Pennsylvania State University
University Park, PA 16802
www.aasl.psu.edu

Analysis Report For:	Copy To:
Compost Technical Services	

LAB ID:	SAMPLE ID:	REPORT DATE:	SAMPLE TYPE:	FEEDSTOCKS	COMPOSTING METHOD	COUNTY
C10485	2018 Mixed Leaves	05/03/2018	Feedstock			

COMPOST ANALYSIS REPORT

Compost Test 1A

Analyte	Results (As is basis) (Weight basis)	Results (As is basis) (Volume Basis*)	Results (Dry weight basis)
pH	5:2	—	—
Soluble Salts (1:20 w:w)	0.06 mmhos/cm	—	—
Bulk Density*	—	27 lb/yd³ BD is likely higher	—
Solids	78.1 %	21 lb/yd³	—
Moisture	21.9 %	6 lb/yd³	—
Organic Matter	70.5 %	19 lb/yd³	90.2 %
Total Nitrogen	1.07 %	0 lb/yd³	1.4 %
Carbon	40.6 %	11 lb/yd³	52.0 %
Carbon:Nitrogen Ratio	37.80	—	37.80

* Volume results are calculated on the basis of laboratory-determined compost bulk density

Figure 4.9. Lab analysis of mixed leaves from Penn State Agricultural Analytical Services Laboratory.

factors. If you look at table 4.5, you'll see that the C:N ratio of the horse bedding that was sampled ranges from 33:1 to 90:1, moisture content is 36 to 73 percent, and bulk density is 235 to 796 pounds per cubic yard. When having an accurate recipe is the goal, it's always a good idea to test your raw feedstocks (but be careful because lab testing can become habit forming).

To have confidence in your analysis, it's important to get an accurate sample. See the *Compost Feedstock Sampling Protocol* sidebar for guidance on taking a representative sample. Figure 4.9 provides an example of a compost feedstock analysis.

Mixed Food Scraps

The feedstock of the day is food scraps. Some people call it food waste, food residuals, kitchen scraps, or garbage. In the industry they are often called source-separated organics (SSO). As I mentioned in the introduction, I make a conscious effort to use the term *food scraps*. That said, all of these terms generally refer to same thing.

Mixed food scraps differentiates commercial, institutional, and residential food scrap sources from food processing residuals, which refers to a typically uniform product that comes from a food manufacturer—for example, potato peels from a potato chip factory.

While food scraps are abundant in most communities, sourcing them is an enterprise in itself. Many composters serve as the processor for one or several food scrap collection services, whereas others do their own collection, a topic that is covered in much greater depth in chapter 12.

As noted elsewhere in the book, food scraps are wet (80 to 90 percent moisture), dense (around 800 to 1,500 pounds per cubic yard), and high in nitrogen (C:N of 15 to 18:1). Since they are mostly water and contain highly available carbon and protein, they are perfectly suited to go anaerobic, which means that managing them effectively requires special care.

Food scraps need carbon, dry matter, and a bulking agent to compost effectively. They also need to be contained; exposed scraps will release odors and attract flies, birds, and other pests.

Figure 4.10. Food scraps in a 64-gallon cart. See the contamination?

Leaves, Grass, and Yard Debris/Trimmings

Organics such as leaves, grass clippings, tree trimmings, and garden trimmings from maintained landscapes, although very different in their individual characters, are generally classified together. This is because they often end up in the same collection streams. Most commonly this stream is called yard waste, though in Hawaii it is often called green waste, and the EPA calls it yard trimmings. I call these feedstocks yard debris, for lack of a better term (and in a conscious effort to remove the term *waste* from the vocabulary when it comes to compostables).

The makeup of yard debris is seasonal. Grass is the primary constituent from April through October, followed by leaves from October through December. Sites that accept yard debris will need to adjust their recipes to seasonal differences and plan their processing accordingly. Many sites only take certain types of yard debris; leaves in particular require the least processing and are a valuable carbon source.

Composting significant quantities of larger-size materials (large particles), such as brush, tree limbs, or stumps, will require chipping or grinding equipment. Once ground, they can be treated as a unique feedstock that is typically similar to chips in consistency but often with a higher percentage of fine particles; analysis I've done of yard debris grindings has come back with a C:N as low as 41:1.

Leaves

Admittedly I don't have a lot of firsthand experience composting leaves on a commercial scale, because most of my operational experience has been rural. That said, leaves make a great compost (I'm stockpiling them for winter behind my shed as we speak).

Leaves are a carbonaceous feedstock, with C:N ratios in the 40 to 80:1 range that vary somewhat

Figure 4.11. Whole leaves (*left*) and shredded leaves (*right*). Note the difference in particle size. The general consensus is that shredded leaves compost faster and are less likely to clump.

depending upon species, and likely based on seasonality. Since leaves typically come in mixed, you'll have to plan around that in your recipe development, either testing a representative sample as best as you can or assuming an average C:N ratio.

Moisture content can also vary widely, so you'll have to use your best judgment as to whether to consider them a wet or dry feedstock at the time of blending. The general consensus is that shredded leaves, which are less prone to matting together, are preferable to whole leaves. Despite their high-carbon credentials, leaves are prone to clump together when they are wet and can become very dense, even anaerobic. Composters have also told me that even relatively dry-feeling leaves aren't immediately absorbent, which makes sense given that leaves are designed to retain water. After a long contact period, they do absorb moisture, but it takes time and mixing.

Common sources of leaves include landscapers, municipal collection, and public drop-offs at both transfer stations and directly at the compost site. In urban and suburban areas, leaves are often both plentiful and accessible. While it's a mystery to me why so many people export their leaves from their properties, their loss is a composter's gain. When it comes to accepting leaves and other yard debris, there is a big difference in terms of contamination between working with the general public and working with landscapers or a municipality; the latter can often provide better controls. Sites that have a good relationship with landscapers might have an honor-system drop box where they can deposit a tipping fee. With a small amount of effort, landscapers and municipalities can be quickly educated about what is and isn't accepted.

In rural areas leaves are typically less available. People just don't seem to care about leaves as much and leave them on the ground or else pile/compost them at home. Leaves that would be available in rural areas are spread out and therefore take more work to collect. Landscapers are probably the most likely source of leaves rurally.

Unless you live in a warm or tropical climate (aloha Hawaii compost 'ohana), leaves are a seasonally available feedstock; typically the cleanest leaves are available from October through December. Fall leaves are going to be fresher and cleaner. Spring leaves are typically the remains of what didn't get cleaned up before winter, so they can be wet and have a higher volume of contamination. Make sure you aren't getting street sweepings. (Most people won't have to worry about this, but in the fall sweepings can be mostly leaves, and from a distance look identical.)

Grass

Many composters ask me about whether or not to accept grass clippings. While virtually every municipal compost site that takes yard debris accepts grass clippings, many community and commercial sites choose not to, due to the wide range of synthetic chemicals used on lawns. Most of the common herbicides that have been studied do appear to break down in the composting process.[39] Still, using grass as a feedstock opens up the gateway to a host of chemical contaminants that would not be present otherwise. There is also the risk that a product won't break down during composting (see *The Threat of Persistent Herbicides*, page 110) or that something new will come onto the market in the future that you really won't want. I rarely dissuade composters from accepting grass, but most operations that I work with choose to avoid it unless they know exactly where it's coming from. Of course, if the clippings come from organically managed turf, which is becoming more and more common, it's a whole different story.

Grass clippings offer several benefits as a feedstock, namely rapid decomposition and the ability to fuel the microbial fire, so to speak. Grass has a low C:N ratio, somewhere in the teens (it's high in protein). When they are fresh, they are also wet, and will become highly anaerobic without carbon, dry matter, and a bulking agent. But given those counterbalances, the sugars and cellulose are readily available carbon sources that drive a fast, active composting process. In the right proportions, leaves, grass, food scraps, and a little bit of wood chips can create a screaming hot recipe. Older grass (such as hay) and thicker grasses like larger ornamental and prairie grasses are going to have a much higher C:N (more like mulch hay or straw), so don't confuse these with grass clippings.

One other upside to grass clippings as a feedstock is that they should at least be free, and you may even want to charge to accept them. The reason for this is related to a couple of other downsides to using them as a feedstock. First, they are nitrogenous, and therefore require carbon, which often comes at a cost. (If you have an N shortage, by all means accept them for free, but many food scrap composters have more than enough N to go around.) Second, they break down to next to nothing; I would estimate a greater than 75 percent volume reduction.

It's important to note that grass is considered a weed pretty much everywhere other than in lawns and meadows, so clippings need to be heat-treated if they are going to be put to horticultural use. This goes for all other feedstocks, too, but grasses can be especially tenacious.

Yard Debris/Trimmings Generation and Capture

Small or micro compost sites can rely on an abundance of leaves and yard debris in practically any urban area. In 2014 the EPA estimated that 61 percent of yard trimmings were diverted to compost facilities.[40] Since these materials are primarily handled by municipal operations at a cost, there may very well be an opportunity to source a small volume periodically—that is, to get a load of leaves delivered to your site by the municipal hauler, by landscapers, or directly from residents. For sites handling only a small portion of local generation, estimated throughput of these feedstocks will most likely be proportional to your demands for processing food scraps or other primary feedstocks.

Larger sites that plan to process the entirety of a town or region's yard debris will require more in-depth analysis. The EPA's waste characterization study reported that yard trimming generation equated to 216.6 pounds per person per year in 2014,[41] which was up slightly from 216.3 pounds per person per year in 2012.[42] These estimates include all of what was composted as well as what went to the landfill or was incinerated, divided by the total US population as reported by the US Census. But the census numbers also include rural areas where few if any leaves end up in the trash, so we can assume that generation in urban and particularly in suburban areas is higher.

The Massachusetts Department of Environmental Protection's *Leaf and Yard Waste Composting Guidance*

Document references several other methods of estimating capture. The most useful metrics include:

- Leaves make up 2 to 5 percent of the total residential waste stream (in Massachusetts).
- Leaf collection equals 150 to 500 pounds per household per year; grass, 400 to 1,000 pounds per household per year (National Data).[43]

Wood Chips

Wood chips are a common high-carbon bulking agent. Their large particle size, structural stability, and light density make them the perfect material with which to add porosity to composting blends. Like leaves, they are relatively abundant, particularly in more populated areas. In some regions chips are in high demand as fuel for biomass energy plants. In those places there may be a cost for chips or similar materials like shredded wood and wood grindings. Still, in most areas small amounts of chips can be sourced for free from tree services and landscapers. There is no shame in chasing down a tree truck. If folks in your area need a little extra motivation, put a cooler on-site and keep a six-pack stocked there to lure them in!

Wood chips are very high in carbon, with reported C:N ratios as low as 200:1 and as high as 1,300:1.[44] Being woody in nature, the C is primarily lignin-based and therefore very slow to break down. Wood chips' recalcitrant nature makes them a great bulking agent, but in terms of making end product, they are on a totally different time line than most other compost feedstocks. Lignin is also low in terms of its biologically available carbon, so while on paper it may seem like it would be great at balancing out high-nitrogen feedstocks, in reality it may supply very little actual carbon where and when it's needed.

Figure 4.12. These wood chips were somewhat moist and stored in a bag for several months. Something sprouted up. Wood chips and yard debris need to be heat-treated to mitigate the spread of invasive species and plant disease.

One of the common recipes that I hear among small composters is 1 part food scraps to 1 part wood chips. Assuming the chips have a C:N of 611, BD of 550, and are 60 percent moisture, and the food scraps have a C:N of 15, BD of 1,000, and are 80 percent moisture, that recipe would have a starting C:N of around 43 and an MC of 73 percent. The C:N would be acceptable—high even—if not for the fact that very little of that carbon is available to bacteria. In reality, what I suspect is happening is that this recipe is a low C:N, high-porosity, and high-moisture mix. The moisture drains, and then it breaks down relatively fast, because there is little available carbon and a decent amount of aeration. In some cases I suspect that this would be okay, particularly in in-vessel systems where nuisances associated with high MC / low C:N are contained. But scale that up or do it in an open system and you are asking for trouble. Also, the vast majority of what's left over will be wood chips. If the food scraps break down completely, but reduce in volume by 80 percent (because they are mostly water), and the chips break down very little (say a generous 25 percent becomes compost, and they also shrink in volume by 20 percent), then you basically end up with somewhere around a 40 percent compost / 60 percent wood chip finished product. Then in most cases the wood chips get screened out, so you lose over half of your finished product and end up with about 20 percent or less of what you started with as fully degraded compost. For all of these reasons, I say wood chips are great, and I highly recommend keeping them at 5 to 10 percent of your overall mix, which is enough to add needed porosity.

A related material that can sometimes be purchased is ground bark. It has a lower C:N ratio and a much larger percentage of small particles. The bark that is produced under freezing conditions is particularly nice because it shatters into tiny pieces. The smaller particle size and lower C:N means it generally breaks down more completely, although it can still take a long time.

One concept that some composters have played with is pre-composting or aging woody materials such as bark, so that their time under active management is reduced and the volume of finished end product from the woody fraction is increased. This is a strategy that could make a lot sense for those with the ability to keep a stockpile on-site. Of course, aged materials contain even less available carbon, so this would not be a feedstock that I would recommend adding at above 20 percent of the total volume; otherwise you'd risk stagnating the process. The same goes for recycled chips or *overs* (oversized particles) that are screened out of the finished product; they are excellent as a bulking agent, but don't count on them as a readily available carbon source.

Livestock Manure and Bedding

Growing up as a composter in farm country made me very partial to livestock manures as a go-to feedstock. Our primary feedstock was food scraps, and we used wood chips and bark to add porosity, but it was bedded animal manures that filled out the bulk of our composting recipe. (I'll just call them manures, but assume that I mean both manure and bedding materials from farm animals, so this also includes livestock urine, which adds both N and moisture.)

The feedstock characteristics of manures vary significantly from animal to animal and farm to farm. Therefore, I'll talk about the most common manures individually following a brief discussion of sourcing manures.

To state the obvious, the source of livestock manures is farms. In urban and suburban areas, this is typically not a source of very much material, but it also shouldn't be ruled out. Given the rise of backyard poultry, chicken bedding in particular comes to mind. Members of my family in the Boston area, for example, are sources of both rabbit and chicken manures. (Yes, the rabbit is a pet, but it's an herbivorous pet.) Small composters who serve residential markets in particular would be well positioned to tap these as a source, and might even be able to charge for the service.

Just outside the city, there are often horse owners or boarding stables. Often urban and suburban livestock owners are hobby farms. While some may be composters themselves, others really need an outlet and are happy to see their manures put to beneficial use. Hobby farms will typically let their manures go for free, and some will pay someone to take it.

In many rural areas the manure options really open up. This is where you can also readily access cow and pig manures.

Cow Manures

Cow manure is truly amazing. It gives finished compost a richness that is hard to replicate. Consider that in many parts of the world people have literally built houses out of it for centuries. It's like glue for soils; we dirt geeks love sticky organic substances that support aggregate formation. I guess there is something to be said for having four stomachs. However, that's the finished compost side of cow manure; the raw feedstock side is a whole other story.

The character of a particular cow manure source can be anything from a pumpable liquid to something that literally looks like sawdust (with a few interspersed surprises). What you get depends entirely on the farm's manure management methods and which animals the manure comes from. To state what may be obvious, you do not want liquid manure.

Most composters look for well-bedded cow manures, which typically come from the calf and heifer barns on dairy farms. Some bedded pack barns will also have nice manure, but often it's old because the pack sometimes sits in the barn for an entire season before it gets cleaned out. Also, many of the pack manures that I've seen are really wet and dense. That said, if the farm gets the bedding right, it could still be a great material, but I digress. With dairy farm manures, my understanding is that calves and heifers are bedded more heavily because they are more susceptible to diseases. In terms of carbon content, they can have C:N as high as many horse manures (30 to 50:1).

Decent compost feedstock from dairy cows, as opposed to calves and heifers, will likely only come from small farms. Small farms often haven't converted to liquid manure systems, and while the manure is still really wet (80 to 90 percent MC), high nitrogen (16 to 18:1), and dense (1,500 pounds or more per cubic yard), it is stackable. In most cases food scrap composters are not going out looking for more N and moisture to add to their recipe, but if you have access to it and can afford it, it adds something to the quality of the end product.

Some medium and large farms have manure separators, which mechanically reclaim the solids from the liquid manure. Separated manure solids, or "press cake," is actually surprisingly suitable as a composting feedstock. It looks dry, light, and fibrous, almost like peat moss, but it's typically wetter than it looks (around 70 percent MC), yet lightweight (500 pounds or less per cubic yard). But the most surprising thing is that the manure press cake I've tested has an almost perfect C:N ratio of about 30:1, so it composts on its own without any other additives. Believe it or not, most farms recover and reuse the manure for bedding, and some put it through in-vessel rotary drum composters for three days to heat-treat and dry it. Agrilab Technologies (see the *Compost Heat Recovery* sidebar in chapter 9, page 269) produces a compost heat recovery system designed to be compatible with these drums in order to recover the abundant heat.

Horse Manures

If you are looking for a feedstock that will reliably help your piles achieve thermophilic conditions, look no further than horse manure. Horses are less efficient at converting feed to energy than cows, so more energy remains in their manure following digestion. (I guess there is something to be said for having one stomach as well.) Not only this, but horses often receive the royal treatment, bedded with high-carbon materials at a much higher rate than their dairy cow counterparts, so their manure is similar to calf and heifer manure. I often joke that we especially like rich horse owners, because they bed at a really high rate.

All of that carbon is like gold to a food scrap composter. Horse manure C:N ratios are reliably 30:1 or higher, and I've seen them as high as 75:1, so it's usually a C source in relation to food scraps. Neutral manures around 30:1 C:N are really great feedstocks for food scrap composters to have on hand, because

Figure 4.13. A good cow manure, like this one, is stackable. Note it's still quite wet, as indicated by the liquid being absorbed at the bottom of the pile. Courtesy of Highfields Center for Composting and Vermont Sustainable Jobs Fund.

essentially you can use as much as you need without throwing off the overall C:N. They can be added to dilute or cover food, which is a really good strategy to control nuisances without impacting the balance of the recipe.

Horse manure does have a few potential downsides, however. One is that manures that are too heavy on sawdust or wood shavings create a compost that resembles the parent material. The solution is not to use too much of that manure—40 percent or less should be good—or increase the N content as needed. I realize that this contradicts what I said in the previous paragraph, but as long as you keep the ratio low enough, both can be true. Second, horse manures often contains de-wormers as well as herbicides that are present in their feed and that can persist both in the horse's gut and through the composting process. This subject is covered in greater depth in *Avoiding and Managing Contamination*, which comes later in the chapter (page 103), but it's worth saying here that if you are composting horse manure, it's very likely that you are also managing persistent herbicides.

Poultry Manures

Chicken and other poultry manure and bedding are common composting feedstocks, typically with low C:N (6:1 plus or minus) and medium to high moisture.[45] They have high levels of volatile N, and often a potent odor that can at least in part be associated with ammonia loss. High pH levels can exacerbate the loss of ammonia. As with other manures, the type of bedding used is a major determinant for its qualities as a compost feedstock. You will notice that the chicken bedding listed in table 4.5 (page 106) has a C:N that is much higher than 6:1, indicating that it was very well bedded.

Hog Manures

Of the four main species of manure described here, hog manure is probably the most challenging as a compost feedstock because it is very wet, high N (9:1 to 19:1),[46] and very odorous. Therefore it should be used sparingly and with adequate control measures in place.

Estimating Manure Generation

Some readers may need to estimate the volume and weight of manure generated by a farm over time. The livestock farming sector has an abundance of information available to guide you through this process, some of which is provided in table 4.4. When estimating manure generation, it's important to account for the addition of bedding, as well as seasonality, which often affects the amount of time the animals are in

Figure 4.14. A well-bedded horse manure. Note the diversity of hay, shavings, and sawdust.

Figure 4.15. Pig shit! Just always wanted a good excuse to say that.

Table 4.4. Livestock Manure Production (Without Bedding)

Animal	Animal Weight (Pounds)	Bulk Density		Manure Production		
		Pounds per Cubic Foot	Pounds per Cubic Yard	Pounds per Day	Pounds per Week	Yards per Week
Dairy cow[a]	1,400	62.5	1,687.5	122.3	856.1	0.507
Lactating cow[b]	1,000	62.0	1,674.0	111.0	777.0	0.464
Lactating cow[b]	1,400	62.0	1,674.0	155.0	1,085.0	0.648
Dry cow[b]	1,000	62.0	1,674.0	51.0	357.0	0.213
Dry cow[b]	1,400	62.0	1,674.0	71.0	497.0	0.297
Beef cow[a]	800	61.4	1,657.8	48.5	339.5	0.205
Swine[a]	135	61.5	1,660.5	11.1	77.7	0.047
Finishing swine[b]	300	62.0	1,674	14.8	103.6	0.062
Sheep[a]	60	62.3	1,682.1	2.4	16.8	0.010
Feeder lamb[b]	100	63.0	1,701.0	4.1	28.7	0.017
Goat[a]	140	62.5	1,687.5	5.8	40.6	0.024
Horse[a]	1,000	61.1	1,649.7	50.3	352.1	0.213
Rabbit[a]	10	28.0	756.0	0.3	2.2	0.003
Layer chicken[a]	4	61.6	1,663.2	0.3	1.8	0.001
Layer[b]	3	65.0	1,755.0	0.2	1.1	0.001
Broiler chicken[a]	2	63.7	1,719.9	0.2	1.1	0.001
Broiler[b]	2	63.0	1,701.0	0.2	1.3	0.001
Turkey[a]	15	63.5	1,714.5	0.7	4.8	0.003
Duck[a]	3	62.4	1,684.8	0.3	2.3	0.001

[a] J. C. Barker and F. R. Walls, *North Carolina Agricultural Chemicals Manual: Livestock Manure Production Rates and Nutrient Content* (Raleigh: North Carolina Department of Agriculture & Consumer Services, 2002), 1–2.

[b] *Manure Management Systems Series: Manure Characteristics*, 2nd ed. (Ames: Iowa State University, 2004), 13.

the barn, where clean manure is easily collected. Manure generation estimates typically involve creating a spreadsheet with separate calculations for each month of the year. Setting up the calculations seasonally provides an opportunity to adjust for the number of hours the animals are in the barn during any particular time of year.

Unique and Challenging Feedstocks

Inevitably most composters are faced with certain inputs that they are not sure how or if they should accept and manage. Knowing how to interpret the character of compost feedstocks gives you many tools in assessing variability in common materials like manures or leaves. Overwhelmingly, these are the same tools needed to assess the unique or challenging materials that will inevitably be coming your way. Challenging materials might include common feedstocks like meat and dairy, paper products, and more recently compostable service ware, but the possibilities are really endless, from apple pomace to zoo manures.

All of these materials can be properly composted, and at times there are benefits to accepting some more challenging products. First and foremost, accepting a wider range of materials from food scrap generators increases the potential diversion in your region and the types of generators you can service.

There are also potential drawbacks. As a composter, you have the final say about what comes into

Figure 4.16. These floral products avoided the landfill, but many composters choose not to accept such material unless they know it came from a clean local source. The cut flower industry is one of the dirtiest in terms of pesticide use. Ever wondered how that rose made it all the way from South America looking so perfect? A veritable bouquet of fungicides might have something to do with it.

Figure 4.17. Covfefe, also known as coffee flavoring. A beautiful and mysterious feedstock that most sites will thankfully never encounter. It spiked a 10-yard test pile to over 180°F in under two weeks. Go figure.

your operation. You are the "gatekeeper of the soil," so to speak, and many composters take that job very seriously. You'll need to work with the generators and/or haulers to let them know about what is acceptable to you, so that you avoid needing to reject unacceptable material that comes to the site. Watch out for materials in the compostable gray area such as paper cups, which are sometimes coated with polyethylene. Similarly, not every compost operation can process compostable plastic clamshells or utensils, so make sure the generator and the hauler are aware of your specifications.

In general, when handling a potentially challenging ingredient such as fish waste or ice cream, dilution is the solution. If a challenging material will overwhelm the system, or throw it off balance, it might not be a good fit. Special accommodations can often be made, but these typically come at a cost. Contaminants such as trash are better avoided altogether, as are persistent chemicals, inorganic materials like rocks, or large particles like whole logs that would take years or expensive equipment to break down.

Meat, Dairy, and Eggs

Can meat, dairy, and eggs be composted? Absolutely, and composters large and small do it around the world. Perhaps the better question is: *Should* you compost them? The fact is that many small composters avoid meat and dairy as a risk reduction measure, and I can't fault them for it. Still, the inability of a composting site or system to reliably handle animal products is a potential disincentive for employing that particular strategy.

Most collection programs will accept all food scraps, so composting animal products is in some ways the benchmark of a system's capacity and versatility. The goal of most food scrap generators is to divert all of their material, which has added benefits: (1) It makes source separation easy—all food can be composted, so less education is required; and (2) it reduces odors from the trash by getting the worst materials out of the waste stream.

So why do some composters avoid these materials? Meat, dairy, and eggs are high in protein, which means that when they are composted, they can generate some very offensive odors if the process is not managed properly. Odors attract animals and other vectors. In essence, composting meat, dairy, and other forms of animal protein increases the risk of issues, which is the main reason why most information about backyard and home composting advises people to leave these materials out of the pile. Following a proper recipe and employing other best management practices controls and breaks down odors.

Animal proteins are also more likely to contain human pathogens than are other types of food scraps, which is why heat-treating all portions of compost at 131°F or warmer for at least three days is an important best management practice. Luckily, active composting can easily reach 131°F without any external heating process (see *Temperature Treatment Standards and the Process to Further Reduce Pathogens*, page 56). Making sure that the compost is fully mature is also critical in reducing potential pathogens.

While there is absolutely nothing wrong with avoiding animal proteins, they can be successfully composted on a small scale. I've worked with numerous schools that use insulated and hardware cloth secured bins to compost 100 percent of the food scraps they generate (see *Bin and Bay Systems*, page 37). If you still have any doubt, consider that many farms effectively compost whole cows, which is considered preferable to other disposal methods. In fact, some of the richest compost I've ever seen was made from cow mortalities. For sites that manage the compost process less than optimally, avoiding meat and dairy is probably good advice, but it's not an answer to controlling odors and vectors in itself. Simply avoiding animal proteins by no means alleviates the risk of issues; best management practices are critical with or without meat and dairy.

Paper Products

Many paper products are an easily compostable carbon input. And there aren't a lot of other materials with reported C:N ratios in the 400:1 to 850:1 range and moisture of 8 percent or more that also have a lot of surface area and are usually free apart from perhaps the cost of collection.[47] Glossy paper should be avoided as well as any mixed paper. Shredded office paper, with the exception of envelopes with plastic windows, is an excellent material. When pondering what to accept, a good question to ask yourself is, *Would I use this as a mulch in my garden?*

Relying on paper too heavily could impact the quality of the process as well as the end product. Some on-site composting systems avoid paper from cafeterias altogether, because it overwhelms the food scraps in terms of volume, throwing off the recipe. If you keep the ratio of paper to other materials low, in my experience it disappears incredibly quickly. Some people who collect compost at home and then bring it to a residential drop-off point like to keep it in a paper bag. By accepting brown paper bags, you can accommodate participants who would prefer this practice.

Compostable Plastics and Service Ware

A growing number of businesses and institutions use compostable service ware. That growth has spawned an increasingly diverse range of compostable products, from silverware, to coffee cups, to single-use coffee pods. There are so many options, in fact, that being a food scrap composter today means having a policy about what, if anything, is acceptable to your operation. Most commonly, these products include a range of paper and compostable or biodegradable plastics. In addition, there are products that span the range from minimally processed plant-based products, like pressed leaf plates and bamboo forks, to purportedly compostable aluminum silverware.

Compostable plastics (such as polylactic acid, PLA) are polymers that should break down under proper composting conditions. They are typically derived from plants such as corn. Many food scrap collection programs utilize compostable plastic bags, which can help encourage participation and address the "ick" factor by creating a barrier between the food scraps and the compost bucket or tote (similar to the brown paper bag). Other programs want to service generators who use compostable food service ware such as compostable plastic cups and silverware.

There are a range of challenges associated with compostable plastics, but manufacturers are getting better at making them actually compostable, so 10 years from now the world of compostable plastics will likely look very different. Currently, many composters find that the conditions they can achieve in their composting process do not adequately break down all of the plastics. If you require a compost system that has this capability and are looking for a commercially available system, this should be one of the first questions you ask a manufacturer. Accept only "compostable" (ASTM D6400 or D6868) products, which is the current industry standard.[48]

Another challenge is that a compostable plastic fork can be hard to distinguish from a petroleum-based plastic fork, making it hard to identify contamination. If the fork doesn't break down, it might then get screened out, which defeats the intent of composting if the overs are then discarded. Composters in some cities are working with food service establishments to use only certified or third-party-approved compostable ware. (This can work well, especially in closed venues such as sports stadiums.) Other composters prefer not to accept them, treating these products as contaminants, which simplifies the process of educating generators.

At the time this book was written, organic agriculture regulation in the United States considers compostable plastics a synthetic. Not all regulating bodies have publicly adopted the recommendation from the National Organic Standards Board that these be on the list of restricted-use products, which has caused some confusion in the industry, but based on recent conversations with organic certifiers, it appears that compost approved for use on organic farms should not contain non-incidental compostable plastics unless your certifier tells you otherwise.

If you choose to accept compostable plastics, expect to deal with a learning curve. Keep these products at the core of your pile to better allow them to break down. Remember that many people view them as "trash"; avoid letting these products spread around the compost site, as this will give visitors the wrong impression. Also, make sure that organic farmers know that your product contains PLA, as they could put their organic certification at risk by using a restricted product.

Figure 4.18. Two compostable forks at different stages of decomposition. One was clearly in the pile, the other on the pile's edge.

Avoiding and Managing Contamination

When non-organic materials make their way into compost feedstocks, it's called *contamination*. Unfortunately, contaminants are one of the challenges that most composters have to contend with at one time or another. Some contaminants are just added work to deal with, like rocks in a manure pile. Other contaminants, such as certain persistent herbicides, are an outright threat to plants.

Figure 4.19. Compostable plastic bags that escaped the pile find their way to all corners of a composting site. While these materials are compostable, when they escape the pile, they should be treated as litter and handled accordingly.

One vital step in assessing a feedstock for use at your operation is to consider potential contamination sources and the cost or risk they might add for the operation. Problems can arrive through numerous and sometimes unexpected pathways, so take your time, do your research, and work to build a relationship of trust with the generator so that there can be ongoing communication about contamination.

Ultimately it's up to each operation to decide:

- What if any contamination is acceptable
- What preventive measures and generator/hauler education are necessary
- What the strategy will be to remove any contaminants and ensure a clean and safe product and work space

Generally contaminants fall into two categories, physical and chemical contamination. We'll discuss both briefly.

Physical Contamination

There is a wide range of materials that composters do not want introduced onto their site and into their product. Unlike chemical contaminants, physical

Anaerobic Digestate as a Feedstock

My counterparts in the anaerobic digestion (AD) industry would not let me live it down if I did not discuss AD's by-products as a potential composting feedstock. There is overwhelming agreement that one of the potential compatibilities between AD and composting is the use of AD as a pretreatment process and composting to add value to the solid component of the end product. Digesters are used for managing many of the same feedstocks as composting: manures, food scraps, food processing residuals, sewage, or some combination of these. (Yard waste is sometimes used in high-solids AD, so it's an AD feedstock, but to a much lesser degree overall.) For the purposes of this book's primary audience, the most appealing feedstocks will be from manure and food residuals digesters (as opposed to from wastewater treatment plants).

The organic by-products following AD are called *digestate*, which can refer to both the solid and liquid outputs from AD. Generally, solid digestate is the most likely composting feedstock, either from a high-solids (dry fermentation) AD system or from solids separated out from liquid digestate, which is often referred to as *press cake*.

In either case, the AD process acts similarly to composting in that there is a reduction in carbon through microbial metabolism. With AD, the C is converted to methane (CH_4) rather than to carbon dioxide (CO_2) as in aerobic composting. The most available forms of C are what methanogens can access. The fibrous remains are the C that comes out in the bathwater, so to speak. The N is largely preserved, although some will be lost as nitrous oxide (N_2O)—and in the case of liquid digestate, much of the N appears to follow the liquid portion post-separation rather than the solid portion.

From the press cakes that I've seen, the end result is solids that are relatively neutral on the C:N spectrum, similar to undigested manure press cake (30:1 plus or minus). Undoubtedly these solids will have slightly less available carbon than undigested manure, but it does seem to heat reasonably well. I would strongly recommend analyzing AD press cake periodically as part of the recipe development process.

It has been posited to me that liquid digestate is a potential composting feedstock, which could be applied or blended into the compost in some way. While I have not run across this as a practice, I could see where the liquid digestate (either separated or not) could be useful where moisture and N were limiting. Yard debris composting sites come to mind. The challenge is, of course, managing a liquid, since most compost infrastructure and equipment is designed for managing solids. Using liquid digestate to add moisture to piles mid-process is also a possible practice, but the site would need to be able to temperature-treat the compost following application, because while AD systems do reduce pathogens, they do not typically meet PFRP.

contaminants are primarily inert, meaning they don't decompose and won't interact chemically with the compost, soil, or plants. In most cases these materials are visible to the naked eye (trash, rocks, and large sticks) and can be picked out by hand or screened out. However, physical contamination might also include particles as small as sand or plastic nano-particles, for which there is no practical method of removal.

Table 4.5. Compost Feedstock Analysis

Feedstock Category	Feedstock Subcategory	Feedstock Description	pH	Soluble Salts (mmhos/cm)
Livestock manure, bedding, and feed	Cow	Dairy—tie stall barn[b]	7.6–8.3	5.8–9.0
		Dairy—free stall barn[c]	7.7	3.9
		Dairy—bedded pack (wood chip bedding)[c]	8.1	5.0
		Dairy—bedded pack (sawdust bedding)[c]	6.6	7.2
		Dairy—bedded pack (straw bedding)[d]	8.1–8.2	18.2–19.9
		Dairy—bedded pack (hay bedding)[d]	6.9–8.3	3.4–17.5
		Calf[d]	7.7–8.2	4.1–7.1
		Heifer[b]	6.6–8.1	6.3–15.9
		Dairy manure solids (press cake)[b]	7.9–8.3	3.1–11.5
		Anaerobically digested dairy manure solids (press cake)[c]	8.4	2.6
	Equine	Horse—very well bedded[e]	6.9–8.2	0.9–3.7
		Horse—well bedded[d]	7.4–7.9	3.0–9.5
	Chicken	Laying hens—very well bedded[c]	7.6	6.2
		Laying hens—average[f]	—	—
	Swine	Pig manure[g]	6.9	—
		Pig manure–average[f]	—	—
	Sheep	Sheep—bedded pack[c]	8.4	8.1–12.1
	Dairy cow feed	Silage[d]	5.6–6.4	6.6–9.1
		Refusals (hay, etc.)[c]	6.7	9.8
	Bedding	Straw—average[f]	—	—
		Sawdust and wood shavings		
Food residuals	Mixed food scraps	Mixed food scraps[h]	—	—
		Mixed food scraps—high meat[c]	—	—
		Mixed vegetable scraps (farmers market)[b]	—	—
	Food Processing	Cranberries[c,j]	2.6–3.1	1.9
		Cranberries—dried sweet[c]	2.9	0.9
		Cranberry sludge by-product[c]	6.8	2.2
		Cranberry trash (leaves, twigs)[j]	3.7–3.9	—
		Niter (sugar sand)[c]	6.3	3.3
		Potato chip processing residuals[c]	4.7	11.7
		Sugarcane plant residues[k]	7.1	—
		Coffee chaff[c]	5.2	3.8
		Cocoa shells[f]	—	—

TEST PARAMETER						
Bulk Density (Pounds/Yard)	% Solids	% Moisture Content	Dry Weight			Carbon : Nitrogen Ratio
			Organic Matter (%)	Total Nitrogen (%)	Carbon (%)	
662–1,591	13.8–27.0	73.0–86.2	65.5–85.7	1.7–3.3	35.7–46.7	14.1–21.0
—	14.8	85.2	86.0	2.5	47.6	18.8
895	29.2	70.8	85.8	1.1	47.2	42.3
582	40.7	59.3	71.3	1.0	37.1	38.0
687	25.8–26.1	74.2–73.9	77.2–80.5	1.9–2.0	38.8–41.0	19.2–21.3
1,000	29.8	70.2–72.6	90.2	1.4–2.7	18.3–47.2	13.4–17.8
430–760	30.6–32.6	67.4–69.4	53.1–95.1	0.8–1.6	27.5–47.5	17.7–56.2
501–1,189	29.9–37.2	62.8–71.7	46.6–89.5	0.6–1.6	13.2–43.3	16.6–31.3
332–610	25.0–32.5	67.5–75.0	91.1–92.2	1.3–1.8	44.0–46.7	25.7–32.9
342	39.7	60.3	94.6	1.5	50.5	33.6
235–680	33.4–63.8	36.2–66.6	70.2–96.9	0.4–0.9	37.6–48.3	50.7–89.9
796	27.3–37.2	62.8–72.7	63.4–87.8	0.9–1.5	32.2–48.9	32.9–34.4
500	—	65.9	—	0.4	13.8	31.3
1,479	—	69.0	—	8.0	48.0	6.0
1,118	15.1	84.9	—	1.8	35.6	19.4
—	—	80.0	—	3.1	43.4	14.0
800	—	39.2	—	0.7	11.4	17.2
316–505	27.6–35.8	64.2–72.4	82.4–86.7	2.3–2.5	42.1–46.3	16.9–20.4
70	57.3	42.7	91.7	3.2	52.0	16.1
227	—	12.0	—	0.7	56.0	80.0
	(See Woody Materials)					
917–1,820[a]	28.3–39.7	60.3–71.7[a]	27.5–93.0	2.4–3.8	36.9–53.2	12.4–18.3
1,740[a]	28.2	71.8[a]	91.6	6.1	53.4	8.7
1,348[a]	20.0	80[a]	—	1.7	46.3	27.2
948	12.2	87.8–90.0	97.9	0.7	43.4	61.8–168.0
1,235	44.9	55.1	98.1	1.5	55.2	36.3
1,211	12.9	87.1	91.5	8.6	47.5	5.5
485	—	63.0–74.0	—	—	—	125.0
2,395	75.3	24.7	69.2	0.2	28.1	166.4
1,366	35.3	64.7	92.3	1.2	48.8	42.3
718	—	36.2	—	1.6	20.0	12.4
360	—	10.4	93.9	2.3	47.0	20.0
798	—	8.0	—	2.3	50.6	22.0

Table 4.5. *Continued*

Feedstock Category	Feedstock Subcategory	Feedstock Description	pH	Soluble Salts (mmhos/cm)
Wood, forestry, and landscaping products	Paper products	Shredded cardboard[g]	7.3	—
		Shredded cardboard (in with food scraps)[c]	6.7	1.6
		Shredded newspaper[g]	3.9	—
		Shredded paper[g]	4.4	—
	Woody materials	Wood grindings[c]	6.9	0.2
		Hardwood bark[c]	6.5	0.3
		Wood chips (mostly softwood)[c]	6.4	0.1
		Wood chips—hardwood (English oak)[l]	—	—
		Sawdust[g]	4.2	—
		Wood shavings[m]	—	—
	Yard trimmings/ debris	Mixed green waste (ground)[c,n,o]	6.2–8.2	1.4
		Fresh grass[c]	6	1.2
		Mixed leaves[c]	5.2	0.1
		Shredded leaves[m]	—	—
Miscellaneous		Peat moss[g]	2.7	—
		Digestate from high-solids AD[p]	8.5	—
		Imperata spp. (cogon grass)[i]	—	—

Note: The green shaded columns are used in recipe calculations.

[a] Food scrap bulk density and MC results are highly variable, likely due to the challenge of getting a representative sample.

[b] 3 lab analyses

[c] 1 lab analysis

[d] 2 lab analyses

[e] 5 lab analyses

[f] Robert Rynk, ed., *On-Farm Composting Handbook* (Ithaca, NY: Natural Resource, Agriculture, and Engineering Service, 1992), 106–13.

[g] "Compostable Raw Materials," Cornell Composting, http://compost.css.cornell.edu/feas.study.tab6.html.

[h] 4 lab analyses

[i] Jan Banout et al., "Investigation of *Imperata* sp. as a Primary Feedstock for Compost Production in Ucayali Region, Peru," *Journal of Agriculture and Rural Development in the Tropics and Subtropics* 109, no. 2 (2008).

There are costs associated with removal of contaminants that can be picked or screened out. For large sites, these can be considerable. Screeners for removal of trash and oversized particles are covered in chapter 6. Prevention is the best long-term strategy, although most operations need to anticipate removal of some physical contamination. In particular, contamination in poorly separated food scraps can pose a significant challenge. Training and education of food scrap generators to prevent contamination through source separation is a topic unto itself, and is therefore covered in chapter 12.

TEST PARAMETER							
Bulk Density (Pounds/Yard)	% Solids	% Moisture Content	Dry Weight			Carbon : Nitrogen Ratio
			Organic Matter (%)	Total Nitrogen (%)	Carbon (%)	
186	—	36.5	95.7	0.4	38.2	107.2
211	—	62.7	92.2	1.3	63.7	48.0
51	—	5.0	—	0.3	39.3	115.5
97	83.4–94.8	5.2–16.6	81.0	0.1–0.3	39.3–41.1	108.0–633.0
365	47.6	52.4	92.7	0.5	40.7	80.9
464	50.4	49.6	92.9	0.6	49.3	80.2
610	41.4	58.6	96.5	0.3	51.0	157.9
359	90.1	9.9	96.5	0.3	49.3	151.0
389	73.1	26.9	—	0.4	44.7	108.7
150	89.3	10.7	84.8	0.1	49.5	402.4
371–887	61.4	34.4–43.5	32.1–65.2	0.7–1.8	33.1–50.4	24.5–40.9
80	10.3	89.7	81.3	4.8	36.5	7.6
27	78.1	21.9	90.2	1.4	52.0	37.8
200	88.1	11.9	85.7	1.1	45.1	42.6
386	—	16.8	91.4	1.2	47.5	38.3
1,331	35.0–39.0	61.0–65.0	44.0–52.0	1.5–1.7	22.7–27.3	15.3
108	—	32.1	—	0.5	50.6	103.0

[j] U. Krogmann et al., "Effects of Mulching Blueberry Plants with Cranberry Fruits and Leaves on Yield, Nutrient Uptake and Weed Suppression," *Compost Science & Utilization* 16, no. 4 (2008): 220–27.

[k] El-Sayed G. Khater, "Some Physical and Chemical Properties of Compost," *International Journal of Waste Resources* 5 (2015): 172.

[l] I. G. Mason, "Design and Performance of a Simulated Feedstock for Composting Experiments," *Compost Science & Utilization* 16, no. 3 (2008): 209.

[m] "Urine as Fertilizer," *Sustainable Agriculture Research & Education*. https://projects.sare.org/project-reports/one14-218.

[n] Paula Robets et al., "In-Vessel Cocomposting of Green Waste with Biosolids and Paper Waste," *Compost Science & Utilization* 15, no. 4 (2007): 274–75.

[o] Frederick C. Michel et al., "Bioremediation of a PCB-Contaminated Soil Via Composting," *Compost Science & Utilization* 9, no. 4 (2001): 274.

[p] Golnaz Arab and Daryl McCartney, "Benefits to Decomposition Rates When Using Digestate as Compost Co-Feedstock: Part I—Focus on Physicochemical Parameters," *Waste Management* 68 (2017): 77.

Chemical Contamination

Although perhaps less costly on a frequent basis than physical contaminants, chemical contaminants are a concern that composters must also take seriously. Because they are not visible to the naked eye, wayward chemicals might find their way into a compost pile completely unbeknownst to the operator or end user. That chemical will then make its way into the soil, which is at best an unfortunate breach of trust between the maker and end user, and at worst might harm or kill someone's gardens. Therefore, you must take adequate precautions to identify and prevent sources of chemical contamination.

There are a number of potential pathways for contamination to occur, including something as simple as a gas spill or blown hydraulic hose. Operations should have spill protocols in place and remove any contaminated material immediately. More challenging to deal with is contamination that comes in the organics. These chemicals might be several times removed from the generator source itself, as is the case with most persistent herbicides, which are discussed in the upcoming subsection.

Part of the challenge with chemical contaminants is in knowing what is and isn't acceptable. It could in fact be argued that composting is a good way to remediate many chemical agents and render them inert. While I personally fall on the side of using natural versus synthesized products, particularly when it comes to foods and land management, there should be no illusions about the fact that compost feedstocks may contain all manner of synthetic substances. It's the back end of a "better living through chemistry" food system, and as have organic food producers, composters have had to make trade-offs about the purity of what goes into the soil. Recycling manures from conventional farms, for example, often involves handling synthetic products. Yet many organic farms would struggle to meet the nutrient demands of their crops without conventional manures, and composters would be severely limited as well.

Food scraps are no different; all of the weird things that we do to our food are what end up at our compost sites. While most of us would like purity in our feedstock, that first involves greater purity in the systems used to produce the foods that we eat. Maybe closing the food-soil loop will be a catalyst for this kind of change. Until then, we'll be faced with the choice of what is and is not acceptable for composting. There is broad consensus that chemical contamination that *can* be prevented *should* be. There is also broad concern about a class of agricultural chemicals that composters call persistent herbicides, because they are not quickly degradable in the composting process.

The Threat of Persistent Herbicides

In 2012 I had the unfortunate experience of witnessing Vermont's biggest food scrap composter, Green Mountain Compost (GMC), go through extraordinary measures to mitigate the damage of an herbicide contamination. An unknown and potent herbicide had made its way through the composting process and had been distributed. By the time all of the damage was done, over 500 gardens were impacted, and the municipal site incurred nearly a million dollars of expenses, compensating their customers, paying for legal fees and testing, losing productivity, and so on.[49] The local composting industry was deeply set back right at a time when food scrap composting was really starting to take off.

At the time, I was a technical assistance provider at the Highfields Center for Composting, and had the job of sitting in on hours of conference calls with the state, then calling just about every small composter in Vermont to explain to them that we really didn't know what was going on, but that it appeared to be an isolated incident. We didn't say it, but we were terrified that the problem might be much more widespread. We knew that this was an acute threat to the robust network of composters throughout the state and beyond that we were working so hard to support and expand.

Fast forward 6 to 12 months. GMC survived. They had outstanding leadership and luckily they serve a very forward-thinking region that is dedicated to composting and waste reduction. Taxpayers ate the cost. Had it been any other compost operation in the state, they probably would not have come out on the other side. The local economy, food scrap collection, and generators would have been set back as well.

Persistent herbicides are herbicides that are still plant-toxic following the composting process. The damage that gave away an herbicide as the culprit is called *epinasty*. These particular herbicides mimic the plant growth hormone auxin and cause a plant's leaves and stems to curl, grow disfigured, and, with enough concentration, die.

After an extensive investigation that would literally make a great detective novel (although nobody who was there would want to relive it), it was discovered that the problem herbicide was aminopyralid (current North American trade names: Milestone and ForeFront). The product had been used on hay nearby and sold as feed to a horse farm. The herbicide persisted through the horse's digestive tract and made it into its manure, which the site had sourced as a feedstock. Although it caused enormous damage, because of the labeling on the herbicide and the fact that it hadn't been classified as a "restricted use" herbicide in Vermont (it now is), there was simply no legal recourse for the site against any of the other parties involved. It was the involuntary cost of doing business as an organics recycler.

The incident was the first time to anyone's knowledge that this particular product had shown its ugly face in compost and been identified, although there is speculation that there have been other instances. It was learned that aminopyralid breaks down very, very slowly in the composting process (it does break down faster in the soil).

We also learned that this herbicide in particular was extremely potent at very low concentrations, which meant that the entire site needed to be treated as potentially contaminated. Unfortunately, there was only one lab in the country that could accurately test for it, and that was at the product's manufacturer, Dow AgroSciences. Despite concerted national efforts on behalf of Dow AgroSciences, the EPA, the US Composting Council, and others, at the time of writing, there is only one known commercially available lab accurately identifying small concentrations of aminopyralid in compost and composting feedstocks like manure and bedding. Unfortunately, these tests cost over $500 apiece, which is simply impractical for most if not all composters.

I tell this story both as a precautionary tale and also because a lot was learned by it. It's unclear, even with the benefit of hindsight, if anything could have been done to prevent the contamination of the compost; there was just so little known about the risks of aminopyralid at the time. An aggressive plant bioassay (growth test) program, however, might have caught this one before the compost was distributed, which is one reason why a plant growth test regimen is so important.

One of the other preventive steps that composters have started to utilize is to chemically bind the herbicides using high-carbon wood ash (HCWA; see the *High-Carbon Wood Ash* sidebar). Aminopyralid is just one of the herbicides composters might come into

High-Carbon Wood Ash

High-carbon wood ash (HCWA) is a fine-particle wood char that is similar to biochar. Composters report that it has the ability to lock up chemicals, including persistent herbicides, rendering them inert until they can be broken down.

Although it has a lower cation exchange capacity than biochar, it is also much cheaper than biochar or activated carbon. Composters have been using it to mitigate the potential for herbicide damage by adding it at a rate of 2 to 4 percent of the total initial batch volume.

It has a high pH (11 to 12) and is very dry, so it will help balance the low pH and high MC of food scraps. While it is high carbon, the carbon is biologically unavailable, so it will impact the final C:N ratio but not increase the amount of available carbon in the compost recipe or finished compost. Therefore, leave it out of C:N recipe calculations.

Make sure to use product that is clean and has been approved for beneficial use. Large agricultural amendment distributors in some parts of the country may be able to source it. Where unavailable, using biochar or activated carbon, or making your own char, are potential alternatives.

contact with. Other known offenders with similar characteristics (and even more likely to show up in compost) include clopyralid, aminocyclopyrachlor, picloram, and bifenthrin.[50] And these will not be the last herbicides of concern to come onto the market. HCWA seems to be an effective insurance policy against the current known offenders, for those who can source it.

Clopyralid seems to be particularly widespread, because it is used in sugar beet production, then makes its way into horse manures via their feed. It could potentially be in other manures as well. Lab analysis of mixed food scraps coming into GMC even came up positive for clopyralid. The EPA's acceptable levels for produce and grains are very high compared with the low concentrations that affect certain garden crops. Other herbicides have been known to show up in yard wastes, so this is not an issue exclusive to livestock manure composters. Clopyralid may be broken down during a prolonged composting process to the point where its effects are negligible, but research trying to determine the exact time frame has had considerably varied results. Composters who are using bioassays sometimes still see effects after nine months, so it's a problem that may be more widespread than most composters realize.

The use of clean high-carbon wood ash or biochar and a solid plant bioassay or herbicide testing regimen (Woods End Laboratories offers these) appears to be the best defense against the threat of persistent herbicides to composters. You can find descriptions of bioassays in chapter 13. Still, composters need to be diligent, keep their eyes open for new threats, and work together to educate and remove threats from their supply chain wherever possible.*

* Dan Goossen, Green Mountain Compost's general manager, reviewed and contributed to this section to ensure its factual accuracy.

CHAPTER FIVE

Processing Capacity and Site Assessment

feedstock
(ˈfēdstäk)
DEFINITION: The raw material used in composting and other organics recycling methods.

In the early stages of site planning, you are developing a clear concept of your project in your mind. I often refer to these as the *Discover* and *Define* phases, where through site assessment and due diligence you are laying the groundwork for successful compost site design, development, and rollout. The initial steps of good compost site and systems planning involve first creating context for the operation in terms of its target scale, then identifying the infrastructure components that will be needed to achieve that target. You'll need to assess the suitability of your potential compost site or sites and become clear on whether your processing needs align with the site, specific to your composting method, as well as to any applicable regulatory factors.

When projects are not well defined based on thorough initial assessment, oversights are made, miscommunication arises, time is wasted, and opportunities are missed. Early due diligence is the basis for more detailed planning, including assessment of permitting needs, revenues, up-front costs, and operational expenses. These *Discover* and *Define* phases should precede the *Design* phase. I recommend investing significant energy in design only once you have defined the site in terms of its components and their scale.

It is also helpful to conceptualize and define the larger goals that are driving the operation's development. Increasing sources of local fertility, remediating denuded land, producing a high-grade commercial compost product, engaging the community in implementing a more independent food system, making a living in a way that gives back to the planet: Any number of factors could be motivating your effort, and these might influence the direction you take in planning, right from the start.

The vision, goals, and objectives can and should change over time, so think of these targets as a starting point. If there are partners or community members involved in the project's implementation, invite those parties into the planning process. Now is the moment to begin identifying potential partners and engaging key stakeholders. It takes time to build trust, and your project will be more successful if the community shares in the vision and goals of the operation.

Even if the project is a solo effort initially, put your goals down on paper and ask the right questions about how to get there. As a college admissions adviser once told me, "I can't tell you what you should do with your life, but once you know, I'll help you get there." It takes time to develop a clear picture of what an operation may look like in the end, and it should be an iterative process. (Not unlike choosing a college major or career.)

One of the goals of this book is to help you identify and answer the key questions that composters face, especially when it comes to scale, systems, and operational components. You may be inclined to jump right into the *Design* phase, but for most readers I recommend starting with a thorough site assessment.*

Compost Site Processing Capacity and System Scale

All aspects of site planning—including permitting requirements, feedstock and equipment costs, infrastructure, and income—depend on one very important variable: processing capacity. As a compost site consultant, one of my most important jobs is helping my clients understand first what their target site capacity is, and second what their actual site capacity is based on current conditions and/or what it will take to get to their target capacity. In early planning stages an exact answer to either question can be elusive for a variety of reasons, which is why it's so important for composters to understand the factors involved in answering these questions.

The following section is designed to guide you through the process and questions involved in defining a target processing capacity. From this target you can calculate the scale of other site infrastructure, as well as value from the operation's services and outputs. It also covers much of the conceptual interplay among site capacity, system phase, pile geometry, processing time, and volume reduction.

First let's start by defining some terminology: processing capacity, throughput, and batch. The *processing capacity* and *throughput* of a site both refer to the volume or tonnage a site can handle in a given time frame, and the two terms can effectively be used interchangeably. Processing capacity is often discussed in relation to a site's annual capacity, but practically speaking site planning involves working in much smaller increments. This is why we are going to use the term *throughput* when referring to smaller and more frequent amounts of material coming onto the site (tons per week, for instance, versus tons per year).

A *batch* of compost is a blend of material of like age that is managed as a unit. It has a distinct start point, where incoming fresh material may be added to the batch, and a distinct stop point after which no new material will be added. *Throughput* also refers to the volume of material coming onto the site in a given time frame, but the term does not necessarily refer to any uniquely trackable unit. To say it a different way: *Throughput* is a unit of flow onto and off the site as a whole; a *batch* is a trackable unit of material within the site, and is often very site- and system-specific. Sometimes these terms can refer to the same quantity if they are representing the same time interval and volume units. They could also be, and very often are, different quantities.

Since this book focuses on food scrap recycling, assume that processing capacity, throughput, and batch size are all relative to the volume of food scraps a site receives. Sizing any food scrap composting operation starts with, or ends with in the case of an existing site, an estimate of throughput.

Estimating Food Scrap Throughput

Those who are planning a food scrap composting operation likely fall into one of several scenarios in terms of estimating the volume of scraps to be processed. Depending upon your scenario, it can be really simple to estimate throughput, or it may involve a fairly significant discovery process. My hope is that one of the following scenarios applies to your situation and will help you clarify the process of estimating throughput:

1. A known volume of food scraps can be delivered to you. Great!
2. You generate a specific volume of food scraps or intend to target a known list of generators, but

---

* Note that the *Site Planning Checklist* in chapter 6 on page 133 outlines the key factors and components involved in compost site assessment.

have never measured the volumes. In this case, visit *Estimating Food Scrap Generation/Capture for Individual Generators* page 324.

3. You already have a target scale in mind in terms of how much finished compost you need to produce. If you already have a specific number in mind, you should then be asking:

    a. How much feedstock do I need to process on a daily/weekly/monthly basis to produce this volume? This will typically be two and a half to three times the volume of the finished compost (see the "Estimating Volume Reduction" sidebar later in this chapter). The percentage that is food scraps will depend on the recipe, but it is typically 20 to 25 percent by volume.
    b. How realistic is it to obtain this much material (see *Collection Service Area and Scale*, in chapter 12, page 316, and its subsections)? What are the costs/revenues associated with that material?

4. You have a site picked out with a limited area and want to know how much compost you can make on that footprint and/or the volume of food scraps that you can process. This is a number that can be backed out of the system sizing equations in the method-specific chapters. A spreadsheet will save you some time. The processing capacity will be very specific to the composting method, with ASP and in-vessel making the most efficient use of space. If this is the scenario, you'll need to pick one or more methods and calculate their capacities based on the available space.
5. You plan to create the capacity to serve as composter to a particular region for a portion of that region's organics. As in scenario three, this involves researching and estimating food scrap generation for the region, then estimating capture rates based on the sectors you intend to target (see *Collection Service Area and Scale*, page 316, and its subsections).

If you intend to charge for food scrap collection or to charge a tipping fee for receiving food scraps, these numbers will be a basis for revenues in these areas. With a target throughput in mind, the next step in site sizing is calculating batch size.

Batch Size

Efficient flow and space usage can be optimized when composting systems are designed based upon a reasonable estimate of batch size. If all of a sudden you try to fit 200 yards of throughput into the space that once fit 100 yards of material, there will be negative process implications. Therefore, a batch's size is ideally formed around a solid estimate of the food scrap throughputs the site needs to be able to handle at peak capacity (see *Estimating Food Scrap Throughput*).*

To arrive at batch size, calculate the total volume, estimating recipe volume based on a particular time frame of throughputs. You can calculate a recipe ahead of time (see *Compost Recipe Development*, page 83) or estimate it based upon a reasonable ratio of additional to primary feedstocks. As a general rule, 4 parts additional feedstocks to 1 part food scraps is a safe ratio, although recipes can go as low as 2:1 or 3:1 and as high as 5:1 or 6:1. Situations that call for conservative recipes, such as operations with odor issues or lots of close neighbors, should plan for a high ratio of secondary feedstocks to food scraps.

The time frame between when a batch starts and when it stops needs to be logical in terms of both the space allowed for the batch and any other general operational or management considerations, such as delivery frequency. Most of the sites that I plan are on a weekly food scrap delivery and blending cycle. Food scraps putrefy rapidly and need to be removed from the generator and processed about once a week.

* Note: Although the terminology is misleading, batch size is a big factor in sizing continuous flow systems because: (1) You need to know how many days or weeks of throughput the system can retain; and (2) when the material exits the system, presumably it will then be stacked with other materials of like age until it's finished, and you need to plan the space around that batch volume.

Therefore the site's throughput is planned on a weekly basis. Assuming there are enough inputs received, one week's material can often be a logical batch size. In other cases weekly mixes might get combined to form a batch every two, three, or four weeks.

The throughput examples that are used throughout this book all assume 10 tons per week of food scraps. That converts to approximately 20 yards of food scraps, which at a ratio of 1:4 food scraps to secondary feedstocks equals 100 cubic yards per week raw feedstock and 80 yards following blending. For the ASP example, batch frequency is one week per batch. For the turned windrow example, the batch frequency is two weeks per batch or 160 cubic yards.

If you think of the batch in units of both time and space, a site should have enough space to fit enough time to manage, mature, store, and distribute a finished product efficiently. In some cases sites might be sized to accommodate smaller batches initially, then accommodate growth later by adding side walls or other means of increasing batch volume without decreasing efficiency. The key is to keep coming back to the system's design capacity and how the site is performing compared with that.

Use step 1 for assistance calculating batch size based on your throughput estimates. The remainder of this section reviews some of the other core concepts and tools used in planning site capacity, including:

- Making quick capacity estimates
- Mass and volume relationships
- Estimating volume reduction
- Pile geometry
- Using spreadsheets

Step 1. Calculate Batch Size

To summarize, you need to know the following to estimate batch size:

- The target throughput of primary feedstock processed in a given time period (tons per week food scraps)
- The ratio of additional feedstocks to primary feedstock by volume
- The frequency with which a batch is created

*1.1 Convert Tons of Primary Feedstock to Pounds**

___ Tons Primary Feedstock per ___ Time Period × 2,000 (pounds per ton) = ___ Pounds per ___ Time Period

1.2 Convert Weight of Primary Feedstock to Volume

___ Pounds per ___ Time Period ÷ ___ Primary Feedstock Bulk Density (Pounds per Cu. Yd.)† = ___ Cu. Yd. Primary Feedstock per ___ Time Period

EXAMPLE

10 Tons ÷ 1 Week × 2,000 = 20,000 Pounds per 1 Week

20,000 Pounds per 1 Week ÷ 1,000 = 20 Total Cu. Yd. per 1 Week

* Note: Skip step 1.1 if you're already using pounds.

† See *Bulk Density*, page 73; *Converting Food Scraps to Tons and Cubic Yards*, page 88; and *Bulk Density: The Relationship Between Mass and Volume*, page 122, for clarification on this step as needed.

1.3 Estimate Volume of Additional Feedstocks

___ Cu. Yd. Primary Feedstock per ___ Time Period × ___ Ratio Additional Feedstock to Primary Feedstock (Typically 3 to 5:1)	=	___ Cu. Yd. Additional Feedstock per ___ Time Period	20 Cu. Yd. per 1 Week × 4:1 Ratio = 80 Cu. Yd. per 1 Week

1.4 Estimate Total Volume of Feedstocks

___ Cu. Yd. Primary Feedstock per ___ Time Period + ___ Cu. Yd. Additional Feedstock per ___ Time Period	=	___ Total Cu. Yd. Raw Feedstock per ___ Time Period	20 Cu. Yd. per 1 Week + 80 Cu. Yd. per 1 Week = 100 Total Cu. Yd. per 1 Week

1.5 Estimate Total Volume of Blended Feedstocks

___ Cu. Yd. Raw Feedstock per ___ Time Period × ___% Blending Shrink Factor (Typically 20%, or Multiply by 0.8) × 27 (Cu. Ft. per Cu. Yd.)	=	___ Cu. Ft. Blended Compost per ___ Time Period	100 Total Cu. Yd. per 1 Week × 0.8% × 27 = 2,160 Cu. Ft. per 1 Week

1.6 Estimate Total Blended Volume Per Batch

___ Cu. Ft. Blended Compost per ___ Time Period	×	___ # of Time Periods per Batch	=	___ Cu. Ft. Blended Compost per Batch	2,160 Cu. Ft. per 1 Week × 1 Week = 2,160 Cu. Ft.

This is your standard "batch size" and can be used with any of the compost system sizing methodologies. For larger systems and especially for both positive and negative ASP systems, use of a biofilter as a capping material is a best management practice and highly recommended. If you are unsure about whether or not you should use a biofilter cap, see the *Capping Piles* sidebar, page 361, and *Capping for Odor Control*, page 268. The estimated volume of capping materials for ASPs is different depending upon whether the system uses positive or negative aeration. If you don't know, positively aerated windrows are the conservative assumption.

1.7 (Optional): Estimate Total Volume/Batch with Biofilter Cap

___ Cu. Ft. Blended Compost per Batch × ___ Biofilter Capping Volume Factor (see table 5.1)	=	___ Cu. Ft. Blended Compost per Batch with Biofilter	2,160 Cu. Ft. × 1.15 = 2,484 Cu. Ft. or 92 Cu. Yd.

Table 5.1. Biofilter Capping Volume Factor

Pile Type	Positive Aeration	Negative Aeration
Bays	1.15	1.07
Windrows	1.25	1.15

Estimating Finished Compost

From your estimates of throughput and batch size, the output of finished compost can be projected with reasonable accuracy. The method for doing this multiplies the estimated throughput by a volume reduction or *shrink factor*. It also typically makes sense to estimate the volume of oversized particles, often referred to as *overs*, that are removed through screening. This percentage can range widely depending upon the proportion of large stable particles in the recipe, the moisture content of the compost that is being screened, and the screen size. Compost that is well suited for screening might have 5 to 10 percent loss to overs, whereas compost that is over 45 percent MC and/or has a significant volume of non-decomposed wood chips might lose well above that. Volume reduction and *shrink factor* are discussed in more depth in the sidebar *Core Concepts in System Scale* on page 122. Step 2, "Estimate Output of Finished Compost," outlines the basic math.

Step 2. Estimate Output of Finished Compost

To calculate finished compost, you need to have estimates of the following:*

- Throughput or batch size (step 1.4)
- Number of throughput unit or batches per year
- Shrink factor

			EXAMPLE
2.1 Calculate Annual Throughput			
___ Total Cu. Yd. Raw Feedstock per ___ Time Period × ___ # of ___ Time Periods per Year	=	___ Cu. Yd. Annual Throughput	100 Total Cu. Yd. Raw Feedstock per 1 Week × 52 Weeks per 1 Year = 5,200 Cu. Yd.
2.2 Apply Shrink Factor			
___ Cu. Yd. Annual Throughput × ___% Shrink Factor (Typically ≥ 60%, or Multiply by ≤ 0.4)	=	___ Cu. Yd. Finished Compost Annually	5,200 Cu. Yd. × 0.4 = 2,080 Cu. Yd.
2.3 Apply Screening Volume Loss Factor			
___ Cu. Yd. Finished Compost Annually × ___% Overs Factor (Typically 5 to 20%, or Multiply by 0.80 to 0.95)	=	___ Cu. Yd. Screened Compost Annually	2,080 Cu. Yd. × 0.9 = 1,872 Cu. Yd.

* If you have not calculated batch size yet, but you know the throughputs of both primary and additional feedstocks, use these to calculate total annual throughput. This may involve converting food scraps from weight to volume and calculating the volume of additional feedstocks if you have not already done so (see the *Compost Recipe Formulas* sidebar in chapter 4, page 85).

Capacity, Methods, and the Phases of Composting

Management and spatial requirements are individualized in terms of both the phase and method of composting; distinguishing each system's requirements leverages those differences to your advantage. Each composting method has unique factors that need to be accounted for when planning capacity, but ultimately every system falls into the basic formula of *throughput > time > output*. In chapter 3 I describe four phases of composting: *primary*, *secondary*, *finishing*, and *curing*. One of the primary purposes of distinguishing these phases is to be able to uniquely plan a site's composting infrastructure in terms of *throughput > time > output* for each phase.

As an example of the ways that phases and methods can be leveraged to create efficiency, let's compare two conceptual systems. In the primary phase, an aerated static pile (ASP) windrow might be stacked 8 feet high and 17 feet wide, whereas a turned windrow pile might be stacked only 7 by 15 feet. Assuming the windrows are the same length, the aerated static pile would hold 20 to 30 percent more material than the turned pile while only adding 13 percent more footprint (this does not ever count the additional footprint required for tractor work space with the turned windrow system). Four weeks of forced aeration might also knock 8 to 12 weeks off the total processing time compared with the turned windrow system. Also, the compost would assumedly shrink at a faster rate, reducing the space required in the secondary and all subsequent phases.

While the ASP in this scenario was used only in the primary phase, it impacts every subsequent phase in terms of volume reduction and total processing time. Using the methods described in the *Core Concepts in System Scale* sidebar, page 122, volume reductions can be applied and pile dimensions adjusted to create efficiencies. The calculations in this book are organized in these four phases to create opportunities to make adjustments along the way.

Figure 5.1. Tightly spaced windrows (*left*) are less efficiently turned, which is a deterrent to implementation of BMPs such as meeting temperature treatment targets. Tightly spaced materials tend to accumulate organic matter and moisture on the compost pad. Sites that maintain planned work spaces (*right*) can be efficiently managed and maintained. Courtesy of Highfields Center for Composting and Vermont Sustainable Jobs Fund.

Quick Capacity Estimates

This book spends a lot of time on processing capacity concepts because I see so many sites getting it wrong. If as an industry we're going to lower the cost of production and maintain product quality, this is something we're going to need to get a lot better at. I've seen sites that planned to stack material 20 feet tall in wide mounds, essentially treating compost as if it were feed silage going into storage. It doesn't take a composting expert to know that that won't go over well. More commonly, I've seen sites where thoughtful effort was put into the designs, but someone just made slightly optimistic assumptions about pile size or processing time. Sometimes people decide to permit the site to the maximum allowable volume under regulations, without regard for physical limitations; then it becomes assumed that the permitted capacity was actually the site's planned design capacity.* The result in each of these scenarios is severely overblown expectations, which eventually lead to overcrowded and unmanageable operations.

There is one very simple technique that can be used to get a rough assessment of a site's capacity, either on paper or while physically on-site. It involves mapping the material on the site in both space and time. Picture a compost site with windrows. Each windrow represents a volume of material that takes up a certain amount of space. Each windrow also represents a certain time frame of inputs. Assuming the flow of material processed by the site is basically consistent, if it takes 2 weeks to construct a windrow, and there are only 5 windrows on a crowded site, we can safely assume that there is close to 10 weeks' worth of material on the site—maybe 12 if one of the windrows was actually two windrows originally that were later consolidated into one. Unless this site possesses some sort of magical microbes, there may not be enough capacity on the site to manage and mature the compost adequately. Unfortunately, this is all too common a scenario.

Whether you're planning or assessing a compost site's capacity or just keeping an eye on a site's capacity as it grows, I highly recommend creating a mental map of material in space and time in this way. It serves as a quick and dirty feedback mechanism, allowing either throughput or capacity to be adjusted before an imbalance becomes a major issue.

Spreadsheets and Calculations

Composting is not rocket science, but good planning and management do involve some math. The most efficient way to do that math is by creating spreadsheets, where you can adjust a few variables and let the formulas do the rest of the work, with fewer errors. As you will see, there are a lot of long-form calculations in this book. Once you know what calculations are useful to you, convert them into spreadsheets.

Compost Site Budgeting and Business Planning

A lot of community composters have no need for a business plan, but even at a micro scale, there are enough labor and material resource needs that budgeting and perhaps some basic cost-benefit analysis about system options is strongly advisable. For those for whom composting is a commercial endeavor and/or a livelihood, comprehensive budgeting, market analysis, and business planning are strongly recommended.

Resource Considerations at the Micro Scale

Micro-scale projects commonly seek to capitalize on shared interests, serendipitous arrangements, and sometimes in-kind donations of labor, time, materials, and equipment. Nonetheless it is advantageous to the overall success of a project to assess its feasibility

* This scenario is actually extremely common and has led to what I believe are probably extremely inflated estimates of capacity in some areas. What worries me is that we may be looking at a gap in infrastructure that is way greater than policy makers realize, which could in fact hold back investment in greatly needed capacity.

through a financial lens, even if money is not the driving factor in the project's implementation or long-term operation. The following questions may be helpful in planning:

Resource requirements. What resources do I have/need? Site, equipment, tools, materials, feedstock?

Organizational structure. Who else is going to be involved in decision making about this project? What will that look like, practically speaking?

Management. How much time will it take to manage this project once it is up and running? Does the team have that kind of time? Does the involvement required make sense in relation to the broader goals? Who else will be involved in the operation of this project, and how? Volunteers? Paid staff?

Reward. What will the rewards be? Do those rewards justify the time and expenses involved? Today? Next year?

At the micro scale it's really easy to underestimate the energy and resources involved, which is why it's so important to learn from and replicate what others are doing successfully.

Financial Assessment and Business Planning for Small to Large Composters

There is no escaping the economic realities faced by composters of all scales. At the time of writing, I can say with confidence that this is a risky business to be in if your goal is to get rich. Still, many composters find a path to economic sustainability.

Without a doubt, a lack of financial planning can hold composters back. Community-scale composters often find themselves in a gray area where, for example, they aren't required to produce a business plan to acquire financing, but not having a clear financial picture can leave them at a disadvantage. I imagine that very few people have ever created a business plan and then thought, *Well, that was a waste of time.*

Collectively, as a community of small composters, better understanding the financial underpinnings of what we do and developing profitable financial models is key to growing this movement. This will undoubtedly involve finding operational efficiencies, lowering the cost of production, and making sure that the communities we serve value the work in fair economic terms.

I recommend that every venture I work with, small or large, develop and maintain a good business plan. This often involves working with a business planner or financial consultant of some kind. A good plan involves more than just numbers; it's an analysis of the factors involved in making those numbers play out to net profit. In a recent *BioCycle* article titled "Postmortem of a Food Scraps Composting Facility," the site's former operator discusses the factors that led to their decision to close, when in theory the numbers could have played out very differently. In his assessment a number of factors led to the operation being unsustainable, including the lack of composting policy incentives (such as an organics ban) and local municipal support/partnership. He also recognizes that they underinvested in marketing, especially early on.[3]

As a primarily technical consultant, I am by no means an expert in financial matters, but I am familiar with the basic revenue and cost areas that compost economics fall into. On the revenue side, these typically include sales of compost and associated products, and tipping fees. Delivery and compost application can be revenue sources as well. Compost made can also offset the cost of purchased compost, and other costs may be offset as well, such as animal feed or energy. Expense categories include start-up and operational costs in five critical areas, none of which can be overlooked:

1. Site/infrastructure/permitting
2. Equipment
3. Labor
4. Feedstocks
5. Sales/marketing/distribution

A good business plan or financial assessment takes all of these revenue and expense categories into account based on well-vetted assumptions and market research.

Core Concepts in System Scale

Let's look at a few factors fundamental to planning or assessing system scale.

Bulk Density: The Relationship Between Mass and Volume

We've talked about the relationship between a material's weight and its volume from a recipe development perspective, but it also can come into play in terms of site capacity planning. When we're taking about capacity, we need to be able to think in space and time. Weight units only correspond to the space a feedstock takes up relative to that specific material. Volume units take up a certain amount of space regardless of the material. Therefore, for capacity purposes, we need to work in volume, not weight. Typically this relationship is described in units of *bulk density* (BD) or the weight of a particular volume of a specific material.

Pounds per cubic yard is the common unit of BD in the United States, which is handy because the volume increments of loader buckets are typically in yards. For capacity planning purposes, the main use for bulk density is in converting feedstocks from weight to volume. For food scraps in particular this is key, but it might come into play with other materials as well. Manure generation, for example, is often estimated by mass rather than volume. For a number of reasons, haulers typically track food scraps by weight. For most readers, though, food scraps are the main feedstock of relevance. It's safe to assume that the BD of food scraps is 1,000 pounds per cubic yard, or 2 cubic yards per ton. On a smaller scale you might track in gallons, which is the unit of most buckets, bins, and carts. Four or 5 pounds per gallon is a fairly safe assumption. It never hurts to cross-check these numbers for your own materials; I would encourage it actually, especially if you receive a tipping fee (also see *Bulk Density* and *Converting Food Scraps to Tons and Cubic Yards* on pages 73 and 88 respectively).

A starting point for most compost system sizing calculations is converting weight-based numbers to volume.

Estimating Volume Reduction

Compost reduces in volume as raw material is processed. In designing compost sites, this *shrink factor* is accounted for in a number of ways. *Blending shrink factor* takes place as different raw feedstocks are combined and blended. Material with smaller-size particles fills in the gaps in materials with larger particles, reducing the total volume by approximately 20 percent. Through a combination of blending and decomposition, finished compost can amount to only 30 to 40 percent of the initial volume of raw materials, depending upon the materials and time involved.

Most methods for compost system sizing account for this volume reduction, and how we apply shrink factor can have a significant impact on facility design. A conservative approach *underestimates* shrink in terms of a site's spatial needs (that is, site capacity and processing time) and *overestimates* shrink in terms of the finished product that's anticipated (revenue).

The main goal at this stage is to introduce this important concept. Studies have shown reductions of as much 43 percent and 69 percent in as little as 15 and 55 days, respectively.[1] Even very small differences in feedstock type can make a noticeable difference, and it's not always intuitive. For example, a volume

Table 5.2. Principles and Rules of Compost Pile Architecture

Principles	Corresponding Rules
1. Taller piles fit more volume on less of a footprint than shorter piles (see table 5.3).	Maximize use of available space by increasing pile height without exceeding the recommended limits (rules 2 and 4).
2. As a pile's height increases, the surface area to volume ratio decreases, which: a. Reduces a pile's ability to passively aerate b. Reduces moisture and heat loss from the pile c. Reduces the distribution of additional moisture through precipitation	a. The maximum recommended height for turned piles is 7 feet (immature). b. The maximum recommended height for ASP piles is 8 feet, depending upon blower capacity. c. Increase pile height to reduce moisture loss and maintain temperature. d. Decrease pile size and flatten piles to increase and distribute moisture.
3. Vertical piles (such as cuboid-shaped bins and bays) fit 50 to 100% more volume on the same footprint as open piles.	When a site has reached capacity with open piles, use side walls to maximize the existing footprint through the use of vertical space.
4. As compost matures, it requires less oxygen to remain aerobic, reducing the requirement for small piles.	Finishing compost can be piled 6 to 8 feet high in tightly stacked windrows. Leave access at the end for occasional turning (once a month or less). Curing piles can be stacked in large masses as needed (this will decrease drying, which may not be ideal). Curing piles should not be stacked more than 8 feet tall, and piles in storage are ideally kept below 10 feet.
5. Triangular and parabolic piles shed moisture whereas flat piles absorb moisture.	a. Form steeply peaked piles to minimize the addition of precipitation and the creation of leachate. b. Flatten piles to absorb precipitation.
6. The cross section of a windrow at the time of formation is essentially two right triangles, which means that the pile width is twice the pile height (give or take a foot or so in either direction). As volume reduces, height reduces but width remains the same.	Turning, restacking, combining piles, and trimming pile edges makes use of space created by pile shrinkage.
7. Due to an increase in biologically active mass and insulation, temperature tends to increase as pile size increases.	A minimum of 1 cubic yard of fresh material (in a bin) is recommended to hit and maintain the common temperature treatment target of 131°F (this is especially true during winter). High temps are easier to hit with larger batch sizes.
8. Plan and maintain pile access ways for both turning and material movement (e.g., turning windrows with a bucket loader from the side is more efficient than turning them from the end).	Leave enough space between piles for efficient access and loader movement.

reduction comparison after 100 days showed a 36.7 percent reduction in wood chips and a 44 percent reduction in bark, despite the bark having a higher C:N (more carbon).[2]

A third volume reduction, or rather volume removal, occurs during screening. Large particles or overs are screened out, and this can account for a significant loss of finished compost overall. One reason why I advocate for using as few wood chips as is necessary to create good pile structure (5 to 10 percent by volume) is to reduce the loss of volume during screening. Volume loss as overs is probably between 5 and 20 percent, but this varies significantly by screen size, the moisture content of the compost being screened, and compost makeup in terms of large particles and contamination.

At the planning stage, precise projections of finished product are impossible. The 20 percent blending shrink factor is an estimate I feel confident in, however, because I've measured it. Especially for food scrap composters who are adding at least 2 or more parts of additional feedstocks, that 20 percent gets applied immediately, so the primary phase infrastructure only needs to accommodate 80 percent of the initial feedstock volume.

When I apply shrink factors as a designer, I find it helpful to break down the composting process into the four phases and apply a shrink factor at each phase. This allows me to be slightly more targeted in my assumptions while accounting for shrinkage and consolidation. In this book's approach, blending shrink factors will be applied when we calculate batch size in step 1 of this chapter. Additional shrink factors are specifically addressed in the sizing methodologies for each method, by phase.

One thing to keep in mind when applying shrink factors for use in capacity estimates is

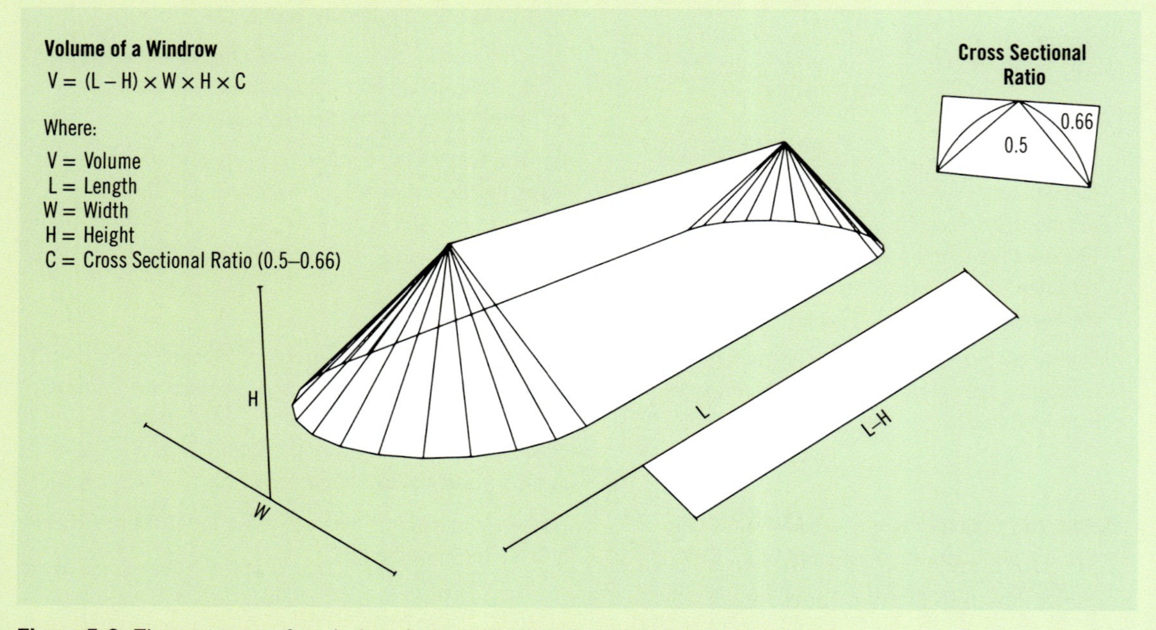

Figure 5.2. The geometry of a windrow in both illustrated and mathematical form.

that volume reduction really only adds space on the site if that space can be utilized by new throughput. Unless piles are aggressively combined and consolidated (which takes time), most operations don't actually see more space on the site as the piles shrink. The width of the pad is the same no matter what, so as the piles shrink, if the pile gets stacked taller, you end up with empty space on the ends, and if you stack longer, the pile gets thinner, but only by a few feet. For example, two batches are formed at 7 feet tall, shrink to 4½ feet, and are then recombined into a new 8-foot-tall pile. One fewer pile creates real and usable space. But if you didn't combine those two piles, the fact that they shrank didn't really add any space for new material—maybe a foot or several over the length of the pad if you went to the effort of restacking the edges. Essentially there is a cost benefit between the value of the additional space and the cost of consolidation. Rather than reducing the size of the pad to account for this shrink factor, take advantage of that extra few feet of space to operate more efficiently. We'll talk about this more in *Geometry and Piles of Detritus*.

It is possible to measure inputs and outputs on a particular site and use that as a feedback loop for future estimates of both throughput capacity and output. This is extremely valuable information. If you are an existing compost site that is planning a capacity increase, it's probably worth going through your records to get a good sense of the ratio of input to output. For new sites in the planning stage, educated guesses are all we have to go on.

Geometry and Piles of Detritus

There is one major paradox that site designers and composting practitioners are constantly managing; let's call it the throughput versus efficiency paradox. There is a natural tendency to want to fit as much material on a site as possible. As I've discussed repeatedly, more material is only a good thing when it can be managed effectively. Piling materials higher and higher, closer and closer, often leads to poor and inefficient management, which is where good design comes in. The shape and size of the compost mass is a primary consideration when it comes to system capacity and effective design. Table 5.2 shows some of the core principles of pile design, as well as some rules that, if followed, help to avoid these problems and lead to a more sustainable process flow.

These are rules that most operators learn over time and adhere to. In addition, some basic compost pile geometry comes in handy (see figure 5.2).

Table 5.3. Volume of a Windrow (Cubic Yards)

Windrow Length (Feet)	Windrow Height (Feet)				
	4	5	6	7	8
10	4.7	6.1	n/a	n/a	n/a
15	8.6	12.2	15.8	19.2	n/a
20	12.5	18.3	24.6	31.1	37.5
25	16.4	24.4	33.4	43.1	53.2
30	20.3	30.6	42.2	55.1	68.8
40	28.2	42.8	59.8	79.1	100.1
50	36.0	55.0	77.4	103.0	131.4
60	43.8	67.2	95.0	127.0	162.7
70	51.6	79.4	112.6	150.9	194.0
80	59.4	91.7	130.2	174.9	225.3
90	67.3	103.9	147.8	198.8	256.6
100	75.1	116.1	165.4	222.8	287.9

Note: Width assumes 2× height.

Finding and Evaluating a Location

In some cases, you will start with a predetermined general location, especially if you are a community garden, urban farm, farm, school, or university. If you are a new enterprise, you may have to look for land. Finding the right location for a new compost site can take a great deal of effort, particularly in urban environments, where farmland is less common.

Visibility may or may not be advantageous to your project, depending on your goals. People are generally more welcoming of small "demonstration" composting systems that might accompany urban agriculture or school and community gardens, whereas larger-scale systems can face more unwelcoming attitudes. It is easier to choose the right location than to have your plans threatened by uninviting neighbors.

Community-scale operations have found homes under bridges, on farms, in abandoned lots, on rooftops, and on or nearby reclaimed landfill sites. What do these locations all have in common? They were all underutilized spaces, for which the operations were able to glean the support of neighbors and/or municipal partners. Look for similar situations for your operation.

Many community compost programs depend on free land and need to work with what they can get. Often a local community organization or community development corporation can partner with you to develop a land-use agreement with their organization or with the city government.

Existing Infrastructure

A primary consideration in site planning is existing infrastructure, which might include buildings, sheds, cement slabs, water, electricity, or roadways. Other site conditions, such as slope, vegetation, and environmentally sensitive areas, must be assessed as well. Few sites are a completely blank canvas, so the task of planning a compost site involves looking at the existing conditions, assessing the pros and cons of what's there through a composting and permitting lens, and then identifying how the site can be best utilized.

Often a site is not perfectly aligned with an operation's goals, so the site, goals, or both must be adapted until a viable plan takes form. Still, I can't overstate the value that existing infrastructure can add to a new operation, especially at the community scale. The start-up phase is hard, and any avoided cost makes a difference.

Environmental Considerations

We talk more about designing conservation measures in site plans in the next chapter, but environmental factors come into play in the pre-design phases as well. Typically for small operations, the major concern is water pollution—both potential surface-water and potential groundwater contamination. The site will need to meet any applicable regulations in terms of setbacks from waterways. There may also be thresholds for distance to bedrock and seasonally high water tables. If no regulations apply to you on this front for whatever reason and you intend to compost a significant volume, safe setbacks are 100 to 300 feet from surface water (depending on slope and risk of runoff) and 3 feet or more above bedrock and the seasonally high water table.

Neighbors and Roads

It is sometimes said that people smell with their eyes. Just the sight of compost, or the impression of garbage, can conjure negative emotions in people. Even the very best and most well-intentioned composter can run into problems with neighbors and NIMBYism (NIMBY = Not In My Back Yard). There may even be a town somewhere that banned composting altogether because an enterprising composter tried to site an operation on a picturesque farmland vista, despite a very good consultant's recommendations to do otherwise. Which is to say, neighbors are a huge consideration in site planning and location.

Conflicts with neighbors, like the one I just described, can stem from irrational objections to composting operations. Such problems are, quite honestly, very difficult to overcome and have caused good composters to pack up shop. However, neighbors who may be impacted by an unfettered odor or an increase in bird activity have legitimate concerns. Some in the industry might disagree with me, but from what I've seen, the latter is more often the case, which is a good thing, because it means that the neighbor issue can probably be mitigated through good management.

From a planning perspective, there are two lines of thinking when it comes to preventing clashes with neighbors. One is to just fly under the radar for as long as possible and hope any neighbors that might take issue don't ever realize what you're doing. On a really small scale, that's probably a fine strategy, but on a larger scale, it's unrealistic. Commercial operations obviously cannot hide what they're doing. Permitted sites may even be required to notify neighbors. So I think that there is a good argument to be made for a second proactive approach, which involves neighborhood outreach as part of your planning process. This will give you a chance take neighbors' concerns into consideration and plan accordingly. Having a relationship with your neighbors can mean the difference between getting a direct call about a complaint or finding out when a regulator arrives.

General rules to follow include siting with as much distance to neighboring residences as possible, while meeting any regulatory setbacks to homes, property lines, and roads. Plan the operation so that raw material has the greatest setback and so that visitors see finished compost first. While at times composting in closer proximity to neighbors is the only practical option, it's always a consideration that needs to be closely weighed. An ounce of prevention in this department is worth at least a few tons of cure, so be a good neighbor first and save the lawyers for a last resort.

Choosing a Composting Method

One of the most critical steps in the discovery process is assessing how different composting methods align with your site, operational, and business requirements. Chapter 2 reviews the common options and provides guidance on the strengths and weaknesses of the different methods. Choosing a method (or multiple) is something that often takes significant research. It's often necessary to run the numbers on the capacity of different systems, or to do some preliminary design and budgeting for different scenarios. One of the best things a composter can do is to visit other compost sites, to see the different methods in action and get operators' firsthand knowledge about working with their system.

Once you have a likely scenario identified (or even before), it's worth reviewing the method-specific chapter/s and other literature on the method, which will help you to more fully define the system concept. Prior to jumping into all of the details of the design, define the following:

- Estimate the system's capacity specific to your site.
  - What is the system's retention time?
  - What are the system's space requirements?
- Identify the system's auxiliary requirements (see chapter 6).
- Outline a budget for the site specific to the system scenario.

Regulatory Assessment and Compliance

Another important early step in site planning is to assess your site options within the context of your local regulatory framework. Permits can be relevant to and sometimes dictate several aspects of site planning, including scale, boundaries and setbacks, stormwater and leachate management, and pad materials, to name just a few. The cost of permitting can be significant,

Producing Compost Suitable For Use on Organic Farms

Confusion often surrounds the use of the word *organic* as it relates to compost. The topic is too large to discuss in full here, but there are a few things that are important to understand. First, compost feedstocks are often referred to as organics, as in "the organics recycling industry," but this is not the same as the organic status of produce you buy at the farmers market.

Organic certifiers, which are distributed across the country, mostly state by state, often review compost products, but these products are not typically "certified organic" in the same way an organic tomato is. This is because in order for a product to be certified, the inputs used to produce it need to be organic, with very few exceptions. While many of the feedstocks composters use are allowed as soil amendments on organic farms, they do not have to be from certified organic sources. Manure from a conventional dairy is just as allowable for amending an organic farm's fields as the manure from an organic dairy. The same is true for composters making compost for use on an organic farm. If all of the feedstock was from certified organic sources, then, as I understand it, a certifier could give compost a CERTIFIED ORGANIC label, but this is very uncommon.

Instead, most compost reviewed by organics certifiers is deemed "suitable for use on organic farms," which means that it has met certain criteria set forth by the National Organic Program (NOP), a branch of the USDA. These criteria include a heat treatment standard similar to PFRP as well as restrictions on allowable and non-allowable inputs.[4] PFRP meets NOP's heat treatment standard, but some certifiers use slightly different language, so it's worth double-checking.

You may be asking: If it's not certified organic, why go through all the trouble? The main reason is that compost that meets organic farming standards can be applied right up to the same day as the harvest (with no harvest interval), while manures or composts that have not met the standards can only be applied 90 days or more prior to harvest of crops that are harvested off the ground (such as tomatoes) and 120 days or more for crops that are in or on the ground (say, leafy greens).[5] Organic farms track all of this and need to be able to show that composts applied without the harvest

as can the cost of meeting certain permit requirements. Above certain scales, there are often additional requirements, such as engineered site plans, greater setbacks, or larger leachate management systems. The impact of applicable regulations can be negligible or large, which is why early and thorough investigation is so critical. Many states have permit specialists who can help identify applicable regulations. Regulators often want to come out and look at a site before you get too far along in the planning process.

How you are regulated and by whom depends on a range of factors including:

- What you're composting
- How much you're composting
- Whether what you're composting was generated on- or off-site

In terms of what you're composting, food scraps, yard debris, livestock manures, and biosolids are all interval have met the standards. Undergoing review allows a composter to put language on their product such as SUITABLE FOR USE ON ORGANIC FARMS, which assures organic farmers that it has already been reviewed and has met the standards. It also allows composters to put the word ORGANIC on their label or website, which helps them stand out in the market. Even though it is not technically organic, quite frankly, no one but an organic farmer would know the difference.

While it may seem contradictory, there are lots of products out there that use the word ORGANIC on their packaging, but most show no signs of being certified or reviewed by certifiers. These companies could potentially face repercussions for misuse of the word. One other thing that is important to know if you are considering producing compost for use on organic farms is that, at least for the foreseeable future, compostable plastics are not allowed as or in feedstocks. The NOP considers the plastics currently on the market to be synthetic and therefore not allowable, aside from incidentals (minor contamination). Unfortunately, when compostable plastics first started to become popular, there wasn't an official opinion on the matter. Certifiers approved composts with plastics in the feedstock, so composters accepted them. I was working in Vermont when word first came that the policy would soon change. Vermont Organic Farmers, bless their souls, got all of the composters and stakeholders together to make sure we had time to adapt. Most composters had never accepted plastics to begin with, but at least one site had to adjust their practices and create a food-scraps-free product for the organic market. Unfortunately, in other cases there did not seem to be any concerted effort to inform composters of the shift, and it has caught some composters off guard when they reapplied with their certifier. One operator told me they lost a half-million dollars in revenue in 2016 because of the shift, which they received zero notice of. When I inquired with the certifier about the change in policy, they told me that this had always been the policy, which seems to contradict the fact that they had approved the very same products in the past. For new composters, this will hopefully not be an issue. Just don't accept compostable plastics if you intend to supply the organic market. Composters that currently accept plastics, and have organic approval, should adapt accordingly. For more on compostable plastics, see *Compostable Plastics and Service Ware*, page 102.

treated slightly differently, and there is a hierarchy for sites that process a combination of materials, where the rules governing the more highly regulated materials take precedence.*

Typically, biosolids have the strictest regulations, followed by food scraps, yard debris, and finally manures (and other ag residuals). Biosolids facilities fall under federal regulations (EPA 503), which are generally implemented at the state level under wastewater laws. Food scraps and yard debris are typically considered solid waste and fall under state regulations, although guidelines for the two are often slightly different, because composting strictly yard debris is less risky in terms of nuisances. Some states

* There are additional categories of materials, such as food processing residuals, animal mortalities, and paunch (livestock renderings that often have their own unique regulatory status), but these are a bit outside of this book's scope.

Common Regulations for Food Scrap Composters

It would be nearly impossible to capture every possible regulation that might apply to every composting situation. Regulations change periodically and are highly case-specific, which is why individual due diligence is so important. Below is an outline of some of the areas of regulation that might apply to different food scrap composting situations across the country.

State Level

- Solid waste permits* (or their agricultural counterparts)
- Stormwater management permits
- Development permits (general, commercial)

Local Level

- Local zoning

* As we move toward zero waste, these will probably be called resource recovery permits or something else along those lines.

- Air quality (for example, California air pollution districts)

Food and Drug Administration: Food Safety Modernization Act (FSMA)

- Biological Soil Amendments of Animal Origin (BSAAO). Standards apply to those producing compost used on farms with over $25,000 in annual revenue. Compost used with zero harvest interval must meet PFRP, and farms are required to have a certificate of compliance for compost not produced on-farm.[6]

Organic Certifiers (Voluntary)

- Local certifiers: approval for use on organic farms
- Local certifiers: organic certification (rare, but possible with all-certified-organic feedstocks)
- Organic Materials Review Institute (OMRI)

provide different regulations for farms (as opposed to non-farm entities) that compost solid waste, but not all do. Composting manure alone is generally regulated by any manure management rules (for instance, regarding setbacks to water), with the exception of commercial manure composting facilities, which might trigger local development rules.

Most state regulations take scale into account and many have a threshold under which composting a small amount of food scraps is exempt. Another common exemption is for sites that generate their own material. Schools often fall under either the on-site or volume exemption.

Some states have written their solid waste regulations to encourage compost site development by creating a streamlined permitting pathway for small sites. Exemptions and small site status can be a strong incentive for composting, so these are options that should be explored, whether you are a composter or a regulator. Using Vermont as an example, food scrap composters need a small composting permit (registration) from solid waste if more than 100 cubic yards per year of total organics are processed (and it contains some food scraps). A permit is required even if all the material is generated on-site. In Massachusetts on-site composting of less than 10 tons per week of food scraps requires notification, but not a permit.

Review applicable regulations and requirements as you choose a composting methodology and design your site.

CHAPTER SIX

Compost Site Infrastructure and Equipment

Compost sites have several infrastructure components and site functions that, in most cases, are useful regardless of system type. Chapters 7 through 11 cover the active composting management areas where the primary composting and secondary processes take place. In this chapter we'll cover auxiliary infrastructure and equipment—that is, the *supporting site components* that are often necessary to process materials successfully from start to finish. These components include spaces where materials are received and managed, such as tipping docks, blending areas, and feedstock storage, as well as curing and storage for finished compost. They include areas for post-processing of compost such as screening and bagging. There is also a significant list of extraneous but essential site elements such as access roads, cover, water, drainage, and treatment for leachate and stormwater. If you are also a hauler (food scraps or otherwise), the site may require dedicated space to meet your collection needs. Some of these auxiliary components are scale-dependent and/or they may not be immediately applicable, but it is wise to anticipate how future growth might impact infrastructure and equipment needs.

Throughout the chapter, we'll discuss the design relationships among these different components. A site's design can either support or diminish operational efficiency and success. Good design also mitigates the potential negative impacts that composting can create both on-site and off-site. With careful consideration, your core operational infrastructure will aid in the development of an optimally functioning composting operation.

psychrophilic
(ˌsī-krō-ˈfi-lik)

DEFINITION: Compost temperature conditions between 32 and 77°F, in which organisms known as psychrophiles are dominant.

Compost Site Infrastructure Components

The purpose of a composting site is to provide a structured system for managing the composting process from raw input through product distribution. Thoughtfully planned infrastructure is not only beneficial, but also absolutely necessary for food scrap composting. This is the part of the book where we move from generalities about composting systems and the planning process into the specific concepts and design of a site's operational components. The chapter may be useful in some aspects of your *Discover* and *Define* phases, but is intended to address the early aspects of actual design. The discussion of infrastructure components follows the natural progression of the composting process as seen in figure

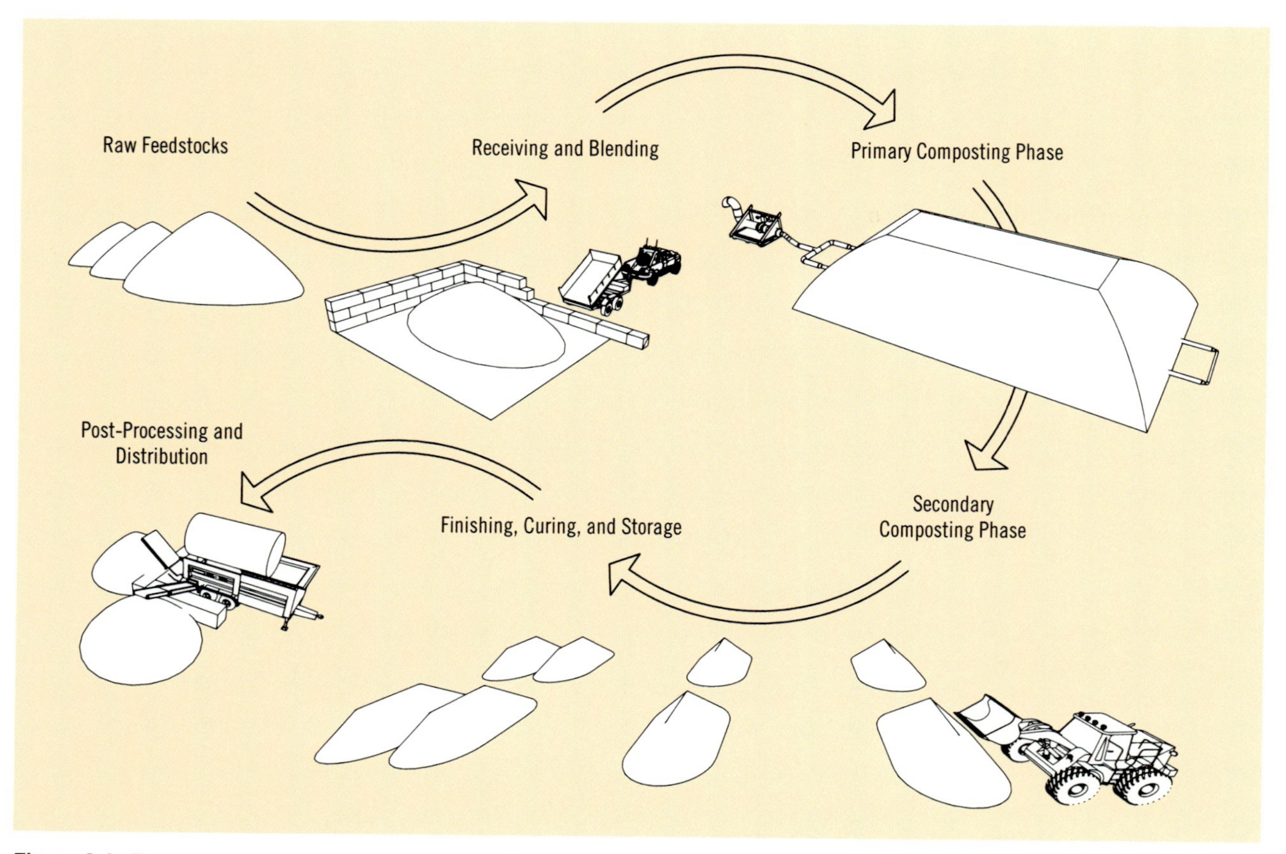

Figure 6.1. Food scrap composting sites typically feature these six elements at a minimum, and often have additional phases, auxiliary infrastructure, and equipment as well.

6.1, starting with receiving and managing feedstocks and ending with post-processing and distribution.

Food Scrap Receiving and Blending Areas, and Equipment

One of the most heavily used site components of food scrap operations is what is often called the *receiving and blending area*. On its most basic level, this is an improved surface on which to receive and blend the operation's primary feedstock as it comes in. In the case of food scraps, it's especially important for this area to remain distinct from the areas where other feedstocks are stored, because food scraps require immediate management upon arrival, unlike manures, chips, and yard debris, which can be temporarily stored.

As a best management practice, food scraps are delivered to the site directly onto a bed of carbon materials, ideally ones that have some immediate absorbency to soak up any delinquent juices. The carbon bed, or *carbon trough*, reduces the mess on the site and immediately begins the process of remediating the putrescent anaerobic state in which scraps typically arrive. Additional carbon materials are then added and blended, following a recipe, to balance moisture content and C:N and create porosity.

The wet nature of food scraps (plus or minus 90 percent moisture) makes receiving and blending on unimproved surfaces challenging at best and unworkable long-term in most cases. Hence, the

Site Planning Checklist

The following checklist outlines the site assessment and planning components covered in chapters 5 and 6. Use the checklist to track applicable site information and refer to related topics in the chapters for more detail and explanation as needed.

Processing Capacity

- ☐ Throughput of primary feedstock ________
- ☐ Ratio of primary to additional feedstock ________
- ☐ Batch size ________
- ☐ Total annual throughput ________
- ☐ Volume of finished compost ________

Existing Infrastructure

- ☐ Existing infrastructure ________
- ☐ Identify footprint available/likely site ________
- ☐ Identify existing equipment ________

Feedstock Sources (note sources, volumes, frequency, seasonality, costs, revenues, etc.)

- ☐ Food scraps ________
- ☐ Leaves ________
- ☐ Yard debris ________
- ☐ Wood chips ________
- ☐ Manure/s ________
- ☐ Sawdust ________
- ☐ Paper ________
- ☐ Other ________
- ☐ Other ________
- ☐ Other ________
- ☐ Other ________

Food Scraps Receiving and Blending Area

BLENDING METHOD:

- ☐ Blending area/pad
- ☐ Total Mix Ration (TMR) mixer

BLENDING AREA PAD MATERIAL:

- ☐ Concrete
- ☐ Packed gravel
- ☐ Reclaimed asphalt
- ☐ Native soil/turf
- ☐ Covered
- ☐ Other ________

Blending area footprint ________

Blending area dimensions ________

Feedstock Receiving and Storage Areas

- ☐ Yard debris/leaf drop-off area footprint ________
- ☐ Feedstock storage area dimensions ________

FEEDSTOCK STORAGE:

- ☐ Open piles
- ☐ Bays
- ☐ Covered

FEEDSTOCK STORAGE AREA PAD MATERIAL:

- ☐ Concrete
- ☐ Packed gravel
- ☐ Reclaimed asphalt
- ☐ Native soil/turf
- ☐ Other ________

Feedstock storage area footprint ________

Feedstock storage area dimensions ________

Primary Composting Area

PRIMARY PHASE COMPOSTING METHOD:

☐ Turned windrows
☐ Aerated Static Pile
☐ In-vessel
☐ Animals
☐ Bins/bays
☐ Vermicomposting
☐ Compost heat recovery
☐ Other____________________

PRIMARY PHASE COMPOST SITE PAD MATERIAL:

☐ Concrete
☐ Packed gravel
☐ Reclaimed asphalt
☐ Native soil/turf
☐ Other____________________

Primary phase compost site footprint:______
Primary phase compost site dimensions: _____

Secondary Composting Area

SECONDARY PHASE COMPOSTING METHOD:

☐ Turned windrows
☐ Aerated Static Pile
☐ In-vessel (typically primary)
☐ Bins/bays
☐ Vermicomposting
☐ Compost heat recovery (typically primary)
☐ Other____________________

SECONDARY PHASE COMPOST SITE PAD MATERIAL:

☐ Concrete
☐ Packed gravel
☐ Reclaimed asphalt
☐ Native soil/turf
☐ Other____________________

Secondary phase compost site footprint: _____
Secondary phase compost site dimensions: ___

Compost Finishing Area

FINISHING PHASE COMPOSTING METHOD:

☐ Turned windrows
☐ Bins/bays
☐ Vermicomposting

FINISHING PHASE COMPOST SITE PAD MATERIAL:

☐ Concrete
☐ Packed gravel
☐ Reclaimed asphalt
☐ Native soil/turf
☐ Other____________________

☐ Finishing phase compost site footprint: ____
____

☐ Finishing phase compost site dimensions:___
____

Compost Curing and Storage Area

CURING AND STORAGE AREA METHOD:

☐ Windrows/open piles (static or turned)
☐ Bins/bays

CURING AND STORAGE AREA PAD MATERIAL:

☐ Concrete
☐ Packed gravel
☐ Reclaimed asphalt
☐ Native soil/turf
☐ Other____________________

☐ Curing and storage area footprint: _______
☐ Curing and storage area dimensions:______

Auxiliary Infrastructure

COVERED/ENCLOSED SPACES:

☐ Feedstock storage

☐ Blending area
☐ Compost storage
☐ Equipment shed/shop
☐ Offices
☐ Greenhouse
☐ Other ______________________

UTILITIES:

☐ Water
☐ Electricity
☐ Heat
☐ Internet

ACCESS:

☐ New roads
☐ Existing roads
☐ Gates
☐ Other ______________________

LEACHATE CAPTURE/TREATMENT:

☐ Vegetative treatment area
☐ Leachate tank
☐ Catchment pond
☐ Sediment pond
☐ Sand filter
☐ Compost socks
☐ Berms
☐ Swales
☐ Other ______________________

STORMWATER MANAGEMENT AREAS:

☐ Diversion
☐ Vegetative treatment area
☐ Storage tank
☐ Catchment pond
☐ Sediment pond
☐ Compost socks
☐ Berms
☐ Swales
☐ Gutters
☐ French drains
☐ Other ______________________

Applicable Regulations/Permits

☐ Solid waste/ag composting permits: ______
____
☐ Stormwater permits: ______________
____
____
____
☐ Commercial/development permits: ______
____
☐ Local zoning/construction permits: ______
____

USE ON ORGANIC FARMS:

☐ Local certifier
☐ OMRI

Equipment

☐ Loader	Existing	Need	n/a
☐ Windrow turner	Existing	Need	n/a
☐ Mixer	Existing	Need	n/a
☐ Screener	Existing	Need	n/a
☐ Bagger	Existing	Need	n/a
☐ Dump truck/s	Existing	Need	n/a
☐ Other	Existing	Need	n/a
☐ Other	Existing	Need	n/a
☐ Other	Existing	Need	n/a

Tools

☐ 18–36″ compost thermometer/s	Existing	Need	n/a
☐ Shovel/s	Existing	Need	n/a
☐ Wheelbarrow/s	Existing	Need	n/a
☐ Rake/s	Existing	Need	n/a
☐ Broom/s	Existing	Need	n/a
☐ Other	Existing	Need	n/a
☐ Other	Existing	Need	n/a
☐ Other	Existing	Need	n/a

Figure 6.2. A load of food scraps is tipped onto the receiving and blending area from the tipping dock at the former Highfields Center for Composting. The load is received on a bed of carbon material and manures, which contains the food scraps, absorbs moisture, and facilitates blending. Courtesy of Highfields Center for Composting and Vermont Sustainable Jobs Fund.

development of an improved surface, ideally a concrete slab, where receiving and blending takes place. In fact, if only one place on the site could be concrete, this would be the spot.

Site the receiving and blending area in a high and dry location. The last thing you need is precipitation from the surrounding landscape flowing into the receiving area and potentially adding moisture to the fresh compost. Work spaces and access around the pad should be gently graded away so that they don't become stormwater catchment. As with the investment in concrete, if only one place on the site could be covered, this would be a significant contender (dry carbon and an office might be pretty high, too). If you are lucky enough to have a deluxe (covered) receiving bay, make sure that the roof sheds water away from pad (or, better yet, into a rainwater collection tank).

It's important to plan the receiving bay with your particular food scrap hauler's access needs in mind. This is especially true if you intend to create a tipping dock.

Tipping Docks

Delivery of food scraps often involves large trucks dumping 5 to 10 tons of material per load, which equates to approximately 10 to 20 cubic yards. Especially for small sites receiving large loads, it can be challenging for trucks to tip without the load spreading all over the place. For trucks, a tipping dock situated above the receiving area can allow the material to be piled higher right out of the truck (figure 6.2), making less of a mess and requiring a smaller footprint. The dock can also double as a short push

Figure 6.3. A gravel receiving area in need of major repair. Ruts trap water and organic matter, which over time makes maintaining a level and firm working surface very hard. Concrete receiving and blending areas are far more durable. Courtesy of Highfields Center for Composting and Vermont Sustainable Jobs Fund.

wall as long as it's tall enough (3 to 4 feet is ideal). On a sloped site you may be able to get enough height by making the concrete pad near level and digging into the slope. Then the ramp can also be raised ever so slightly to meet the height of the dock.

On one compost site, when designing the receiving bay and tipping dock, we weren't sure what the final collection vehicle was going to be. At the time, the solid waste district was doing the hauling with a box truck and swapping collection carts (leaving a clean cart when they picked up a dirty one). But there was a likelihood that eventually a truck that would tip carts in the truck bed, then tip the load from the bed, would want to deliver there. The solution was to build a tipping dock that a truck could back up and tip from, but that the box truck could also back up to and unload totes onto. The dock was built at the height of the box truck so that the two would marry nicely with only a small extension ramp.

Food Scrap Receiving and Blending Area Sizing

Managing incoming food scrap materials immediately is critical. The last thing you want is to have a bottleneck in the very first step of the composting process. There are numerous ways you could approach calculating the size of a receiving and blending area (file under "things compost technical geeks say"), but to my knowledge this is not a design algorithm that has been well outlined in other technical guides. I based the calculations represented in table 6.1 largely on my past experience designing, working in, and watching others work with systems

Design Considerations for Tipping Docks and Receiving and Blending Areas

General Recommendations

- Size the pad according to batch size or throughput in relation to blending frequency (that is, how much material you intend to accumulate there prior to blending).*

$$\text{Volume of Primary Feedstock (e.g. Food Scraps)} + \text{Volume of Other Feedstocks (Typically 3–5} \times \text{Volume of Primary Feedstock)} = \text{Volume on Pad}$$

- A concrete or packed gravel pad surface is highly recommended, even for sites only composting small volumes (packed gravel will have only a two- to four-year life with heavy use).
- Locate as close to feedstocks and the primary composting area as possible, to minimize the distance traveled when pile blending and building.
- Locate in a high and dry spot.
- Divert stormwater and snowmelt if needed with grading and swales.
- Consider locating under cover.
- For covered pads, contain and absorb leachate in the receiving and blending area when possible.
- For uncovered pads, design with a slight slope to move precipitation and runoff to an approved catchment or treatment area.

* For more on batch size and throughput, see chapter 5; for more on pad sizing, see *Food Scrap Receiving and Blending Area Sizing* on page 137.

- Wood chip and compost berms, "compost filter socks," vegetated swales, and vegetative treatment areas can be utilized in combination to treat leachate and runoff if allowable. See *Managing Moisture On-Site*, page 152.
- Ensure that the design meets all applicable regulations.

Tipping Docks

- Tipping docks should be designed with at least 2 to 4 feet of height above the receiving and blending area.
- If you're using concrete blocks for the tipping dock, set the blocks into the ground or on concrete footing.
- Place a safety "bumper" or "curb" at tipping edge of the ramp for truck safety.
- Safety comes before efficiency in design.

Push Walls

- Design back and/or side push walls to facilitate material blending and removal.
- The loading dock may double as push wall (higher docks will work better for this).
- While many receiving and blending areas are designed as a "bunker," with three walls, single- and two-sided push walls are often preferable, as they leave access from multiple angles, which adds efficiency when blending.
- "Waste" concrete blocks are often used for push walls (these come in varying sizes and styles from different concrete plants; find out what's available locally before designing).

of all different scales in relation to the volume of material that they process.

The following example walks through the sizing of a 10-ton-per-week pad for receiving and blending food scraps. I have seen plenty of comparable sites that are able to operate on smaller receiving bay footprints, but I can guarantee you that all of them would prefer a larger one if given the chance. As with other compost site infrastructure components, space = flexibility = efficiency. In reality, a pad that is 25 to 50 percent larger would probably be the most ideal, but I wanted to provide a midsized pad as an example.

When sizing a receiving and blending area, first you need to have a comfortable estimate of the volume of food scraps that will be delivered in a load as well as the frequency that the site will receive loads, and how frequently the material will be blended and removed. It's ideal to have enough space to hold at least one week's worth of blended material, even if it comes in two batches (as is the case in the example). It's not necessary to have enough space to hold an entire batch if it will take several weeks to create it; rather, consider the load delivery size and how many loads you want to be able to manage in the space before the material must be removed.

Second, consider the height of the pile once the additional recipe materials are added and as it's being blended. It's easier to get a good blend on shorter and smaller piles for a couple of reasons: (1) The high surface area to volume ratio of a shorter pile makes layering in new feedstocks more efficient—in other words, shorter, wider piles are better when blending materials; and (2) with large piles, it can be hard to access and effectively blend the pile's innards (the back and corners), which is also why I like to have access from as many angles as possible. Of course, what is considered a small pile is very relative to the size of the loader; blending a 6-foot pile is easy for a front-end loader with a 3- to 4-yard bucket. The pile's dimensions are relative to its height. Consider what pile height you can work with effectively and the trade-offs among pile height, footprint, and efficiency. Last, remember that table 6.1 only provides the footprint of the compost itself, not the work space that's required to manage and access it.

Steps 1 through 1.5 walk you through using table 6.1 and arriving at a receiving and blending area pad size.

Step 1. Size Receiving and Blending Area (Pad)

Use table 6.1 to estimate the footprint for the compost in the receiving and blending area based on the estimated height of the materials as they are being blended. From the compost footprint, find a logical pad width and length (round multiples), and then add work space. Work space for most loaders is between 15 feet (for a very small tractor) and 35 feet in width. It's also good to have room for the material to be moved around as it's blended. Additional space may be in one, two, three, or four directions depending upon the design. I would recommend adding work space in either two or three directions (with the other sides being the tipping dock and/or push walls). Some or all of the work space may be concrete.

1.1 Identify Batch/Load Size

Identify the total volume of food scraps you want the space to handle at any given time. This is your load size.

1.2 Estimate Average Pile Height

Identify the average height that materials will be stacked in the receiving bay.

1.3 Identify Rough Compost Footprint

Using table 6.1, find the square footage that corresponds with your batch/load size and average pile height. This is the rough area that the compost will occupy in the receiving and blending area.

1.4 Find Compost Footprint Dimensions

Find two round numbers (factors) whose product, when multiplied, is close to the footprint of the compost. These are the length and width of the compost footprint.

1.5 Add Loader Work Space

Identify how much area is required surrounding the compost to operate the loader (or for people to work). The width of the work space should be added to the compost footprint in each applicable direction. These are the final receiving and blending area dimensions.*

* Do not forget space for truck access and/or the tipping dock.

Table 6.1. Food Scrap Receiving and Blending Area Sizing for Loader Blended Sites (Does Not Include Workspace)

Batch Size			Rough Compost Footprint (Square Feet)							
Tons Food Scraps	Cubic Yards Food Scraps	Cubic Yards in Batch (Assumes 4:1 Ratio Feedstocks to Food Scraps)	1' Pile Height	2' Pile Height	3' Pile Height	4' Pile Height	5' Pile Height	6' Pile Height	7' Pile Height	8' Pile Height
0.25	0.5	2.5	80	53	48	n/r	n/r	n/r	n/r	n/r
0.5	1	5	153	94	80	n/r	n/r	n/r	n/r	n/r
1	2	10	295	172	137	125	122	n/r	n/r	n/r
2	4	20	575	321	245	213	198	n/r	n/r	n/r
3	6	30	853	467	348	296	270	n/r	n/r	n/r
4	8	40	1130	612	450	377	339	319	n/r	n/r
5	10	50	n/r	n/r	n/r	456	406	378	n/r	n/r
6	12	60	n/r	n/r	n/r	534	472	436	416	n/r
7	14	70	n/r	n/r	n/r	n/r	536	493	467	n/r
8	16	80	n/r	n/r	n/r	n/r	600	549	518	n/r
9	18	90	n/r	n/r	n/r	n/r	664	604	567	545
10	20	100	n/r	n/r	n/r	n/r	727	659	616	590
12	24	120	n/r	n/r	n/r	n/r	851	767	713	678
14	28	140	n/r	n/r	n/r	n/r	975	874	808	765
16	32	160	n/r	n/r	n/r	n/r	1,097	979	902	851
18	36	180	n/r	n/r	n/r	n/r	1,218	1,084	995	935
20	40	200	n/r	n/r	n/r	n/r	1,339	1,188	1,088	1,019

Note: n/r = not recommended

Example

1.1 Identify Batch/Load Size

Identify the total volume of food scraps you want the space to handle at any given time. This is your batch or load size.

Site receives 10 tons of food scraps per week in two 5-ton batches on Tuesdays and Thursdays. Therefore the site processes about 100 cubic yards per week of total feedstock.

1.2 Estimate Average Pile Height

Identify the average height that materials will be stacked in the receiving bay.

Plan for up to 5-foot-high piles during blending.

1.3 Identify Rough Compost Footprint

Using table 6.1, find the square footage that corresponds with your batch/load size and average pile height. This is the rough area that the compost will occupy in the receiving and blending area.

10 tons of food scraps per week or 100 cubic yards per week at a 5-foot pile height would require 727 square feet. However, since the material comes in two loads, the site has the option to process the week's material in one or two batches. If the first load can be blended and removed before the second load comes in, then the receiving area can be half as big. In this case we'll go with a smaller receiving area, which is sometimes advantageous due to limited space and funding. The drawback is that there is less flexibility should the timing to move that material along be a problem, but some

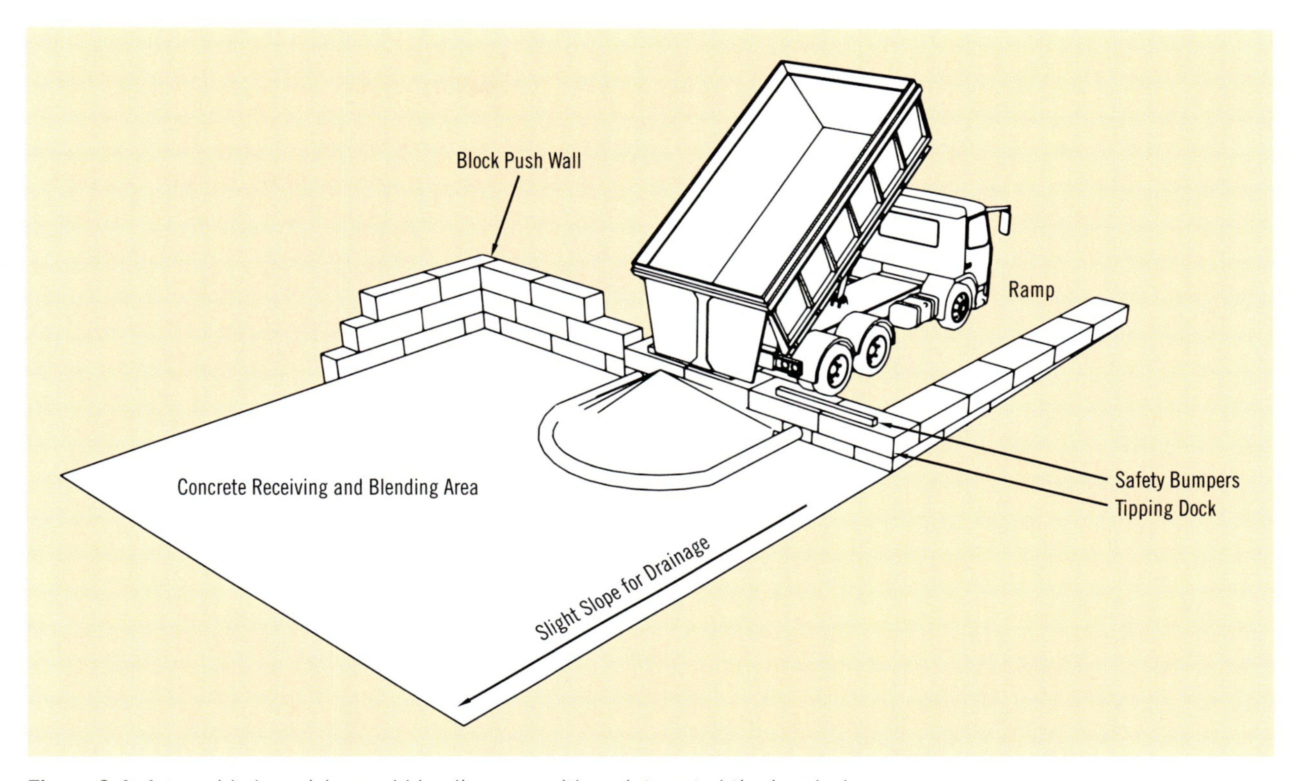

Figure 6.4. A two-sided receiving and blending area with an integrated tipping dock.

flexibility will be accounted for when we add work space (work efficiency will be sacrificed in that case). At 5-ton and 50-cubic-yard batches at a 5-foot pile height, the compost will require around 406 square feet.

1.4 Find Compost Footprint Dimensions

Find two round numbers (factors) whose product, when multiplied, is close to the footprint of the compost. These are the length and width of the compost footprint.

There are several dimension options to meet these needs:

- 15 feet × 25 feet = 375 square feet
- 20 feet × 20 feet = 400 square feet
- 15 feet × 30 feet = 450 square feet

We can go a bit over or under the recommended 406 square feet since we will be adding another pad anyway for loader operation and material movement.

1.5 Add Loader Work Space

Identify how much area is required surrounding the compost to operate the loader. The width of the work space should be added to the compost footprint directions in each applicable direction. These are the final receiving and blending area dimensions.

Assume:

- An 80-horsepower tractor loader with a turning radius of around 25 feet
- Tipping dock, push wall, and open loader access on two sides
- A concrete pad surrounded by packed gravel

A safe work space would be 25 feet in two directions, so the 15 × 25-foot dimension from the previous step is a logical option, making the final pad dimensions 40 feet × 50 feet or 2,000 square feet. I would also add at least 15 feet of packed gravel roads in all access directions, which can double as a work space buffer.

Total Mixed Ration Mixers

Many larger composters use agricultural feed mixers, known as *Total Mixed Ration* or *TMR mixers*, to efficiently and thoroughly blend feedstocks. Getting a really good blend of your feedstocks can have significant and immediate advantages, such as odor mitigation, fast breakdown of food scraps, ease in handling, and some immediate volume reduction. It should be noted that mixers replace the blending function of the receiving and blending area, but not the receiving function. So in designing a site with a mixer, it is still critical to have a designated place where material can be delivered (and covered if not immediately blended), then loaded into the mixer along with the other feedstocks.

Mixers are especially beneficial to those utilizing the aerated static pile (ASP) method, and many people in the industry don't advise food scrap composting with ASP without a mixer. This is because with ASP, having a very homogeneous blend, especially in terms of porosity and moisture, increases the uniformity of the aeration and decreases the likelihood of preferential air channeling.

Some mixers allow an operator to add water or leachate in precise doses to accurately achieve target moisture content. Some sites have conveyers or systems of conveyors that move the blended materials on to the next location, while others make a small discharge pile out the side. It's good to have a small bunk or push wall at the discharge point. Many mixers also have small knives on the mixing augers

Figure 6.5. A TMR small mixer in action. Courtesy of Renee Whittick.

that help to reduce particle size, which is especially helpful when mixing hay and straw (some can even shred a whole round bale). In cold climates we often have to deal with "food-cicles" (pronounced like Popsicles, also called "tote-cicles"): large frozen blocks of food that come out of buckets and carts in a solid cube. Most mixers can break up food-cicles and get them blended without a problem. The alternative with food-cicles, assuming you can't thaw them out somehow, is trying to crush the blocks with a loader bucket (this has varying degrees of effectiveness depending on the size of the loader and the hardness of the block). If that fails, you can put the block into the windrow whole, cover it really well, and wait for it to thaw. Turning will get it blended, but obviously this is not great for an ASP that's static for four weeks. For large operations in extremely cold climates, you can imagine how beneficial a mixer is going to be.

Vertical, as opposed to horizontal, auger mixers are the style of choice for most composters. New self-powered mixers cost upward of $30,000, while PTO-driven tractor-powered mixers cost less. Black Earth Compost, located north of Boston, powers their mixer with a car engine. There are also very small mixers (Jaylor is the brand that I'm most familiar with) that are more suitable for small community composters. Big Reuse Compost in Queens, New York (see the *Space and Place* sidebar about Big Reuse Compost, page 210) has an awesome system with a stand-alone tote tipper that tips food scraps into a Jaylor mixer, which can then be driven over to the compost pile and dispensed before being formed into an ASP windrow. These small mixers start at around $10,000 new for a two-cubic-yard capacity. Many composters also find that they can purchase a used mixer for a reasonable price. Mixers do not typically hold huge volumes, so in most cases a "batch" of compost in the mixer will not equate to an entire "batch" in the field. Sizing a mixer is a matter of balancing the cost with the number of times you need to fill and empty it in a given period. You'll also want to be sure it can handle larger chunks of materials, such as round bales, square bales, and food-cicles, if that is a requirement.

Figure 6.6. Manure spreaders are a popular method for mixing compost. I don't recommend them for food scraps, however, because of the cleanup involved. Also, these are not well suited for mixing in below-freezing temperatures. When they are working well, though, they do a great job of mixing, and I've even seen them used as a mobile chicken feeding troughs. Courtesy of Highfields Center for Composting and Vermont Sustainable Jobs Fund.

Blending on the Primary Composting Pad

The main alternative to having a receiving and blending area where food scraps are received is to have loads of food scraps delivered directly to the primary composting area. For example, there may be times when operators cannot access their receiving area for whatever reason, so instead they build a bed of carbon at the end of a windrow, then blend and form the pile in place. On an unimproved or even a gravel pad, I would strongly urge against this as a permanent strategy, because all the moisture and blending of fresh material destroys the work surface very quickly. On the other hand, there is a certain logic to receiving, blending, and forming the pile basically in place; it's quite efficient. I bring this up because a compost site could build a concrete receiving and blending area directly on a turned windrow pad. I'm actually surprised that I've never seen this anywhere. I'm sure that such a design exists, because I honestly can't think of any reason why the design and management challenges that would arise from this couldn't be overcome. Basically the primary pad would have one section of concrete that's wide and long enough to fit at least one and ideally two or more windrows. Loads would be delivered, blended, and formed into piles in sequence until a "batch" or windrow is created.

For direct pad delivery to work as a strategy, fresh material would have to be moved along fairly quickly to make room for subsequent loads. This doesn't align well with the fact that with fresh food scrap blends, you often want to give them a couple of weeks before turning aggressively. For this reason, enough room for two or more windrows (at least four weeks of material) would be ideal, so that the delivery spot can be rotated among the batches and you don't have to turn material that's only a week old if it might cause a nuisance. If a batch will eventually be made up of four loads over two weeks, for example, you can start to turn and move the first two loads while the next two are building, knowing that they will be combined when the later material is ready.

Another caveat with this scenario is that it would be more challenging to do well with ASP due to the pile architecture it requires (wood chip plenum as base). In most cases this wouldn't even come up, because ideally the material would be blended in a TMR mixer, so the receiving and blending would happen in distinct locations. That said, there are definitely creative ways that a primary-phase ASP could be designed in combination with or in very close proximity to a receiving area, but again, I have yet to see this.

Finally, locating a tipping dock in logical relation to the pad may or may not make sense depending upon your site's access, slope, and material flow. If the receiving pad is large enough, a tipping dock is not as necessary; you'll just need to clean up and consolidate the food scrap load.

Drop-Off Areas for Leaves and Yard Debris

Many compost sites accept yard debris and trimmings (not *waste*), including leaves, branches, wood chips, and so on. Drop-off areas should be designed for clear and safe access, and should allow you as the operator to have control over the material that comes in. When working with landscapers and municipalities, you'll want to be able to talk to them and make sure you can trust them not to bring you contaminated loads or feedstocks that you may not want (such as grass clippings or stumps).

Use signs to direct people where to go and, if you accept multiple types of materials, how to separate those materials where you want them. Many small sites will receive and store yard debris in the same location, while larger sites are likely to have a segregated drop-off area, a processing area (typically involving a grinder), and a separate storage area for the processed materials. Dry materials are often an especially valuable resource to food scrap composters and may be worth covering, either in a feedstock shed or with a tarp. Concrete pads are by no means a necessity for leaf and yard debris drop-offs, but they should be high and dry, with adequate room for loader operation. Push walls and divider walls are often advantageous as well. See *Leaves, Grass, and Yard Debris/Trimmings* in chapter 4, page 93, for more on accepting and managing these materials.

Feedstock Storage and Management

Another necessary component of composting sites is space to effectively receive, manage, and store feedstocks (raw organic material inputs). For some sites, feedstocks might be kept in open piles near the receiving and blending area. Other sites prefer to keep feedstocks in bays with sides and back push walls. In both cases, it's important to be able to both differentiate between and track feedstocks on the site, from arrival through end product. If a feedstock became a problem, you'll want to know which composts it was used in. Tracking feedstocks is discussed in more depth in *Feedstock Tracking and Inventory*, chapter 13, page 354.

From a site management perspective, you'll want the ability to keep distinct piles of feedstocks, both for tracking purposes and because you'll need to be able to distinguish between materials in terms of source as well as age. Anticipating what types of feedstocks you'll be composting may be important to the design of feedstock storage. For example, dry matter (sawdust, really well-bedded manure, leaves) is often highly valuable to food scrap composters, making covered storage for your driest feedstocks a huge asset. When you are considering potential sites, consider the feasibility of using any existing sheds or barns that might be on the site in terms of their size, location, and accessibility. Some materials other than food scraps have odor potential as well (dairy manure or silage, for example). Obviously, you will want to be conscientious about the location of any materials that might be a nuisance to neighbors and manage/use them accordingly.

Like other critical site infrastructure components, sizing feedstock storage capacity should be of primary concern. Here are some feedstock storage and management design considerations, as well as calculations for sizing:

- Feedstock storage should be located in close proximity to food scraps receiving and blending area/s.
- Plan access for safe and efficient delivery.
  - Design feedstock sheds for truck tipping access (roof height, bay width).
- Covered areas keep dry matter dry.
  - Utilize existing structures where applicable.
  - Cover materials with tarps if needed.
- Plan the site to move runoff (stormwater) away from feedstocks.
- Manage uncovered materials to shed moisture and minimize contact with precipitation.
  - "Peak" piles and stack them neatly to create a smooth surface.
  - Tall piles reduce the surface area to volume ratio.
- Maintain distinct feedstock piles by source (and age when possible).
  - Use bays or divider walls (particularly where space is limited, to make use of vertical space).
- Push walls behind feedstocks assist with loader bucket operation.
- A water source should be available for sites with large, dry feedstock piles (piles around 30 to 45 percent moisture and over 20 feet tall).

Figure 6.7. A beautifully constructed feedstock storage shed (*left*) at the Hudak Farm compost site in Swanton, Vermont, is situated directly adjacent to the receiving and blending area (*right*). The tipping dock is under construction. The shed is used for storing high-carbon dry materials such as leaves, chips, and straw. Courtesy of Highfields Center for Composting and Vermont Sustainable Jobs Fund.

Step 2. Sizing Feedstock Storage

Use the following process to estimate your feedstock storage needs. These calculations assume food scraps are your primary feedstock and are not stored on-site like additional feedstocks would be. If you consider another feedstock your primary ingredient and store it temporarily as batches are being created, treat that material as an additional feedstock for sizing purposes.

2.1 Identify Volume of Additional Feedstocks Required per Time Period

Based on your target processing capacity, what volume of additional feedstocks will you process in a given time period? This will provide the cubic yards of additional feedstock per time period in step 2.4. See *Batch Size* in chapter 5, page 115, if you are unsure about this.

2.2 Identify Feedstock Storage Duration

How many months or weeks do you anticipate storing feedstocks on average? This will provide the storage duration for step 2.4.*

2.3 Identify Number of Unique Feedstocks

How many distinct feedstock types do you anticipate receiving?† This will provide the number of unique feedstock piles for step 2.5.

2.4 Calculate Total Volume of Feedstock in Storage

___ Cu. Yd. Additional Feedstock per Week × # of Weeks Feedstock Storage Duration × 27 (Cu. Ft. per Cu. Yd.) = ___ Cu. Ft. of Feedstock in Storage

2.5 Calculate Volume of Feedstock per Unique Pile

___ Cu. Ft. of Feedstock in Storage ÷ # of Unique Feedstock Piles = ___ Cu. Ft. per Feedstock Pile

2.6 Identify Feedstock Storage Design

Will feedstocks be stored in open piles or in storage bays? You may also use a combination of both open piles and bays. Use option A and/or B below based on your feedstock storage design.

EXAMPLE

2.1: 80 yards per week additional feedstocks

2.2: Up to eight weeks.

2.3: Six distinct piles.

2.4: 80 Cu. Yd. × 8 Weeks × 27 = 17,280 Cu. Ft.

2.5: 17,280 Cu. Ft. ÷ 6 = 2,880 Cu. Ft.

2.6: Half open piles (3 piles), half storage bays (3 bays).

---

* The feedstock storage duration will be different depending on the material, but in most cases an average time period will be adequate. If you have one material that you need to store for a long time (leaf and yard debris is a common example), size its storage area separately using the same methodology.

† In order to track and maintain a consistent recipe, many sites find it beneficial to keep similar materials distinct if they come from different sources (say, dairy manure from two different farms) or are of different ages.

2.7 (Option A) Calculate Footprint of Open (or Windrow-Shaped) Feedstock Piles

Note that the geometry of open piles can vary, so you may want to adapt the formula to more closely represent a particular pile formation. The formula below will provide a conservative estimate of the space you'll need.

2.7.1 (A) Identify Pile Width (or Length) and Height

What is the height and width of the feedstock piles?

Height = ___ feet
Width = ___ feet

Height = 7 feet
Width = 15 feet

2.7.2 (A) Calculate "Adjusted Feet"

___ Cu. Ft. Feedstock Pile × ___ Height of Pile × ___ Width of Pile ÷ 0.66 Cross-Sectional Area* = ___ Length of Pile – Height

2,880 Cu. Ft. × 7′ × 15′ ÷ 0.66 = 42′

2.7.3 (A) Calculate Pile Length

___ Length of Pile – Height + ___ Height of Pile = ___ Length of Pile

42′ + 7′ = 49′
(Rounded to 50′)

2.7.4 (A) Calculate Width of Feedstock Storage Area

___ Width of Pile × # of Distinct Feedstock Piles = ___ Feedstock Area Width

15′ × 3 = 45′

2.7.5 (A) Calculate Feedstock Area Footprint

___ Feedstock Area Width × ___ Length of Pile† = ___ Sq. Ft. Feedstock Area Footprint

45′ × 50′ = 2,250 Sq. Ft.

If you also have feedstock storage bays, continue with step 2.7 (option B); otherwise you can move on to step 2.8.

2.7 (Option B) Calculate Footprint of Bay-Type Feedstock Storage

Calculate bay size based on the cubic foot capacity required in each bay (steps 2.1 through 2.5).

2.7.1 (B) Calculate Bay Footprint

Calculate bay dimensions by using a simple formula below. Alternatively, refer to the methodology in chapter 7, *Step 3. Calculate Primary Bin Size,* (page 178), which uses sizing tables for suggested dimensions based on pile height.

---

* If the pile is wider with a flat top, you can use a higher cross-sectional area. If you cut a cross section of the pile, what percentage of a rectangle would be filled? This is your cross-sectional area (figure 5.2, page 124).

† Add 2 feet for back push wall if applicable.

BAY DIMENSION FORMULA

Where:

V = bay volume = ___ cubic feet
H = pile height = ___ feet
W = bay width = ___ feet
D = bay depth = ___ feet

$V \div H \div W + (\frac{1}{2} H) = D$

V = 2,880 cubic feet
H = 7 feet
W = 14 feet
D = 32.89 feet (rounded to 34 feet)

2,880 cubic feet ÷ 7′ ÷ 14′ + (0.5 × 7′) = 32.89′

The dimensions 14 × 34 feet are also recommended for a volume of 2,970 cubic feet or 110 cubic yards in table 7.2.

2.7.2 (B) Calculate Width of Feedstock Storage Area

___ Bay Width × # of Distinct Feedstock Piles = ___ Total Width of Bays

14′ × 3 = 42′

2.7.3 (B) Adjust Width for Walls

Bay walls add to the overall footprint and should be accounted for accordingly. Add together the width of the walls and sum their combined widths with the combined widths of the bays.

___ Total Width of Bays + ___ Combined Wall Width = ___ Feedstock Area Width

42′ + 8′ (2′ Wall) = 50′

2.7.4 (B) Adjust Depth for Back Wall

___ Bay Depth + ___ Wall Width = ___ Feedstock Area Depth

34′ + 2′ (2′ Wall) = 36′

2.7.5 (B) Calculate Feedstock Area Footprint

___ Feedstock Area Width × ___ Feedstock Area Depth = ___ Sq. Ft. Feedstock Area Footprint

50′ × 36′ = 1,800 Sq. Ft.

2.8 Add Work Space

Along with the footprint of the feedstocks themselves, you will need to allot enough space for access by trucks and loaders where applicable. All access areas on the site ideally provide other uses as well. Feedstock access might double as work space for your receiving and blending area, for example. You may add that into your calculation now by multiplying the feedstock area width by the length of the tractor and truck work space needed (typically 20 to 25 feet, more for large trucks and trailers). It may also be easier to add this in on paper as you lay out your site.

The width of my feedstock storage areas is 50 feet plus 50 feet; 100 feet total. To give myself at least 30 feet of access, I will look at where there is space in the layout, and minimize the amount of a space that is solely dedicated to this purpose. For example, maybe there is a corner of the site where I could lay out the piles in an L-formation, or perhaps they could run parallel facing each other, so that the access to each row overlaps.

Active Composting Areas (Primary and Secondary)

Typically, once the feedstocks are blended, they are moved directly to the site's active compost management areas. Active management starts immediately following blending and ends when a batch cools and requires less frequent attention. I subdivide the active period of composting into primary and secondary phases for purposes described in *The Phases of Composting*, chapter 3 (page 58).

The *primary phase* is the hottest period in the process, when temperature treatment targets are reached and significant management activities are conducted. The *secondary phase* takes place following the primary phase, and ends as the compost cools off significantly and raw materials are no longer distinguishable. Management is required during the secondary phase, but the intensity reduces greatly over time, as the material requires less oxygen and less frequent mixing to remain aerobic.

During the primary and secondary phases of composting, one or a combination of active compost management methods will be utilized. An overview of all these methods is provided in chapter 2. You will need to refer to the method-specific chapters (7 through 11) for planning primary- and secondary-phase infrastructure in detail.

While each composting methodology is quite distinct, they often share design components. Some of the common site components include:

- Compost pad improvements
  - Compost pad materials
  - Compost pad slope
- Covered work and storage areas
- Leachate collection and treatment
- Stormwater diversion and treatment
- Access, roads, and travel lanes
- Access to water, electricity, and utilities

We'll discuss these common infrastructure and design components in the following sections. Infrastructure improvements can be beneficial in multiple areas of the compost site, but they are especially important in the active compost management areas where efficiency is critical and nuisance and pollution potential are high.

Compost Pad Improvements

Composters often improve the working surfaces of compost management areas, which often are called *compost pads*, to improve the general workability of a site and to meet regulatory requirements. While existing pad conditions at some sites may already be suitable for some applications, this is the rare exception. In most cases both the grade (slope) and the pad surface need to be improved to shed moisture and make it stable to work on and maintain. Composting pads can be constructed out of any of a wide array of pad materials, as discussed in table 6.2.

Managing moisture on the site is one of the most critical considerations in terms of keeping the site workable year-round, while protecting ground- and surface water. The grade and levelness of the site, as well as the quality of the materials used in pad improvements, will significantly impact management. Compost pads function best with a slight slope (2 to 4 percent) so that moisture moves and does not pool. Windrows should always be oriented with the slope, not against it, to allow for water to move between windrows. In addition, both built and natural environments surrounding the pads require careful consideration.

General pad improvement and design considerations include:

- Divert clean water up-grade of the compost site (before it reaches the management areas).
- Consider pad materials.
- Vertical separation distances to groundwater and bedrock should be 3 feet or more.
- The pad surface should drain (be permeable and/or create flow across).
- The surface should remain firm following usage and moisture.

Figure 6.8. Windrows oriented with slope to allow moisture to flow to vegetative treatment areas. The upper pad is for mature compost, and although there is a vegetated swale that treats and moves water around the lower pad, this configuration prevents contamination of mature compost with leachate from fresh compost, were runoff to overflow. Courtesy of Highfields Center for Composting and Vermont Sustainable Jobs Fund.

- Grade site:
  - The slope should be 2 to 4 percent, if possible.
  - Terrace to reduce slopes.
  - Break up long slopes (over 150 feet) with lateral diversion swales—creating multiple pads—or multiple grades to reduce the size and increase the effectiveness of leachate treatment areas.
- Orient windrows downslope to prevent ponding at the pile's base.
- Stage finished compost upslope of fresh compost to prevent contamination.

The following sections, along with table 6.2, provide some information about common compost pad materials.

Concrete

Concrete is an ideal surface to work on because it is firm even in the wettest conditions (although it may get slippery). Working compost with a tractor on concrete is highly preferable to all other options. The main drawbacks are the upfront cost of concrete and concrete's impermeability, which means an increase in the intensity of leachate and stormwater runoff management systems downgrade. Concrete is, for all intents and purposes, impermeable to liquids. Rain and leachate hit it and flow in whichever directions they can, so concrete compost pads need to be planned with liquids capture and management in mind. While the capture of these potentially nutrient-rich liquids is appealing, leachate from fresh compost is usually extremely odorous (especially from food scraps) and can potentially contain pathogens. Therefore, leachate should be handled in such a way that it does not contaminate your heat-treated compost, crops, or clean water of any kind.

Packed-Gravel Pad

Good packed-gravel products can create a firm and reasonably smooth surface. Operating a tractor on packed gravel requires care to avoid rutting and digging during bucket operation. But while it can be frustrating at times, packed gravel is certainly preferable to working on unimproved surfaces. Maintaining the level of the pad and keeping it clear from organic matter are critical to maintaining the longevity of a gravel pad. The gravels that work best as compost pad surfaces have a high ratio of fine particles; they are well graded, moistened, and then packed with a vibrating roller.

Table 6.2. Common Compost Pad Materials

Pad Material	Description/Application	Pros	Cons	Cost per Sq. Ft.[a]
Native soil and sod	Using existing soil/sod works for small sites, for hand-turned sites, and for finishing/curing areas, as weather allows.	Minimal cost, pervious	Challenging to work on in moist conditions, hard to maintain clean surface and grades, careful machinery operation required	—
Packed gravel	Typically a ¾" or smaller gravel is used with a high percentage of fines. Products range by distributor; I've seen it called Staymat or Sure Pack. A depth of at least 6" is recommended. Remove the top layer of native soil; this will work its way up through the gravel and degrade the pad over time. Often a filter fabric base is installed beneath the gravel, which keeps the native soil from migrating up.	Low cost, semi-pervious, locally sourced materials	Requires consistent maintenance (grading, removing organic matter); 5- to 10-year resurfacing/replacement; can rut, so it requires careful machinery operation	$0.50–2.00
Reclaimed asphalt paving (RAP)	Asphalt grindings that have been removed from roadways are more and more commonly reused for other applications, including for composting pads and access ways. If applied evenly and compacted with a roller, this can make a very effective working surface.	Lower cost than asphalt or pavement, long lasting, hard and uniform surface, some water percolation, utilizes recycled material	I have been told that compost acids can degrade asphalt over time, but have not heard any complaints about this from composters. RAP materials can be variable—make sure to find material that will create a flat and well-bonded surface. Ruts and potholes may form. May need to be resurfaced periodically.	$1.00–3.00
Asphalt	It's uncommon to see compost sites that were newly paved using asphalt. More common are sites that are using existing asphalt (say, abandoned lots). That said, this is an option, and is slightly less expensive than concrete.	Very solid working surface, long lasting (about 20 years with maintenance), easy to maintain and repair	Costly up front. I have been told that compost acids can degrade asphalt over time, but have not heard any complaints about this from composters. Ruts and potholes may form. Requires maintenance and may need to be resurfaced periodically. Impervious—increases runoff and treatment requirements. Energy-intensive.	$3.00–5.00
Concrete	Concrete pads are common on large sites, particularly in heavy-use areas such as blending areas and primary composting pads.	Very solid working surface, long lasting (about 30 years), low maintenance	Costly up front, hard to repair, impervious—increases runoff and treatment requirements. Energy-intensive.	$4.00–6.00

Source: Adapted courtesy of Compost Technical Services, LLC.

[a] Does not include prep, grading, or base materials. The cost of an excavator and operator typically runs $1,000 a day or more, plus transportation.

Native Soil and Sod

It is certainly possible to compost on the native soil of your site, and unimproved surfaces can be entirely adequate on a really small scale, in dry climates, or on highly permeable soils. The challenge with composting on bare soil or sod is that moisture and organic matter tend to make mud, and over time this creates unworkable conditions. Systems operated by human power as opposed to tractor power will hold up longer. Another strategy used by farms that have enough space and don't want to put concrete or gravel down on good soil is to move the composting area around the farm. This has the added benefit of building the soil somewhat as you go (or disturbing it, depending upon how well it's treated).

Managing Moisture On-Site

It is important to ensure that the compost site does not negatively impact its surrounding environment and that the design complies with applicable regulations. One of the most important aspects of this is planning for moisture management at your facility, including moisture from both the compost and compost management areas, as well as precipitation. With every aspect of the site design, think (1) where will water come from, and (2) where will it go?

A general principle in best management practices for compost sites is to manage clean water separately from water that's come into contact with compost and composting pads. The phrase *Keep clean water clean* is a popular one. In practice this means:

- Diverting precipitation or runoff upslope from compost and composting areas, and managing it separately
- Minimizing contact between stormwater and compost/organic matter:
  - Clean pads.
  - Orient piles with the slope.
  - Cover and cap piles.

Water that comes into contact with compost and composting areas should be treated thoughtfully and by approved methods where specific regulations apply. Having been trained in permaculture design, I am partial to capturing and treating water on-site and as high in the watershed as possible to minimize and disperse flow. This can be achieved by breaking up the footprint of pad areas that shed to any one point, then moving water slowly through vegetation to infiltrate moisture, remove sediment, and absorb nutrients.

Vegetative Treatment Areas

Vegetated water treatment systems are referred to by many names, and there are all kinds of design variations of differing complexity (including vegetative buffer strips, rain gardens, and constructed wetlands). I typically call simple systems *vegetative treatment areas* or VTAs. In my opinion, for most small sites, vegetative treatment in combination with wood chip berms and compost socks is the preferable method. These are low-tech and low-cost, can often be well designed and constructed without outside expertise, and very often take advantage of existing site resources (such as a big field). Even where adequate vegetation is currently lacking or impractical to utilize, vegetation can often be planted.

Sizing is of course critical when it comes to runoff and leachate treatment. Vermont's regulations for small sites (under around 18 tons per week of food and 5,000 cubic yards of total organics) require a VTA equal in size to the compost management areas and equal in length in the downslope direction. Without being an expert in the design of this type of infrastructure, I have found this to be a good reference point, although with an effectively designed system, this high a ratio of vegetation to pad is probably more than is necessary.

Ideal vegetation for VTAs is going to be regionally specific. In the Northeast the recommendation is for a mix of warm- and cool-season perennial grasses, due to their ability to synthesize nutrients across multiple seasons. Grasses are also relatively easily maintained. Sites might consider incorporating a mix of native and/or non-invasive species, adapted to meadow-type habitat. There would certainly be nothing wrong with adding some woody or other favorite species as well.

Figure 6.9. Organic hay fields located slightly downslope from a prospective food scrap composting site. It's the perfect "existing infrastructure" for treating runoff (not to mention utilizing it) from the composting pad. Don't be discouraged if your VTA doesn't look like this one; few do!

Other Treatment Methods

Due to scale, regulation, spatial constraints, or design needs, many sites will require other methods of stormwater or leachate treatment system. Common alternatives to VTAs include sand filters; sediment, retention, and evaporation ponds; and leachate storage tanks. Other than leachate management systems for the aerated static pile method (covered in *Leachate Collection and Management*, chapter 9 on page 236) and compost socks and wood chip berms (following), these are not covered due to their higher degree of technicality (most often an engineer will need to be involved).

Wood Chip Berms and Compost Socks

One of the simplest ways to slow, infiltrate, and treat site moisture and nutrients is to use organic matter itself as a physical barrier. In particular, compost sites often use high-carbon composts and wood chips, which are permeable enough to slow but not

Figure 6.10. This compost sock was probably nicked by the loader over winter. Note the fungal activity. The compost can be left in place (after removing the sock) when the sock is no longer needed or replaced. Courtesy of Highfields Center for Composting and Vermont Sustainable Jobs Fund.

stop flow. *Compost socks* are mesh tubes typically filled with specified compost blends. The tubes retain the filtration media so that they don't blow out in high flow. Filtrexx, one of the leading brands of compost socks, is a great source of information about their use. Wood chip berms serve a similar function, although they are less reliable under heavy flow.

One of the benefits of compost and chips is that they are biologically active, even at very low temperatures, when vegetation would not be.

Redundancy

While some treatment is better than no treatment, redundant treatment is the best treatment of all. Large treatment systems can be costly to construct and maintain, and are not always 100 percent effective. Incorporating a combination of compost socks, berms, and vegetation, even when larger catchment and treatment areas are involved, can be a logical or even necessary way to ensure optimal effectiveness.

Figure 6.11. Two wood chip berms at NYC Compost Project hosted by the Brooklyn Botanical Garden at Red Hook Community Farm catch sediment and nutrients off the compost pad. The rich sediment at the first berm shows that it was both needed and effective.

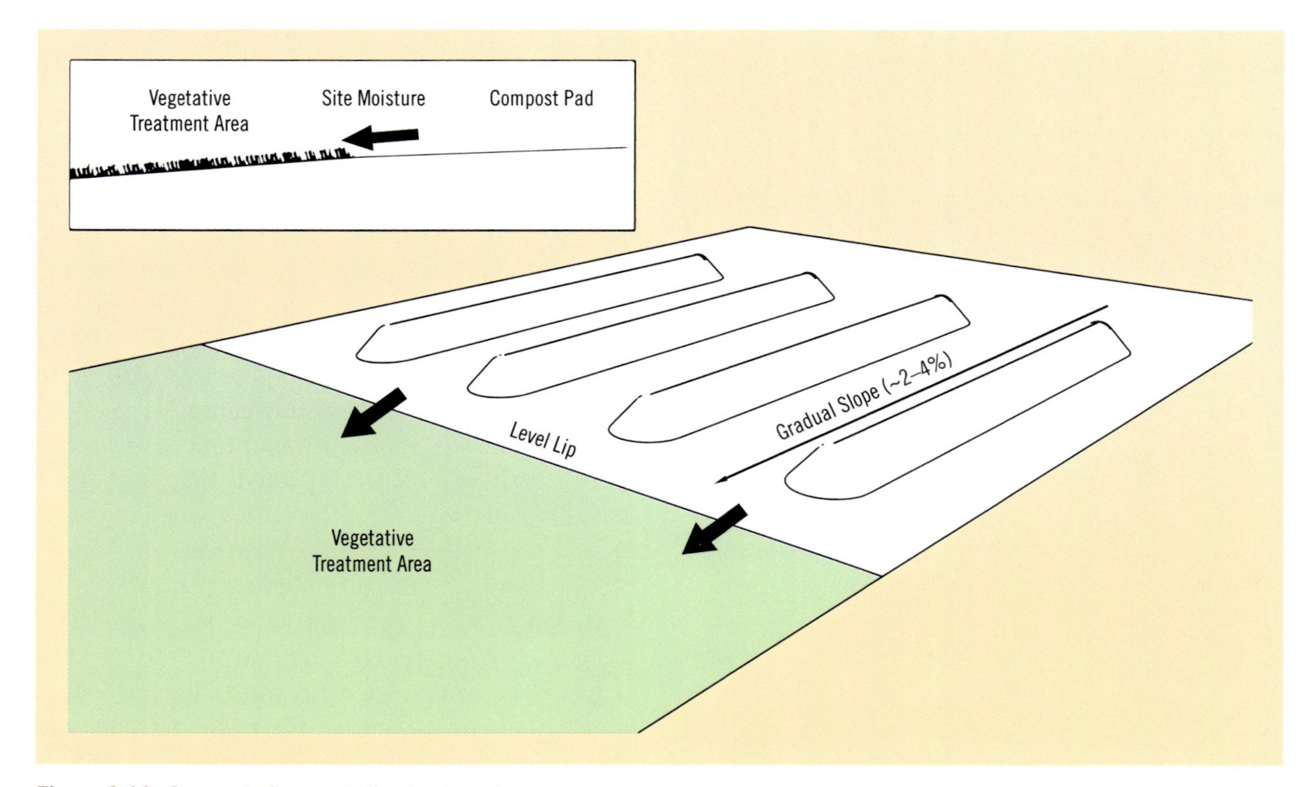

Figure 6.12. Compost site vegetative treatment areas.

Compost Pad Maintenance

Due to its criticality, throughout the book I'll continue to reiterate that pad maintenance is a huge factor in moisture management. Ruts, organic matter buildup, and ponding can all prohibit good drainage and create a cycle of pad degradation. See *Pad Maintenance* in chapter 13, page 374, for more on compost pad best management practices.

Access

Access roads are a critical piece of infrastructure, especially if you frequently receive materials from off-site. To operate safely and efficiently, while at the same time protecting the environment, plan appropriately for the level of traffic. Roadways may be covered under stormwater permits, in which case plans for roadways will be included in your stormwater pollution prevention plan (SWPPP or similar) and need to include approved best management practices for mitigating pollution.

Recommendations for planning and developing access to your compost site include:

- Improved roadway surfaces such as gravel, concrete, or asphalt
- Graded roadways that evenly shed moisture:
  - Avoid pooling and channeling moisture.
  - Slow runoff as it exits the roadway and direct it to vegetated infiltration zones.
- Provide adequate space for trucks to maneuver and safely load and unload.
- Provide signage to direct traffic:
  - At larger sites plan retail traffic and large truck traffic separately.
- Plan for snowplowing in winter:
  - Avoid plowing snow onto feedstocks and compost.
  - Avoid blocking drainage for snowmelt and rain.
- Leave as little unvegetated land surrounding your access ways as possible.

Finishing, Curing, and Storing Compost

After the primary and secondary phases (or the active phases), the composting process requires time and space to finish. Most composters also require additional capacity to cure and store their product until it is distributed. The management, infrastructure, and space needs are quite different for compost in these latter phases, so sizing and planning are done distinctly from the active management areas in order to save space and cost.

Some sites might plan one area for everything following secondary composting, while others might find it beneficial to calculate storage areas separately from finishing and curing or split up all three. In reality, all of the phases might take place in one contiguous space, but by planning them separately there is an opportunity to factor in assumptions that would be challenging to model into one equation. The phases of composting and reasoning behind phased capacity calculations are covered in much greater depth in *The Phases of Composting*, chapter 3, and *Compost Site Processing Capacity and System Scale*, chapter 5 (pages 58 and 114 respectively).

Infrastructure differences between the active and later phases often include less intensive leachate treatment systems and lower-cost pad materials. We're assuming that compost that has progressed to the finishing phase has met the required heat treatment standards during active composting. For this reason, runoff from more mature compost is no longer of high pathogen concern, and nutrients are stabilizing. Therefore moisture coming from these areas may be treated with less intensive methods than from areas where raw compost is present. (This would only be a concern for sites that are required to have more intensive leachate management systems in place.)

In addition, oxygen demand is significantly reduced, so piles can remain relatively aerobic with only passive airflow—in other words, less turning. A site that utilizes concrete pads in the primary and secondary phases might go with gravel for finishing, curing, and storage. During finishing and curing,

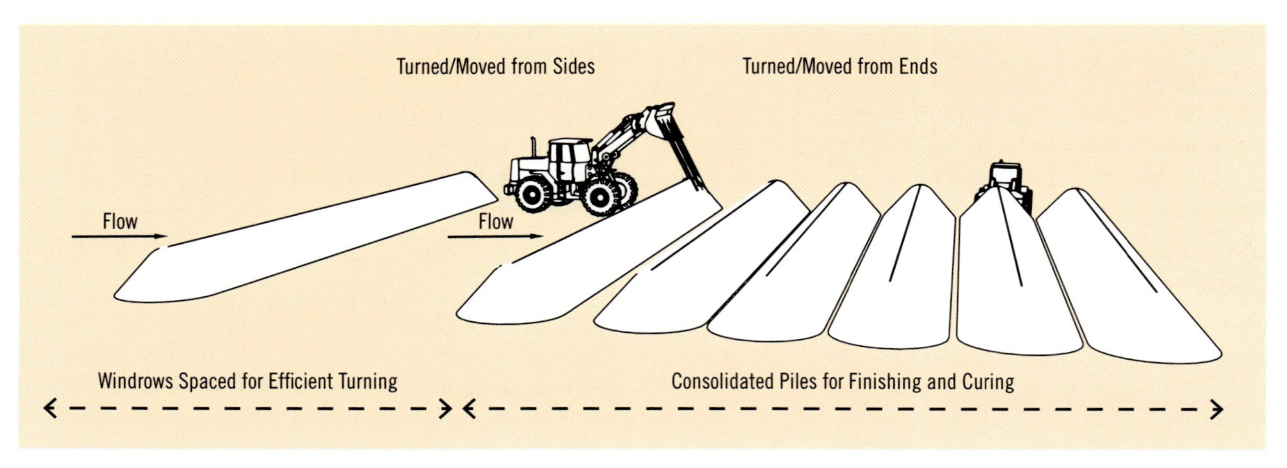

Figure 6.13. Stored compost typically leaves the site in large volumes at least twice a year during sales seasons (spring and fall), leaving empty space. Following these periods, that space can be accessed to efficiently turn the finishing and curing piles, until it gets crowded again.

piles can be stacked toe-to-toe in order to conserve space, without work lanes in between them for turning, as long as the compost is adequately porous and the windrows are relatively small.

Finishing and curing piles can still benefit from turning, but this is often a management decision based on the feel and homogeneity of the pile, not a necessity as it was for meeting temperature requirements. Larger piles typically require more frequent turning.

Compost finishing, curing, and storage area design considerations include:

- Finishing, curing, and storage areas can be sized individually or in combination using the calculations below.
- Pile size can differ depending on phase:
  - Finishing compost is ideally 7 feet or less in height.
  - Curing compost is ideally 8 feet or less in height (6 feet or less is suggested in the *On-Farm Composting Handbook* for highest quality[1]).
  - Stored compost is ideally 10 feet or less in height (*On-Farm Composting Handbook* states should not exceed 12 feet[2]).
- Improve finishing, curing, and storage pads (see *Compost Pad Improvements*, page 149). Gravel is often adequate; just be sure that you can still access the pads year-round.
  - On farms, compost maturation could take place in a dispersed manner, near fields where the compost will be used. Access for turning is still beneficial.
- Mitigate pre-heat-treatment leachate and compost from intermingling with post-heat-treatment compost.
- To encourage drying, consider cover (sheds, tarps/covers, coverall structure/barn).
- Plan adequate access (you may not need space for turning compost laterally) so you can access from windrow ends.

Calculating Finishing, Curing, and Storage Areas

Calculating the footprint required to finish, cure, and store compost uses a method similar to that of sizing the active turned windrow compost pad. The key differences are in the assumptions you make about pile height, work alleys, and other space for access. Also, the volume in each batch can be adjusted based on material shrinkage and pile combining.

Step 3: Calculate Finishing, Curing, and Storage Areas

The following assumes windrow-shaped piles. If you intend to use storage bays, refer to *Sizing Bins and Bays*, chapter 7, page 176, to reference the appropriate formulas.

3.1 Identify Sizing Assumptions

You need to know the following in order to make this calculation:

a. Batch frequency
b. The processing time or retention capacity from start to finish during:

- Finishing
- Curing
- Storage

c. Will finishing, curing, and storage areas be sized distinctly?
d. Pile combining strategy (if applicable)
e. Pile height
f. Pile width (2 times the height plus 1 foot)
g. Width of work alleys (tractor work space)
h. Width of perimeter access

3.2 Adjust Batch Size for Shrink Factor

___ Cu. Ft. Compost per Batch in Secondary Windrows × ___% Shrink Factor (Example: For a 20% Reduction, Multiply by 0.8)* = ___ Cu. Ft. Compost per Batch After Shrinkage

3.3 Estimate Windrow Dimensions

Based upon what you know about your site location (likely pile size and batch volume), estimate the size of your windrows. Windrow width is typically twice windrow height. If you are not using windrows, use alternative calculations to find the pile dimensions.

* In estimating the shrink factor going into the finishing, curing, and storage phases, take into account how much volume reduction has been subtracted thus far, as well as whether any additional shrink factor will be subtracted in later calculations. In the given example, 20 percent was subtracted for blending, and then 30 percent was subtracted for shrinkage during the primary phase, which left 56 percent of the original volume. By calculating in an addition 20 percent volume reduction from the secondary, that brings the total to 45 percent of the original volume, which would account for shrink in the secondary as well as all subsequent phases. While in reality, total shrink may be 60 percent or more (or 40 percent or less of the original throughput volume), that reduction continues on throughout the finishing, curing, and storage process, so it makes sense to account for some but not all of it when sizing this pad.

EXAMPLE

a. Every four weeks, from secondary
b. 20 weeks
4 weeks
16 weeks (assumes a four-month window from November through February when no compost is sold and stored compost accumulates. Salable inventory must be distributed in spring by June and in fall by the end of October.)

TOTAL = 40 WEEKS

c. No
d. Windrows are already incorporating four weeks of throughput, so no additional combining is necessary, although this would always be an option if deemed advantageous by the operator.
e. 7 feet
f. 15 feet
g. No work alleys. Piles can be turned from the sides when there is space following the sales season and from the ends when the pad gets crowded.
h. access 10-foot perimeter

Step 3.2

6,048 Cu. Ft. × 0.8
= 4,838 Cu. Ft.

3.3.1 Calculate Windrow Length Minus Height

Formula		Result	Example
___ Cu. Ft. Blended Compost per Batch After Shrinkage ÷ ___ Height of Windrow ÷ ___ Width of Windrow ÷ 0.66 Cross-Sectional Area (Assume 0.5 if Conservative)	=	___ Length of Windrow – Height	4,838 Cu. Ft. ÷ 7′ ÷ 15′ ÷ 0.66 = 69.8′

3.3.2 Calculate Windrow Length

Formula		Result	Example
___ Length of Windrow – Height + ___ Height of Windrow	=	___ Length of Windrow	69.8′ + 7′ = 76.8′ (Rounded to 80′)

3.4 Estimate # of Windrows

Formula		Result	Example
___ Processing Time on Compost Pad ÷ # of Time Periods per Batch	=	# of Windrows on Compost Pad	40 Weeks ÷ 4 Weeks = 10

Keep the time period consistent across all calculations. I typically use weeks for food scrap calculations.

3.5 Calculate Pad Width Required for Windrows

Formula		Result	Example
# of Windrows on Compost Pad × ___ Width of Windrow	=	___ Combined Windrow Width	10 × 15′ = 150′

3.6 (Optional) Calculate Width Required for Work Alleys

Formula		Result	Example
___ # of Work Alleys on Compost Pad × ___ Width of Work Alleys	=	___ Combined Work Alley Width	n/a

You may want to use this step if you're sizing a finishing pad that is frequently turned.

3.7 Calculate Width of Compost Pad

Formula		Result	Example
___ Combined Windrow Width × ___ Combined Work Alley Width + ___ Width of Perimeter	=	___ Width of Compost Pad	150′ × n/a + 20′ = 170′

3.8 Calculate Length of Compost Pad

Formula		Result	Example
___ Length of Windrow + ___ Width of Perimeter (Assumes Two Perimeter Edges)	=	___ Length of Compost Pad	80′ + 20′ = 100′

3.9 Calculate Area of Compost Pad

Formula		Result	Example
___ Width of Compost Pad × ___ Length of Compost Pad	=	___ Sq. Ft. Area of Compost Pad	170′ × 100′ = 17,000 Sq. Ft.

3.10 (Optional) Size Curing or Storage Pad
If you're calculating a curing or storage pad separately from the finishing pad, repeat steps 3.1 through 3.9 for those areas.

Finalizing Site Scale

Albeit a detailed and often circuitous process, scaling your site does have an end point once all of the site infrastructure components have been identified and sized. If you are reading this, then hopefully you've arrived at this point.

In order to move from system scale into the details of the design and then on to site development, it's helpful to lay out the site to scale both on paper and while actually standing on it with a measuring tape and stakes. Most designers will have already done some of this, but I want to emphasize that the scale cannot be considered finalized until each component has a verified home both on paper and as actually staked out on the site. It's just too easy to overlook the little things that can make a design unworkable.

Critical mistakes to avoid include:

- Undersizing access and work space
- Underestimating feedstock storage needs; lack of feedstock and carbon storage in particular can lead to everything from pile stagnation to major odor issues
- Undersizing compost storage needs
- Allowing excessive moisture accumulation on pads and roadways; keep pad sizes reasonable, use existing grades to your advantage, and create drainage as needed to maintain a workable site
- Accepting more than the intended throughput or overestimating the system's throughput capacity
- Not identifying setbacks
- Underestimating the need for water

Additional Infrastructure

While this chapter covers the critical infrastructure needed by most community-scale composting operations in reasonable depth, there are several components that at least deserve a mention even though they won't be handled in depth. These include cover, water and compost watering systems, storage and space for maintaining tools and equipment, office space, facilities such as bathrooms, and greenhouses and growing space (growing plants in your compost is the best way to ensure and improve its quality). Many sites would benefit from demonstration and classroom space for workshops and other educational purposes. Screening, blending, and bagging finished products also takes up considerable space and ideally happens under cover. Bagging ideally happens in an enclosed and possibly even a heated space in cold climates, because it often is conducted during the colder months.

Compost Site Equipment

The following equipment is commonly used for producing compost. As I've said elsewhere in the book, I am not an equipment guy. I don't even change the oil in my own car (although I have). Therefore this section discusses the function of the equipment from a composting perspective rather than the specifics of what to look for in one piece of equipment or another.

Tractor Loaders

A veteran farmer I knew once said, "You can't farm without a tractor." What he meant was, "You can't farm with a skid steer." Tractors are built for many farming jobs that other loaders simply aren't, and tractors with loaders can also make compost. Just as a loader is the essential multifunction tool for composters, tractors are the essential multifunction tool for on-farm composters. From a composting perspective, tractor loaders offer few benefits over other loaders, aside from maybe cost. Tractors with loaders are typically less agile, slower, harder to get the bucket perfectly level, and so on, than other loaders. But if you are farming and only get to pick one piece of equipment, get a tractor with a loader, because you just can't farm without one. On the other

Figure 6.14. A loader with a several-yard bucket piles material into a windrow at Green Mountain Compost in Williston, Vermont. It will be watered to increase its moisture content in between phases of aerated static pile treatment.

hand, if your tractor is needed elsewhere on the farm most of the time, it might actually be necessary to invest in another piece of equipment, in which case a skid steer or other type of loader with different advantages might be beneficial in other areas of the farm as well.

Skid-Steer Loaders

Anyone who has ever driven both a tractor and a skid steer will tell you that they are night and day in terms of their speed, versatility, and ease of use. If you've ever watched an experienced operator trying to get a job done fast, it's kind of like watching a stunt car driver doing doughnuts. Some skid steers can move like helicopters. Some have tracks, some wheels that are typically much wider than tractor tires, which increase traction and reduce compaction. Skid steers are also very powerful in terms of their loader capacity because their main purpose is lifting and moving stuff. Also, skid steers are often compatible with lots of powerful attachments, from hole augers to backhoes to mowers. There are several useful attachments for composters such as graders and sweepers for maintaining pads and roads.

There are a few drawbacks to skid steers, however. First, they can cost a pretty penny, and parts and maintenance are also expensive. Second, due to their tight turning ability, skid steers can rip up surfaces that aren't concrete or asphalt, unless they are driven really carefully (that is, not like a skid steer). Last, the largest-size skid loaders have a limited lift capacity, so to scale up to larger-size buckets, you really need to go with a traditional loader.

Bucket Loaders

Traditional bucket loaders come in a huge range of types and capacities. They are built for moving material quickly. If what you need is a material mover and turner, not a tractor or smaller skid steer that can serve all sorts of other functions (to be fair, some loaders do have other attachment options), a bucket loader is the tool for you.

Telescopic Handlers (Telehandlers)

Some material handlers have a telescopic or extending arm. These are typically known as telehandlers, and although they are most often used with pallet forks, they can just as easily be fitted with a loader bucket. The significant reach of telehandlers facilitates a great variety of composting applications, from turning a windrow without having to even move your tires, to mixing, to stacking compost or feedstocks in much larger piles than would be possible with a traditional loader without driving on it. When operating a telehandler, you seek to minimize the use of the telescoping action to mitigate wear and tear, but having it there when it's needed is an amazing advantage. The main drawback of telehandlers, as with skid steers, is their cost, both up front and in terms of service and maintenance.

Excavators

While excavators are rarely the first equipment that comes to mind when people think of composting, they are versatile tools with lots of potential advantages in pile turning. But let's start with the main drawback of excavators in composting applications (other than their cost): They are not built for moving volumes of materials over any distance. They move from place to place slowly, so even with a large bucket, they aren't efficient unless you are moving material within their immediate radius. What excavators are good at is digging and site development work, precise and fast bucket movements (within their reach), and turning and stacking large piles of materials. They also require a lot less room for maneuvering between piles, because they can move laterally and rotate in place as they work. This means alley space requirements are minimal, and, because they typically move on tracks rather than tires, there is less rutting of the pad. For sites or farms that require a lot of site earthworks, having an excavator could be a real advantage. Just keep in mind that excavators can't really do the job of a loader, so a loader is almost always going to be the first priority.

Total Mix Ration Mixers

TMR mixers are used for blending materials, and were therefore covered earlier in the chapter.

Conveyers

Belt conveyers are often used in combination with other equipment (screeners, grinders) to increase operational efficiency. For example, the discharge point from a small screener might fill with compost

Figure 6.15. One of the tools of choice for turning piles at the Vermont Compost Company (VCC), in East Montpelier, is an excavator. VCC has the ability to quickly reverse the bucket, which they do for most turning applications. In this picture they are using the excavator to build an ASP pile that supplied heat via the Agrilab Compost Heat Wagon to the farm's greenhouse.

quickly, forcing to the operator to both load and unload compost simultaneously. A conveyer with a hopper could catch the compost as it comes out and stack it away from the screener, allowing the operator to load faster and remove the compost less frequently.

Chippers and Grinders

Woody materials and yard debris are often processed through chipping or grinding equipment. Chipping and grinding reduces particle size, which in turn increases uniformity, carbon availability, the ability to absorb moisture, and, in turn, the rate of decomposition. Chippers are designed for branches and make a clean chip, while grinders can handle a wider variety of materials, including leaves, and create more of a mix of shredded debris and fine particles.

I know one composter who actually uses his grinder as a mixer, combining food scraps with his yard debris. It's not a practice that I have enough experience with to endorse, but I can see the appeal.*

Chippers and grinders come in a variety of sizes and price points. Chippers cost much less and are often purchased as a tractor PTO attachment. There are even mini chippers that cost under $1,000 that can handle small saplings. Chippers are often on farmers' and composters' wish lists, but the process of chipping is extremely time consuming in relation to the yield of chip. For this reason, sites that accept yard debris often accumulate the material for a year or more, then rent a grinder to process it all at once.

* On a related point, I am often asked about methods of chopping up food scraps for composting. Mixers with knives achieve this to some degree, but generally I tell people that grinding food is not a necessary step. In fact, to my knowledge, none of the composters that I've ever worked with perform this additional step. Obviously smaller food particles will break down faster, and it probably is worth chopping up a watermelon with whatever tools you have available (such as a shovel or machete). Some of the older composting literature talks about grinding food scraps, so it might be worth researching further if it's something you're looking for.

In most cases, I see owning a chipper as only being practical on a micro scale (small community sites with volunteer labor, for instance) or where the chipper can be used for other things (say, if you own a tree service). At a large site, grinders are the way to go, but they are extremely expensive, both to purchase and to maintain. Most community-scale sites are going to be sourcing chip elsewhere. Luckily in most places there are a plethora of options.

Screeners

While some sites sell unscreened compost, more commonly, composters use screens to remove large particles or *overs* (oversized particles) to create a more uniform end product. Most micro sites make their own screeners with hardware cloth or another wire mesh.

Many sites, particularly those in start-up stage, outsource screening, either by renting one and screening themselves or by paying someone to come to the site and screen for them. Screeners are sometimes available from gravel pits, demolition sites, and equipment rental companies.

Trommel screens are the preferred tool for most composters, with rotating cylinders of screen that the compost moves through. Deck screens, which are essentially vibrating platforms, are the other common style and are the most likely rentable option.

Most large screeners cost $40,000 or more. There are a few lower-cost options that are worth mentioning. Sittler makes several small screeners in the $15,000 to $25,000 range. HotRot sells two small trommel models, one of which can be loaded with a small bucket loader. I recently learned about DeSite deck screeners, which are sold for as little as $5,000. New options come on the market all the time. As with any equipment purchase, do some digging and talk to people who have used them.

The ability to reuse overs, which are typically great bulking agents, is highly desirable. Disposal of contaminated overs is costly and wasteful. Although there are now screeners that remove contamination from the overs to produce a reusable product, these

Figure 6.16. A self-built trommel-style screen constructed with hardware cloth and bike tire rims at Compost With Me (Falmouth, Massachusetts).

Figure 6.17. The Sittler trommel screen at the Highfields Center for Composting screened about 25 cubic yards per hour with the help of a conveyer. Courtesy of Highfields Center for Composting and Vermont Sustainable Jobs Fund.

are expensive. A lot of composters rig up fans to blow out light contamination (plastic, paper) and use gravity to sort heavy contaminants such as rock from their compost.

Baggers

Operations that bag their finished compost often purchase bagging equipment. The simplest bagging operations might consist of a small funnel/hopper used to fill the bag and a hand stitcher for sealing woven poly bags. Above this DIY scale there are a variety of different options and price ranges. It's not equipment that I have a lot of experience purchasing beyond a small amount of online research. Most of the small systems that I've encountered run in the $10,000 to $20,000 range, which would be the starter scale that most small to medium sites would probably be planning. Larger systems get expensive quickly. I ran across one very simple mulch bagging system from Weaverline that ran only $5,000. It is basically a hopper and bag doser that can be filled with a loader and would speed up the process of manually filling bags.

Hacking Equipment, Infrastructure, and Functionality

I wrote this book at a time when the word *hacking* had become cliché, but the hacker/maker/DIY movement is growing more popular for a reason. People have a desire to create stuff, and as we become more and more embedded in the digital world, the draw to the physical is amplified. In the non-criminal sense of the word, hacking simply means finding a creative solution to a problem using the tools available. It's a practice that both farmers and composters are extremely familiar with.

As a child of the 1980s, I thought the television character MacGyver was pretty much the coolest guy alive, so hacking systems and equipment is one part of composting that I am really drawn to. (One of the first compost bins I ever made was constructed with bamboo and wire in a Belize rain forest.) I know I'm not unique in this; composters and the composting industry at large have borrowed tools and equipment from other industries and repurposed them for our own utility over and over. Livestock feed mixers become mixing wagons. Blowers, ductwork, and timers all serve other purposes, but can be used to create an ASP. Existing farms, warehouses, and parking lots can all be redesigned as compost sites. I am constantly observing my surroundings both on and off compost sites and looking for tools that would help a composter do a job cheaper and better.

Since the latter half of the 20th century, the composting industry has grown to the point that large-scale production of specialized composting equipment is justified. But there remains a need for creative solutions, particularly when it comes to tools for working at the community scale. Better yet, we need tools that are scalable, and low enough in cost that they will work at any scale. One of the ways that I propose that we improve the tools of composting is to open-source "Ⓒompost" information, designs, and tools to the greatest extent possible. By increasing replication and having a shared design process, we can open up a world of ideas and avoid making the same mistakes over and over. It's especially important for veteran composters and designers to share their experience with the younger generation. A wealth of knowledge could be lost if it isn't documented and shared.

As a community of compost practitioners we have an opportunity (and I would argue a responsibility) to create systems for capturing and managing information such that it can be most useful, both internally and externally. I have heard the term *information infrastructure* used to describe this important but often overlooked component of our composting systems. Information infrastructure could be anything from pile monitoring and receiving logs, to food scrap generator data and maps, to this book. Like other tools, it can be time consuming and clunky or it can be efficient and user-friendly.

At a minimum, composters need to track what's going on at their operation from day to day. Internal feedback loops are critical to meeting targets and making a quality product. In many cases that information is also necessary to show regulatory compliance.

But I also think that it's worthwhile to envision what forms compost information capture and presentation could take in the future and how the functions of those systems might improve the lives of their users and the larger community.

CHAPTER SEVEN

Bin and Bay Composting Systems

Along with open piles and windrows, bin and bay systems are a very common composting method. Many people are familiar with backyard bin systems, yet we also see the bin concept utilized on a larger scale. Some systems integrate forced aeration (chapter 9) and even highly engineered systems for turning, such as agitated bed systems, utilizing side walls. As the scale of the system grows, I often refer to a bin as a *bay*, simply because the word conveys a larger capacity than I associate with *bin*. That said, the two words are interchangeable for our purposes; some readers may call a small bin a bay or vice versa. Bins and bays are categorized here as composting contained by at least two sides. Bays are more likely to be open on the top and front, while bins are often fully contained within a six-sided cube or cuboid (rectangular cube).

Walls contain material vertically, increasing the volume that can be retained within a given footprint compared with a freestanding pile or windrow. Simply by adding side walls, volume can increase by 50 to 100 percent on the same footprint. The main disadvantage of adding walls is that they add costs. Also, walls can make turning less efficient because they limit the options for access and material movement.

This chapter is devoted to the application and design of bin and bay composting systems. The chapter starts with a discussion of common applications of both small and large bins and bays, then covers both common and uncommon design strategies and materials, such as using insulation in bins, which mimics composting at a much larger volume and can help small batches reach thermophilic temperatures. Next, it covers the flow of materials in bins, and finally calculations and strategies for properly sizing bin and bay systems.

If you read this book all the way through, you might notice that I have a bias against bin systems. This bias comes from their frequent mismanagement, which I believe stems in part from limitations inherent in common bin design. In this chapter I discuss the limitations I have observed in standard bin layouts and propose some design modifications that are intended to overcome them.

carbonaceous
(ˌkär-bə-ˈnā-shəs)
DEFINITION: An organic material consisting primarily of carbon compounds.

Common Applications for Small Bin Composting Systems

In backyards, schools, and community gardens, bins are common due to their simplicity, efficient use of space, and relatively low installation cost. Compared with open piles or prefabricated in-vessel options, bins have a lot of potential benefits:

Figure 7.1. An O2Compost Micro-Bin System produces compost on the rooftop farm at the Brooklyn Grange Rooftop Farm. The system is primarily used to compost the farm residuals that are produced on-site.

- Walls help contain odors, and physical barriers such as hardware cloth can keep out vectors.
- They are relatively low cost compared with other contained systems.
- They can be designed and constructed by community members.
- Often they are built with recycled or locally sourced materials (such as pallets on a small scale, locally sourced lumber, and even straw bales).
- They have a smaller footprint than open piles.
- They are often aesthetically pleasing.
- They organize material flow.
- Both passive and forced aeration can easily be added.
- Insulation can be easily added to assist smaller volumes of material in hitting temperature treatment targets.

There are also some common pitfalls with bins that you should do your utmost to avoid. In my experience the most common issues stem from the fact that their walls make it difficult to efficiently turn the material. Once the pile grows above a cubic yard or two, it's hard to turn the compost without removing all of the material and then re-piling it. Underturning is common with open piles as well, so you can't place the fault entirely on the bins, but I do think they add a layer of difficulty. When compost goes unturned for long periods of time, it increases the likelihood of pests taking up residence, even if the bins are well secured. Unturned piles also decrease the quality and speed of composting.

Because of these potential pitfalls, you may sense some ambivalence on my part toward bin systems when it comes to school and community garden applications. For home composting, bins are great: A few pallets or bales of straw and voilà, you have a bin. Beautiful and creative home composting systems inspire me, and I have absolutely nothing bad to say about them. But at the scale of most schools, it's just not practical to turn bins as frequently as they should be turned without a tractor. Nevertheless, as of right now, there are few other options (depending on who you talk to) that can handle a couple hundred pounds of food scraps a week, that will heat year-round in the cold, and that are affordable for schools ($5,000 to $6,000 is many schools' upper limit).

My opinion is that bins are less than perfect for composting at the micro scale, though good operators can make them work. To all of the inventor types reading this: Please come up with a better system that can handle 50 to 500 pounds of food a week, at a relatively low cost.

Common Applications for Larger Bay Composting Systems

Scaled-up bins or bays are a method for making better use of limited site space. Particularly with aerated static pile (ASP) systems, bays can add capacity vertically without reducing efficiency. With turned piles, I think that bays are usually more of an aesthetic choice, and that compared with open windrows, they actually reduce operability.

Bays are used in all types of larger-scale applications, from farms to universities to commercial operations, especially where space is limited. Bays are also a very practical choice for storing feedstocks and finished compost, as we discussed in the previous chapter.

Standard Bin and Bay System Design

While bins and bays are simple, there are still a lot of design variables that need to be thought through. The rest of this chapter walks through these topics.

Bin and Bay Construction Materials

One of the great things about bins is that there are so many different possibilities for construction materials. The choices are endless. I've built bins with bamboo in Belize and Costa Rica, chicken wire in Hawaii and California, straw bales in Arizona. Most of the bins I've designed for schools in Vermont have been built with local rough-cut, rot-resistant lumbers such as hemlock or cedar. Larger bay systems are typically built with large concrete blocks, the kind that concrete plants form when their trucks come back with extra concrete. The key is to choose materials that will work for your desired application. In places where rodents and other vermin are a big concern, a simple pallet bin system is probably not the right fit. In warm climates bins as small as 1 cubic yard or less will heat reliably, but in colder climates like my native New England, higher volume and/or some insulation are often required.

The following sections cover both common and uncommon materials for constructing the bin's skeleton and walls, along with leachate control methods, and materials for rodent control, insulation, and cover.

Rodent Barriers

Regardless of the other materials you use, ensuring the bins mitigate rodent access is absolutely critical in on-site composting applications. The use of hardware cloth, concrete pavers, and plastic bins are the main methods for controlling rodents and other potential vectors.

Hardware Cloth

Hardware cloth is a steel mesh that is widely available from hardware and building supply stores. Get the highest gauge available (¼-inch spacing), as I've heard stories of rodents penetrating hardware cloth. These rolls go for about $0.50 to $1.00 per square foot, depending upon how much you buy and how wide the roll is. Make sure that all sides and cracks are secured.

Pavers

A DC-based company called Urban Farm Plans has plans for a composter they call the Compost Knox.

One of the things that I like about the plans is that they call for concrete pavers as a floor. The benefits of using pavers is that they are essentially a rat-proof barrier as long as they are closely spaced, and they are a nice surface to work on with shovels. If the bin floor is at ground level, pavers create a rodent barrier between the compost and dirt that is simpler than pouring concrete. They are also permeable, so liquids can flow down. If the bin floor is off the ground, the pavers help preserve any wood below them.

Thick Plastic Bins

Thick plastic bins like the polyethylene tubs and tanks discussed briefly later in this chapter are rugged enough that on their own they create a really good rodent barrier.

Insulation

In Vermont we've found that small bins with 2 to 4 inches of rigid insulation added to their walls can reliably heat to 131°F or higher year-round. The design we've tried called for insulation on three sides, with wooden slat walls on the fourth side. Insulating the top, bottom, and fourth side would improve this even further, especially for smaller bins. The smallest bins we've tried were 4 × 4 × 4 feet, so they could hold just over 2 cubic yards of material. Larger bins are better from a heating perspective, even with insulation. Make sure that insulation is installed on the interior of the walls so that it doesn't come in direct contact with the compost. Maybe in the near future, natural forms of insulation, such as mushroom-based materials, will be more widely available. On that note, straw bales provide incredible insulation as long as they stay dry.

Bin and Bay Walls

On a small scale bin walls can be and have been constructed out of a large range of materials. If you are a purist, you'll want to avoid compost coming into contact with materials like pressure-treated lumber or foam insulation that may leach chemicals into the compost over time. Larger bays have a more limited number of options for wall materials, since they need to be structurally sound. I've found that even 8-foot tall concrete-block walls want to lean, especially if they aren't perfectly square or if they sit on an unstable base. Here are some of the most likely wall materials:

BIN WALLS

- Rot-resistant lumber or bamboo
- Heat-treated or kiln-dried pallets*
- Pressure-treated lumber (not recommended for contact with compost)
- Recycled plastic wood board
- Hardware cloth
- Concrete cinder blocks
- Straw bales (temporary)
- Polyethylene tubs or tanks

BAY WALLS

- Concrete blocks (2 feet or more wide) on a level and stable base
- Formed concrete
- Jersey barriers with walls or fence on top
- Wood

Bin and Bay Floors/Pads

Having a solid base to work on is key when composting food scraps in the same place over and over. Some floor and pad options are provided here, along with a brief discussion on passive aeration using raised bins and leachate management.

BIN FLOORS

- Perforated rot-resistant lumber
- Heat-treated or kiln-dried pallets*
- Concrete
- Pavers
- Hardware cloth (be careful working on hardware cloth; only use flat plastic shovels)

BAY FLOORS/PADS

- Concrete
- Gravel

* Pallets that are heat-treated or kiln-dried are marked *HT* and *KD*, respectively. Avoid pallets marked *MB*, which stands for "methyl bromide." *DB* just means "debarked."

Figure 7.2. Bin components. Adapted from Highfields Center for Composting, courtesy of Vermont Sustainable Jobs Fund.

Passive Aeration

Bin systems, when raised off the floor, allow air to flow as convection draws fresh air into the pile passively, which is a great benefit. To get this aeration effect, the bin floor needs to have air conduits such as slots between boards or drilled perforations. Because of the tendency for holes and cracks to clog, I would recommend drilling ¾- to 1-inch holes every 9 to 12 inches. Make sure that these holes are covered with hardware cloth to keep out rodents. For older piles, I tend to think that direct contact with the dirt is good, but for active piles the added airflow is a welcome benefit. See also *Passively Aerated Static Pile,* page 39.

Leachate Management

Even with a great recipe, wet food scraps are often a source of leachate (liquid discharge). Depending upon the surface that you are working on, you may or may not ever see the leachate. On impermeable surfaces (concrete, plastic) leachate may build and create stagnant conditions under the pile, or drain out from under the pile. If raised bins are draining into the ground, over time it may create odor and fly issues, and it will certainly increase the likelihood of attracting unwanted visitors. There are a few approaches you can take. Management-wise, building a thick (12 inches or more) base layer of carbon materials below food scraps in the bin will minimize most if not all drainage and stagnation issues. Alternatively, you can capture the leachate, either in a tank or in a separate bed of carbon materials. Applicable leachate management and treatment options are covered in much more depth in *Managing Moisture On-Site*, chapter 6, and *Leachate Collection and Management*, chapter 9 (pages 152 and 236 respectively).

Polyethylene Tubs/Tanks

While commercially available home composting systems are too small for most applications beyond demonstrations, other types of plastic containers have been repurposed as bins by small-scale composters. The NYC Compost Project hosted by the Lower East Side Ecology Center uses insulated ice container pallets as bins. These are available up to at least 40 cubic feet; some have drains and other entry points that could be used to connect aeration or drainage pipes. The benefit to these is that they come insulated. However, they only open from the top, so emptying them requires tipping them over with a tractor, or fabricating a side opening. Collapsible bulk containers are similarly sized and have removable sides, but aren't insulated.

Figure 7.3. At the New York Compost Project hosted by the Lower East Side Ecology Center, they utilize insulated pallet tubs as their primary composting bins.

Intermediate Bulk Containers (IBC tanks) also have repurposing potential. These are fairly easily available used and are the frame that Green Mountain Technology's Earth Cube is built around. IBC tanks require both insulation and access fabrication. It's also worth noting that not all of these plastic tub/tank options are necessarily aesthetically pleasing, but they provide a great opportunity for a design/art project.

Lids, Roofs, and Structures

As a general rule, bins that are utilized for vector security purposes have lids, while larger bays do not. Wooden lids can be heavy and therefore hard to open, especially for kids. Lighter materials such as sheet metal can work well, as can counterweight and pulley systems like the ones in figure 7.4. Either might also

Figure 7.4. A rock acts as a counterweight on large bin lids at the Craftsbury Outdoor Center, Craftsbury, Vermont (*left*). Pulleys are used on the bins at the Charlotte Central School (*right*). In some cases, multiple pulleys might be needed to pull the lids higher while keeping the rope out of the way of the lid. These are both good engineering experiments for kids at schools! Courtesy of Highfields Center for Composting and Vermont Sustainable Jobs Fund.

have a roof to shed moisture, and other structures for added protection and aesthetic purposes. I've heard concerns that bin roofs dry out piles, and structures certainly add cost. But overall, I think the added moisture control is a net positive. It certainly reduces the potential for leachate and runoff and makes siting easier in terms of nutrient and other pollution concerns. Also, in northern climates, removing snow from and around the composting areas can be a big time sink, adding to the energy it takes to make composting work. In a guide we created at the Highfields Center for Composting, we talk more about design of A-frame roofed structures, like the beautiful one in Charlotte, Vermont (figure 1.8, page 22).[1] Other options include composting in greenhouses and small hoop structures, like those you see at public works facilities to cover sand and salt.

Materials Management and Flow in Bin and Bay Systems

An essential first step in designing bin and bay composting systems is planning how materials will move through the site. Addressing material flow starts with an understanding of the following:

1. How much primary feedstock will be handled?
2. How will the primary feedstock be managed as it enters the system?
   a. Will you be blending feedstocks? If so, where will this take place?*
   b. Will you be layering them directly in the bins?
3. How much additional feedstock (high C and others) will be needed?
   a. Where will this be stored?
4. How frequently will a batch be created?
5. How long do you want to retain materials in each bin or bay?
   a. How many times do you want to transfer materials from bay to bay or otherwise turn materials?
6. Will compost be cured and stored outside of the bays?

* Even in small bins, blending is highly advantageous. One option is to use a large wheelbarrow as a mixing system. Mixing small batches is easier, even if it means dealing with multiple small batches.

As in the other system-specific chapters, the discussion of bins and bays focuses on what happens in the primary, secondary, and possibly the finishing phases of composting. For planning of compost feedstock storage, receiving and blending, compost curing and storage, and other auxiliary site elements, please refer to chapter 6. Much of this chapter is devoted to bin and bay system sizing, for which you need to have a good estimate of the volume of materials (food scraps or otherwise) that you intend to handle. Strategies for estimating food scraps and other feedstocks are described in chapters 4, 5, and 12.

Static Bins and Bays

I make no secret of my low opinion of unturned or static (non-aerated) compost piles, at least where food scrap composting is concerned. Especially at schools and other institutions, unturned piles are associated with too many risks. I always wonder how much the compost is valued when the plan is essentially not to manage it. In all of the systems described in this chapter and book, assume that a managed approach including some turning and aeration is recommended.

Turned Bin and Bay Flow

Like windrows and ASPs, bins and bays are batch systems. Typically one bin is filled, constituting a batch, then managed by turning it, either to another bin or in place. Three-bin home composting systems are the classic example; typically Bin 1 gets filled with fresh materials, when it's full it's moved to Bin 2 to make room for new material, then Bin 2 moves on to Bin 3 once Bin 1 fills up again. There is a logical succession, from bin to bin, with one bin receiving deposits, one bin constituting the "bank," so to speak, and the third bin used for finishing and withdrawals.

This same type of deposit-bank-withdrawal flow is the goal at larger scales as well, including schools and small compost sites. However, from what I've seen, the flow that works well for backyard three-bin systems doesn't scale up well. The main reason is that it's best to give a batch time to cook without having to move it immediately once the first bin is full. If you add to Bin 1 every week, the newest material will only be a week or two old and still raw. It's fine to turn it, but not ideal to have to turn it just to make room. It's nice to allow some time for raw food to break down before that first turn. That means that you need two or more "Bin 1s," so that one can be filled while the other is breaking down somewhat prior to turning.

I think of these Bin 1s as primary bins, where temperatures are typically highest, although this depends largely on the retention time. With ASP bins and bays, temperature treatment standards can easily be met in Bin 1. For turned piles, meeting temperature treatment targets in Bin 1 (see *Temperature Treatment Standards and the Process to Further Reduce Pathogens* on page 56) in most cases requires you to turn the piles in the bins, or remove the pile from the

Figure 7.5. These pallet collar hinges stack, allowing bin walls to be easily constructed, deconstructed, and stored as needed.

bin, then return it to the same bin, which provides a more thorough turn. This is not particularly efficient, but it's what people would have to do to reliably meet turning and temperature treatment requirements. More often what happens is that piles get turned only to make room for new material. Though piles will likely have climbed well above 131°F by this time, they may not climb that high again following turning from Bin 1 to Bin 2, so temp treatment is more similar to an ASP or in-vessel than to a turned pile. I don't love this and encourage more frequent turning in the primary phase to ensure that all material has maintained a temp rise to 131°F or more for at least three days, but accept that it may not be realistic for hand-turned bins in all cases. I realize that composting proponents may not love me for saying this, but bins are simply impractical to manage, which makes holding them to the standards of other composting systems challenging.

I hope that small-scale practitioners will experiment with alternative material flows and designs. No doubt there are ways to add efficiency to hand-turned bins in terms of more frequent turning. One such strategy that could make pile turning easier is to have longer bins/bays that can be turned from one end to the other and back if desired. It could be opened up on both ends so that older material could be pulled out of the far side, making room so that newer material could more easily be rolled backward. Longer bins could have a divider wall in the center with slats that could be removed, essentially making one bin into two.

Figure 7.6. Charlie Bayrer and Marisa DeDominicis of the NY Compost Project hosted by Earth Matter on Governors Island are proponents of collapsible bin design. This 4 × 4 × 4-foot cube is functional in its simplicity. It's a versatile design that could be used right on the edge of a farm field.

Figure 7.7. A deconstructable 1-cubic-yard bin, ideal for home composting applications.

Deconstructable bins are another bin design strategy that I've come across on several occasions. When the pile is ready to turn, the sides come off. The bin gets reconstructed and the pile is moved to a nearby location. Depending upon the design, it may be beneficial to have two or more deconstructable bins, so that you always have an empty bin on hand to fill. One variation on this design that I have been testing is the modular bin concept illustrated in figure 7.8, which involves stacking and re-stacking layers of wall in succession and leverages gravity for fast turning. I have been using this system in my backyard with great success, and early reports from a pilot at Hazen Union High School (Hardwick, Vermont) suggest turning a 2-cubic-yard pile takes about 10 minutes. The pallet collar hinges shown in figure 7.5 are perfectly designed for creating stackable sides. You only need one bin per batch with this system, but it seems to work best with at least two hardware cloth bottoms that are used in rotation.

Another option I'd like to pilot is using small, insulated tumblers like the Jora and locating them directly above insulated bins. Using an insulated tumbler for mixing and heat treatment over a few weeks would be ideal, as long as there was adequate retention time for each batch. Then the material would be emptied into the bin directly below it. The tumbler and bin could even both be housed within one small insulated structure, which would help the tumbler stay active in the winter.

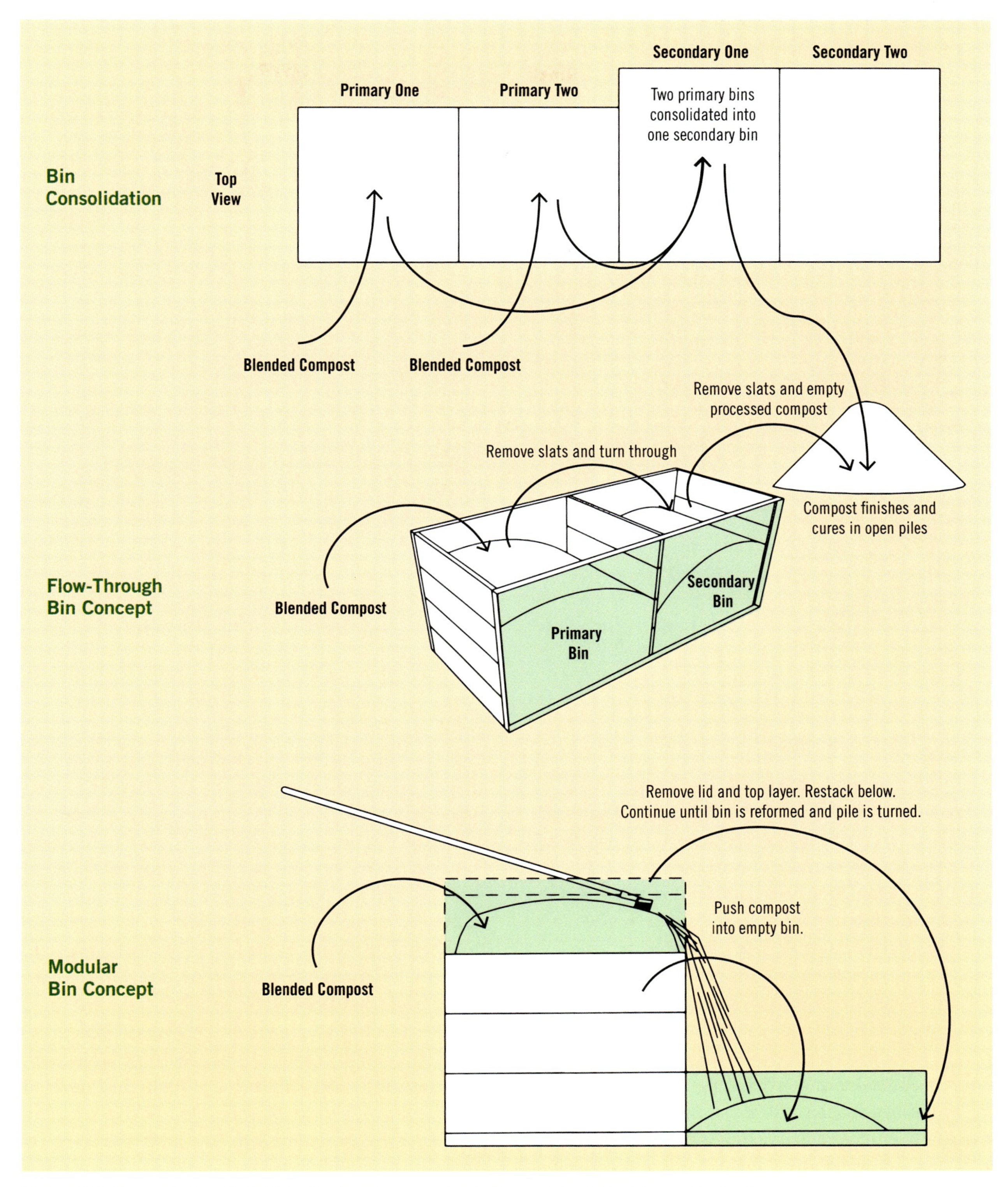

Figure 7.8. Conceptual material flow in bin and bay systems (beyond the typical three-bin flow).

How Much Material Is Required for Hot Composting?

Those planning to compost on a micro scale may be asking themselves if they have enough material to make a hot pile. It's widely taught that 1 cubic yard is the minimum volume required to hit the 131°F heat treatment target. In my experience, the volume of food scraps that my family generates on a weekly basis—roughly 25 pounds of food (5 gallons)—plus leaves, shredded straw, and grass clippings, is more than enough to hit 131°F, even during the coldest periods of winter in Arlington, Massachusetts. The minimum volume of material that I've seen exceed 131°F is about ⅓ cubic yard.

It's important to emphasize that no two systems are identical, though, and there are a variety of factors that make my system heat through the coldest spells, even without insulation. First, I make the mix very dry (50 to 55 percent moisture content) in the winter, which provides more insulation and lowers the specific heat of the mix. Second, I only add room-temperature food scraps—nothing frozen—and I blend everything well. Third, I use cardboard to create a circular pile shape inside my wooden bin (32 × 32 inches), which adds more height, lowers the surface area to volume ratio, and provides a small amount of additional insulation. I turn and add fresh material every week or two for the first six to eight weeks (ideally it would be more like four weeks, but this would require more bins than I currently have).

While it is clearly possible for very small volumes of compost to heat, it's important to remember that moisture content, pile density, age of material, mixing, and pile shape are all major factors. Adding more volume and external insulation is a good insurance policy when composting smaller volumes particularly in cold climates.

Sizing Bins and Bays

As with any compost method, correctly sizing bin systems for your specific throughput is a critical step. Bins are typically batch systems with a known volume of fresh material accumulating over a set period of time. Planning shorter fill times is beneficial, because each batch is ideally made up of material of similar age. You also want to size your bins to have enough volume to heat consistently; for smaller systems this means balancing adequate volume with maintaining a relatively short fill time, and also not having too many bins/batches to manage simultaneously.

Planning capacity around combining bins when turning from the primary to the secondary phase is a common strategy and a key factor in properly sizing some systems. Prior to sizing the system, you should think through the flow of materials from bin to bin, as well as the retention time in each bin (see *Materials Management and Flow in Bin and Bay Systems*, page 171).

Because the applications of bins and bays can be so drastically different in terms of scale, I will provide two examples. The first will be more relevant to schools and other micro sites, while the second example will be for 10 tons per week and be consistent with the other chapters.

Note that the sizing methodology below assumes you have a target processing capacity in mind over a set period of time. Other infrastructure such as feedstock storage, mixing areas, compost curing, and finished compost storage are probably necessary in addition to the active composting areas. These auxiliary components are covered in chapter 6.

FOUR-SIDED BIN

Step 1. Identify Key Assumptions

You need to know the following to size your bin:

a. The target food scraps to be processed in a given time period
b. The ratio of additional feedstocks to food scraps by volume
c. The frequency with which a batch is created
d. The processing time from start to finish in the primary, secondary, and finishing composting phases as applicable
e. Flow from bin to bin: Will bins be combined? What is the retention time in each bin?

Step 2. Calculate Batch Size

To calculate batch size, you will use the *Compost Site Processing Capacity and System Scale* (page 114). The batch size formula from chapter 5 is included here as a smaller-scale example.

*2.1 Convert Tons of Food Scraps to Pounds**

___ Tons Primary Feedstock per ___ Time Period × 2,000 (pounds per ton) = ___ Pounds per ___ Time Period

2.2 Convert Weight of Food Scrap to Volume

___ Pounds per ___ Time Period ÷ ___ Primary Feedstock Bulk Density (Pounds per Cu. Yd.)† = ___ Total Cu. Yd. Primary Feedstock per ___ Time Period

2.3 Estimate Volume of Additional Feedstocks

___ Cu. Yd. Primary Feedstock per ___ Time Period × ___ Ratio Additional Feedstock to Primary Feedstock (Typically 3 to 5:1) = ___ Cu. Yd. Additional Feedstock per ___ Time Period

* Note: Skip step 2.1 if you're already using pounds.

† See *Bulk Density*, page 73; *Converting Food Scraps to Tons and Cubic Yards*, page 88; and *Bulk Density: The Relationship Between Mass and Volume*, page 122, for clarification on this step as needed.

BIN EXAMPLE

200-Student Elementary School On-Site Composting Scenario

a. 226 pounds per week (200 students @ 1.13 pounds per student per week)
b. 2.5:1
c. Every four weeks
d. 37 weeks or more
e. There are two primary bins and three secondary bins.
Two primary bins combine into one secondary bin.
Each primary bin has four weeks or more of retention per bin (five to eight weeks total retention).
Each secondary bin has 8 weeks or more of retention (12 to 16 weeks total retention).
Compost that is 17 to 24 weeks old will be finished, cured, and stored outside, or additional bins will be created if needed.

Step 2.1

0.113 Ton per 1 Week × 2,000 = 226 Pounds per 1 Week

Step 2.2

226 Pounds per 1 Week ÷ 1,000 = 0.226 Cu. Yd. per 1 Week

Step 2.3

0.226 Cu. Yd. per 1 Week × 2.5 Ratio = 0.565 Cu. Yd. per 1 Week

2.4 Estimate Total Volume of Feedstocks

			Example
___ Cu. Yd. Primary Feedstock per ___ Time Period + ___ Cu. Yd. Additional Feedstock per ___ Time Period	=	___ Total Cu. Yd. Raw Feedstock per ___ Time Period	0.226 Cu. Yd. per 1 Week + 0.565 Cu. Yd. per 1 Week = 0.791 Total Cu. Yd. 1 Week

2.5 Estimate Total Volume of Blended Feedstocks

			Example
___ Total Cu. Yd. Raw Feedstock per ___ Time Period × ___% Blending (and/or Bin*) Shrink Factor (Typically 20%, or Multiply by 0.8) × 27 (Cu. Ft. per Cu. Yd.)	=	___ Cu. Ft. Blended Compost per ___ Time Period	0.791 Total Cu. Yd. per 1 Week × 0.8 × 27 = 17.08 Cu. Ft. per 1 Week

2.6 Estimate Total Blended Volume

					Example
___ Cu. Ft. Blended Compost per ___ Time Period	×	___ # of Time Periods per Batch	=	___ Cu. Ft. Blended Compost per Batch	17.08 Cu. Ft. per 1 Week × 4 Week = 68.3 Cu. Ft. (Rounded to 68 Cu. Ft.)

This is your standard "batch size." Keep the time period constant though all of your calculations; I typically use weeks (or days for very large operations).

Step 3. Calculate Primary Bin Size

Now we can use your operation's standard batch size to calculate your primary bin size. Step 4 will give you the option of resizing your secondary bins after accounting for shrinkage and/or pile consolidation. Calculating bin size should be relatively intuitive. Based on the number of cubic feet capacity required for a batch (step 2), you can divide to find the inner dimensions of your bins, or look at table 7.1 for suggested bin dimensions.

If you load the bins by hand, don't make the pile higher than 4 feet, unless you have a plan to dump into it safely from above the bin. A 4-foot-tall compost pile in an insulated bin is enough to get really active.

In larger systems the bays will be designed to load and unload with a bucket loader. These bins could be piled 6 or 7 feet in height for non-aerated systems and up to 8 feet for ASP systems. The width of the bays and the height of the shed ceiling will need to accommodate loader bucket movements.

Some shrink factor can be accounted for in the bin. There will be more shrink with longer retention.

* "Bin" shrink factor would account for shrinkage that is happening in the bin as it's being filled. There will be more shrinkage with longer retention.

Table 7.1. Bin Sizing: Minimum Dimensions by Volume

Batch Volume		MINIMUM BIN DIMENSIONS (FOUR WALLS)									
		3′ Height		4′ Height		5′ Height		6′ Height		7′ Height	
Cubic Yards	Cubic Feet	Width	Length	Width	Length	Width	Length	Width	Length	Width	Length
1	27	3	3	n/r	n/r	n/r	n/r	n/r	n/r	n/r	n/r
2	54	4	5	3	5	n/r	n/r	n/r	n/r	n/r	n/r
3	81	5	5	4	5	n/r	n/r	n/r	n/r	n/r	n/r
4	108	6	6	4	7	4	5	n/r	n/r	n/r	n/r
5	135	n/r	n/r	5	7	5	5	n/r	n/r	n/r	n/r
6	162	n/r	n/r	5	8	5	6	n/r	n/r	n/r	n/r
7	189	n/r	n/r	6	8	5	8	5	6	n/r	n/r
8	216	n/r	n/r	6	9	5	9	6	6	n/r	n/r
9	243	n/r	n/r	6	10	6	8	6	7	n/r	n/r
10	270	n/r	n/r	8	8	6	9	6	8	6	6
12	324	n/r	n/r	8	10	6	11	6	9	6	8
15	405	n/r	n/r	8	13	8	10	6	11	6	10

Note: Round up dimensions as logical for construction purposes.
n/r = not recommended

FOUR-SIDED BIN DIMENSION FORMULA

Where:

V = batch volume = ___ cubic feet
H = pile height = ___ feet
W = bin width = ___ feet
D = bin depth = ___ feet

$V \div H \div W = D$

Note that H, W, and D can all be used interchangeably depending upon which dimensions you know.

V = 68 cubic feet or 2.5 yards per batch
H = 4 feet
W = 4 feet
D = 4.25 feet, or 4 feet rounded, assuming some shrinkage over the four-week fill time

68 cubic feet ÷ 4′ ÷ 4′ = 4.25′ (Rounded to 4′)

Or use table 7.1, "Bin Sizing: Minimum Dimensions by Volume."

Step 4 (Optional). Calculate Size of Secondary Bins or Bays

Based on the flow of materials from bin to bin, it is often logical to size your secondary bins differently than your primary bins to account for shrinkage and batch consolidation (combining older batches).

4.1 Assume Primary Phase Shrink Factor

Estimate the total amount each batch will shrink prior to being moved to the secondary bins or bays. Earlier in the chapter I suggest 3 to 5 percent volume reduction per week as a conservative estimate. If you're combining bins, you can use a slightly higher shrink rate, because the first batch that goes into the secondary bin will continue to shrink before the next batch is added. If you have already subtracted shrink factor in step 2, be sure to account for that in your estimate. Also, keep in mind that total shrink including during blending is typically around 60 to 70 percent including 20 percent during blending, of which about 40 to 50 percent happens in the first two to four months.

___ Cu. Ft. Blended Compost per Batch in Primary × ___% Bin Shrink Factor (Typically 20 to 30%, or Multiply by 0.8 to 0.7)	=	___ Cu. Ft. Compost per Batch After Primary Shrinkage	68 Cu. Ft. × 0.7 = 47.6 Cu. Ft.

This is the adjusted volume for your secondary bays if you intend to keep each batch distinct. For those who are combining batches, follow step 4.2. If not, skip to 4.3.

4.2 (Optional) Combine Volumes from Primary into Secondary

If you intend to combine two or more batches into one secondary bin or bay, multiply the volume after shrinkage by the number of batches. You can also test the volume of combined bins here to see if combining makes sense. In cases where the bins need to be really large, combining piles in the bins might not make sense, especially for a hand-turned system.

___ Cu. Ft. Compost per Batch After Primary Shrinkage × ___ # of Batches Being Combined	=	___ Cu. Ft. Compost per Combined Batch in Secondary Bin	47.6 Cu. Ft. × 2 = 95.2 Cu. Ft. (or 3.5 Cu. Yd.)

This is the volume for your combined secondary bays.

4.3 Repeat Step 3 to Find Dimensions of Secondary Bays and Bins

Once you have calculated the volume of space required for the secondary bins or bays (4.1 and 4.2), repeat step 3 to find logical dimensions based on their adjusted volume. Depending upon the configuration of the site, you may want the depths of the bins in the primary and secondary bins to be the same, to make construction simpler. It may take some trial and error to find a combination of dimensions that suits both the primary and secondary stages.

If you plan on having a third or fourth set of bins, repeat steps 4.1 through 4.3 as necessary.

V = about 95 cubic feet or about 3 cubic yards per batch
H = 4 feet
W = 5 feet
D = 4.75, rounded to 5 feet

120 cubic feet ÷ 4′ ÷ 5′ = 5′

Note that width and depth are interchangeable, but in order to minimize the width of the front panels, the design chooses the shorter of the two. A deeper bin also leaves room to turn the first batch of material that goes into the secondary bin, prior to it getting filled by the second batch.

THREE-SIDED BAY

Step 1. Identify Key Assumptions

You need to know the following to size your bays:

a. The target food scraps to be processed in a given time period
b. The ratio of additional feedstocks to food scraps by volume
c. The frequency with which a batch is created
d. The processing time from start to finish in the primary, secondary, and finishing composting phases as applicable
e. Flow from bin to bin: Will bins be combined? What is the retention time in each bin?

Step 2. Calculate Batch Size

To calculate batch size, refer to the *Compost Site Processing Capacity and System Scale* in chapter 5 (page 114), then resume with step 3.

Step 3. Calculate Primary Bay Size

Now we can use your operation's standard batch size to calculate your primary bay size. Step 4 will give you the option of resizing your secondary bays after accounting for shrinkage and/or pile consolidation. Calculating bay size should be relatively intuitive. Based on the number of cubic feet capacity required for a batch (step 2), you can divide to find the inner dimensions of your bins, or look at table 7.2 for suggested bay dimensions.

If you load the bays by hand, don't make the pile higher than 4 feet, unless you have a plan to dump into it safely from above the bin. A 4-foot-tall compost pile in an insulated bay is enough to get really active.

In larger systems the bays will be designed to load and unload with a bucket loader. These bays could be piled 6 or 7 feet in height for non-aerated systems and up to 8 feet for ASP systems. The width of the bays and the height of the shed ceiling will need to accommodate loader bucket movements.

Some shrink factor can be accounted for in the bay. There will be more shrink with longer retention.

THREE-SIDED BAY DIMENSION FORMULA
(ACCOUNTS FOR SLOPED PILE EDGE ON ONE OPEN SIDE)

Where:

V = batch volume = ___ cubic feet
H = pile height = ___ feet
W = bay width = ___ feet
D = bay depth = ___ feet

$V \div H \div W + (1/2\ H) = D$

BAY EXAMPLE

On-Farm ASP Composting in Bays

a. 10 tons per week
b. 4:1
c. Every one week
d. 12 or more weeks
e. Flow from bin to bin: Will bins be combined? What is the retention time in each bin?
There are four primary ASP bays and four secondary ASP bays.
Two primary bays combine into one secondary bay.
Primary bays have four weeks or more retention.
Secondary bays have eight weeks or more retention.

Step 2

92 yards per batch, including a biofilter cap

Step 3

V = 2,484 cubic feet or 92 yards per batch rounded to 90 yards
H = 6 feet
W = 13 feet (assumes the bay is two 6-foot blocks wide plus a 1-foot gap in the center for an aeration pipe)
D = 35.34 feet (rounded to 36 feet)

2,484 cubic feet ÷ 6′ ÷ 13′ + (0.5 × 6′)
= 34.84′ (Rounded to 35′)

Note that H, W, and D can all be used interchangeably depending upon which dimensions you know. Also, with three-sided bays that are open on the top, you can round by a few yards or feet if needed to make round numbers. Piling the compost a few inches higher can easily make up this difference.

Table 7.2. Bay Sizing: Minimum Dimensions by Volume

Batch Volume		MINIMUM BAY DIMENSIONS (THREE WALLS)									
		4′ Height		5′ Height		6′ Height		7′ Height		8′ Height	
Cubic Yards	Cubic Feet	Width	Length	Width	Length	Width	Length	Width	Length	Width	Length
5	135	5	9	5	8	n/r	n/r	n/r	n/r	n/r	n/r
6	162	5	10	5	9	n/r	n/r	n/r	n/r	n/r	n/r
7	189	6	10	5	10	5	9	n/r	n/r	n/r	n/r
8	216	6	11	5	11	6	9	n/r	n/r	n/r	n/r
9	243	6	12	6	11	6	10	n/r	n/r	n/r	n/r
10	270	8	10	6	12	6	11	6	10	n/r	n/r
12	324	8	12	6	13	6	12	6	11	n/r	n/r
15	405	8	15	8	13	6	14	6	13	6	12
20	540	8	19	8	16	8	14	8	13	8	12
25	675	8	23	10	16	8	17	8	16	8	15
30	810	n/r	n/r	10	19	8	20	8	18	8	17
40	1,080	n/r	n/r	10	24	10	21	10	19	10	18
50	1,350	n/r	n/r	12	25	12	22	10	23	10	21
60	1,620	n/r	n/r	12	30	12	26	12	23	10	24
70	1,890	n/r	n/r	14	30	12	29	12	26	12	24
80	2,160	n/r	n/r	14	33	12	33	12	29	12	27
90	2,430	n/r	n/r	16	33	14	32	12	32	12	29
100	2,700	n/r	n/r	16	36	14	35	14	31	14	28
110	2,970	n/r	n/r	18	36	16	34	14	34	14	31
125	3,375	n/r	n/r	18	40	16	38	14	38	14	34
150	4,050	n/r	n/r	20	43	18	41	16	40	16	36
175	4,725	n/r	n/r	20	50	20	42	18	41	16	41
200	5,400	n/r	n/r	24	48	20	48	20	42	18	42
225	6,075	n/r	n/r	24	53	24	45	20	47	18	46
250	6,750	n/r	n/r	24	59	24	50	24	44	20	46
275	7,425	n/r	n/r	24	64	24	55	24	48	24	43
300	8,100	n/r	n/r	24	70	24	59	24	52	24	46

Note: Round up dimensions as logical for construction purposes.
n/r = not recommended

Step 4 (Optional). Calculate Size of Secondary Bins or Bays

Based on the flow of materials from bay to bay, it is often logical to size your secondary bays differently than your primary bays to account for shrinkage and batch consolidation (combining older batches).

4.1 Assume Primary Phase Shrink Factor

Estimate the total amount each batch will shrink prior to being moved to the secondary bins or bays. Earlier in the chapter I suggest 3 to 5 percent volume reduction per week as a conservative estimate. If you're combining bins, you can use a slightly higher shrink rate, because the first batch that goes into the secondary bay will continue to shrink before the next batch is added. If you have already subtracted shrink factor in step 2, be sure to account for that in your estimate. Also, keep in mind that total shrink including during blending is typically around 60 to 70 percent including 20 percent during blending, of which about 40 to 50 percent happens in the first two to four months.

___ Cu. Ft. Blended Compost per Batch in Primary × ___% Bin Shrink Factor (Typically 20 to 30%, or Multiply by 0.8 to 0.7)	=	___ Cu. Ft. Compost per Batch After Primary Shrinkage	2,484 Cu. Ft. × 0.8 = 1,987.2 Cu. Ft.

This is the adjusted volume for your secondary bays if you intend to keep each batch distinct. For those who are combining batches, follow step 4.2. If not, skip to 4.3.

4.2 (Optional) Combine Volumes from Primary into Secondary

If you intend to combine two or more batches into one secondary bin or bay, multiply the volume after shrinkage by the number of batches. You can also test the volume of combined bays here to see if combining makes sense. In cases where the bays need to be really large, combining piles in the bays might not make sense, especially for a hand-turned system.

___ Cu. Ft. Compost per Batch After Primary Shrinkage × ___ # of Batches Being Combined	=	___ Cu. Ft. Compost per Combined Batch in Secondary Bay	1,987.2 Cu. Ft. × 2 = 3,974.4 Cu. Ft. (Rounded to 3,975 Cu. Ft.)

This is the volume for your combined secondary bays.

4.3 Repeat Step 3 to Find Dimensions of Secondary Bays and Bins

Once you have calculated the volume of space required for the secondary bins or bays (4.1 and 4.2), repeat step 3 to find logical dimensions based on their adjusted volume. Depending upon the configuration of the site, you may want the depths of the bays in the primary and secondary bays to be the same, to make construction simpler. It may take some trial and error to find a combination of dimensions that suits both the primary and secondary stages.

If you plan on having a third or fourth set of bays, repeat steps 4.1 through 4.3 as necessary.

Note that width and depth are interchangeable, but in order to minimize the width of the front panels, the design chooses the shorter of the two. A deeper bin also leaves room to turn the first batch of material that goes into the secondary bay, prior to it getting filled by the second batch.

V = 3,975 cubic feet
H = 8 feet
W = 13 feet (assumes the bay is two 6-foot blocks wide plus a 1-foot gap in the center for an aeration pipe)
D = 42.22 feet (rounded to 42 feet)

3,975 cubic feet ÷ 8′ ÷ 13′ + (0.5 × 8′) = 42.22′ (Rounded to 42′)

With three-sided bays that are open on the top, you can round down V and D by a few yards or feet if needed to make round numbers. Piling the compost a few inches higher can easily make up this difference.

CHAPTER EIGHT

Turned Windrow Composting Systems

Composting in long piles, known as windrows, is a very common method on farms and in other larger-volume scenarios. Piles are usually built out in succession over time; new material is added to the end of the pile as it is comes in, although windrows can be formed all at once as well. Over a short period, ideally under a month, a batch is formed.

Volume for volume, the shape of the windrow provides more "edge" than a large round or rectangular pile would. The higher surface area to volume ratio of windrows increases critical compost/air interactions in a passive manner.

The long edge of the windrow also allows for easy access when turning the pile. Windrows are most commonly turned with a bucket loader, although equipment designed especially for turning windrows is used as well. At the community scale hand-turned windrow systems have gained popularity. Later in this chapter David Buckel describes the New York Compost Project hosted by the Brooklyn Botanic Garden at the Red Hook Community Farm; there, the absence of specialized equipment facilitates a community building and educational mission, while minimizing the environmental footprint.

The focus of this chapter is the planning and management of turned windrow composting operations. In concept these systems are quite simple, and on a small scale they can require relatively minimal planning, especially in rural settings. The chapter will guide you through common applications and management factors such as meeting the temperature standards specific to turned windrows. Then we'll discuss turning methods and equipment before moving on to site design components, layout, and flow, as well as sizing for targeted site capacity.

convection
(kohn-VECK-shun)
DEFINITION: The movement of warm-up and cool-down in a gas or liquid.

Common Applications for Turned Windrow Compost Systems

Due to the straightforward nature and low start-up cost of turned windrows, they are often the go-to method for operations that are testing the waters or in the early stages of development. Turned windrows are not typically the most space-efficient system (although windrow turner vendors would probably debate that point), but for a farm that already has a tractor with a bucket loader, or for a composting operation that would need to buy a loader to move material regardless, utilizing the loader for pile aeration as well as material movement is a no-brainer. Simply put, turned windrow sites get the job done with very little fuss.

Due to its wide applicability, turned windrow composting is by far the most commonly used method. In a 2017 survey of composting operations in the

United States, 63 percent were windrow operations (23 percent were also listed as static piles, so I have to assume that these were primarily unturned windrows).[1] When you look at the operation at Red Hook Community Farm, you can see why. The site was built on an existing and underutilized asphalt lot, so an excellent pad already existed. The operation tried composting in bins at first, but found that the sides got in the way of efficient material movement. They eventually scrapped the bins and went with hand-turned windrows, which only required shovels and volunteers. Since the site has an educational mission, applying human labor was a pro rather than a con.

Another relevant story is that of Kingdom View Compost. Operated since 2005 by Eric Paris and family, the site serves one of the most rural regions of Vermont, the Northeast Kingdom. The site has operated on an approximately 40,000-square-foot gravel pad on the family's organic dairy farm, and has scaled up from a couple tons of food scraps a week to about 15 tons per week over 10 years using the turned windrow method. I have been very lucky to work with the farm over the past five years or so in planning their scale-up. The site is a case study on the reasons why operations, when under pressure to grow, often start with tractor-turned windrows, then move on to more advanced technology. As more material is processed, space and labor become more and more limiting, and efficiency is lost. The farm piloted a low-cost aerated static pile (ASP) system in 2018, and will likely add additional ASP units until its full permitted capacity can be efficiently handled. Turned windrows allowed the farm to scale up slowly, with little cost. Now ASP is allowing them to expand the volume they process while minimizing labor and maintaining their existing footprint.

On the other end of the size spectrum, large commercial and municipal yard debris and biosolids composting operations frequently use the turned windrow method. Often these sites use large windrow turners, which can turn a lot of material very quickly. Sometimes yard debris composting facilities contract someone to come in with a large turner several times a year, rather than purchasing one outright. Of course, once you invest in a specialized piece of equipment like a windrow turner, you're fairly locked into that as a method. Since loaders can move material as well as turn piles, they are a necessity either way, so moving on to ASP or another method doesn't waste a piece of equipment. I recently ran into this exact scenario, where a small windrow turner is currently lying idle on a site that had switched to ASP. They simply didn't have the space to work with 4½-foot windrows, which was the equipment's maximum turning height.

One of the challenges with composting food scraps in turned windrows is that turning the pile early in the process exposes food scraps, increasing the potential for attracting pests and releasing odors. With a tractor-turned windrow system, the frequency of turning is typically less than with a windrow turner. Exposed food scraps can be re-capped with compost if need be, but this is something sites typically want to minimize. With a windrow turner, piles are typically turned more frequently, exposing raw materials, which can make it hard to take advantage of a turner's full capacity early in the process. Some large sites use covers that are lifted and repositioned by the turner as it passes through, which is a great system, but a very large investment. That said, on a smaller site covers are a good solution if you have the labor.

Turned windrows are ideal for the secondary and finishing composting phases for a number of reasons: (1) Food scraps and odors will be mostly broken down later in the process, making turning the pile less risky in terms of nuisances; (2) demand for oxygen is greatly reduced later in the process, so the material stays more aerobic passively, and with less aeration than is required earlier on; and (3) investments in composting technology such as ASP or in-vessel are best made to handle material in the early stages of the process, so you can save money on infrastructure by going low-tech with turned windrows for the later stages (although secondary ASP is often still a good investment following primary ASP). To get a quick overview of how windrows and other methods are commonly used in the different phases of the composting process, see table 2.2 on page 48.

Meeting Temperature Requirements

The temperature treatment standards of different composting methods have important distinctions (see *Temperature Treatment Standards and the Process to Further Reduce Pathogens* in chapter 3, page 56). Heat treatment using the turned windrow (or turned open pile) method is typically more challenging to achieve than the standard for aerated static pile (ASP) or in-vessel. Most standards, even the National Organic Program's (NOP) standards, are based on the Environmental Protection Agency's process to further reduce pathogens (PFRP), which was designed for biosolids composting. PFRP using the turned windrow method requires maintaining temps of 131 F or higher for at least 15 days with at least five turnings during that period.[2] The goal is essentially for all material to hit 131°F or more for a minimum of three days (in between turnings). State regulations may vary slightly, but most stick to this standard pretty closely.

While I have yet to come across a case where a windrow facility has been penalized or shut down due to not precisely meeting regulatory temperature standards, it is definitely more work to "hit temps" with turned windrows than with ASP or in-vessel, which only require three days at 131°F.

Methodology for pile monitoring and documentation that should meet most regulatory requirements is covered in *Compost Pile Monitoring and Management* in chapter 13, page 360. Methods and equipment for turning windrows are covered in the following section.

Pile Turning Methods and Equipment

Bucket loaders are the standard piece of equipment for most composting operations, because they can both move material and also turn piles. Many farms already have a tractor equipped with a loader, so purchasing new equipment may not be necessary to get started in composting. Most of this chapter focuses on loader-turned windrows because this is the most common scenario, but loaders are by no means the only equipment you can use.

There are respectably sized operations where volunteers turn windrows with shovels; there are operations that prefer to turn with an excavator; and there are many sites that use specialized windrow turning equipment. Before getting into site design and sizing, we'll discuss some of these options from a composting perspective. Keep in mind that my focus is on the function of the tool in the application, not on the mechanics of the tool. I am a firm believer in purchasing equipment where expertise and parts for that equipment are readily available locally. Even for those experienced with mechanics, this is undoubtedly a plus, especially for large equipment. While there is always something to be said for self-sufficiency in equipment maintenance and fixes, if you have a $50,000 loader, a Bobcat for example, be sure that you have an expert take a look at it frequently or you will almost certainly regret it later.

Human-Turned Windrows

There is also something to be said for avoiding the headaches of equipment altogether. Few people know more about the technique involved in moving and turning significant volumes of material better than David Buckel, which is why I asked him to share his experience in his piece about the Red Hook Community Farm (page 191). I had always assumed that hand-turned windrows would be turned from the side like their loader-turned counterparts, but David uses a method called the "walk-turn" (figure 8.4, page 194), which moves the pile from the ends, turning it simultaneously.

As David points out, while using manual labor to do the job a machine could do may seem inefficient, it provides ample opportunity to find efficiencies, which is how the walk-turn came about (combining the move and turn). There are several excellent videos on YouTube in which David outlines the method to the madness in meticulous detail. I highly recommend watching them; the thought and care that he uses in performing the simple act of turning compost is something most composters would strive to emulate.

Loader-Turned Windrows

Compost can be turned with a bucket loader of any size or kind, but some work better than others. Most of the windrows I have turned were with a lovely old 80-horsepower Zetor, which is a Czechoslovakian make. I can still feel its slow bucket work in my shoulder and its loose clutch in my back. It was a challenging piece of equipment to work with, but it did get the job done (about 97 percent of the time). In reality, we probably paid more in breakdowns and lost productivity than we ever did for the tractor. It's always hard to know exactly what the cost-benefit breakdown of any particular piece of equipment is going to be, but I would almost always advocate for the value of reliability.

One key factor for bucket loaders is a powerful engine. Most of the job your tractor does is with the loader bucket. The engine powers the hydraulics on the bucket, so more power equates to faster, more powerful loader movement. If you are processing more than 300 cubic yards a year of raw material, at least 80 HP is highly advisable.

Of course, the loader's lifting power is also crucial. Loaders are rated to a certain lift capacity. Lift capacity and bucket size limit the volume of material you can effectively lift or move. Loader buckets come in all different sizes; your bucket size should correspond to the lift capacity of your loader.

Figure 8.1. A windrow turned with a loader from the side at the old Highfields Center for Composting site in Hardwick, Vermont. Courtesy of Highfields Center for Composting and Vermont Sustainable Jobs Fund.

Most of the time loaders come with the right size bucket to handle the density of your typical feedstocks (dairy manure can weigh upward of 1,700 pounds per cubic yard), but you wouldn't want to put a 2-cubic-yard bucket on a loader that only has a 1,500-pound lift capacity, as a cubic yard of finished compost usually weighs around 1,000 pounds, unless you were only going to lift lighter materials like wood chips. Loaders that have quick-connect buckets are fantastic; some even have buckets that can be connected and disconnected from the seat of the loader. Many operations use different buckets for different purposes—say, a 2-cubic-yard bucket for handling raw materials and a clean 1-cubic-yard bucket solely for loading customers.

Windrow Turners

Compost windrow turners represent a third option, alongside human power and loaders. These turners are specialized pieces of equipment that straddle the windrow. The turner is either pulled through the compost by a tractor or is self-propelled. As it moves, a spinning drum with paddles or flails mixes and aerates the compost. The motion and shape of the turner re-forms the windrow behind it. Windrow turners are good at efficiently turning piles and thoroughly mixing the material. For example, the smallest model produced by one leading brand has a self-propelled windrow turner that can turn a 14-foot-wide, 7-foot-tall windrow at a maximum rate of 3,900 cubic yards per hour. How does that compare with a loader-turned pile? Depending upon the size of the loader, how fully the pile is turned, and how much space your site has for efficient maneuvering, I would estimate you could turn anywhere between 75 and 500 cubic yards per hour with a loader.

Typically, windrow turners move the compost a bit each time they turn, so they can be used in a "continuous flow" manner if new materials are added at one end and slowly moved to the other where the older compost is removed. This would work for sites that handle smaller volumes of material. More typically, my understanding is that compost is made in "batches," and the direction the turner moves through the windrow alternates so the pile moves only slightly back and forth on the pad.

Since turning increases the homogeneity of the finished compost, it stands to reason that the efficiency of windrow turners creates a more uniform end product, especially when compared with more static management styles. The criticism sometimes leveled against management with windrow turners is that aggressive and frequent agitation disturbs the microbial communities, which could be detrimental to the end quality, particularly in the maturation stages when higher-level organisms are establishing. It's not unlike a no-till or low-till philosophy in agriculture. This is less of a concern, however, if maturing piles are consolidated and infrequently turned with a loader.

On several occasions I've consulted with composters who started out with their minds set on either buying or building a windrow turner. Certainly in many instances that makes perfect sense, but in the majority of cases, windrow turners are an expansion strategy rather than a start-up strategy. Similar to ASP, when compared with a loader-turned windrow, turners have the advantage in theory of speeding up the process and reducing the footprint and labor required to produce a set volume of product. But most composters in start-up don't yet know exactly how quickly they will be scaling up and at what point they can financially justify another equipment purchase beyond the loader (which is definitely required if you're at the scale where you're considering a turner).* If you know that you need technology to produce more compost on a smaller footprint, I highly recommend comparing turners to aerated static pile, and if you are on a really small scale (less than 5 tons per week), in-vessel systems are also worth researching.

* A related pitfall is focusing on and investing in processing equipment and infrastructure while ignoring marketing the end product. Most operations that are on a scale that can ultimately justify a windrow turner are making a significant volume of compost and should prioritize marketing and adding value to their product over processing technology, although both are ultimately critical to a sustainable operation.

There are two main styles of windrow turner: self-propelled turners, which are stand-alone, and tractor-pulled turners.

Self-Propelled Turners

Large composting operations are often managed using massive and powerful windrow turners that are both tractor (they are driven) and turner in one. I've only seen these in operation and have never had the pleasure of operating one. Some facilities (yard debris facilities in particular) contract turners to be brought in periodically. I mention self-propelled turners mainly to illustrate the full range of equipment out there, but this is not the technology most people envision when they think about community-scale composting.

Tractor-Pulled Turners

Tractor-pulled windrow turners are the most likely choice for composters operating at the community scale. These are PTO-driven, pull-behind tractor attachments, so they require a tractor with adequate power. In addition, the tractor needs to have a low gear, "turtle gear," or hydrostatic transmission, which allows the engine to supply power to the PTO while keeping the tractor's travel speed low enough to slowly turn the pile. In other words, not every tractor is compatible with windrow turners, so this is a factor that you need to consider when researching tractor/turner options.

Figure 8.2. A tractor-pulled turner, custom-built based on open source designs by the Highfields Center for Composting (then the Highfields Institute) and Vermont Technical College. The designs are available online and have been built by operations around the world. They require some modifications from the original designs, however. A cool feature of this turner are the sprayers that can be turned on to water the pile while it's being turned if needed.

Aside from the compatibility of your equipment, there are other limiting factors with tractor-pulled turners that need to be considered. Turning a tractor that's towing a windrow turner requires a much larger footprint than would a tractor alone, so the work space at the end of the windrows needs to be designed with the turner and tractor in mind. Another important factor is the relationship between pile height and volume, which affects both site capacity and seasonal pile management. Pile size is limited by the size of the turner (as it is limited by loader size with loader-turned windrows). But small piles require exponentially larger footprints per volume, so while you gain in processing time and turning efficiency, you lose vertical capacity. The ideal application for tractor-pulled turners is for sites with large footprints—for example, a farm composting in rotating field locations or a site composting on a very large reclaimed lot. The volume of a pile also affects its ability to heat, and small piles (less than 5 feet tall typically) are going to have a much harder time maintaining heat in cold climates. The outside of small piles can even freeze solid, in which case forget turning it with the turner.

Other Turner-Style Equipment

In addition to tractor-pulled and self-propelled windrow turners are a number of other equipment types that perform similar turning functions. *Elevating face conveyers* convey materials up and over a large belt, usually moving the windrow as it's turned. There are loader attachments with hydraulically powered auger buckets that turn a windrow as you push the bucket through it. *Agitated bed systems* turn a long bay filled with compost, moving it from one end to another with each pass, typically as a continuous flow system.

NYC Compost Project Hosted by Brooklyn Botanic Garden at Red Hook Community Farm

by David Buckel, senior organics recovery coordinator

Brooklyn is home to the largest community-based compost site in the United States run entirely by renewable resources: solar power, wind power, and human power. The NYC Compost Project hosted by Brooklyn Botanic Garden operates the site at the Red Hook Community Farm. The farm occupies an entire city block of NYC parkland, with the portion devoted to composting comprising about 20,000 square feet. The Department of Sanitation funds the NYC Compost Project, which features locations and events across the city, with different hosts in each borough.

Brooklyn Botanic Garden's site at Red Hook Community Farm processes 150 to 200 tons per year from the landfill with the help of up to 2,000 volunteers each year, hot or cold, rain or snow, nice or not. The incoming material is roughly 2 parts carbon-based and 1 part nitrogen-based (mostly food scraps).

Figure 8.3. Brooklyn Botanic Garden's community composters, David Buckel (*left*) and Domingo Morales (*right*). Courtesy of Terry Kaelber.

Windrows, aerated by hand or solar-powered blowers/pipes, are the chosen composting vehicle, replacing bins because the latter were inefficient. The processes of building and turning windrows are unencumbered by a structure (that is, the bin walls), and windrows are more friendly for large groups of volunteers because there is more space in which a large team can maneuver, and each individual can experience meaningful work. The more volunteers, the more compost.

In an urban setting, rats are a top concern. Surprisingly, the windrows control rats better than bins when combined with improved protocols for materials management:

- Using large rat-proof Brute barrels for materials storage
- Moving stored materials periodically to prevent rat habitat formation
- Timely processing of fresh nitrogen-based inputs
- Placing windrows in open space to better expose rats to insecurity
- Thickly sealing windrows with more mature compost so rats encounter heat if they penetrate far enough to reach food, and are repelled
- Turning windrows often to maintain high levels of heat while material is still of interest to rats and to prevent habitat formation (especially in winter)
- *Sweeping, sweeping, sweeping*, to leave nothing of interest to rats on the ground

The previous system of bins failed on rodent control even though they were covered in hardware cloth—top, sides, and bottom—and placed away from fences on flat asphalt. The asphalt is a barrier to rats forming habitats below the bin (for warmth) or burrowing upward to access food. But New York City rats are professionals, and spent whatever time they had through the night clawing for access. By contrast, a properly constructed/turned/maintained windrow, combined with all other protocols to frustrate rodents, has left the site without a single rat for over six years, even though they are visible on nearby blocks.

Restrictions on inputs further help with rat and odor control, essential for an urban setting with neighbors across the street (a K–12 school on one side, a block-long IKEA storefront on the other). The site does not accept meat, dairy, or any post-consumer waste. Meat and dairy contain dense protein and post-consumer waste contains oils, both of which are slow to break down. Those items would keep a windrow of interest to rats for a longer period, meaning the windrow would need a thick seal of developed compost for more turns than would be needed without those items present.

By honoring the needs of an urban site with neighbors (controlling rats and odors), Brooklyn Botanic Garden can fulfill the goals of a community-based compost program, including processing the community's organics locally for the benefit of greening the community. The finished product amends the soil for growing food at a dozen farms in Brooklyn. The finished product also supports food gardens in neighborhood public schools, street tree pit maintenance, and Brooklyn's community gardens.

By maintaining a nice site, Brooklyn Botanic Garden advances another goal of community composting: fostering environmental stewardship. An appealing site helps draw in the public for opportunities to develop and use the knowledge and skills to divert organic material from landfill, make compost, and rebuild soil for a greener city. That focus on public education, reducing waste, and rebuilding neighborhood soil is at the core of the NYC Compost Project's mission.

But an appealing site is not enough to motivate volunteers to become environmental stewards in compost and expand their ranks. Work at the site must be meaningful in order to motivate them. This is addressed in two ways at the Red Hook Community Farm: Volunteers' work really matters to the operation, and their work is part of an exciting national model for environmental friendliness. Both are connected to the rule that the operation runs entirely on renewable resources. Ten kilowatts of solar power run two blowers for a few actively aerated windrows, as well as two heat lamps to keep volunteers' gloves warm in the winter. A small wind turbine helps in the winter when there is less sun to power the solar. But most of the power is human power, with pitchforks, shovels, and wheelbarrows for building, turning, and sifting mounds. It's done by hand: artisanal compost. Without the volunteers, the site would not be the largest compost site in the United States that runs entirely on renewable resources, so we can honestly and glamorously conclude every work session with a "thank you for making the world a better place."

These motivational components help sustain volunteer events every Friday and Saturday, year-round and through bitter winter winds and sweltering summer heat, in addition to many annual volunteer events for groups like students or corporate employees.

All that said, not all volunteers flourish, despite a daily creative struggle to find easier ways to do things by hand. As is true in many fields, a fair number of folks fail to connect lofty ideals for sustainability with the work required to achieve those ideals. In addition, composting by hand is hard work. Some volunteers experience shock over the dissonance between what they talk about with friends on the exciting topic of sustainability and what sustainable practices require of both body and mind in the real world. This is especially stark at a site that promotes an ethic of keeping in motion even while talking. That ethic applies not only to regular volunteers but to walk-in surprise visitors (or groups who want a tour), for whom a deal is struck that they either work while they talk to a worker, or move with the worker as the work gets done. Absent that ethic, very little would get done, as many folks want to come and just look and talk.

This challenging dynamic became a monster after Hurricane Sandy surged 2 feet of ocean water across the farm, salting the soil beyond repair and otherwise piling up anything not bolted down, including a 200-pound worm bin that floated half a city block away. Cleanup required skid steers and large bucket loaders, parked across the farm. On compost days those gas-powered machines greatly tested volunteers, because it was clear that machines could do the composting in minutes rather than hours (although with less quality and more damage to the environment). But the whole point of the work, as was reemphasized, is not only to divert local waste to local resources, but to do it in the most environmentally friendly way, thus maximizing public participation to foster environmental stewardship and greening the community.

A common theme since the machines went away has been that sustainable practices are hard work. If they weren't, everyone would already be committed to them. So a meaningful commitment to sustainable practices will form

Figure 8.4. Volunteers turning two windrows using the walk-turn. Courtesy of David Buckel.

only with the experience of their real-world challenges. Community-based composting is hard, but well worth doing.

At the same time, our goal is not to make composting harder than it has to be. Our site operators constantly look for efficiencies. For example, the site is configured so that material enters one end and, as it is processed, flows to the other. Operators prioritize what are called "walk-turns," whereby volunteers aerate a windrow by turning it just a few feet, as part of a longer journey of more turns toward the curing and sifting area. This allows for a quick turn because of the short distance, in contrast with instances when the windrow must be loaded up in wheelbarrows and "roll-turned." As another example, operators place new windrows where carbon-based material and cover seal material are close at hand. In winter, hot mounds move close to where fresh builds will occur so the new mounds get kick-started with material from the hot mounds and do not freeze.

Other operational efficiencies include micro management and equipment choices, such as:

- Set up any task to allow for the fewest steps to move material from one point to the next.
- When moving a windrow, try to fix any dis-uniformity with material held in the shovel rather than just pushing the windrow around with the shovel. (The material has to move anyway, so use that motion to do a fix, meeting two goals with one motion rather than two.)

- Build sifting tables to allow for sifting directly into wheelbarrows for immediate transport.
- Place sifting leftovers directly into barrels for incorporation into builds, rather than mounding them.
- Get rid of all pneumatic tires in favor of flat-frees to avoid needless wheelbarrow maintenance.
- Favor tumblers as transitional devices for small amounts of incoming nitrogen-based material, but favor tumblers under which a wagon/wheelbarrow can collect contents to save the step of shoveling off the ground.
- Configure materials storage so that there is an easy visual code for determining what is older/newer (say, signs that move as the receptacles enter or leave the inventory).

Developing the most efficient practices is part of developing general best practices for community-based composting, and that is of added importance for low-income communities that cannot afford fossil-fueled machines or are looking to create jobs that are meaningful and foster good health for workers.

One curveball when sourcing materials from agricultural settings is when they deliver huge volumes of a single type of material (such as weeds when a section is cleared, or large collard stems when several beds are cleared) that would overwhelm a build of a windrow. Too much of one narrow type of material can lower the quality of the compost recipe and also require extraordinary time to reduce to an appropriate particle size by hand. To handle these situations, we invented the "J row." The *J* stands for "just weeds," which encompasses both weeds and spent crops that are considered weeds once they are no longer wanted on the fields. The material is layered with wood chips into a long boxy mound, and then gradually introduced to the compost pipeline as appropriate for the recipe of any new windrow.

Composting at the Red Hook Community Farm has scaled up rapidly since 2009, with each year bringing new protocols to replace old ones that were once best practices. The site has nearly reached full scale in terms of achieving the NYC Compost Project's mission of creating environmental stewardship through public education, reducing waste, and keeping resources in the community for the benefit of rebuilding soil and greening that community.

In the midst of the success, one special feature stands out. In community composting there is a physical connection to the world of recycling that is difficult to find with other resources like paper, metal, plastic, or glass. With composting, participants can put their hands and bodies to work, see the steam rising from the hot windrows, touch the beautiful compost that once was raw food scraps, and glance out at the fresh produce growing for the neighborhood or at the happy trees in the community. For some, helping the environment has an added plus of getting a really good workout without being shut up in a gym, what they call "motion with meaning."

But for all, the relish in being present is in slowly rolling out the tools, folding into the task with the entire body, and methodically working the material for hours as part of something really important. In this way, what happens feels like what happens with the Slow Food movement. Slow Food is partly about accepting that most of the meaning with food comes from loving preparation that takes time and work, and a product that is to be savored for what it adds to our lives. Community composters are part of Slow Compost.

Windrows in Space and Time (Planning Your Turned Windrow Operation)

The remainder of this chapter is devoted to planning turned windrow composting operations. Regardless of the method of turning, the components of an optimally functional windrow composting system will be laid out in a logical order, sized to adequately process the target volume of material in a reasonable amount of time without causing environmental harm or public nuisances. And of course, the site needs to meet any applicable regulations. You do not need to be an expert to plan a windrow composting facility; to do it well simply requires due diligence and a solid understanding of the functionality of your site's components.

Turned Windrow Site Infrastructure

The defining component of turned windrow infrastructure is usually one or more active composting pads. There are typically several other site components that need to be planned for. These might include:

- Access roads
- Feedstock storage
- Food scrap receiving and blending areas
- Stormwater and leachate catchment and treatment
- Compost curing and storage

Chapter 6 covers compost site design factors that are non-compost-method-specific and should be referenced in conjunction with the turned-windrow-specific information that follows. These components are mentioned here as a reminder to think about the site holistically. Feedstock other than food scraps usually comes onto a site in trucks and may need to be stored until later use. These materials are then blended with food scraps in a designated area, ideally a concrete pad or bunk, as fresh food scraps are received. Even for very small sites, an improved pad will be a useful piece of infrastructure, because it will keep the wet material from creating a stinky and mucky mess. Blended material then goes out to the primary composting area, where it is actively managed. For larger sites there are often several pads, and windrows may live out their active phase on one pad, or move to a specific secondary pad, following heat treatment. Material that has matured to the point where it no longer needs frequent turning can be finished, cured, and stored in consolidated manner. The curing and storage areas have some unique factors, so I often recommend sizing and planning them separately from the actively managed compost pads (see *Calculating Finishing, Curing, and Storage Areas* in chapter 6, page 156).

All sites require a thoughtful and compliant approach to managing stormwater and leachate as they flow through and off the composting areas. Because turned windrow sites typically have a much larger footprint than would other composting methods of like capacity (ASP or in-vessel, for example), they often require larger treatment infrastructure.

Work Space for Windrow Management

It is absolutely critical to plan for *work space* surrounding the windrows. Work space is where pile management activities, such as loader and operator movement, take place. Turning windrows with a loader is most efficiently done by *rolling* the piles from their sides. While windrows can be turned from end to end, it is not an efficient way to frequently aerate or mix the windrow with a loader, although it is the preferred method for turning windrows with human power at the Red Hook Community Farm, where they pioneered the "walk-turn."

Windrow turners and excavators require less work space than a loader-operated site of a similar capacity. No matter what pieces of equipment you're using, you'll want to have a very good sense of pile size and work space requirements for that equipment before designing the site. Tractor-pulled windrow turners need a lot more space to turn around at the end of the windrows, for example, than do self-propelled windrow turners. For loader-turned operations, plan the alley (or aisle) space between windrows with your loader's turning radius in mind. Designing work alleys that are twice the length of the loader is reasonable for most loaders—typically 20 to 30 feet, depending upon the equipment. Alleys for sites managed with windrow turners are going to be narrower, and the vendor can give you the specifications. I call the space around the perimeter of the site or in between rows of windrows travel lanes. These need to be a single lane wide, just enough for the equipment to travel down, which is about 8 to 10 feet for typical loaders.

The active management areas for turned windrows are going to have as many work alleys as windrows if they are sized properly. Some site designs plan irregular spacing in an attempt to achieve spatial efficiencies, such as turning windrows toward each other, then combining them. In my experience, this can lead to site flow that is hard to manage and sustain. You can and will find efficiencies as you operate the site, but I highly recommend keeping the layout simple at the outset.

Site Flow and Layout

The layout of a turned windrow operation should be simple in terms of material flow, minimizing the number of times and distance that any material needs to be handled aside from turning. There is an early window of time in the first one to three months of the process, which is often called the *active phase*. This period of intensive management is critical, but it also creates a challenge in terms of flow, because not all of the windrows need to be turned at the same frequency. For sites with windrow turners and excavators, this is not an issue, because closely stacked material can still be turned efficiently. For loader-turned windrows, give yourself plenty of space for flexible management during the active phase.

One technique I use to conceptualize the flow of materials as I design is to think about the process in phases. As described in chapters 3, 5, and 6, I break the active stage into the *primary* and *secondary phases*. The primary phase is where temperature treatment is met. The secondary phase is where the

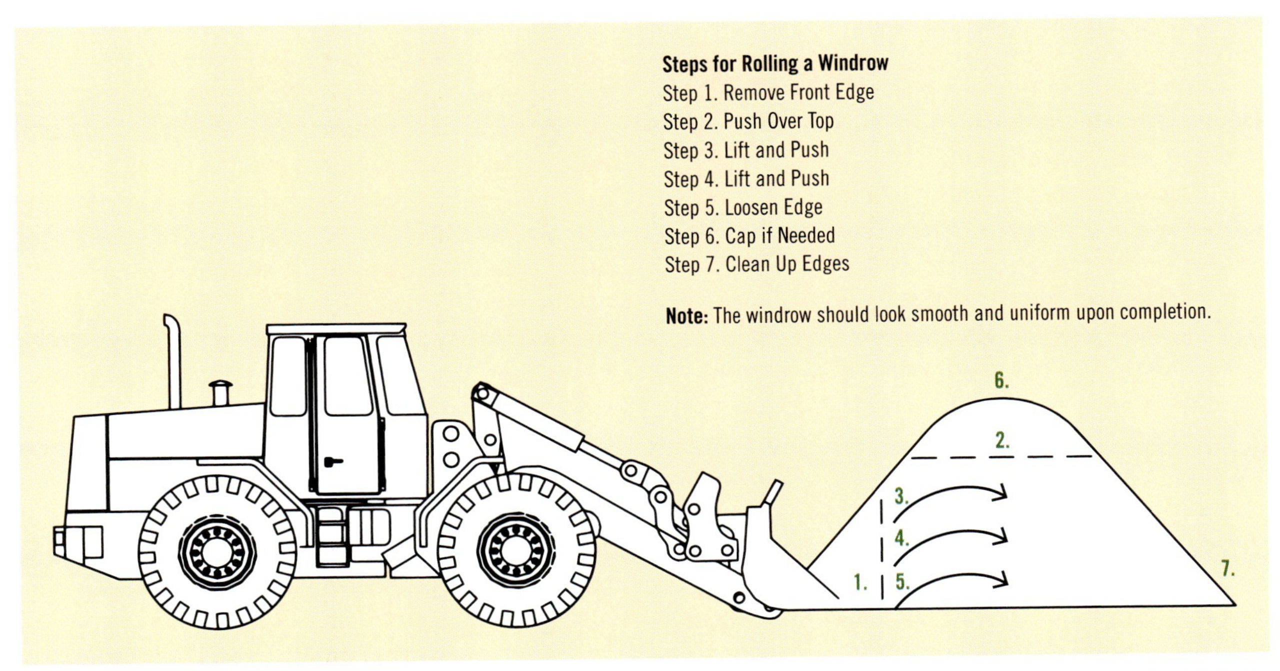

Figure 8.5. Rolling a windrow is a style of turning that infers turning from the side and sometimes only turning a third to half of the width of the pile.

material starts to cool and stabilize. Both involve frequent turning, but during temperature treatment, a pile might be turned five times in two weeks, while in the secondary, it might be only once every one or two weeks.

Some sites might utilize windrows only from the secondary phase on, because the primary phase is handled using aerated static pile or in-vessel systems. There can be a lot of different variation from site to site, so the key is to understand the level of management that is required for any given material in any given area, and look for strategies to utilize space more efficiently throughout the process. For example, two primary windrows might be consolidated to form one secondary windrow. Conceptually, this allows the designer to account for shrink factor (see *Estimating Volume Reduction* in chapter 5, page 122) during the primary phase, and spatial efficiency in re-forming larger piles when sizing the secondary pads.

After the secondary phase, material management decreases dramatically and piles can finish and cure in consolidated areas (figure 6.13, page 156), which are out of the way of active material and runoff from active pads.

In concept, an ideal flow of materials for loader-turned piles is from one end of the pad to the other, with the piles being moved as they are turned. Once the compost reaches the end of the pad, it is ready to be moved to a curing pad. In my experience, however, what often ends up happening is that fresh piles move toward the older piles at a faster rate than the older material needs to move, crunching the work alley between the piles and forcing the operator to move the older piles as well. Eventually the fresh piles can get too close to the older piles, and they can no longer be turned from that side. You then have to move the pile from the ends or at a super-tight angle with almost no work space, which equates to lost efficiency.

Operators who have the time and forethought deal with this by moving the older piles preemptively to create room. The only alternative I've found is to leave enough space on either side of a fresh windrow to turn it in one direction, then turn it back so that it's not pushing the other material over unnecessarily. If a site has enough space, a pile could theoretically live out its whole life in one zone, surrounded by enough work space to be managed as needed, at least for the primary and secondary phases. To achieve this type of site layout would require an extra windrow width of pad for each windrow, which may not be realistic for all of the windrows, but might be possible for some portion of your newest windrows, say the first one to three months' worth of material.

The takeaway is to give yourself extra room in sizing the pads in order to maintain flexibility and efficiency. When sites are undersized, the options are either some elaborate and time-consuming system of site flow, or simply taking the static approach whereby piles are stacked tightly and managed infrequently. Keep this in mind when you size your site, which is the topic of the next section.

Sizing Turned Windrow Composting Pads

The following sizing steps and equations can be used to calculate the footprint of primary- and/or secondary-phase turned windrow pads.* The calculation for pad width assumes a single row of windrows running parallel. The layout of the site can be split into multiple rows; for example, 16 windrows could be split into 2 rows of 8 windrows or 4 rows of 4 windrows. In your calculations, make sure that you adjust the pad width and length to account for additional perimeter and travel lanes accordingly. Sketching the site will help to visualize the layout.

Tracking and managing each windrow as its own distinct entity or batch is a key best management practice. Effective batch performance requires that all of the material in the batch be of *like age*. A new windrow should be started every one to four weeks, six weeks at an absolute maximum. These calculations assume that you have a throughput, batch size, and batch frequency identified. If you do not yet have

* Infrastructure other than the active composting pads such as feedstock storage, mixing areas, compost finishing, curing, and storage are covered in chapter 6.

an estimated batch size, revisit *Compost Site Processing Capacity and System Scale* in chapter 5, page 114.

Another important factor to consider in determining pad size is windrow dimensions, which may in part be determined by how high your equipment can stack material. In general, keeping windrows less than 8 feet tall is recommended. Large piles increase the risk of the pile's core going anaerobic. Windrows five to seven feet tall at the time of formation are ideal. Windrow width is typically twice the height of the windrow, plus a foot or two depending upon how vertically your mix stacks.

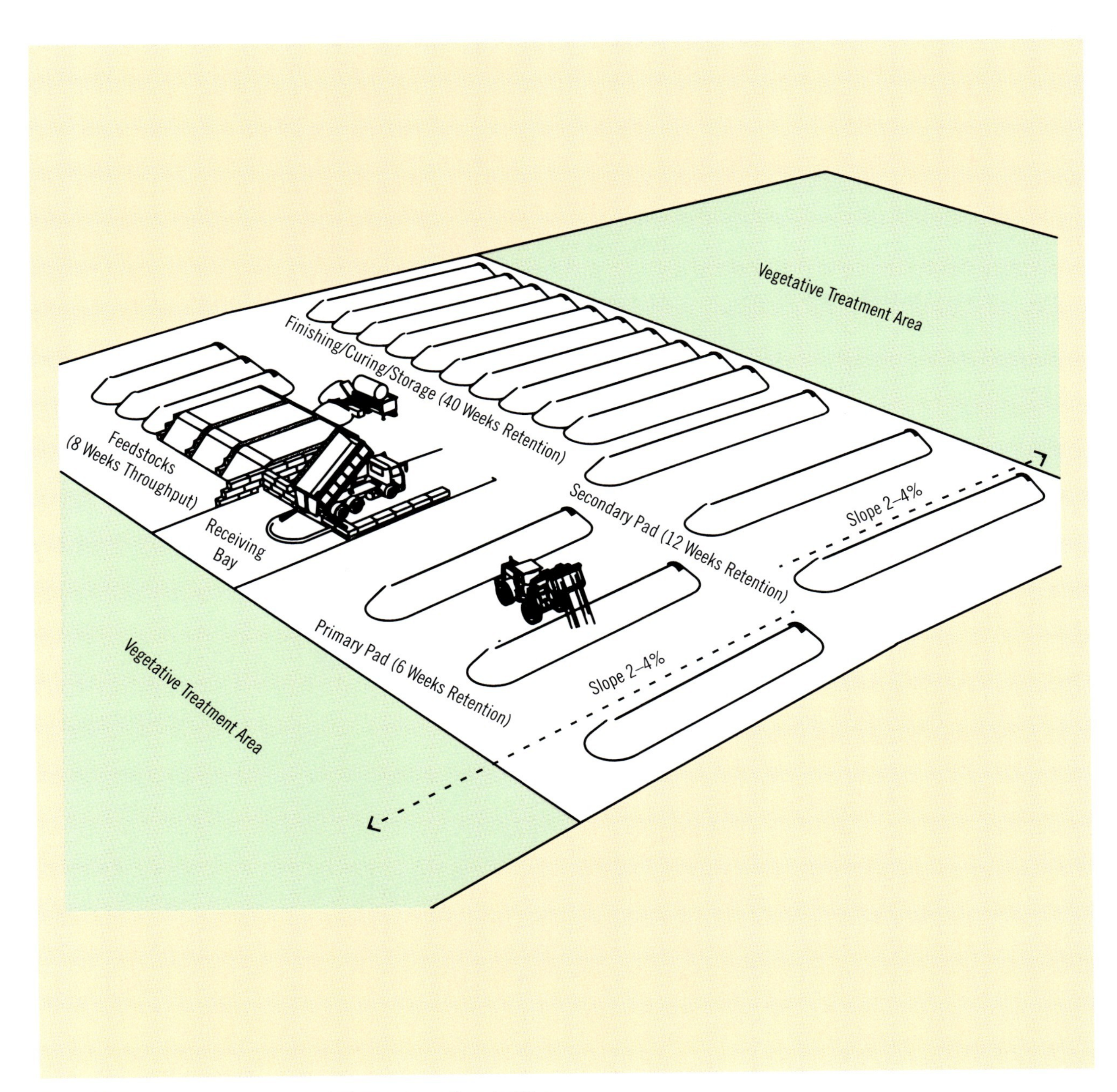

Figure 8.6. A conceptual turned windrow composting facility.

Step 1. Identify Key Assumptions

You need to know the following in order to size your windrow pads:

a. The target food scrap throughput
b. The ratio of additional feedstocks to food scraps by volume
c. Batch frequency
d. The processing time from start to finish in the primary and secondary phases
e. Pile combining strategy (if applicable)
f. Pile height
g. Pile width (2 times the height plus 1)
h. The width and layout of *work alleys* (tractor work space, typically equal to the total number of windrows)
i. The width and layout of *perimeter* access and travel lanes

EXAMPLE

a. 10 Tons/Week
b. 4:1
c. Every 2 weeks
d. 6 weeks
   12 weeks
e. Two primary windrows are combined to form one secondary windrow (one secondary windrow = four weeks' throughput).
f. 7 feet
g. 15 feet
h. 25-foot work alleys
i. 10-foot perimeter

Step 2. Calculate Batch Size

To calculate batch size, refer to *Compost Site Processing Capacity and System Scale* in chapter 5 (page 114), then resume with step 3.

Step 3. Estimate Windrow Dimensions

Based upon what you know about your site location (likely pile size and batch volume), estimate the size of your windrows.

3.1 Calculate Windrow Length Minus Height

___ Cu. Ft. Blended Compost per Batch*
÷ ___ Height of Windrow
÷ ___ Width of Windrow
÷ 0.66 Cross-Sectional Area
(Assume 0.5 if Conservative)
= ___ Length of Windrow – Height

Calculations for Primary Phase Pad

4,320 Cu. Ft. ÷ 7′ ÷ 15′ ÷ 0.66 = 62.3′

Calculations for Secondary Phase Pad

6,048 Cu. Ft. (Secondary Combined Batch) ÷ 7′ ÷ 15′ ÷ 0.66 = 87.3′

3.2 Calculate Windrow Length

___ Length of Windrow – Height
+ ___ Height of Windrow
= ___ Length of Windrow

Primary: 62.3′ + 7′ = 69.3′ (Rounded to 70′)

Secondary: 87.3′ + 7′ = 94.3′ (Rounded to 95′)

* For those calculating primary and secondary pads separately, shrink factor will be applied only when calculating the secondary pad width. If you're calculating one pad for all active management, you may want to adjust the batch size slightly to account for this (10 to 25 percent, depending upon processing time on the pad). See the *Estimating Volume Reduction* in chapter 5, page 122.

Step 4. Estimate # of Windrows

Calculation		Result	Example 1	Example 2
___ Processing Time on Compost Pad ÷ ___ # of Time Periods per Batch	=	___ # of Windrows on Compost Pad	6 Weeks ÷ 2 Weeks = 3	12 Weeks ÷ 4 Weeks = 3

Keep the time period consistent across all calculations. (I typically use weeks for food scrap calcs.)

Step 5. Calculate Pad Width Required for Windrows

Calculation		Result	Example 1	Example 2
___ # of Windrows on Compost Pad × ___ Width of Windrow	=	___ Combined Windrow Width	3 × 15′ = 45′	3 × 15′ = 45′

Step 6. Calculate Width Required for Work Alleys

Calculation		Result	Example 1	Example 2
___ # of Windrows on Compost Pad × ___ Width of Work Alleys	=	___ Combined Work Alley Width	3 × 25′ = 75′	3 × 25′ = 75′

Step 7. Calculate Width of Compost Pad

Calculation		Result	Example 1	Example 2
___ Combined Windrow Width × ___ Combined Work Alley Width + ___ Width of Perimeter	=	___ Width of Compost Pad	45′ + 75′ + 10′ = 130′*	45′ + 75′ + 10′ = 130′*

Step 8. Calculate Length of Compost Pad

Calculation		Result	Example 1	Example 2
___ Length of Windrow + ___ Width of Perimeter	=	___ Length of Compost Pad	70′ + 20′ = 90′†	95′ + 20′ = 115′†

Step 9. Calculate Area of Compost Pad

Calculation		Result	Example 1	Example 2
___ Width of Compost Pad × ___ Length of Compost Pad	=	___ Sq. Ft. Area of Compost Pad	130′ × 90′ = 11,700 Sq. Ft.	130′ × 115′ = 14,950 Sq. Ft.

* Assumes one perimeter edge; the other edge was counted as a work alley.

† Assumes two 10-foot perimeter edges at the ends of the windrows.

Step 10 (Optional). Size Secondary Pad

If you're calculating a secondary pad separately from the primary pad, repeat steps 3 through 9 after adjusting the batch size for pile combining and shrink factor.

10.1 Adjust Batch Size for Shrink Factor

Calculation		Result	Example 1	Example 2
___ Cu. Ft. Blended Compost per Batch × ___% Primary Shrink Factor (Example: For a 20% Reduction, Multiply by 0.8)	=	___ Cu. Ft. Compost per Batch After Primary Shrinkage	4,320 Cu. Ft. × 0.7 (or 30%) = 3,024 Cu. Ft.	N/A

10.2 Adjust Batch Size for Pile Combining

Calculation		Result	Example 1	Example 2
___ Cu. Ft. Compost per Batch After Primary Shrinkage × ___ # of Batches Being Combined	=	___ Cu. Ft. Compost per Combined Batch in Secondary Windrows	3,024 Cu. Ft. × 2 = 6,048 Cu. Ft.	N/A

Use this adjusted windrow volume when repeating steps 3 through 9.

Step 11 (Optional). Calculate Combined Area of Primary and Secondary Compost Pads

Calculation		Result	Example 1	Example 2
___ Sq. Ft. Area of Compost Pad (Primary) + ___ Sq. Ft. Area of Compost Pad (Secondary)	=	___ Sq. Ft. Area of Active Compost Pad	N/A	11,700 Sq. Ft. + 14,950 Sq. Ft. = 26,650 Sq. Ft. The active pad provides 18 weeks of processing capacity with spacious work alleys for flexible turning.

As you make your calculations, it is useful to make a sketch of the windrows and pad on graph paper. It's critical that you check your calculations against an actual design and adjust the layout as needed before finalizing anything. Chapter 6 covers compost site design factors that are non-compost-method-specific (finishing, curing, and storage pads, feedstock storage, and so on) and should be referenced in conjunction with the turned-windrow-specific information in this chapter.

CHAPTER NINE

Aerated Static Pile Composting Systems

One of the foundational pillars of managed composting is the consistent provision of oxygen to the compost, or rather, to the microbes feeding on the compost. When fresh oxygen is directly supplied to compost by fans, without agitating the compost by turning, it is known as the aerated static pile (ASP) or forced air method. In the 1970s researchers in Beltsville, Maryland, developed this method for use in the biosolids composting industry. Another name for this method is the Beltsville process.[1] ASP composting, as we'll call it, works by pushing or pulling a fresh supply of air through ductwork connected to perforated aeration channels beneath the compost.

The primary goal in the design and operation of ASP systems is to supply oxygen consistently and evenly throughout the compost, without relying on constant agitation. In theory, by seamlessly meeting the microbes' demand for oxygen, or *biological oxygen demand* (see *Biological Oxygen Demand*, page 53), the active composting phase is rapid and hot. Compost can be processed quickly, in many cases in half to a third as much time as turned windrow composting, cutting the processing time from raw material to usable product. The efficiencies that accompany this reduction are hard to precisely quantify, but the shorter processing time and other benefits of ASP composting have made this a very common method across the spectrum of composters.

interstices
(in-ˈtər-stəs)
DEFINITION: Small spaces in between particles of matter.

In a 2017 survey by *BioCycle*, just below 10 percent (107) of the responding compost sites reported utilizing ASP.[2] That said, I'm seeing an increasing number of community-scale composters, in both rural and urban regions, moving to ASP systems.

The subject of ASP system design and management is large and nuanced; therefore I have devoted a significant amount of energy to this chapter. Starting with a description of the overall concept, benefits, drawbacks, and applications of ASP systems, the chapter then guides you through the two primary types of systems, positively and negatively aerated, and their applications. We'll look at likely site components, layout, and material flows where ASPs are utilized, as well as the sizing and proper design capacity of ASP systems. Methods for sizing and designing aeration systems are covered in depth, including aeration rates, fans (blowers), ductwork, manifolds, and air channels. Included in the chapter are the unique rules for designing ASP systems, as well as options for system controls and odor mitigation. In addition, we'll discuss the many complementary ways in which ASP composting is integrated with the other methods of composting covered throughout the book.

Common Applications for ASP

When considering the ASP method, weighing the benefits and drawbacks as they apply to your own situation will hopefully clarify whether ASP is a good investment for your operation. Situations where land and neighbor relations are at a premium are likely candidates for ASP. Urban composters and farms, for example, have to use space very efficiently and need to keep tight control over odors and vectors.

Turned windrow operations that have maxed out their processing capacity and can't expand the overall footprint of their facility are also good candidates for ASP composting, whereby ASP infrastructure represents an expansion strategy rather than a start-up strategy. In my experience working with community-scale food scrap composters, urban operations and those that are scaling up are the most common users of ASP compost systems.

The use of ASP systems by rural farms that compost food scraps, particularly farms that intend to produce compost that is "suitable for organic production," could also become more common. Because farmers are so busy, and composting is often a side enterprise, it does not always get the full attention that it deserves, resulting in anaerobic pile conditions and all of the issues that go along with this. ASP systems can maintain aerobicity and meet the temperature treatment requirements in organic standards with less attention than turned windrows.

ASP Benefits and Drawbacks

Even operators accustomed to other methods of composting grow to appreciate the efficiency and control that a well-managed ASP system offers. There is an operational learning curve, however, and with ASP composting in particular, there are a lot of little things that can make or break the functionality of the system. We'll start by going over some of the benefits of ASP systems and then look at some of the drawbacks. Note that both benefits and drawbacks listed here assume a well-designed ASP system. Systems that do not follow best practices in design may not reap the benefits and are likely to experience exacerbated drawbacks.

Benefits

The benefits of well-designed ASP systems include the following.

Reduced Processing Time

As described in the introduction of this chapter, ASP systems can speed up the oxidation of compost, reducing the time to compost maturity. I have personally witnessed six weeks of ASP composting reduce a 10-month turned windrow process by 3 to 4 months (producing a finished and cured compost). There are many types of systems available in the industry, with claims of three to four months (of constant forced aeration) to usable product. Note that the definition of maturity across systems is variable and dependent on end use.

Reduced Site Footprint

By reducing the processing time from raw material to end use, ASP composting can reduce the total footprint required for a given capacity compared to turned piles. In addition, by design, ASP composting can take advantage of spatial efficiencies; vertical space can be better utilized with taller piles (8 feet tall and even 10 feet with adequate fan capacity). With the addition of walls, piles can be formed as a rectangular block, which can hold almost double the volume on the same footprint as a triangular windrow. At a minimum piles can be stacked side by side, with access from the ends, reducing the need for work spaces between turned windrows; using the *extended aerated static pile* (EASP) method, the efficiency of a cuboid-shaped pile can be achieved without walls. Estimates of spatial efficiency by ASP and EASP composting are said to be as high as one-tenth of the footprint of turned windrow composting in some cases.

Phase II Site Expansion Strategy

ASP composting reduces the required site footprint for a given capacity compared with turned piles;

Figure 9.1. A four-channel (pipe) extended aerated static pile system (EASP) at the NYC Compost Project hosted by Earth Matter on Governors Island, New York City.

therefore, ASP can increase the capacity of a turned windrow compost site without increasing the footprint. Once a site has maxed out its physical capacity using other methods such as turned windrows, adding ASP is a logical strategy for expanding processing capacity without expanding footprint.

Reduced Loader Operation

The *static* in aerated static pile means that the compost remains unturned for the majority of the ASP phase, although turning at around three or four weeks is common, and sooner may be beneficial given an efficient method to do so.

Run the math for your operation and see how it pencils out. A loader and operator may cost your operation anywhere between $50 and $150 an hour, depending upon the size of the equipment. To manage a 100-cubic-yard windrow in the first month might be a two- to eight-hour job just in turning (again, depending upon the size of the equipment), maybe costing you $100 to $400. To run a 200-cubic-feet-per-minute (CFM) blower on that same 100-cubic-yard pile for 15 minutes out of every hour for a month would only cost $7.06 (58.86 kWh at $0.12 per kWh). Double the aeration and even double it again and you're still talking about a small percentage of the cost.

Efficiency in Meeting PFRP

Typically, regulations on heat treatment with ASP composting require three days at 131°F or greater. You should verify this based on the specific rules your operation falls under. Three days at 131°F or more unturned is quite easy to achieve for most ASP operators, given the prime aerobic conditions. By comparison, turned windrow systems are typically required to hit 131°F or warmer for at least 15 days with at least five turnings. Meeting the standard may require daily monitoring of the piles during the treatment period as well, which is fairly time consuming. The shorter ASP heat treatment period can therefore be more efficient and consistent, which can be a big plus for many operations. Temperature treatment standards are explained in more depth in chapter 3.

Odor Control

The best way to prevent the escape of nuisance odors is to contain and treat them biologically as they are created. ASP systems typically remain unturned for the early period of the compost's life, which is when strong odors pose the most risk. In an unturned or "static" pile, raw and potentially stinky materials can be contained within a blanket of compost, which acts as both a barrier and filter for fetor.

The two primary strategies for capturing and treating odors with ASP systems work by the same principle of biological remediation and achieve the same outcome when properly deployed. The first strategy is to simply contain odors within a thick blanket of compost or other suitable feedstock, which acts as a *biofilter* on the face of the compost pile (see *Capping for Odor Control* on page 268). The second strategy involves negative aeration working with an external biofilter. Air is pulled out of the compost into ductwork and can then be channeled through carbon-rich materials such as moist wood chips. This is the method more commonly associated with the term *biofilter*. As fresh air is pulled into the pile, exhaust is pushed through the biofilter. Odors attach themselves to the moist carbon-rich surfaces of the biofilter medium, where they can be broken down.

Capture Heat, CO_2, and Nitrogen

Compost piles constantly interact with their environment. Early on in the composting process, a significant volume of thermal energy, carbon dioxide (CO_2), and nitrogen are lost to the surrounding environment. Systems to recover these resources have been designed and tested over the past 50 or so years, although using heat from manure dates much earlier.

Bruce Fulford and his team pioneered the use of ASP systems to move heat, vapor, CO_2, and N to recovery systems directly in greenhouses.[3] Later innovations heated water utilizing heat exchangers.[4] Compost heat recovery systems with varying degrees of technology and complexity can be built utilizing ASP composting as the platform; this practice is sure to develop in coming years.

Drawbacks

Even well-designed ASP systems pose some drawbacks, including the following.

Higher Start-Up Costs

Although some purveyors of ASP composting systems might dispute this point, the initial cost of developing ASP infrastructure is typically higher compared with infrastructure for the turned windrow method. The actual cost per ton of processing capacity varies hugely depending on the system design or manufacturer, as well as the existing infrastructure and conditions on-site.

Logic dictates that investment in a technology like ASP pay for itself in reduced operational costs and site footprint over time, and based on the numbers I've run, this seems to be the case. Nevertheless, it is a real challenge to fairly compare one method of composting system to another on an economic basis alone. Rather, the deciding factor is usually some combination of considerations such as space, labor, or management priorities. For this reason, this chapter looks at several different designs that vary widely in cost and labor efficiency.

Preferential Air Channeling

While providing uniform airflow throughout the compost is the goal of good ASP system design, air behaves like a fluid, finding the path of least resistance. Compost is not uniform, and the air moving through it will find any crack, avoiding the denser portion of the mix, just as water moves very evenly through sand and pools or moves around clay. Low-pressure pathways, or *preferential air channels,* form over time no matter how perfectly the system is designed, and as they form the aeration process opens and dries out the material surrounding the pathways, exacerbating the problem. Denser portions of the pile receive much less airflow, which reduces the effectiveness of the ASP method.

Homogeneity, particularly in terms of porosity, is especially important in composts made in ASP systems, but even with mixing equipment, raw compost doesn't stay completely uniform. For this reason, reblending the pile at least every four weeks is the usual protocol. Removing the pile from the primary ASP phase remixes the compost, breaking up preferential pockets and creating a more homogeneous mix. This is also the most effective time to re-wet the material if it is getting dry.

Excessive Drying and Challenges with Re-Wetting

With ASP systems, many operators find that the benefit of continuous aeration can lead to the problem of excessive drying. Particularly in pockets where preferential air channels form, dryness can eventually lead to non-uniform degradation. In summer months the challenge of the compost overheating can require a higher rate of aeration just to cool the pile, which in turn leads to excessive drying.

A lack of efficient and effective re-wetting techniques creates an additional challenge for composters. Dealing with and mitigating overdrying is unfortunately a necessary learning curve for most ASP operators (and many non-ASP composters as well), with recipe/pile moisture content, pile temperature, weather and seasonal conditions, and aeration rates being the main variables. Working with a wetter recipe as the starting point can be a huge benefit, but it requires an added level of diligence and monitoring to mitigate anaerobic conditions and odors. A shorter period of unturned aeration with watering is the main strategy for mitigating overdrying.

Remediating excessive drying will require the addition of water, but this is easier said than done. Evenly watering 100 cubic yards of compost is a lot of work. For example, to bring 100 cubic yards of compost that has a moisture content of 45 percent up to 60 percent requires 21 cubic yards of water, which is approximately 4,240 gallons, distributed as evenly throughout the pile as possible. That's a lot of water!

Co-locating water and sprinkler systems nearby the ASP system is ideal. Having the ability to water as the pile is turned or moved is the best strategy to evenly distribute water throughout the pile. Some operations use water trucks to spray windrows after the compost is removed from the ASP. While some operations water their compost with leachate, it's important to use water that is free of potential pathogens once the compost you are watering has been heat-treated (or heat treatment will need to be re-met).

Figure 9.2. A cross section of a pile following ASP. There are three distinct zones in the pile, each at a slightly different stage and level of moisture content. The veins of dryness extending into the wetter material are known as preferential air channels.

Uneven Degradation

Different portions of material decompose at different rates, which is especially noticeable in a static pile. The reasons already mentioned—preferential air channeling, drying, and differences in density and porosity—all cause uneven degradation within the pile. In addition, the outside of the pile tends to dry out and be cooler, which causes it to degrade much more slowly.

One of the goals of turning is to move the outside of the pile to the more active core, but in a static pile this happens less frequently. All of these factors mean that what comes out of the ASP may not be as uniform as with methods that are more frequently agitated. For this reason, long static periods (typically greater than 4 weeks) should be avoided. Another strategy is to use pre-composted materials as a *cap* or *skin* on the surface of the pile. This material serves as a biofilter, but also as a pre-heat-treated insulation, so that all of the fresh material is in the core of the pile where it can heat and biodegrade.

Overheating

ASP composting can be so effective at initiating microbial activity that temperatures can quickly skyrocket far beyond the ideal range. While it may seem counterintuitive, the most common response is to cool the pile by increasing the aeration rate, either by increasing the duration of the fan's on-time or by increasing the fan speed. It can take just over a day or so to correct the issue. Where a more aggressive response is needed, soaking the hot spot with water can quickly cool it. Of course when all else fails, remove the pile from the ASP.

Leachate Capture and Handling

Many, though not all, ASP systems are designed to capture and reuse leachate. But the ability to recycle leachate from compost is not necessarily the blessing that it might seem at first. Leachate collection and storage systems are expensive to get right, and even then they can be prone to issues, especially in cold climates where leachate pipes can freeze. Once leachate is captured, the liquid needs to be applied to the pre-PFRP compost or utilized in another approved manner (such as being land applied).

At a very small scale, this may not be a huge deal; leachate could even be absorbed with sawdust and then used in the next batch of compost. At a large scale, however, captured leachate can become a managerial cost. For those who apply a lot of water to add moisture to their compost, applying leachate might not end up being a huge added expense, but in such cases it's not necessarily the solution to dryness, either, because leachate represents only a fraction of the water required and can only be applied to pre-heat-treated material. Leachate is also highly odorous and can create nuisance issues.

Leachate generation varies significantly depending on recipe, but measurements taken over several years at the Highfields Center for Composting showed between 1 and 2 gallons of leachate per cubic yard of raw compost, most of which was released in the first two weeks of food scrap composting as cell walls degrade.

Electricity Needed

While reduced tractor use is a huge benefit for both the operation and the environment, running blowers to aerate compost is not without its costs, and these can vary considerably between setups and seasons (summertime, for instance, when the fan is on more). That said, using fans for aeration will save money over time, so fan use is not a drawback compared with tractor usage (see *Reduced Loader Operation* on page 205). The bigger drawback can be the cost of running electricity to remote locations (higher start-up costs).

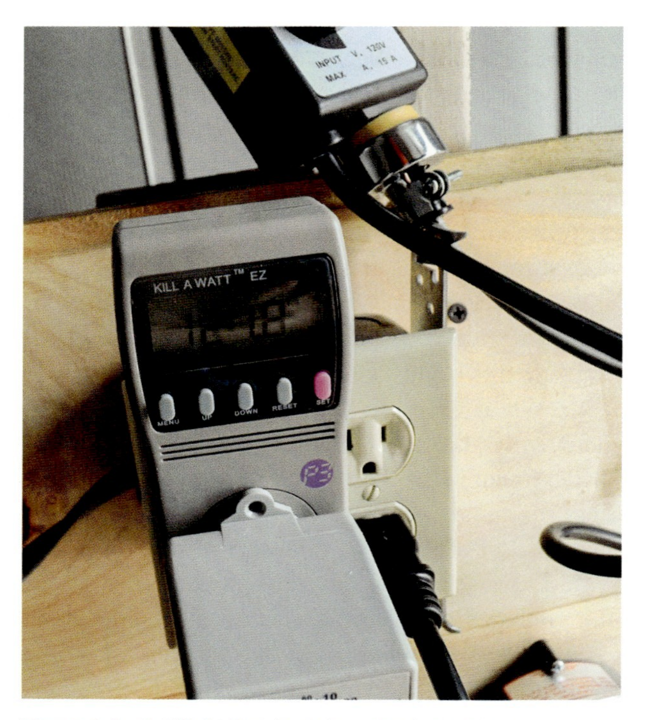

Figure 9.3. A Kill A Watt is an inexpensive tool that can be used to track energy usage and the cost of ASP fans (or any appliance, for that matter).

Challenges with Negative Aeration

Negative aeration systems have several drawbacks that are not associated with positively aerated systems. Many operations struggle with getting the design, components, and operation of their system right. Commonly reported issues include corrosion and high energy use.

Exhaust that is pulled out of the compost is highly corrosive, filled with moisture, ammonia, and organic acids. Fans that would work just fine for positive aeration can wear out in no time. Fans that hold up better to these exhaust compounds come at a much higher cost, but still there is little guarantee how long they will work over the long term. PVC ductwork, which is the standard for positive aeration, is not made for high composting temperatures and even CPVC, which is rated to 200°F, is not designed for compost exhaust's low acidity, so the much higher-cost HDPE pipe is the recommended material for conducting compost exhaust.

With regard to energy use, negative ASP systems typically pull air through the pile and then push the exhaust into a biofilter, which creates resistance on both sides of the fan. Resistance equals work and work equals electricity. Negatively aerated fans typically draw more electricity than positively aerated fans for the equivalent air volume.

Locating the fan on the opposite side of a condensate trap, heat recovery system, or biofilter won't reduce the added resistance of negative ASP, but it should extend the fan's life, effectively capturing much of the heat, vapor, and corrosive chemicals before entering the fan. Of course, with compost heat recovery systems, that extra energy use can be earned back.

ASP System Strategy and Design

The core design concepts for effective ASP composting all come down to one really simple idea: controlling the composting process by pushing or pulling air through the compost with a fan.

One of the advantages of forced aeration is that it lends itself to so many different strategies, scales, and design scenarios for deploying this concept. For example, compost piles can be formed in standard windrow formation, as a large block (often called extended ASP or EASP), or in hybrid formations. Aeration can be added to bays or in-vessels. While the number of options might seem overwhelming at first, developing a system strategy comes down to logic and geometry (cross-checked by cost and biology).

For engineers and others competent in the design of our built environment, the concepts here are probably quite straightforward. However, I know that I am not the only one who has seen systems that either are overbuilt or missed some of the most basic concepts, like adequately sizing air channels to minimize friction. To make this technology more accessible to the wide spectrum of practitioners doing this work, I have done my very best to translate the technical knowledge into a simplified form, without losing the critical nuances involved in designing a customized system.

I strongly recommend using the *Aerated Static Pile Design Specification Checklist* sidebar (page 214) as you move through the different design steps, as it's a helpful way to keep track of all of your design information in one place. Also, it's never too early to get out a pad of graph paper and begin to sketch out how the different ASP options might fit on your site.

You'll need to start by considering how the piles will be set up. Then you can move on to specifying the system details.

Pile Configuration

ASP batches can be formed in a number of different configurations. These include:

Windrows. Batches are formed into parabolic windrow shapes, where the edges of each batch closely abut one another, yet remain distinct. Compost is added and removed from the pile ends.

Bays. Each batch has a distinct composting bay (a large bin), with two, three, or four walls. The

(continued on page 214)

Space and Place: Adapting to the Constraints of Urban Composting

by Shakara Petteway, Leah Retherford, and Erik Martig

The NYC Compost Project hosted by Big Reuse is one of seven compost sites funded by the New York City Department of Sanitation (DSNY). All seven NYC Compost Projects (NYCCP) have varying programmatic directions, but share a mission of rebuilding the city's soil by providing New Yorkers with the knowledge, skills, and opportunities they need to produce and use compost locally. The projects also assist DSNY as the department looks to grow organics collection and composting as a system of waste management for the city. The NYC Compost Project hosted by Big Reuse fulfills this mission in three ways: stewardship, food scrap collection, and organics processing. Our operation is located not far from residential housing. This intersection of programming needs and location has created unique and interesting challenges that have come to shape our overall operations.

Our Model

Our three main efforts of stewardship, organics collection, and processing allow us to further the mission of both NYCCP and DSNY: improving the city's soil health while diverting organic materials from landfill. Through stewardship programs, we help residents connect their food scraps—commonly perceived as trash—to the utility of compost. We do this with tree care events and compost distribution to community greening projects around the city. Food scrap drop-off sites, where residential participants bring their food scraps, are important sites for communicating information while providing outreach about DSNY's curbside collection and recycling and sustainability initiatives. Finally, processing allows the NYC Compost Project hosted by Big Reuse to recycle diverted food waste materials and use them in our

Figure 9.4. Charles Lynch, operations manager at Big Reuse, hand-leveling a newly formed ASP pile at their site under the Queensboro Bridge, Queens, New York City.

stewardship programming: a full circle. Our operations are continually growing because of increased participation in organics recycling.

Our processing site is situated in a unique location: under the Queensboro Bridge linking the boroughs of Manhattan and Queens. The space is owned by the New York City Department of Parks and Recreation (DPR). They allow us to operate on a portion of the compound. Just north of the lot are the high-density residences of New York City Housing Authority's Queensbridge community. As can be imagined, many considerations must be taken to ensure an amicable and healthy relationship with our neighbors. We strive to operate our processing site in a way that minimizes pests and odors while maintaining a clean, welcoming, and educational community composting site.

The Challenges

The majority of materials processed by the NYC Compost Project hosted by Big Reuse are collected at our 14 weekly food scrap drop-off sites, as well as by our community partner GrowNYC, an environmental organization that runs several of New York City's farmers markets. On average, we receive approximately 20,000 pounds of food scraps per week. Wood chips and leaves—our primary bulking agents and source of carbon—are sourced from DPR and other partners.

When we first started in 2009 under the name Western Queens Compost Initiative, we operated five food scrap drop-off sites and processed the materials at six small-scale sites in Queens. In 2013, after we moved our operations to a single site beneath the Queensboro Bridge, we were processing almost 310,000 pounds of organic materials, including the browns that we were mixing in. In 2017 we processed more than 1.3 million pounds, and fully expect that we will continue to increase our intake as more residents participate in the program.

As our organics intake and finished compost output increase, we must continue to address

Figure 9.5. Volunteers bag unscreened, finished compost for use in street tree care while a GrowNYC driver unloads his food scrap pickup from the truck. Space constraints on the site make for a tight squeeze, but everyone manages. Courtesy of Renee Whittick.

the issues that come with space constraints and our proximity to the surrounding residential community. Incoming browns and greens, outgoing finished compost, space for trucks and other equipment to maneuver, and storage of toters—the containers we use for food scrap collection—are just a few of the concerns that our operations manager contends with on a weekly basis. Add to this a desire to process quickly and efficiently while continuing to produce a high-quality product, and it becomes clear how much the challenges of urban composting have shaped our operations.

Our Operations

At the NYC Compost Project hosted by Big Reuse, we adopted our current practices over years of trial and error in our effort to address the constraints and challenges we face. We use innovative small-scale technology that has been incrementally and collectively developed. In this time we have found a system that works within our boundaries: dividing our windrows into phases, mechanizing much of the process, and instituting careful consideration of how, when, and where each aspect of the process operates.

Figure 9.6. A Phase 1 batch at Big Reuse consists of four to six weeks' worth of throughput, managed with a Gore-Tex-covered ASP system. Courtesy of Erycka de Jesus.

Our site is host to five phases overall: Phases 1 through 3 and the curing and finished phases. Unlike other operations, we distinguish our phases by the length of decomposition. Phase 1 consists of the first four to six weeks, when we add to the pile weekly. The four- to six-week window dictates when we move materials to Phases 2 and 3. Phase 2 consists of two windrows, approximately 40 feet long, 6 feet tall, and 10 feet wide. These are managed for the next four to five weeks by aerating weekly with a fork attachment on a skid steer. Phase 3 is a single windrow that is approximately 40 feet long, 10 feet high, and 12 feet wide and consists of the next four to five weeks of the process. Phases 2 and 3 continue to reach temperatures above 131°F. The curing phase follows; this stage is no longer aerated, though it is still very active. After four to six weeks of curing, we begin distributing the finished compost to local community greening groups and projects. Prior to this distribution, we evaluate the maturity of the compost using a Solvita test. From start to finish, the compost is between four and five months old by the time of distribution.

This method means that we are moving piles between phases every four to six weeks. While moving is resource-intensive and time consuming, it thoroughly remixes and aerates the material while helping us to make room within the site for greater maneuvering. Box trucks that are 18 or more feet long must enter and turn around to deliver food scraps. This maneuverability is also important for the equipment we use in our daily operations: Processing is highly mechanized and utilizes several small-scale technologies to increase the efficiency of the decomposition process while maintaining quality.

For instance, Phase 1 is an aerated static pile (ASP) in combination with a Gore-Tex cover, a system designed by Sustainable Generation. Originally we built the windrow 9 feet high using only shovels and pitchforks, but we found the method impractical, especially as our organics intake increased. Now we use a skid steer with a bucket attachment to build windrows and move piles to the next phase. Mixing our carbon and nitrogen sources with the bucket, however, did not give us the best results. So we began utilizing a vertical feed mixer, the Jaylor A100, with the assistance of a toter tipper to empty the food scraps into the mixer. Using the skid steer, we load wood chips and leaves into the mixer as well.

This mechanization has greatly improved our efficiency and our ability to counteract issues such as odors, leachate, and aesthetics. Using the mixer allows for better incorporation of moisture into the browns, reducing overall leachate and allowing us to reach the necessary temperatures sooner. The Gore-Tex cover combined with ASP gives us fast processing, mitigates odors, sheds stormwater, prevents pests, and gives us the ability to monitor oxygen and temperature conditions in our Phase 1 windrow, helping us to ensure the quality of our final product. Overall these technologies allow us to address many of the challenges and concerns inherent to fulfilling our mission in a dense urban location.

Additional strategies in our operation and layout further mitigate issues. NYC Compost Project staff are on-site on a daily basis to perform operations and keep an eye out for potential problems. Staff operate at the site during weekday work hours when most people are not home and cannot be disturbed by the sounds—or smells—of our processing. We've also laid out the site such that the Phase 3 and curing piles—the most finished phases—are closest to our residential neighbors while Phases 1 and 2 sit farther away. Also, by not accepting meat, dairy, or oil at our food scrap drop-off sites, we avoid any additional odor and pest issues that come with processing these materials. While it is unfortunate that by not accepting these items we are also not diverting them from the landfill, the health, happiness, and goodwill of our neighbors remains the more pressing concern. These solutions are very simple, yet they go a long way in ensuring that we continue to be good neighbors to the surrounding community.

Ever-Adapting

The way in which the NYC Compost Project hosted by Big Reuse operates has greatly been determined by the demands of our mission and the constraints of our location. As we educate New Yorkers about the importance of compost and empower them to participate in its creation and use, we are also sensitive to the needs of the communities in which we operate. Our growth in capacity has required significant adjustments to maintain the quality of our finished product while addressing odors, pests, noise, and cleanliness. By dividing our piles into phases, mechanizing much of the process, and developing best practices in the layout and operation of our site, we have been able to meet these constraints. These solutions have improved our operations over the years, but we still face issues such as a general lack of space, the breakdown of equipment, and the ability to maintain high standards while dealing with those issues. As we have done in the past, the NYC Compost Project hosted by Big Reuse continues to seek solutions and adapt as necessary.

ability to stack material vertically against the walls considerably reduces the footprint required to compost a given volume of material. Compost is added and removed from the front or back.

Extended aerated static pile (EASP). Compost batches are consolidated into a large pile, typically trapezoidal in shape and open on the front and sides. Parallel batches are added over time, forming one large mass instead of distinct piles. Often the mass is aerated by a single fan, with air channels connected to the same fan manifold. Aeration pipes are added and taken away as the pile gets built out to keep them out of the way of loaders, and dampers or valves are used to turn off channels that aren't being used.

Hybrid EASP/bays. Multiple batches stacked side by side in one really wide bay with a back and end side walls is another common configuration. I've never heard a name used for this specific type of configuration. Although this is not typically called an EASP, they are similar in that multiple batches are in direct contact with one another. They

Aerated Static Pile Design Specification Checklist

As you plan your ASP, it may be helpful to organize all of your design specs in one place. The following checklist covers all of the key system choices and calculations.

System Capacity and Configuration

- ☐ Pile type
  - ☐ Windrows
  - ☐ Bays
  - ☐ EASP
  - ☐ EASP/bay hybrid
  - ☐ Enclosed/containerized
  - ☐ Semi-turned
- ☐ Target scale (cubic yards or tons per time period) ____________
- ☐ Batch frequency ____________
- ☐ Retention time
  - ☐ Primary ____________
  - ☐ Secondary ____________
- ☐ Number of ASP zones
  - ☐ Primary ____________
  - ☐ Secondary ____________
- ☐ Pile dimensions
  - ☐ Primary ____ (L) ____ (W) ____ (H) ____
  - ☐ Secondary _ (L) ____ (W) ____ (H) ____
- ☐ Batch size
  - ☐ Primary ____________
  - ☐ Secondary ____________

Aeration System

- ☐ Target blower capacity
  - ☐ Air exchanges per hour____ (steps 5–5.3)
  - ☐ CFM ____________ (steps 5–5.3)
- ☐ Ductwork
  - ☐ Header duct material____________
  - ☐ Header duct diameter/dimensions____ ____________ (step 6.1)
  - ☐ Header duct length ____________
  - ☐ Distribution duct material____________
  - ☐ Distribution duct diameter/dimensions ____________ (step 6.2)
  - ☐ Distribution duct length____________
- ☐ Air channels
  - ☐ Aboveground
  - ☐ Belowground
  - ☐ Air channel materials
    - ☐ Formed concrete
    - ☐ HDPE
    - ☐ PVC
    - ☐ Other ____________

differ from EASP in that batches are typically removed individually, whereas all of the compost is removed simultaneously with EASP. Also, each batch is often individually zoned, with its own dedicated blower/controls. The key distinction between this and bays is that there is no physical separation between batches. Compost is added and removed from the front.

Fully enclosed or containerized ASP. One variation on bays is the fully enclosed or containerized system (see figure 10.10, page 285). These are often referred to as *containerized in-vessel* systems with forced aeration. There are several commercially available containerized ASP systems.

Semi-turned ASP. Compost can be configured in such a way that it can be efficiently turned early in the ASP process, while you also maintain the ability to statically aerate the material. As a strategy, *semi-turned ASP* aims to yield the benefits and avoid the drawbacks of both ASP and turned windrows. One way that this is achieved is to design the air manifold so that a windrow turner can drive over it.

- ☐ Air channel diameter/dimensions__________ ____________________
- ☐ Air channel length __________
- ☐ Air channel perforations
  - ☐ Length of perforated zone __________
  - ☐ Diameter __________
  - ☐ Spacing __________
- ☐ Static pressure in system
  - ☐ Static pressure from header duct __________ ____________________
  - ☐ Static pressure from distribution ducts ____________________
  - ☐ Static pressure from air channels __________
  - ☐ Static pressure from compost __________
  - ☐ Total static pressure in system __________
  - ☐ Optional: static pressure from negative aeration and biofilter __________ ____________________
- ☐ Fan selection
  - ☐ Fan CFM at static pressure __________
  - ☐ Fan model (or options) __________
- ☐ Fan controls
  - ☐ Cycle timer
  - ☐ Heat-controlled system
  - ☐ Oxygen-controlled system
  - ☐ Variable speed controller (variable frequency drive or VFD)

Leachate Management System

- ☐ Leachate collection
  - ☐ Air channel collection
  - ☐ Gutter
  - ☐ No collection
- ☐ Leachate storage
  - ☐ Tank
  - ☐ Pond
- ☐ Leachate reuse
  - ☐ Applied to compost
  - ☐ Land-applied

Systems Footprint

- ☐ Primary phase
  - ☐ Pile footprint __________
  - ☐ Blowers, manifold, mechanical housing ____________________
  - ☐ Leachate management __________
  - ☐ Other (ASP-specific) __________
- ☐ Secondary phase
  - ☐ Pile footprint __________
  - ☐ Blowers, manifold, mechanical housing ____________________
  - ☐ Leachate management __________
  - ☐ Other (ASP-specific) __________ ____________________

☐ Total ASP systems footprint

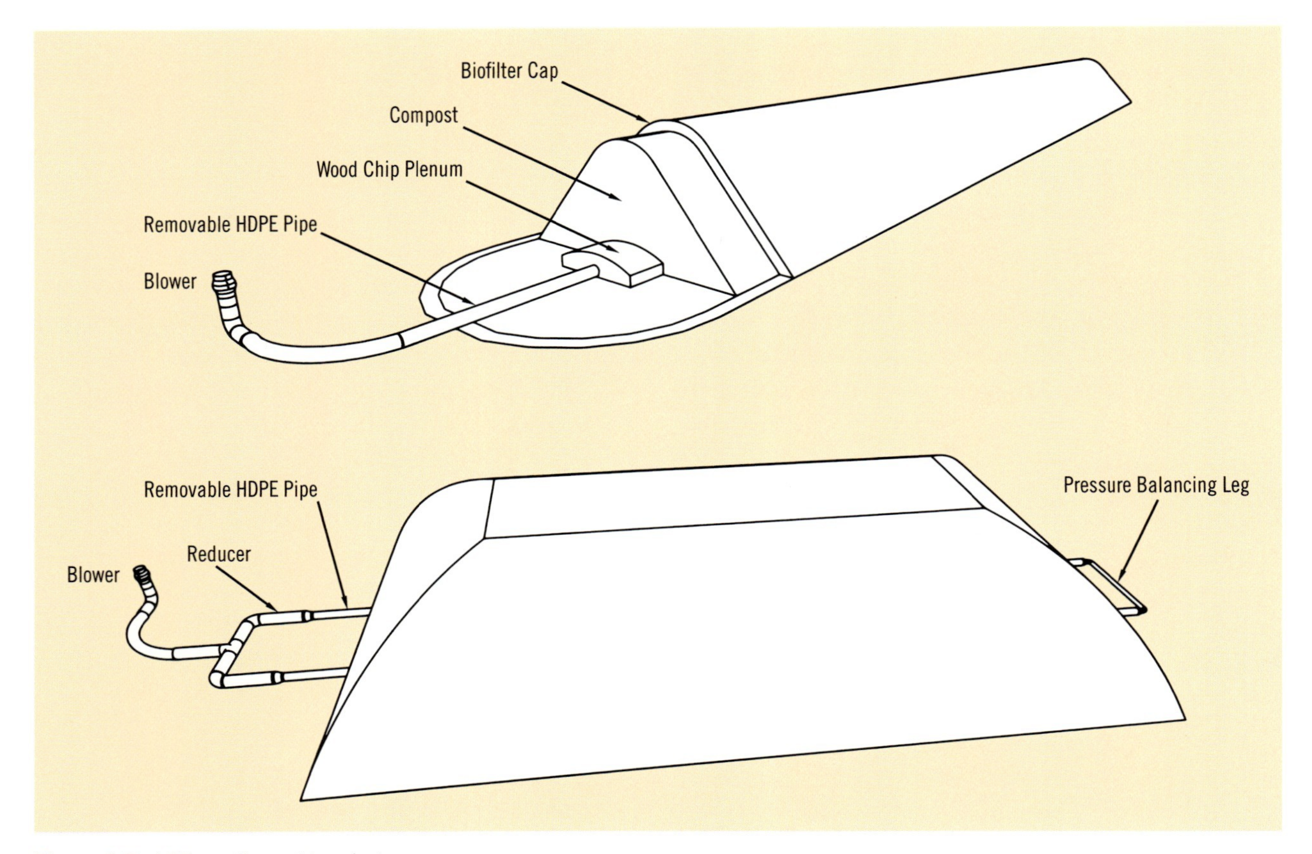

Figure 9.7. ASP configured to windrows.

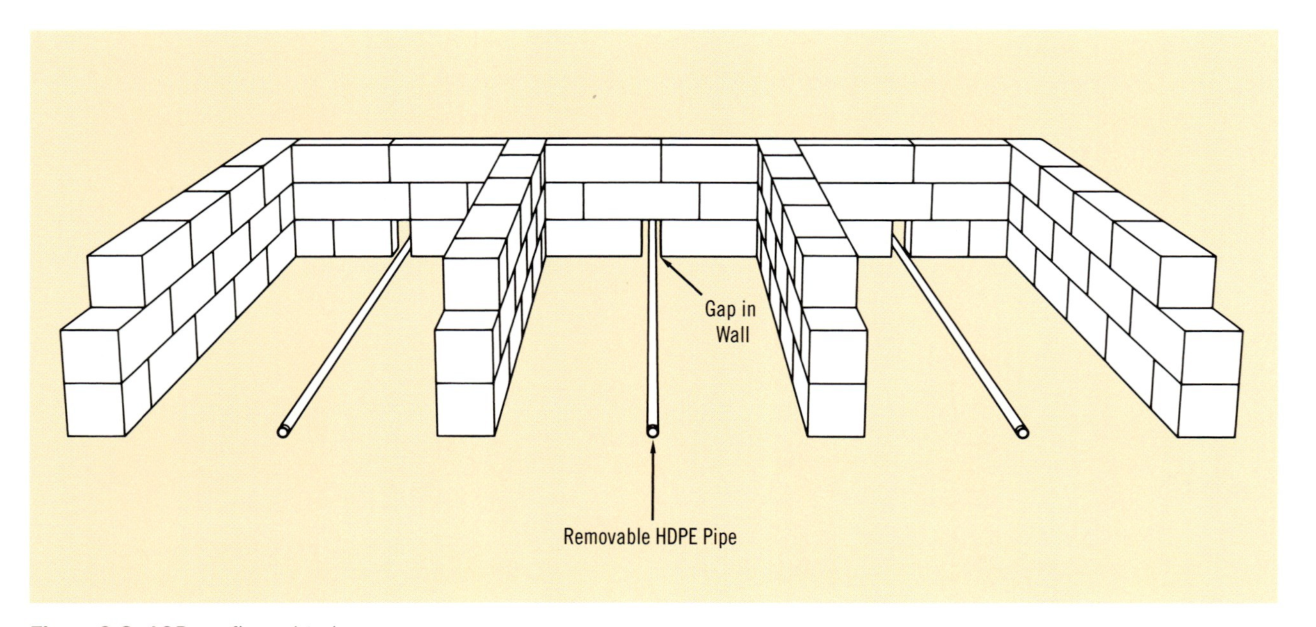

Figure 9.8. ASP configured to bays.

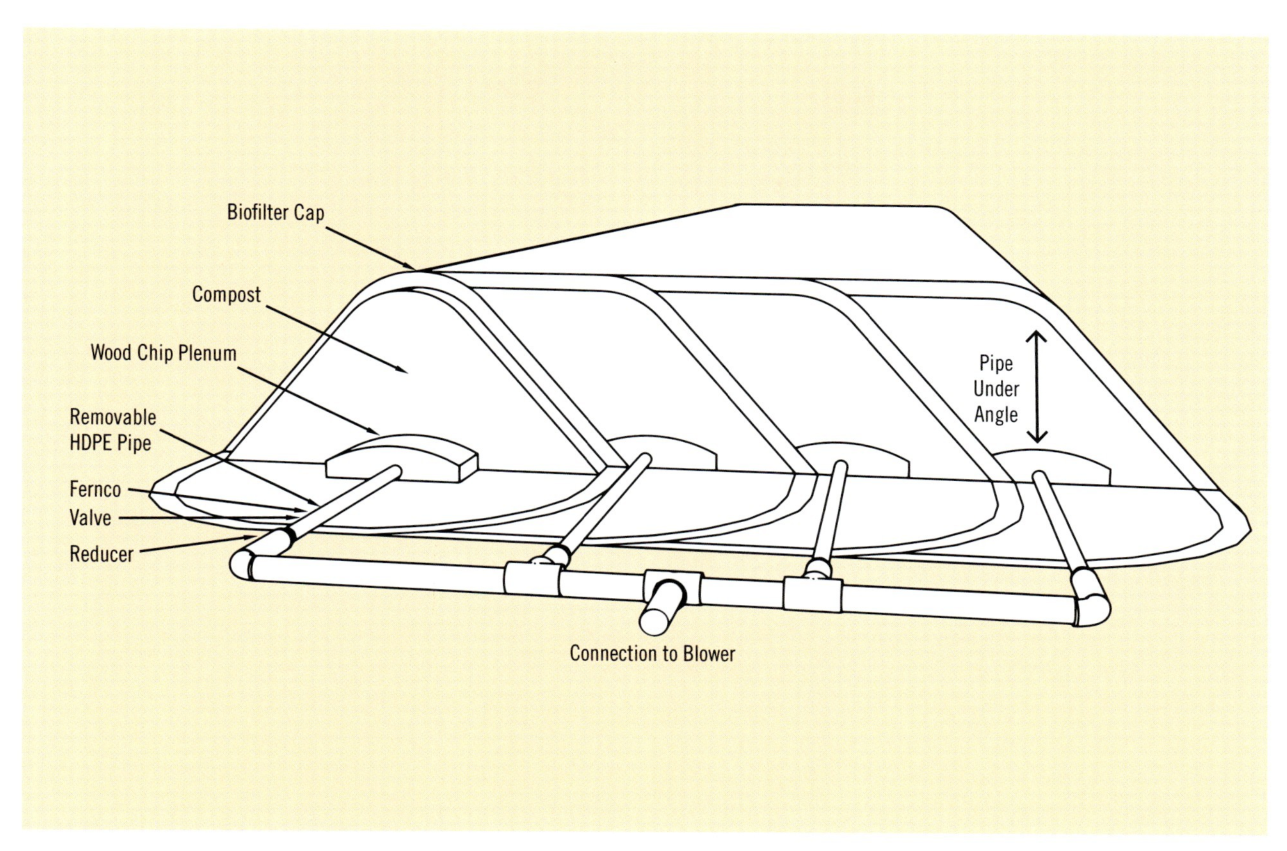

Figure 9.9. Extended ASP (EASP) configuration.

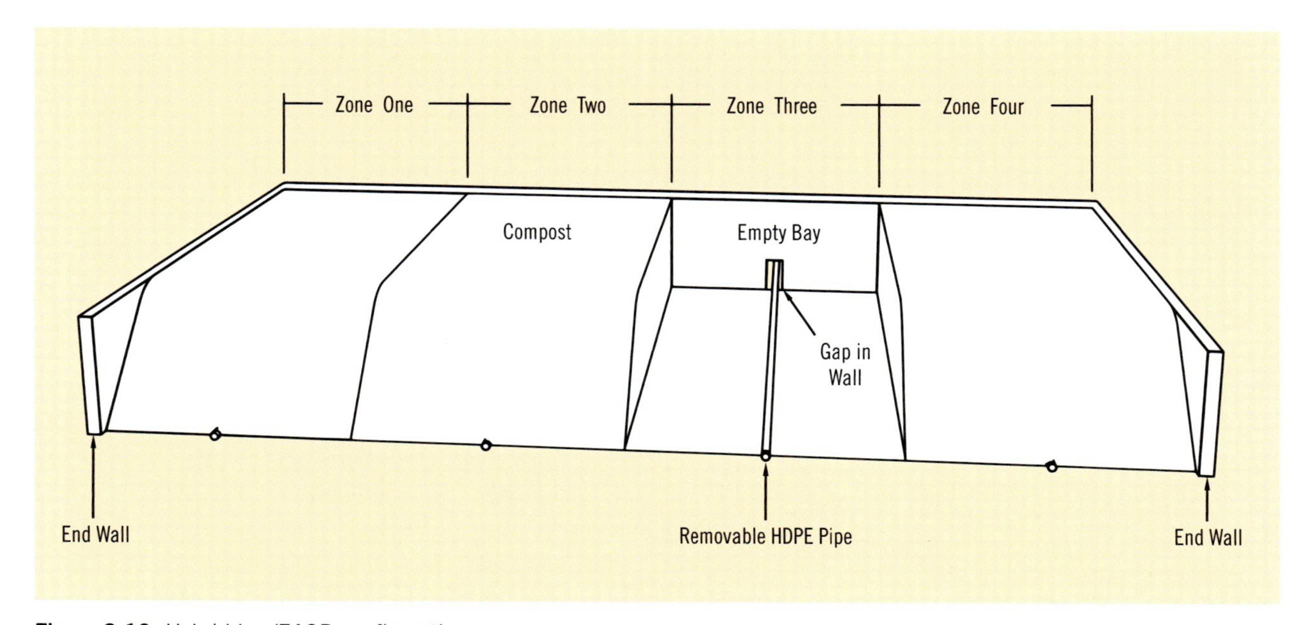

Figure 9.10. Hybrid bay/EASP configuration.

ASP Batch Size

With very few exceptions, ASP composting is a batch composting method. A compost pile is formed as a unit that is processed through the ASP system as a distinct and trackable mass. Typically the ASP system designates an "aeration zone" for each fresh batch of compost as it is created (with the exception of EASP). Plan the batch size for your target maximum processing capacity, remembering that if you only receive half of the target amount at first, you can simply take twice as long to create a batch.

From both a production and economic standpoint, you eventually want to hit a realistic sweet spot, where a batch fills an aeration zone once every one to two weeks. This should be your target batch size. Often there are regulations about how long food scraps can be left before they are incorporated into a compost blend; as a best management practice, this is less than 24 hours. That said, creating a compost mix and filling the ASP are two different things, and you should be able to receive food scraps, blend and cap them, then manage them as you would any other compost without removing the compost from the receiving bay, until enough material comes in to create an ASP batch. An ideal batch frequency at the community scale for ASP is usually weekly and no less frequently than every other week. Larger operations create daily batches, but this is uncommon at smaller scales. Once the frequency has been identified, you should be ready to calculate your batch volume.

Now we can use your operation's standard batch size to calculate pile dimensions.

Step 1. Calculate Batch Size

For step 1, refer to *Compost Site Processing Capacity and System Scale* in chapter 5 (page 114), then resume with step 2.*

* Note that methods for estimating a target throughput capacity are covered in depth in *Estimating Food Scrap Throughput*, chapter 5 (page 114). Refer to that section before designing an ASP system if you feel unclear about the operation's target inputs (feedstocks).

Example

Step 1. Calculate Batch Size

The example given in chapter 5 calculated 92 cubic yards per batch. I'll round up to 100 cubic yards per batch for planning purposes.

The Phases of ASP Composting and Batch Retention Time

The capacity of an operation to retain a batch of compost on site is called its *retention time*. Retention time is basically a unit of capacity and could refer to the whole site or to a particular phase of the composting process. Forced aeration is typically used in the *active phase* of composting, which for design purposes I like to subclassify into *primary* and *secondary phases*. These are when the most oxygen is required to maintain optimal microbial activity. After the primary phase—usually three to four weeks utilizing the ASP system—the compost has been heat-treated and demand for oxygen has been greatly reduced.

Following the primary phase, some operations then move the compost on to a secondary ASP phase, while others manage the subsequent phases in turned windrows until it matures to the point of curing and then storage. The addition of a secondary ASP phase will further reduce processing time, as well as the site's footprint compared with loader-turned piles. Keeping the compost actively aerated following the primary phase allows some operations to cut total production time to around three months using ASP. The design of systems for the secondary ASP phase usually accounts for volume reduction during the primary phase as well as reduced oxygen demand, and therefore reduced aeration rates.

The *ASP Benefits and Drawbacks* section (page 204) discusses the challenges of prolonged static periods, such as preferential air channeling and drying, in great depth. My hope is that one of the things you take away from this chapter is that there is a happy medium between turned and unturned, generally found right in that three- to four-week time period, although finding ways to efficiently turn even more often is probably advantageous in terms of the finished quality.

Step 2. Decide ASP Retention Time

2.1 Choose Primary ASP Retention Time
How long will a batch be processed in the primary ASP?

2.2 (Optional) Choose Secondary ASP Retention Time
How long will a batch be processed in the secondary ASP?

Keep the time period constant throughout all of your calculations, typically weeks or days for very large operations.

EXAMPLE
Four weeks. After four weeks each batch will be moved to the secondary ASP.

Eight weeks. After eight weeks each batch will be moved to the curing pad.

Primary and Secondary ASP Zones

Based on the batch frequency and the batch retention time in the primary ASP, you can now calculate the number of aeration zones needed to handle your target rate of incoming material. We'll refer to these as zones. The underlying principle is that the aeration within each zone can be controlled distinctly from other zones, and therefore each batch can be distinctly managed. A zone could be designed to aerate a bin, a bay, a container or vessel, a windrow, or really any specific portion of material. Calculating the number of ASP zones is a logical calculation: Divide the batch retention time by the batch frequency. For example, a batch retention time of four weeks in the ASP with a weekly batch frequency will require four ASP zones. The oldest batch gets emptied weekly in order for the new batch to get loaded weekly, creating a constant throughput that maximizes the value of the infrastructure to the operation.

Step 3. Calculate Number of ASP Zones

3.1 Calculate Number of Primary ASP Zones

__ Primary Batch Retention Time in __ Time Period
÷ __ Primary Batch Frequency in __ Time Period
= __ Primary ASP Zones

4 Weeks ÷ 1 Week(s) = 4

3.2 Calculate Number of Secondary ASP Zones

__ Secondary Batch Retention Time in __ Time Period
÷ __ Secondary Batch Frequency in __ Time Period
= __ Secondary ASP Zones

8 Weeks ÷ 2 Weeks = 4

Keep the time period constant throughout all of your calculations, typically weeks or days for very large operations.

Pile Dimensions

In order to lay out the operation and design the aeration system, you will need to calculate your pile/batch dimensions and then account for the number of batches in the ASP at any given time.

Step 4. Calculate System Dimensions

Since the pile formations are identical to other methods, refer to those sections as applicable to your design:

- Windrows (See *Core Concepts in System Scale* on page 122)
- Bins (See *Sizing Bins and Bays* on page 176)
- Containerized systems and bays (See *Sizing Bins and Bays* on page 176)

For EASP (see figure 9.9, page 217) the first pile is a windrow, then each extension has a trapezoidal shape, with the base width being half of the base of the original windrow, or approximately equal to the height of the pile.

If the ASP has a secondary phase, you will need to calculate those dimensions separately. Account for approximately 30 to 40 percent shrinkage after the typical four-week primary phase.

Aeration System Design

With the basic pile layout identified, now the aeration system can start to be sketched out. The core of the ASP design comprises three components:

- A fan, also called a blower
- A system of air ducts, often called the manifold
- Perforated ducts below the compost, sometimes called air or aeration channels

The Relationship Between Airflow and Static Pressure

When you're designing an ASP system, it is helpful to understand the dynamic relationships among air volume, airflow, and pressure loss. Picture a pipe with a fan connected at one end. The fan can move a set volume of air through the pipe at any given time. The volume of air that moves through the pipe in time is the *flow rate*. In the United States flow rate is typically represented in cubic feet per minute (CFM). The speed at which that air is traveling through the pipe is the *velocity*, and the velocity is typically represented in feet per minute (FPM). If you increased the flow rate through the same-size pipe, say by using a larger fan, what would happen to the velocity of air moving through the pipe?

It doesn't take a rocket scientist to know that more flow through the same-size pipe will increase the air's speed. What may not be entirely obvious, though, is that the increase in speed creates friction and turbulence inside the pipe, which causes less efficient flow. The friction and turbulence actually build pressure within the system, known as *friction loss*. Fans work harder when moving air through a system that has more friction and hence higher pressure. The smaller the pipe, the greater the pressure loss. Also, the longer the pipe run, the greater the friction loss (more surface area), which equates to greater pressure in the system. While more pressure might seem like a good thing in terms of creating uniform aeration, it's not if it's restricting efficient flow (CFMs).

In simple terms, the pressure within an ASP duct system is known as *static pressure* and is measured in inches of water column

(WC). Fans are rated to generate a specific CFM at a specific static pressure. As the static pressure in the system increases, the fan's performance decreases, eventually to a point of complete inefficiency.

To illustrate what efficient airflow looks like, let's go back to the fan-tube scenario from before. For example:

- 360 CFM of air moving through a 6-inch PVC duct will travel at a velocity of approximately 1,834 FPM. The pressure loss would be around 0.8 inch of WC per 100 feet of pipe. If the pipe is 50 feet long, the blower would need to move 360 CFM at 0.4 inch of WC.

Now double the airflow. Adding a larger blower, for example, increases the flow, which increases the velocity, which increases the static pressure loss.

- 720 CFM of air moving through a 6-inch PVC duct will travel at a velocity of approximately 3,667 FPM. The pressure loss would be around 2.82 inches of WC per 100 feet of pipe. If the pipe is 50 feet long, the blower would need to move 720 CFM at 1.41 inches of WC.

In this example, doubling the CFM more than tripled the static pressure. The 1.41 inches of WC alone would not be bad, but when you start to add additional sources of static pressure, you will see that 1.41 inches of WC from the pipe alone might be an unnecessary inefficiency. The pressure created by increasing the flow is similar to what happens with an ASP system. The friction of the pipe and of the compost itself, along with the perforations, pipe size reductions, and so on, create pressure loss within the system, constraining the flow.

At the simplest level, the task of designing the ASP aeration system involves (1) targeting the air supply to meet the specific demands of the compost; (2) sizing the air distribution system to supply that airflow efficiently; (3) estimating the pressure within the system at the target aeration rate; and (4) selecting a blower that can supply the target airflow at the static pressure estimated within the system.

As was previously stated, blowers are rated to deliver a specific airflow at a given static pressure. The relationship between airflow and static pressure is typically presented in what's known as a *fan curve*. As the static pressure increases, the CFM delivered by the fan decreases at a predictable rate, eventually to a point of real inefficiency. Actual static pressure in an ASP system can be measured using a *manometer*. Once the system is built, a static pressure measurement can be used to estimate the actual CFMs being delivered based on the fan curve, which in turn can help to calibrate fan run times (see *Testing ASP Design Performance* on page 261).

Designing for optimal efficiency within your ASP system can feel like a game of whack-a-mole, because every variable affects the others. The more complex the system, the more carefully you'll have to account for the dynamics among different parts of the system. Double-check, cross-check. In all honesty, it might feel like overkill, but I think that the process that I've fleshed out will help you design a system you can have confidence in. And by designing your aeration system, you'll hopefully understand the logic and assumptions that went into it, which is helpful later when it comes to system optimization.

These, along with electricity, shelter for the blower, and controls (such as a timer), are the only absolutely necessary elements of an ASP system. The following sections will guide you through designing an aeration system that is customized for your particular scenario. There are six steps to sizing the fans, ductwork, and air channels, starting with calculating target blower capacity. Sizing happens parallel to the actual design of the manifold and air channels, with the design informing the sizing and vice versa.

Target Blower Capacity

There has been a significant amount of research related to target aeration rates for forced air systems, although across the research and industry, the recommendations vary significantly. I have made a study of aeration rates and methodologies for sizing blowers and have ultimately developed a methodology for sizing ASP fans, being careful to cross-check it against other methods (see table 9.2, page 226).

The method that I use is similar to how you might size an energy recovery ventilator for a building, designing for an aeration capacity that will replace the air in a given space a prescribed number of times in an hour. When you look at a well-made food scrap compost pile, remember, you're looking at somewhere between one-third and one-half air. Based on an estimate of the percent *free air space* (*FAS*), it is easy to calculate the approximate volume of air in that pile. For example, the batch of compost in the example from step 1 was 100 cubic yards or essentially around 33 cubic yards (891 cubic feet) of free air space (at 33 percent FAS). To replace those gases

Positive and Negative Aeration

Forced air (ASP) systems either push air or pull air through compost. When pushing air, the system is called *positive aeration* (sometimes called *forced draft*). Pulling air is called *negative aeration* (*induced draft*). Some systems are reversible (*push/pull* or *balanced draft*), but this is less common. Negative aeration is used to capture gases from the composting process. These gases can then be filtered using a biofilter, and the heat, CO_2, and nitrogen compounds can be captured and repurposed.

There is broad agreement that positive aeration is simpler, and without a doubt it is less costly. Haug explains that velocity pressure loss works in favor of even airflow distribution across the air channel perforations with positive aeration.[5] Although I can't say that I understand the physics as he describes them, I can infer that positive aeration is more naturally suited to have more uniform airflow across the base of the pile (that is, more even flow through each individual perforation), which makes intuitive sense to me.

When deploying negative aeration, fans need to be suited to handle the corrosive exhaust, and the manifold pipes need to be sloped to drain and capture condensate. Positive air manifolds also need to drain, but negative systems create more condensate that is also corrosive. The manifold for negative aeration systems is ideally made out of HDPE, which is resistant to both high heat and corrosion. Due to cost, many people use PVC for negative aeration, but it will have a shorter life span than HDPE. Finally, negative aeration systems require a biofilter, typically with a condensate trap as well. Biofilters add to the footprint and cost of the site and add to the static pressure in the system. See *Designing Biofilters* on page 251.

with fresh air 20 times an hour would require a fan that could move 891 cubic feet every 3 minutes, or about 297 CFM blower at maximum capacity.

I call this the "volumetric air exchange" method, and it represents my spin on more technically complex methods that I find unapproachable as a non-engineer. It comes from my early research into ASP systems design, where I heard the concept of air exchanges mentioned. It is a concept that has stuck with me, because it puts airflow rates (CFM), fan timing, and pile volume all into a uniform specification. Most other methods are based on the weight of volatile solids, which is highly variable, even within the same operation, and also a hard sell to laypeople. Also, comparing the multiple methods and formulas that I've researched, like so many other things across the industry, is like comparing apples and oranges. I have done my best to translate the oranges into apples, in order to compare what one system does with another on relatively even terms. This is not to minimize the in-depth work that has been done in optimizing aeration rates, but simply to translate it for the lay practitioners who are ultimately carrying out much of the organics recycling at the community scale.

Interestingly, although Haug's methods were totally different, he does talk about the concept of "exchange volumes" in his book *The Practical Handbook of Compost Engineering*. (In hindsight this is probably the root of what initially brought this concept to my attention.) He actually describes a methodology for calculating the total number of air exchanges required to meet *stoichiometric oxygen demand* (the oxygen required during aerobic decomposition from start to finish). For a pile of mixed yard waste, he calculates 1,510 air exchanges would be required in total before oxygen demand is met.[6]

The methodology that I describe here is different, as it offers the number of air exchanges per hour, as opposed to in total, as a way to arrive at a rate of aeration. At a minimum that rate is designed to meet oxygen demand, but it also needs to remove excess heat during peak activity. As a thought experiment, consider how quickly different aeration rates would replace the air in a pile 1,510 times. At a rate of one air exchange per hour it would take 63 days. At the highest aeration rates that I propose, of say 30 to 40 air exchanges per hour, it would only take one and a half to two days, except that those rates are well above the rate at which microbes can utilize that oxygen, so from an oxygen demand perspective most of that air would be wasted. Coker and O'Neill suggest peak aeration rates that would (assuming free air space of 33 percent) equate to approximately 20 to 40 air exchanges per hour in days 1 through 10, 10 to 20 air exchanges per hour in days 10 through 20, and 3 to 10 air exchanges per hour in days 20 through 40.[7] These rates align very logically with what we know about oversupplying air for heat removal and the time line for meeting oxygen demand. (They also provide aeration rates in cubic feet air per yard of compost, which is just about the closest thing to English that I've found in the literature as far as aeration rates goes.)

As you will see in tables 9.1 and 9.2, there are low, medium, and high aeration rates that correlate with the various methodologies described in the literature. I have translated these different methodologies into estimated air exchanges per hour (AE/H) by pile volume, based on the assumption of 33 percent free air space.

Determining where your operation fits into the spectrum of potential aeration rates is a personal design choice, but I would strongly recommend that larger batches of compost be paired with a medium to high aeration capacity, because at a large scale it is critical to be able to remove enough heat to cool the pile during the peak of thermophilic activity. It is estimated that removing enough heat from the pile to cool it requires anywhere from 10 times[8] to as much as 20 or more times the amount than just to maintain oxygen in the pile.[9] Large ASP batches can reach extreme temperatures (over 180°F), which can set back the composting process significantly. At a smaller scale, simply meeting oxygen demand may be acceptable, but those using lower aeration rates may find that when a pile gets hotter than ideal (more than 160°F) the fan does not have the capacity to fully cool the pile. If it's a small pile this will probably only be a problem during the hottest times of year. This may be an acceptable trade-off for the lower start-up and operating costs of a smaller system.

There will also be times when low aeration rates are preferable, such as during cold spells or compost maturation, or if the system is recovering heat. Remember that these calculations are based on aeration during peak demand (when the fan is constantly on at full blast). Simply reducing the on-time on the timer or reducing the fan speed, if your system has that capacity, can dramatically reduce the aeration rate.

Based on your batch size, refer to table 9.1 to identify a target blower capacity. The actual fan you use should be in this general range, ideally within 10 percent of the target CFM. Note that the estimated fan CFM provided by the fan's manufacturer is relative to the estimated pressure loss in the system. Pressure loss is typically represented in inches of water column (for example, 3″ WC), and fan manufacturers provide a fan curve that shows the estimated CFM of a fan at a given pressure.

If you buy an ASP system from a vendor, they will have their own fan sizing methodology. You can ask them what CFM their fan typically operates at for a given volume of compost and compare that to the sizing information in tables 9.1 and 9.2 to get a sense of its capacity relative to other systems.

Table 9.1. Volumetric Air Exchange Blower Sizing

System Scale	Low Aeration Rate		Medium Aeration Rate		High Aeration Rate	
Batch Volume (Cu. Yd.)	**CFM@5 AE/H**	**CFM@10 AE/H**	**CFM@15 AE/H**	**CFM@20 AE/H**	**CFM@30 AE/H**	**CFM@40 AE/H**
5	4	7	11	15	22	30
10	7	15	22	30	45	59
20	15	30	45	59	89	119
30	22	45	67	89	134	178
40	30	59	89	119	178	238
50	37	74	111	149	223	297
60	45	89	134	178	267	356
70	52	104	156	208	312	416
80	59	119	178	238	356	475
90	67	134	200	267	401	535
100	74	149	223	297	446	594
120	89	178	267	356	535	713
150	111	223	334	446	668	891
200	149	297	446	594	891	1,188
250	186	371	557	743	1,114	1,485
300	223	446	668	891	1,337	1,782
350	260	520	780	1,040	1,559	2,079
400	297	594	891	1,188	1,782	2,376
450	334	668	1,002	1,337	2,005	2,673
500	371	743	1,114	1,485	2,228	2,970

Note: CFM = cubic feet per minute, AE/H = air exchanges per hour

Source: Adapted courtesy of Compost Technical Services, LLC

"Volumetric Air Exchange" Blower Sizing Methodology

Table 9.1 provides a reference for fan sizing based on the "volumetric air exchange" methodology utilized by Compost Technical Services. The example below shows the math behind fan sizing for the 100-cubic-yard batch size from step 1, following this methodology.

2,700 Cu. Ft. Blended Compost per ASP Batch × 0.33% Estimated Pore Space (Multiply by 0.33)	=	891 Cu. Ft. Estimated Pore Space (FAS)
60 Minutes per Hour ÷ 20 Target Air Exchanges per Hour	=	3 Minutes per Air Exchange
891 Cu. Ft. Estimated Pore Space (FAS) ÷ 3 Minutes per Air Exchange	=	297 Cubic Feet per Minute (CFM) Blower Capacity

Step 5. Calculate Target Blower Capacity

To estimate the target blower capacity needed to aerate the batch of compost calculated in step 1, use table 9.1. All aeration rates assume 100 percent fan on-time at maximum fan speed—in other words, the fan's capacity during peak demand. Sizing the fan in this way is a precaution for extreme overheating situations. Following the period of peak air demand, reduce the aeration rate by decreasing the fan on-time and/or the fan speed.

5.1 Locate Batch Volume

On the far left column of table 9.1, find a batch volume closest to your own batch size from step 1. Note that if you are using one fan to aerate more than one batch, the fan will need to be sized for the batches' combined volume.

5.2 Choose Target Blower Capacity

On the top row of table 9.1, choose a low, medium, or high aeration rate. This can be a range.

5.3 Find Corresponding Target Blower Capacity

Find the estimated blower capacity that corresponds with your batch size and AE/H, given in cubic feet per minute (CFM). In step 6 you will select an actual blower that will fall within that range, based upon the estimated static pressure in the system.

EXAMPLE
for 100-Cubic-Yard Windrow

100 cubic yards per batch.

Medium aeration rate of 20 AE/H.

297 CFM target blower capacity.

Table 9.2. Fan Sizing Methodology Comparison

Methodology	Yards of Material	Blower CFM	Air Exchanges per Hour	Notes
***OFCH*—Fan On 100%**	100	117	8 (low)	Based on 233 lbs. of volatile solids per yard
***OFCH*—Fan On 50% / Off 50%**	100	198	13 (med)	Based on 233 lbs. of volatile solids per yard
***OFCH*—Fan On 33.3% / Off 66.7%**	100	291	20 (med)	Based on 233 lbs. of volatile solids per yard
Engineered System	100	600	40 (high)	Converted from 35–40 cubic meters per hour per square meter for a 3-meter-high pile
Practical Handbook of Compost Engineering	100	432	29 (high)	Assumes a 2:1 mix of biosolids to wood chips
Composting Aeration Floor Functions and Designs	100	300–600	20–40 (med–high)	Peak blower capacity in days 1–10. Aeration rate of 3–6 cu. ft. per cu. yd.
Composting Aeration Floor Functions and Designs	100	150–300	10–20 (low–med)	Peak blower capacity in days 10–20. Aeration rate of 1.5–3 cu. ft. per cu. yd.
Composting Aeration Floor Functions and Designs	100	50–150	3–10 (low)	Peak blower capacity in days 20–40. Aeration rate of 0.5–1.5 cu. ft. per cu. yd.
Wales Research Study: Oxygen Demand	100	89	6 (low)	Converted from: 5.5×10^{-4} m^3 s^{-1} of air supplied for every m^3 of waste. Assumes an organic waste with a density of 500 kg per m^3, composting at the rate of 100 g CO_2 per kgVS per day.

Sources: Craig Coker and Tim O'Neill, "Composting Aeration Floor Functions and Designs," *BioCycle* 58, no. 6 (2017): 28; Roger Tim Haug, *The Practical Handbook of Compost Engineering* (Boca Raton, FL: Lewis Publishers, 1993); Robert Rynk, ed., *On-Farm Composting Handbook* (Ithaca, NY: Natural Resource, Agriculture, and Engineering Service, 1992); G. Hewings, T. Griffiths, and K. Williams, "Calculating Airflow Requirements for Forced Aerated Composting Systems," presentation at the 22nd Solid Waste Technology and Management Conference, Philadelphia, March 17–21, 2007; Michael Bryan-Brown and Jeff Gage, "Lessons Learned in Aerated Static Pile (ASP) Composting," presentation at the 20th Annual National Compost Conference, Ottawa, September 22–24, 2010.

Designing and Sizing the Aeration Manifold

The ASP manifold starts with a header duct pipe that connects directly to the blower.

From the header, the manifold may take a number of different forms. No matter how simple or complex the configuration, the ducts should provide a fluid path of travel from blower to compost. To achieve this, all of the ducts need to be adequately sized for your estimated airflow. Splits, reductions, and perforations need to supply uniform pressure and airflow across the base of the compost.

A rough layout of the manifold is needed in order to size the ducts, so now's the time to start putting lines and measurements on paper. Steps 6–9 walk through the sizing of the ductwork and selection of the blower based upon your target aeration rate, but they do not specifically spell out the design process. See the *Manifold Design Rules* sidebar, page 231, and review figures 9.7 through 9.12 for some common manifold configurations, and when in doubt keep the design simple. Other design factors and recommendations are covered throughout the text going forward.

Header and Manifold Ducts

A number of variables play into the design of ASP ducts and manifolds, and as with other parts of ASP

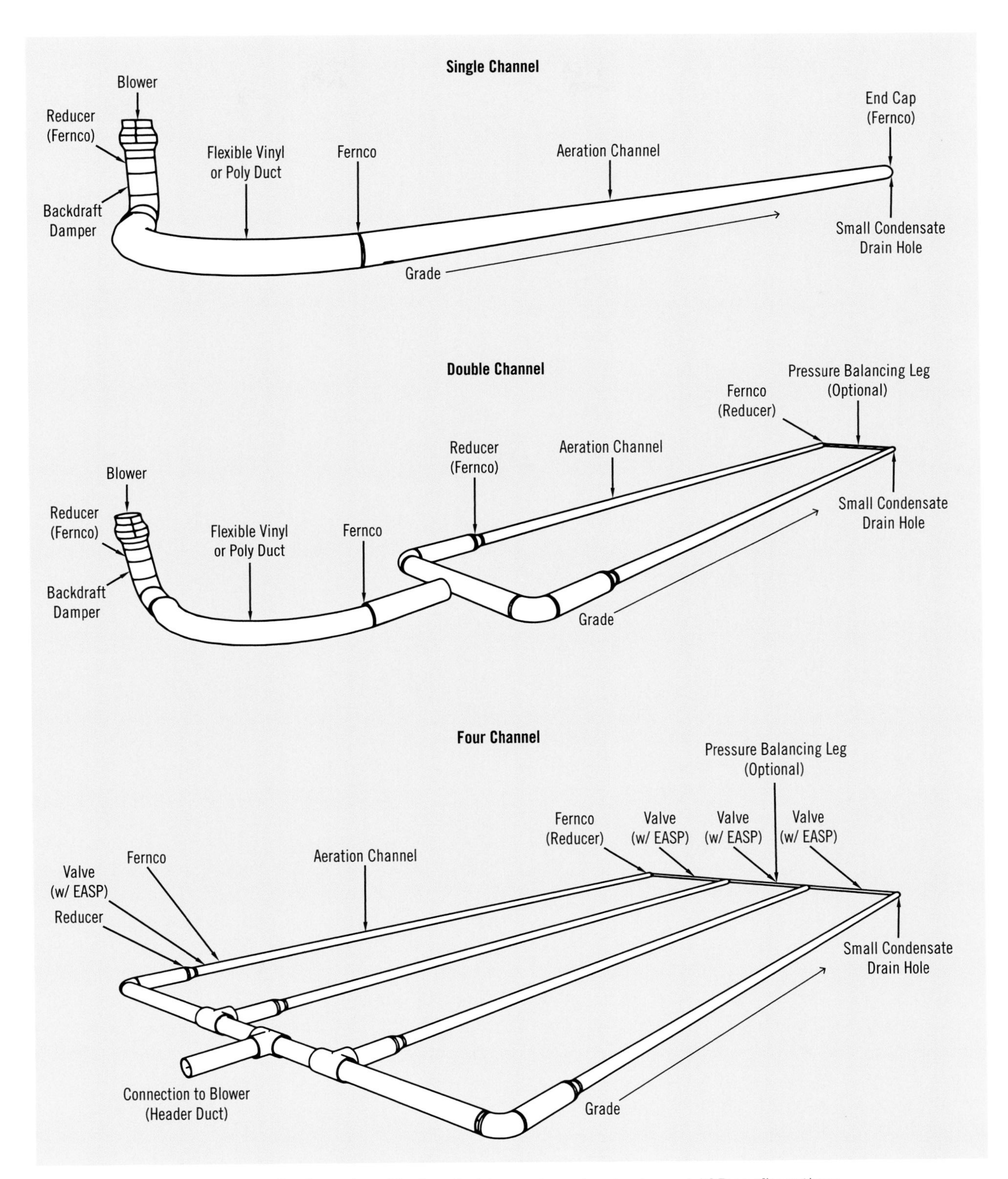

Figure 9.11. Aeration Manifold Configuration. Single-, double-, and quadruple-channel ASP configurations.

system design, there are important correlations among the different components. Step 6 walks you through the sizing of the manifold ducts.

You will need a rough design, as well as the target blower capacity identified in step 5. Designing in order from the blower to the compost, we'll call the first pipe in the system the *duct header*. Sizing this pipe is very straightforward. To size both round and square/rectangular ducts, refer to tables 9.6 and 9.7. Alternatively, you can use online calculators or apps; I use an app on my iPhone called Duct Sizer. These tools tell you the velocity and static pressure loss of a specific flow rate (CFM) of air through a specific size duct.

To minimize pressure loss for smaller systems, I typically target duct airflow velocity to within a few hundred feet per minute (FPM) of 1,000 FPM. Some sources recommend maximum velocities as high as 2,800 to 3,000 FPM.[10] As you will see in tables 9.6 and 9.7, the pressure loss increases dramatically as duct diameter decreases (or conversely as airflow increases). Unless it is a large system, velocities between 2,000 and 3,000 are inefficient and should be avoided if possible.

Manifold Materials

Throughout the composting industry are ASP manifolds designed with a wide array of materials; just about any type of duct that an HVAC system can be built with, ASPs are also built with. You also see materials that would be common in agriculture, such as PVC. Hybrid combinations are also common.

Cost is a primary consideration in material choice, and one of the greatest factors in material cost is the size of the manifold. Materials such as PVC and galvanized ducts are comparable in cost at pipe diameters of 4 inches or less, but PVC goes way up after 6 inches compared with galvanized and flexible ducts. Pipe fittings and valves go up in price especially fast at larger diameters. Your choice of materials will likely be different if you are negatively aerating your compost, as the gases that are pulled out of compost piles are extremely hot, wet, and corrosive, and the ducts need to be rated for these types of conditions in order to last.

For positively aerated systems, I have typically used SDR-35 sewer pipe (non-perforated) for the ducts/manifold. It is a lightweight, low-cost PVC product that is widely available and surprisingly durable; but beware, it is not compatible with your typical Schedule 40 PVC without an adapter coupling. Rubber Fernco fittings, or similar couplings, are practical for connecting the blower, air channels, and any components that may be periodically decoupled. Ferncos are widely available, are flexible, and quickly attach/detach. They also come as reducer couplers.

When designing a manifold it helps to have some knowledge of plumbing/ductwork, but it's not absolutely necessary. The best thing to do if you are having a hard time picturing how all of the parts fit together is to go to a store that carries large plumbing components and get familiar with the different components and how they fit together. There are also large online suppliers that may have more components than the average store, but if you're unfamiliar with which parts are compatible, you might feel lost initially starting online.

Table 9.3 lists common air manifold materials and their pros and cons.

Table 9.3. ASP Aeration Manifold Material Choices

Manifold Duct Material	Pros	Cons	Notes
PVC (polyvinyl chloride)	Widely available Good for collecting condensate	High cost at larger diameters Can be challenging to work with (especially at large diameters) Not rated for above 140°F (works but becomes brittle) Expensive to insulate	SDR-35 PVC is thinner and cheaper than Schedule-40 PVC, and it's quite durable. It is my choice for PVC duct, but if SDR-35 fittings are not easily available in your area, Schedule-40 may be a better option.
CPVC (chlorinated polyvinyl chloride)	Designed for high temps up to 200°F	More expensive than PVC Not as widely available as PVC Expensive to insulate	
PE (polyethylene)	Corrosion-resistant / non-reactive with compost exhaust Good for negative aeration	Heat-welded Not widely available Challenging to work with Expensive to insulate	Haug recommends this in *The Practical Handbook of Compost Engineering*, but I have never used it or seen it in use.
HDPE (high-density polyethylene)	Corrosion-resistant / non-reactive with compost exhaust Good for negative aeration Most rugged pipe	Heat-welded Not widely available Often comes in 40-foot lengths Challenging to work with Expensive Expensive to insulate	This is the recommended material for creating aboveground removable air channels. Can be easily attached with Fernco fittings rather than heat-welded. Make sure to get straight sticks (not the flexible or corrugated kind).
Galvanized duct	Lower cost at larger diameters Widely available Reasonably easy to work with	Not recommended for negative aeration Expensive to insulate	Typically connected with small metal screws, then sealed with duct tape or aluminum tape. Can be sealed with cement as well.
Flexible duct	Very low cost Available insulated at low cost Easy to install	Not designed for draining condensate (because it's not rigid and therefore condensate will pool)	While this is by far the cheapest insulated duct option, unfortunately it's not great for negative aeration where you would most likely want insulated duct. It could work for vertical or near-vertical runs, where condensate can collect and cause the pile to sag or clog. More rugged flexible duct may work just fine in some negative applications.

Although some ASP systems consist simply of one header feeding a single air channel, often the header duct splits into several ducts leading to multiple air channels, which together make up the "manifold." We'll call these the *distribution ducts*. Depending upon how the system is controlled, these distribution ducts may flow to the same or different aeration zones. When the flow from a single header duct gets split between several distribution ducts, then the size of the ducts can be reduced to save cost and maintain velocity. The flow rate of the single header duct is comparable to the sum of the flow rates through all of the distribution ducts that are connected to the header duct, assuming that the system is designed for even distribution. In other words, 300 CFM moving through one pipe that splits evenly into 4 ducts comes out to 300 CFM ÷ 4 ducts, or 75 CFM per duct. These ducts should be sized in the same manner as the header duct, using the estimated flow rate of an individual distribution duct and tables 9.6 and 9.7 (pages 242 and 243) to find the duct size that is ideal for your scenario.

Figure 9.12. Deep-formed concrete air channels in a Phase 1 ASP bay at Green Mountain Compost in Williston, Vermont. The perforated 4-inch HDPE pipe inside the channel provides uniform airflow and pressure distribution throughout the system and can be removed to clean out the channel as needed. There is a separate leachate trap at the end of each air channel that connects to a tank.

Air Channels and Pipes

Air channel is the term that I use to describe the perforated pipe/duct that is located beneath the compost (it was the preferred term of my technical mentor, Tom Gilbert). Air channels should be sized by the same method as ducts, although they may differ somewhat from the distribution ducts; for example, the distribution ducts might be round pipes, going to rectangular concrete air channels. The type of air channels you use needs to be compatible with the material you use for the pad or floor of your ASP. The *Air Channel Design Rules* sidebar, page 234, lists critical rules to follow when designing for uniform pressure and airflow across the base of the compost. This section looks at the most commonly used forms of air channels.

Figure 9.13. Plastic grating covers formed concrete air channels in a Phase 1 ASP bay at Green Mountain Compost in Williston, Vermont. The grating does not provide back pressure (see figure 9.12).

Formed Concrete Channels

ASP systems that have air channels formed below grade, typically in concrete, are permanently installed and out of the way of loader operation.

Manifold Design Rules

Follow these rules to maximize airflow efficiency and to create uniform distribution using the ASP method.

1. The flow rate (CFM) of a single header duct is comparable to the sum of the flow rates through all of the distribution ducts that connect to the header duct, assuming that the system is designed for even distribution.
2. Header ducts that split to multiple distribution ducts should be centered in between those distribution ducts to avoid preferential flow. For this reason, an *even* number of distribution ducts is ideal.
3. Size all ducts to minimize static pressure by keeping air velocity below 2,800 feet per minute and ideally around 1,000 FPM. This will keep friction and pressure loss at a reasonable level, which is critical to efficient aeration and energy use (see step 6 and tables 9.6 and 9.7).
4. The header duct immediately following the blower should be straight for a length three times or more times the diameter of the blower outlet. This is to allow adequate velocity to accumulate.
5. Following any transition in pipe diameter or direction (such as a reducer or elbow), the pipe should be straight for a minimum of three times the diameter of the pipe.[11]
6. Slope ducts slightly (2 percent or more) to shed condensate to low points where moisture can be trapped and removed.

These are typically viewed as the top of the line as far as design goes. There are two main variations: One utilizes a perforated pipe set into the channel, which all but guarantees uniform pressure and distribution of airflow from the pipe; the other is simply a formed concrete duct. With a pipe in the channel, the channel cover can be very porous, as it is not responsible for pressurizing the system, whereas without the pipe, the channel cover must be adequately perforated (not overperforated) in order to pressurize the system. With either design, leachate can be collected into the air channels, then piped on to the collection and management system. These air channel design options are shown in figures 9.13 and 9.23.

The main drawback with formed concrete channels is the labor involved in cleaning them and in managing collected leachate. One operator who has both belowground and pipe-on-grade channels told me that the labor involved in cleaning the belowground channels outweighed the value of them being out of the way in his mind. Also, these are considerably more expensive than pipe-on-grade systems.

Gravel Pad/Gravel Air Channel Floor

One of the air channel designs in which I see promise, but have yet to try, is to form individual concrete channels into a packed gravel pad. Gravel is a semi-pervious surface, so it has the benefit of infiltrating a significant volume of leachate. A concrete pad, on the other hand, is impervious, so it concentrates leachate wherever it collects and then that leachate needs to be managed.

The benefit with this concept is that with the channel formed below grade into the gravel pad, only a very small percentage of leachate will end up in the air channel. The channel itself could have a coarse gravel bottom, so that it infiltrates liquids, and/or an absorbent layer such as wood shavings could be

Figure 9.14. Using metal for channel covers is a common mistake. Most metals corrode in the wet and acidic composting environment. These were also too heavy to be functional given how often they need to be removed.

spread on the bottom of the channel to absorb any liquid that does collect.

Full disclosure: I have yet to test this design in practice, so I can't vouch for its 100 percent effectiveness; it may very well have some drawbacks. The concept came about as the result of working on several other much more costly systems and searching to find a lower-cost alternative that would combine the operational efficiencies of both the higher- and lower-cost systems. The concept removes two of the most expensive components of ASPs—the full concrete pad and the leachate collection system—and is more efficient to work with than pipe-on-grade channels, which need to be worked around and removed. There would be added operational efficiencies, because leachate management can be expensive and time consuming.

On clay soils the channels may be more likely to heave, so creating a foundation that has good drainage would be advisable. Sites with a high water table or bedrock should choose a different system or raise the pad to have at least a 3-foot buffer in order to avoid groundwater contamination.

Air Channel Covers

With the use of formed air channels, you'll need to choose a material to use as channel covers. Rot-resistant wood (such as hemlock or cedar) and plastic grating are two good options. The choice will depend on whether the channel cover itself provides the back pressure, as with the wooden covers, or if a pipe inside the channel provides the back pressure as with the plastic grating. I've seen systems designed with a pipe in the channel, then gravel filled to level with the pad surface, but this is not a preferable option in my opinion, because it would clog over time and be laborious to clean. Whatever material you choose, make sure the channel cover sits below the lip of the concrete, so that the loader's bucket doesn't catch and destroy it.

Removable Air Pipes (Pipe-on-Grade)

When composters want to save on the upfront costs of a belowground system, or are first piloting ASP systems, they will often use aboveground air channels, also called pipe-on-grade or POG. These are typically pipes that have been perforated, then laid with perforations at 5 and 7 o'clock and covered with a wood chip plenum (see *Creating the Wood Chip Base or Plenum* on page 265). The compost pile is then constructed over the plenum. Removing the pipes can happen before or after the compost is removed.

POG is a common method for extended ASP systems (EASP systems).

HDPE PIPE

High-density polyethylene, or HDPE, is the pipe of choice for aboveground ASP air channels. It is extremely rugged and non-reactive with the corrosive compounds released during the composting process.* For this reason it can be used over and over

* Note that this is not the flexible HDPE used for drains. The stuff you want comes in a stick (it's rigid), like PVC, and is not pre-perforated or corregated.

Figure 9.15. A pulling eye or towing head (*left*) is a tool that fits into the end of a thick-gauge pipe and allows a tractor or winch to pull the pipe from beneath a pile without first removing the compost. A pipe chain wrench is used to grab and hold pipes as they are moved (*right*).

Figure 9.16. A heat-welded HDPE cap with a U-bolt embedded for pulling the pipe.

without much damage. Heavy-walled HDPE (DR 17 or 11) is recommended.[12]

The drawbacks of HDPE pipe are that it can sometimes be hard to source, is expensive compared with PVC, typically only comes in 40-foot lengths, and is heat-welded (as opposed to glued). Most systems will not need lengths greater than 40 feet for air channels, and Fernco fittings (flexible rubber) work just fine for connecting the air channels to the manifold, so you will be able to avoid the heat-welding process.

PVC PIPE

Aboveground ASP air channels made out of PVC are another common option. Since PVC comes in various thicknesses (SDR-35, Schedule 40, Schedule 80, etc.), the more solid the PVC gauge, the longer it should last, at least theoretically. But heat and the chemistry of the compost degrade the PVC, making it even more brittle than it already is. For this reason, PVC is generally considered a "throwaway" material, as far as air channels go. People who are extra careful with it might make it last more than one cycle (for hand-operated systems it's fine, for example), but PVC is not well suited to applications where it comes into direct contact with both compost and equipment.

Figure 9.17. Aboveground HDPE in a second-phase ASP. The pipes fit through small gaps in the back push wall, where they connect to the aeration manifold.

Figure 9.18. While PVC is not ideally suited for use with compost, for small manually operated systems, it's common and does just fine.

Air Channel Design Rules

Follow these rules to maximize airflow efficiency and to create uniform distribution using the ASP method.

- Shorter air channels maintain pressure efficiently. Channels have a maximum efficient length of 50 to 75 feet before excessive energy is required to maintain pressure.[13]
- Center the channel, or channels, to the base of the pile.
- A channel spacing of 4 feet apart is standard.
- Keep a minimum of 3 to 4 feet between the air channel and any side walls. Walls can become the site of preferential airflow (this is known as the coanda effect, or the tendency of a fluid to attach to a surface).[14]
- The channel perforations should end an equal distance from the end of the pile as the pile's approximate height. Ideally, no edge of the pile should be closer to the air channel perforations than any other; otherwise, there will be a short circuit in pressure, and air will escape at that point, creating uneven airflow.
- The perforations in an air channel should be sized correctly based on the length and cross-sectional area of the air channel. This is critical to providing uniform airflow along the length of the channel (see tables 9.4 and 9.5, pages 236 and 237 respectively).
- Size air channels to minimize static pressure by keeping air velocity below 2,800 feet per minute[15] and ideally around 1,000 FPM. This will keep friction and pressure loss low, which is critical to efficient aeration and energy use (see step 6 and tables 9.6 and 9.7, pages 242 and 243 respectively).
- If possible, connect the ends of air channels to balance pressure and flow between air channels (see *Air Channel Balancing Legs*, page 236).
- Plan for constructing a wood chip plenum each time a pile is constructed (see *Creating the Wood Chip Base or Plenum*, page 265).
- Design for compost and leachate interactions, so that pipes don't get clogged with liquids and debris (see *Leachate Collection and Management*, page 236). Channels should be sloped to the leachate/condensate removal system.
- Channels should be accessible for cleanout; this means removable covers, end caps, and so on.

Valves and Dampers

There are places in ASP design where duct valves and dampers might be useful. For example, I often recommend putting a back draft damper directly after the fan, to prevent vapors from migrating up the tube and into the fan when the fan is off. The force of the fan opens the damper when it's on, and in off mode it closes automatically. Other simple design instances include manual valves that can be opened or closed to open up an aeration zone or adjust airflow to a specific area. Sometimes these valves are automated, which is an area where I do not have a lot of design experience. From what I've seen, pneumatically (air pressure) controlled gate valves are a preferable choice for automated valves and dampers. Automated control systems are available from several vendors. Manually controlled gate valves are also common, especially for larger-size

Air Channel Perforations

The size and spacing of perforations in the air channels depend upon the channel size and length. There is contradictory information on the exact design basis for air channel perforations in the composting literature. The *On-Farm Composting Handbook* (*OFCH*) recommends that the combined area of perforations be twice the pipe's cross-sectional area.[16] The common recommendation I have used is that the combined area of perforations in an air channel should be slightly less than the cross-sectional area of the air channel, which is more conservative in terms of maintaining pressure. I have used this with good success in the early systems I designed and it has not led to excessive pressure in the system.

More recently, I have used calculations for creating a simplified sparger described by Hartsock, Crotteau, and Gage.[17] Tables 9.4 and 9.5 provide perforation sizes adapted from those calculations. Perforations assume 12-inch spacing based on your air channel diameter (round pipes) or cross-sectional area (rectangular channels) and length of the perforated zone. Refer to step 6 if you are unsure about your air channel sizing. Round up in terms of pipe length as needed to avoid oversized perforations.

Typically the perforations are drilled into the channel in a zigzag pattern for better distribution and to maintain the strength of the channel. For aeration pipes, *OFCH* recommends drilling them on the bottom of the pipes at 5 and 7 o'clock.[18] Refer to the *Air Channel Design Rules* sidebar, for guidance in designing to maintain uniform air distribution and to avoid preferential air channeling.

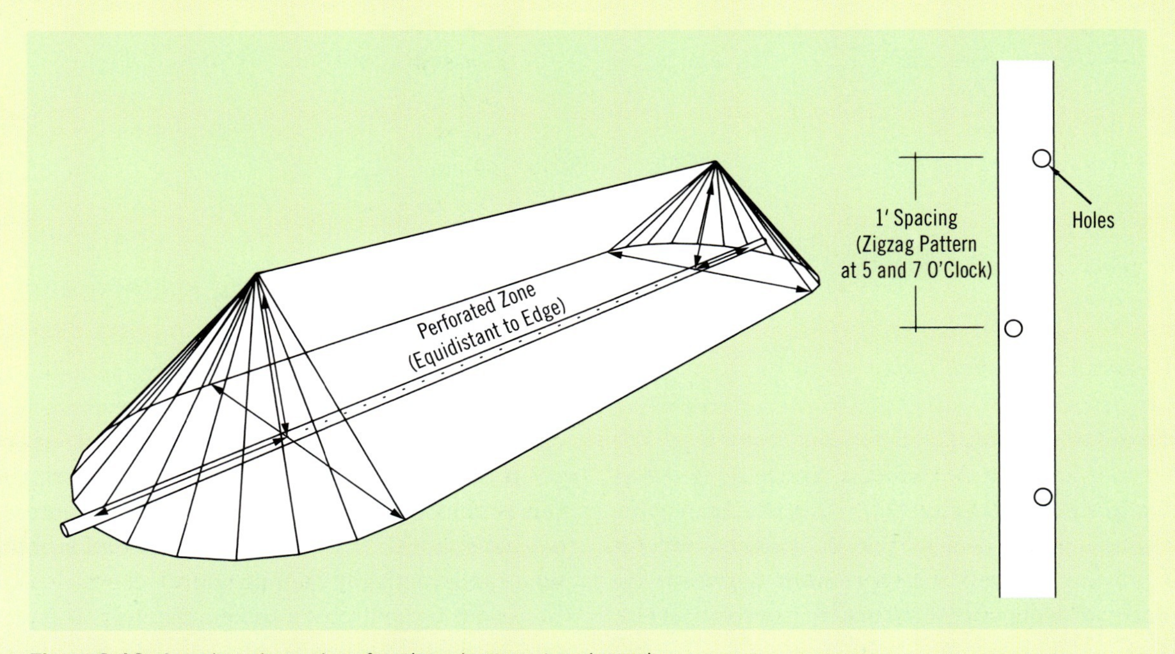

Figure 9.19. Aeration channel perforation placement and spacing.

Table 9.4. Maximum Perforation Size (≤ Inches) for Round Pipe Air Channels at 12-Inch Spacing

Length of Perforated Zone	Pipe Diameter				
	2"	3"	4"	6"	8"
2′	4/5"	n/r	n/r	n/r	n/r
4′	5/8"	1"	n/r	n/r	n/r
6′	5/9"	4/5"	1"	n/r	n/r
8′	4/9"	5/7"	1"	n/r	n/r
10′	3/7"	5/8"	5/6"	n/r	n/r
15′	n/r	1/2"	5/7"	1"	n/r
20′	n/r	4/9"	3/5"	8/9"	n/r
25′	n/r	2/5"	5/9"	5/6"	n/r
30′	n/r	n/r	1/2"	3/4"	1"
40′	n/r	n/r	4/9"	2/3"	7/8"
50′	n/r	n/r	2/5"	3/5"	7/9"
60′	n/r	n/r	n/r	5/9"	5/7"
70′	n/r	n/r	n/r	1/2"	2/3"

Note: Perforation diameters were calculated using the Knaebel simplified sparger design calculations described in Douglas R. Hartsock, Gerry Croteau, and Jeff Gage, "Uniform Aeration of Compost Media," presented at the National Waste Processing Conference Proceedings ASME, 1994: 217.

Perforation sizes may not correspond perfectly to available drill bit sizes. If this is the case, find the closest drill bit size available without going over the recommended maximum size.

n/r = not recommended

pipes, where ball valves get extremely expensive and hard to open and close.

Air Channel Balancing Legs

One logical design feature that I often recommend is the aeration channel *balancing leg* (from *The Practical Handbook of Compost Engineering*).[19] As shown in figure 9.11 (page 227), this is a small pipe that connects the ends of the air channels for the sole purpose of balancing pressure between the channels. For aboveground pipes these also act as spacers. Balancing legs do not need to be the full diameter of the air channel, because there shouldn't be flow through the pipe.

Leachate Collection and Management

Leachate is the liquid that is often released from compost as cell walls break down and moisture becomes mobile. Compost (food scrap compost especially) tends to release a flush of moisture at some point during the first two weeks of composting. For this reason, leachate is an especially important consideration for the first phase of composting, which is when ASP is typically deployed.

Leachate is high in nutrients, full of volatile organic acids that smell really strongly, and it potentially carries pathogens (*E. coli*, *Salmonella*, and more). It is both a pollution and a health concern,

9.5. Maximum Perforation Size (≤ Inches) for Rectangular Air Channels at 12-Inch Spacing

Length of Perforated Zone	Air Channel Cross Sectional Area								
	4"	6"	9"	12"	16"	20"	25"	30"	36"
2′	8/9"	n/r	n/r	n/r	n/r	n/r	n/r	n/r	n/r
4′	5/7"	7/8"	1"	n/r	n/r	n/r	n/r	n/r	n/r
6′	3/5"	3/4"	8/9"	1"	n/r	n/r	n/r	n/r	n/r
8′	1/2"	2/3"	4/5"	8/9"	1"	n/r	n/r	n/r	n/r
10′	1/2"	4/7"	5/7"	5/6"	1"	1"	n/r	n/r	n/r
15′	2/5"	1/2"	3/5"	2/3"	4/5"	8/9"	1"	n/r	n/r
20′	n/r	3/7"	1/2"	3/5"	2/3"	7/9"	6/7"	1"	1"
25′	n/r	n/r	4/9"	5/9"	5/8"	5/7"	7/9"	6/7"	1"
30′	n/r	n/r	3/7"	1/2"	4/7"	5/8"	5/7"	7/9"	6/7"
40′	n/r	n/r	n/r	3/7"	1/2"	5/9"	5/8"	2/3"	3/4"
50′	n/r	n/r	n/r	n/r	4/9"	1/2"	5/9"	3/5"	2/3"
60′	n/r	n/r	n/r	n/r	2/5"	4/9"	1/2"	5/9"	3/5"
70′	n/r	n/r	n/r	n/r	n/r	3/7"	4/9"	1/2"	5/9"

Note: Perforation diameters were calculated using the Knaebel simplified sparger design calculations described in Douglas R. Hartsock, Gerry Croteau, and Jeff Gage, "Uniform Aeration of Compost Media," presented at the National Waste Processing Conference Proceedings ASME, 1994: 217.

Perforation sizes may not correspond perfectly to available drill bit sizes. If this is the case find the closest drill bit size available without going over the recommended maximum size.

n/r = not recommended

and also poses a significant odor risk. Therefore, it needs to be managed with care.

Leachate is also a potential resource that can be captured and returned to the compost, as long as that compost will be heat-treated after its reapplication.

In terms of ASP design, you'll need to decide whether or not you need to capture leachate from your system, and if you do, how you will reuse the captured leachate. The answers to these questions will depend on your scale, existing infrastructure, and environmental factors. They may also depend upon how your state defines leachate and how it regulates leachate management based upon the size of your operation. Typically larger operations receive more scrutiny.

Compost infrastructure generally falls into one of two leachate collection scenarios: (1) Leachate is allowed to infiltrate into the ground, usually involving packed gravel or native soils and an open-air on-site treatment area, or (2) leachate is contained, collected, and stored for reuse, usually involving an impervious or semi-pervious composting pad and leachate storage tanks or ponds. Systems might involve some hybrid of these two scenarios as well, but these are the most common.

As described in other sections of this chapter, leachate management is a primary consideration in the type of compost pads and air channels you design. Other than the aeration system, the next most critical design component of ASPs is their leachate collection system. If you're planning aboveground air channels, you might be in luck. Since the channels are aboveground, liquids will flow around them, and therefore your leachate collection and aeration systems do not need to be integrated. Likewise, the *Gravel Pad/Gravel Air Channel Floor* described on page 231 will not need additional leachate collection. You will still need to plan a leachate and stormwater management strategy, but you will do this as part of the overall site plan, so see *Managing Moisture On-Site* (chapter 6, page 152).

For systems with concrete pads and belowground air channels, leachate collection will need to integrate into the ASP pad and channel design. If you ignore or underplan for leachate, you *will* be sorry! Believe me, the last thing you want is to be lying in bed dreaming about leachate clogging your air channels.

The main components of an integrated air channel leachate collection system are: (1) collection from the channel; (2) leachate drainpipes; (3) air lock; (4) leachate tank or storage; (5) leachate pump; and (6) leachate reuse strategy. The next section describes strategies and considerations for integrating these components into your ASP design.

Integrated Air Channel and Leachate Collection System

With the belowground formed concrete channel design, the channel itself acts as a leachate gutter that transports liquids either directly into a drainpipe or into the aeration manifold then into a drainpipe.

The ASP pad and air channels need to be designed to move leachate to a drain. The designs we did at the Highfields Center for Composting (figure 9.22) are a good example of a sloped pad and channel design that moves liquids across the pad, to the channels, then out of the system. Figure 9.20 shows how the formed concrete channel abuts a 6-inch SDR-35 PVC pipe, which connects the aeration and leachate collection system. The Highfields designs did not have aeration pipes in the channels; leachate went directly to the air manifold, then down to the leachate tank.

For systems that have aeration pipes in the concrete channels, a separate leachate drainpipe will need to be formed into the concrete. The air channel is effectively two layers, an inner pipe layer that transports gases and an outer layer that transports liquids. The channel will still need to be sloped to move liquids to the drainpipe.

Leachate Drainpipe

From the leachate collection system in the ASP pad (air channels or alternative), liquids should gravity-flow through a drainpipe into the leachate tank. The drainpipe should have a good slope if possible, because a fast flow will reduce the amount of sediment and risk of clogging or freezing. One-and-a-half- to 2-inch PVC or sewer pipe is a good option for small systems.

Air Lock

When the leachate collection system is connected directly to the air channels, it will need to be designed to maintain pressure within the aeration system. It will also need to trap gases in the leachate tank because the drainpipe is a potential escape for odors. One of the best ways that pressure can be sustained is to install a p-trap in the leachate drain. P-traps are what plumbers use to create a barrier for sewer gases in drainpipes (look under your kitchen sink if you're not familiar). They create a low spot in the drainpipe that holds a small amount of water, creating an air lock. The first time you turn on the fans, you will need to pre-charge the p-trap with water to create a block. Eventually this water will be replaced with leachate.

An alternative to the p-trap is simply dead-ending the drainpipe into an airtight leachate collection tank, which will maintain pressure in the aeration pipe but not trap gases in the tank. Having tried this before, it's not necessarily the approach I would recommend. With this design you have to be sure the fans are turned off when you're emptying the leachate tanks; otherwise the tank can become pressurized and spew leachate everywhere (not something you ever want to experience). In addition, operators of one such system reported some random leachate odors. It seemed likely to be coming from the tanks, because the drainpipes were not sealed off from the aeration pipes.

Figure 9.20. A 6-inch PVC (SDR-35) air duct formed into the back wall of a concrete ASP bay. The PVC abuts a 4 × 4-inch formed concrete air channel that starts 4 feet from the back wall. This allows leachate to flow from the channel into the PVC, where it can drain into a storage tank. Courtesy of Highfields Center for Composting and Vermont Sustainable Jobs Fund.

Leachate Tanks

There are a wide array of tank options that can be used to store leachate until it can be reapplied. Everything from a septic tank for larger systems to a recycled 55-gallon drum for smaller systems could suffice. A wide access point on the top of the tank where a pump can be inserted is a good feature to have. You'll want as much flexibility in how you empty the tank as possible, particularly with larger operations. The size of the tank you use will depend on how frequently you plan on emptying it, so the tank, the pump, and the reuse strategy all need to be compatible.

Compost releases moisture at an uneven rate, with the vast majority of free moisture loss happening in the first two weeks of composting. After this heavy release period, the leachate slows considerably unless moisture is re-added (irrigated or heavy precipitation).

The total volume and flow rate of leachate depends on the recipe, so when handling a particularly wet batch, expect more leachate. I was unable to locate any quantitative information about leachate generation from food scrap compost. While operating an ASP at

Figure 9.21. PVC air ducts are formed into a poured concrete wall footing. The manifold will be on the back side of the wall in a chase. Courtesy of Highfields Center for Composting and Vermont Sustainable Jobs Fund.

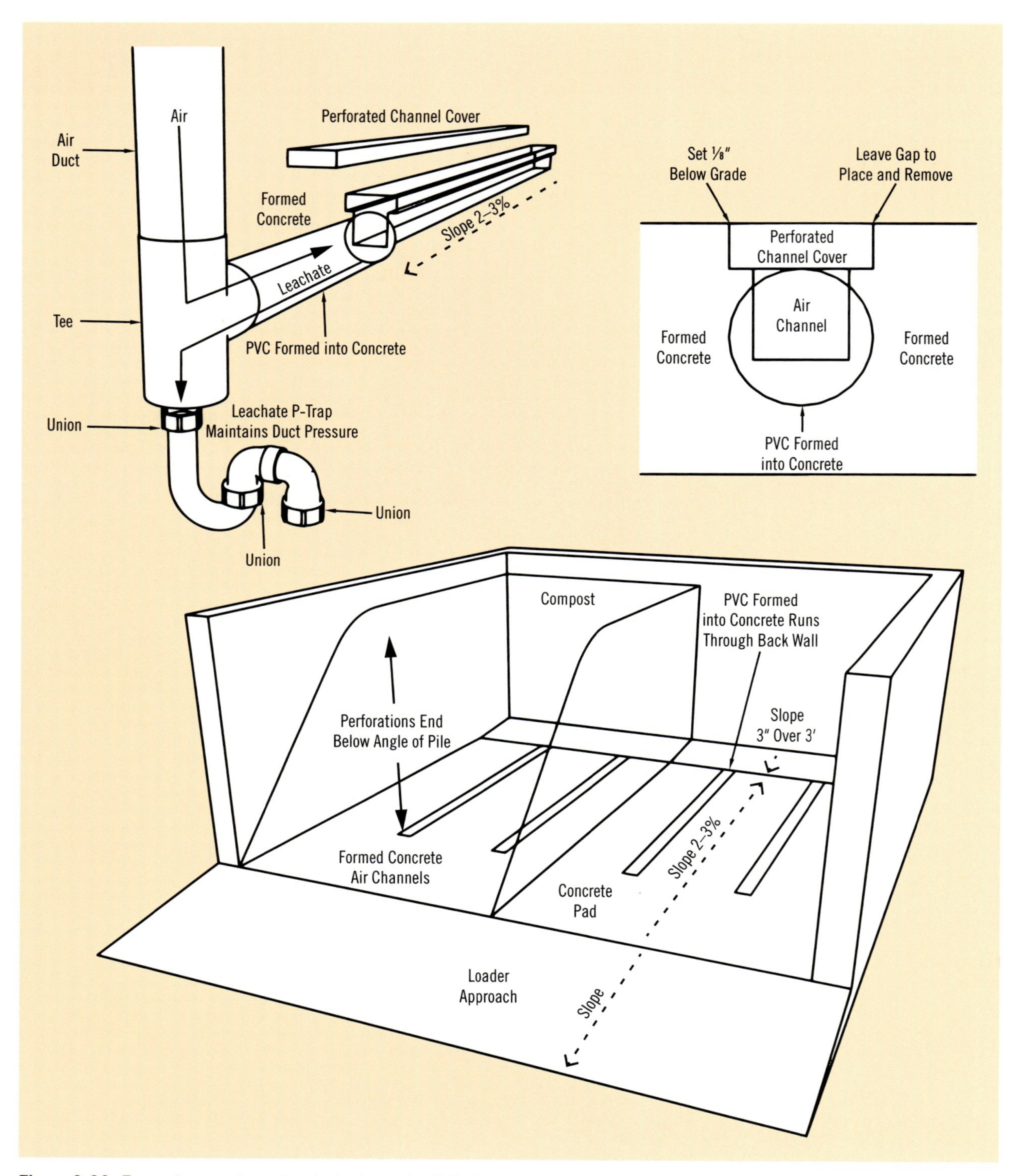

Figure 9.22. Formed concrete pad and air channels. PVC ducts gravity-flow to p-trap drains. Adapted from Highfields Center for Composting, courtesy of Vermont Sustainable Jobs Fund

the Highfields Center for Composting we collected data on how much leachate we captured from our concrete ASP bay. We found that total leachate generation from an 80-cubic-yard batch of compost ranged from 80 to 160 gallons over the first two weeks of composting or 1 to 2 gallons of leachate per cubic yard of freshly blended food scrap compost. (The moisture content of our mixes ranged between 60 and 63 percent.)

Freezing

In cold climates frozen leachate in air channels and drainpipes can be a major challenge to contend with. Directly beneath the compost, the pipes are insulated enough that freezing is unlikely, but moisture moving through pipes outside of the pile is very likely to freeze if it's exposed to cold. For this reason, it makes sense to insulate the pipes and possibly even to keep them in a heated space, such as a chase (see *Mechanical Housing*, page 243). Heat tape on the pipes and even waterproof heating wires in the pipes are also possibilities. It may be easiest, however, to simply send the drain below the frost line, as long as you maintain an air lock and the ability to clean the pipe if need be.

Where to Reapply Leachate

Systems that capture leachate in a tank or pond will need a plan for its consistent utilization. Commonly, leachate is added back into a compost batch that will then be heat-treated, which is a best management practice. Sites that require moisture in the initial mix might add leachate during blending. Sites that can't afford the dry matter in blending might apply the leachate later, after the pile has dried out, assuming there is still adequate pile activity to then hit heat treatment targets.

As far as I know, utilizing leachate on-site to add moisture to compost is universally allowed as long as best management practices are followed. Allowable alternatives to compost application differ state-to-state and site-to-site. Large sites often truck it off-site, where it is then land-applied or processed at a water resource recovery facility (formerly wastewater treatment plant). It's possible that a digester might take it. Off-site options are generally going to be more costly then on-site utilization.

For more on leachate treatment options, see *Managing Moisture On-Site*, page 152.

Step 6. Size Ductwork and Air Channels to Minimize Static Pressure

You need to know the following in order to size your ductwork:

- Target blower capacity (CFM) (from step 5)
- General concept for aeration system
  - Single versus multiple air channels
  - Round versus rectangular air channels

6.1 Size Header Duct

Target Blower Capacity = ___ CFM

Use tables 9.6 and 9.7 to find the duct size that is ideal for your scenario.

___ CFM through ___″ Round or ___″ × ___″ Rectangular Header Duct = ___ FPM @ ___″ WC per 100′

EXAMPLE

297 CFM (from Step 2)

297 CFM through 6″ round = 1,512 FPM @ 0.57″*

* The CFM was rounded from 297 to 300 to estimate static pressure based on table 9.8, page 245. Velocity was calculated in the Duct Sizer app.

Table 9.6. Sizing Circular Ducts and Air Channels

Airflow Rate (CFM)	Duct Diameter at ≤ 1,500 FPM	Pressure Loss in Inches Water Column per 100 Feet	Duct Diameter at ≤ 2,000 FPM	Pressure Loss in Inches Water Column per 100 Feet	Duct Diameter at ≤ 3,000 FPM	Pressure Loss in Inches Water Column per 100 Feet
5	2″	0.08″	n/r	n/a	n/r	n/a
10	2″	0.26″	n/r	n/a	n/r	n/a
20	2″	0.89″	n/r	n/a	n/r	n/a
30	3″	0.26″	2″	1.85″	n/r	n/a
40	3″	0.43″	2″	3.10″	n/r	n/a
50	4″	0.16″	3″	0.65″	n/r	n/a
75	4″	0.34″	3″	1.35″	n/r	n/a
100	5″	0.19″	4″	0.56″	n/r	n/a
125	5″	0.29″	4″	0.84″	n/r	n/a
150	5″	0.39″	4″	1.17″	n/r	n/a
175	6″	0.21″	4″	1.55″	4″	1.55″
200	6″	0.27″	5″	0.66″	4″	1.98″
250	6″	0.41″	5″	1.00″	4″	2.98″
300	8″	0.14″	6″	0.57″	5″	1.40″
350	8″	0.18″	6″	0.76″	5″	1.85″
400	8″	0.23″	6″	0.97″	6″	0.97″
450	8″	0.30″	8″	0.29″	6″	1.20″
500	8″	0.35″	8″	0.36″	6″	1.46″
600	10″	0.17″	8″	0.50″	6″	2.05″
700	10″	0.22″	8″	0.66″	8″	0.66″
800	10″	0.28″	10″	0.28″	8″	0.85″
900	12″	0.15″	10″	0.35″	8″	1.06″
1,000	12″	0.18″	10″	0.43″	8″	1.28″

Note: Pressure loss assumes PVC or HDPE pipe. Uncommon pipe sizes such as 3″ and 5″ may not be available, in which case it is recommended that you choose the next largest available pipe size to avoid creating excessive friction.

Pipes greater than 4″ may be heavy and cumbersome to move if used for pipe-on-grade systems. Consider splitting from one larger air channel to two or more smaller channels.

n/r = not recommended; n/a = not applicable

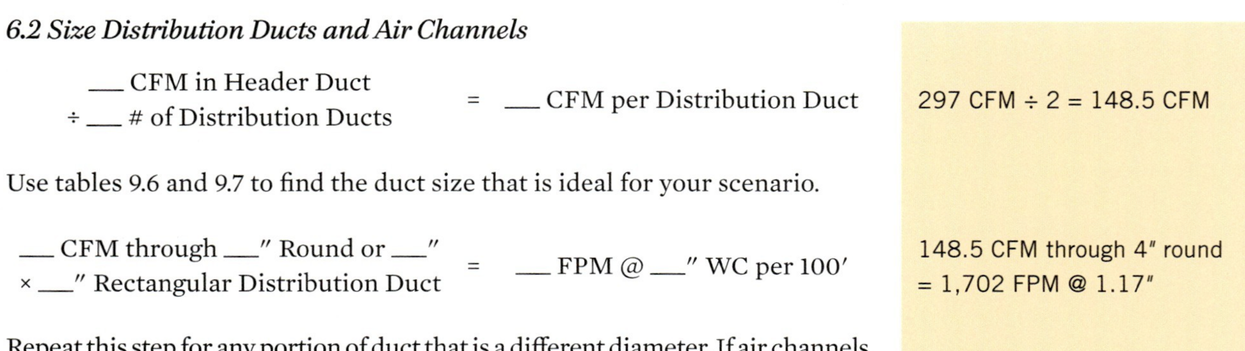

6.2 Size Distribution Ducts and Air Channels

___ CFM in Header Duct
÷ ___ # of Distribution Ducts = ___ CFM per Distribution Duct

297 CFM ÷ 2 = 148.5 CFM

Use tables 9.6 and 9.7 to find the duct size that is ideal for your scenario.

___ CFM through ___″ Round or ___″ × ___″ Rectangular Distribution Duct = ___ FPM @ ___″ WC per 100′

148.5 CFM through 4″ round = 1,702 FPM @ 1.17″

Repeat this step for any portion of duct that is a different diameter. If air channels are different from distribution ducts, size separately using the same method.

Table 9.7. Sizing Square and Rectangular Ducts and Air Channels

Airflow Rate (CFM)	Duct Cross-Sectional Area (Square Inches) at ≤ 1,000 FPM	Pressure Loss in Inches Water Column per 100 Feet	Duct Cross-Sectional Area (Square Inches) at ≤ 2,000 FPM	Pressure Loss in Inches Water Column per 100 Feet	Duct Cross-Sectional Area (Square Inches) at ≤ 3,000 FPM	Pressure Loss in Inches Water Column per 100 Feet
5	4	0.07″	n/r	n/a	n/r	n/a
10	4	0.25″	n/r	n/a	n/r	n/a
20	9	0.12″	n/r	n/a	n/r	n/a
30	9	0.25″	4	2.12″	n/r	n/a
40	12	0.21″	4	3.74″	n/r	n/a
50	12	0.33″	6	2.03″	n/r	n/a
75	16	0.33″	9	1.50″	n/r	n/a
100	20	0.33″	9	2.65″	n/r	n/a
125	24	0.32″	12	1.95″	n/r	n/a
150	32	0.23″	12	2.80″	n/r	n/a
175	32	0.31″	15	2.17″	n/r	n/a
200	40	0.24″	18	1.80″	n/r	n/a
250	48	0.24″	20	1.98″	n/r	n/a
300	50	0.28″	24	1.78″	20	2.85″
350	60	0.24″	28	1.66″	24	2.43″
400	70	0.22″	32	1.56″	28	2.16″
450	75	0.24″	35	1.46″	28	2.73″
500	84	0.20″	40	1.28″	32	2.42″
600	96	0.21″	50	1.07″	35	2.58″
700	120	0.15″	60	0.94″	40	2.52″
800	128	0.16″	70	0.86″	50	1.90″
900	144	0.16″	75	0.93″	60	1.44″
1,000	160	0.15″	84	0.78″	60	1.77″

Note: Pressure loss assumes typical formed concrete.
n/r = not recommended; n/a = not applicable

Mechanical Housing

Many ASP systems include a space called a *chase*, an enclosure that abuts the ASP and provides space for ducts, blowers, controls, and other mechanical systems. It's also a space that allows easy access for operation and maintenance as needed.

Typically, the chase is located behind the back push wall of the ASP. Pipes run through the wall and connect to the aeration and leachate manifolds. Often the chase is lower than the grade of the ASP pad, which creates drop for leachate flow and also puts the ductwork at a more convenient working height. Heating and insulation may be necessary where freezing is an issue. Blocking PVC from sunlight extends its life, because PVC solar degrades rather quickly.

Many sites with existing infrastructure are able to use a shop or barn for housing ASP equipment, in which case proximity is key. In other cases mechanicals are placed in the open air with nothing more than a shed roof, serving the same function as a chase, for less cost. Small systems might be able to fit several blowers in a small shed, on a covered pallet, or even into a large tool box (like a truck box).

Figure 9.23. Magnehelic static pressure gauge (manometer). The range of 0–10″ water column is typical for ASP systems.

Estimating Static Pressure Within the System

Before we can select a blower, we need to form a reasonable estimate of what the static pressure will be within the system. For our estimates, we're going to use: (1) static pressure from the ductwork; (2) static pressure from the compost; and (3) optional: static pressure from the biofilter for negative ASPs. Two areas that are intentionally not factored into our equation for simplicity's sake are static pressure loss from fittings (elbows, tees, and so on) and loss at the perforations. As long as you follow the other design rules, these factor in only minimally. When you select a blower, if you are conservative and oversize it slightly you will be in great shape.

Step 4 provided us with an estimate of the static pressure loss per 100 feet of duct. In this next step, we will use those factors to calculate the total loss in the ducts. You will need to have a solid design concept for the manifold before finalizing your static pressure estimate and selecting a blower. We will adjust the static pressure from the compost for negative ASP systems to account for the added losses associated with those systems.

Static Pressure From Compost

As you might have anticipated, compost itself creates resistance to the flow of air from the blower. This resistance should be accounted for in selecting your blower, just like the other sources of resistance in the system. To do this, we're going to make a ballpark estimate based on two factors: (1) your estimated pile height, and (2) the number of air exchanges per hour your fan is sized for (step 5).

The methodology for this calculation was adapted from Haug, who created specifications for compost based on Darcy's law of fluid flow through porous material. These equations are actually relatively straightforward, but as I've done with other topics, I've simplified this step further by creating table 9.8. Unless you are sizing a really large system (more than 300 cubic yards per batch) and can do some legwork to understand the porosity of your actual mix, you are not going to get much more accurate than this, and really don't need to. (That said, if you are going to build a large system or just want to understand the math behind this, refer to *The Practical Handbook of Compost Engineering*.)

Pressure drop from compost in negatively aerated systems is greater than with their positively aerated counterparts. The pulling of hot moist air toward the air channels has a wetting effect on the compost directly around the channels, which increases density and resistance. In some wet mixes, this can even become a "hard pack," severely impacting the performance and efficiency of aeration. In addition, many negative aeration systems push air through a biofilter on the opposite end of the fan, creating additional resistance. You'll need to adjust static pressure for these two factors if you plan on operating a negatively aerated system. Adjusting static pressure from compost for negatively aerated systems is covered in step 5, and static pressure from your biofilter is covered in *Designing Biofilters* on page 251.

Step 7. Estimate Static Pressure from Compost

To summarize, you need to know the following to estimate the static pressure created by the compost:

- Your estimated pile height
- The number of air exchanges per hour (AE/H) your fan is sized for

7.1 Locate Your Air Exchanges per Hour

On the top row of table 9.8, locate the number of air exchanges per hour (AE/H) your fan is sized for.

7.2 Locate Your Pile Height

On the left-side column of table 9.8, locate your assumed pile height.

7.3 Find Corresponding Static Pressure

Find the estimated static pressure from compost that corresponds with your AE/H and pile height, which will be given in Inches WC. This number will be used in step 8 to calculate the total static pressure in the system.

EXAMPLE
for 100-Cubic-Yard Windrow

Medium target aeration rate of 20 AE/H or 297 CFM.

Trapezoidal-shaped windrow—pile dimensions of 35 × 20 × 7 feet (L × W × H). See figure 9.7 for the two-channel configuration.

0.3666″ WC from compost.

Table 9.8. Static Pressure from Compost

SYSTEM SCALE	Low Aeration Rate		Medium Aeration Rate		High Aeration Rate	
Pile height (feet)	5 AE/H (Inches WC)	10 AE/H (Inches WC)	15 AE/H (Inches WC)	20 AE/H (Inches WC)	30 AE/H (Inches WC)	40 AE/H (Inches WC)
3	0.0040	0.0110	0.0199	0.0304	0.0551	0.0841
4	0.0092	0.0255	0.0463	0.0707	0.1284	0.1959
5	0.0178	0.0492	0.0893	0.1363	0.2474	0.3776
6	0.0304	0.0841	0.1526	0.2330	0.4228	0.6454
7	0.0478	0.1323	0.2401	0.3666	0.6653	1.0154
8	0.0707	0.1959	0.3556	0.5428	0.9851	1.5037
9	0.1000	0.2770	0.5028	0.7674	1.3928	2.1259
10	0.1363	0.3776	0.6853	1.0460	1.8985	2.8978

Note: These static pressures were calculated using equation 15.24 and compost media permeability coefficient for a 1:1 mix of wood chips and biosolids as described in Roger Tim Haug, *The Practical Handbook of Compost Engineering* (Boca Raton, FL: Lewis Publishers, 1993), 525–26.

Step 8. Estimate Static Pressure Within System

You need to know the following to estimate the total static pressure in the system:

- The dimensions of each section of ductwork
  - Length
    - Header
    - Distribution ducts/air channels
- The static pressure associated with each section of ductwork (step 6)
- The static pressure from the compost (step 7)
- Optional: the static pressure from the biofilter. See *Designing Biofilters* on page 251.

EXAMPLE

Header = 20 feet

Distribution ducts/air channels = 60 feet

8.1 Calculate Static Pressure per Foot from Header Duct

___″ WC per 100′ from Header Pipe ÷ 100 = ___″ WC/FT from Header Pipe

0.57″ ÷ 100 = 0.0057″

8.2 Calculate Total Static Pressure Loss from Header Duct

___′ Length of Header Pipe × ___″ WC/FT from Header Pipe = ___″ WC Static Pressure Loss from Header Duct

20′ × 0.0057″ = 0.114″

8.3 Calculate Static Pressure per Foot from Distribution Ducts and Air Channels

___″ WC per 100′ from Distribution Ducts/Air Channels ÷ 100 = ___″ WC/FT from Distribution Ducts/Air Channels

1.17″ ÷ 100 = 0.0117″

8.4 Calculate Static Pressure Loss from Distribution Ducts and Air Channels

___′ Length of Distribution Ducts/Air Channels × ___″ WC/FT = ___″ WC Static Pressure Loss from Distribution Ducts/Air Channels

60′ × 0.0117″ = 0.702″

Repeat steps 8.3 and 8.4 for each unique duct section (any duct with a variable diameter or airflow).

Ducts A. 4-inch distribution ducts (60 feet at 1.17 inch WC per 100 feet) = 0.702″ WC

8.5 Determine Total Static Pressure Loss from Distribution Ducts and Air Channels

Add the loss from each duct section together to calculate the total loss from all distribution ducts and air channels.

___" WC Static Pressure Loss from Distribution Ducts/Air Channels (Ducts A) + ___" WC Static Pressure Loss from Distribution Ducts/Air Channels (Ducts B) + ___" WC Static Pressure Loss from Distribution Ducts/Air Channels (Ducts C)	=	___" WC Total Static Pressure Loss from Distribution Ducts/Air Channels	0.702" + n/a" + n/a" = 0.702"

8.6 Sum Pressure Loss from Header Duct, Distribution Ducts, Air Channels, and Compost

___" WC Static Pressure Loss from Header Duct + ___" WC Total Static Pressure Loss from Distribution Ducts/Air Channels + ___" WC Static Pressure Loss from Compost	=	___" WC Estimated Fan Operating Pressure for Positive ASP	0.114" + 0.702" + 0.367" = 1.183" Or roughly 1.2-inch WC operating pressure.

*8.7 (Optional) Adjust Static Pressure for Negative Aeration**

___" WC Estimated Fan Operating Pressure for Positive ASP × 2	=	___" WC Pressure Adjusted for Negative ASP	1.183" × 2 = 2.366"

8.8 (Optional) Add Static Pressure from Biofilter

___" WC Pressure Adjusted for Negative ASP + ___" WC Pressure from Biofilter	=	___" WC Estimated Fan Operating Pressure for Negative ASP	2.366" + 0.541" = 2.907"

See *Designing Biofilters* on page 251.

* Note: This step is a rough estimation. Due to the nature of negative aeration, pressure loss is a moving target that varies significantly based on factors such as pile moisture and density/porosity that are hard to accurately adjust for. With more frequent turning, the actual operating pressure can be minimized.

Selecting a Blower

Once you have an estimate of the pressure loss within the ASP system, you can research and select a blower. The fan needs to supply the target aeration rate at the estimated pressure loss. Fans—or blowers, as we call them—have a "blower curve" that the manufacturer developed for exactly this purpose. The blower curve will give you the fan's estimated flow rate (CFMs) at a given static pressure loss.

To find the right blower, you'll have to do a little bit of research yourself. A few blower manufacturers that you'll see at compost sites are New York Blower and Fantech, for medium and larger systems, and Koala for smaller systems. Yes, the Koala blowers are for inflating those giant bouncy houses (just imagine the static pressure in those things!).

In addition, before selecting a blower, you'll need to know (1) if you are providing negative or positive aeration; (2) if you want a variable speed drive (also called variable frequency drive or VFD); and (3) the basic fan type you are looking for. If you are doing advanced controls, the fan will need to be compatible with the control system.

Step 9. Select Blower

You need to know the following to select a blower:

- Target blower capacity (CFM)
- Estimated static pressure within the system
- Positive or negative aeration
- Variable speed drive or not
- How the fan will be controlled

9.1 Identify Target Blower Capacity and Estimated Operating Pressure

Target blower capacity: ___ CFM

Estimated fan operating pressure (positive aeration): ___ inches WC

or

Estimated fan operating pressure (negative aeration): ___ inches WC

9.2 Research Blowers
Work with local distributors or search online. Locate blower performance curves and find models with the capacity to provide the desired CFM at the estimated operating pressure. For negative ASP, make sure to source blowers that are designed to handle corrosive exhaust.

Example

9.1 Identify Target Blower Capacity and Estimated Operating Pressure

Target blower capacity: 297 CFM

Estimated fan operating pressure (positive aeration): 1.183 inches WC

or

Estimated fan operating pressure (negative aeration): 2.907 inches WC

9.2 Research Blowers
Work with local distributors or search online. Locate blower performance curves and find models with the capacity to provide the desired CFM at the estimated operating pressure. For negative ASP, make sure to source blowers that are designed to handle corrosive exhaust.

Figures 9.24 and 9.25 show blower curves for fans capable of providing about 300 CFM at ~1.2 inches WC and about 2.9 inches WC respectively.

Note that you will need to perform steps 5 through 9 for each unique "zone" within the system. For example, for multiphased systems, the aerations requirements and ductwork are typically different between the primary and secondary phases.

Controlling Your ASP

Composters have a wide variety of options to choose from to control the aeration rates of ASP systems. Unfortunately, the systems that provide the most

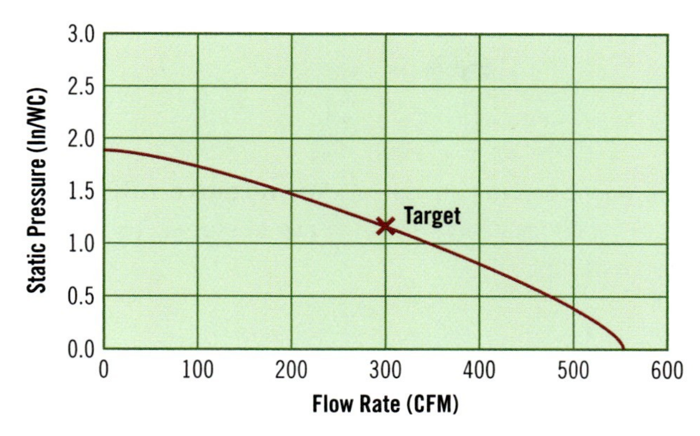

Figure 9.24. A theoretical fan blower curve that can provide approximately 300 CFM at around 1.2 inches WC.

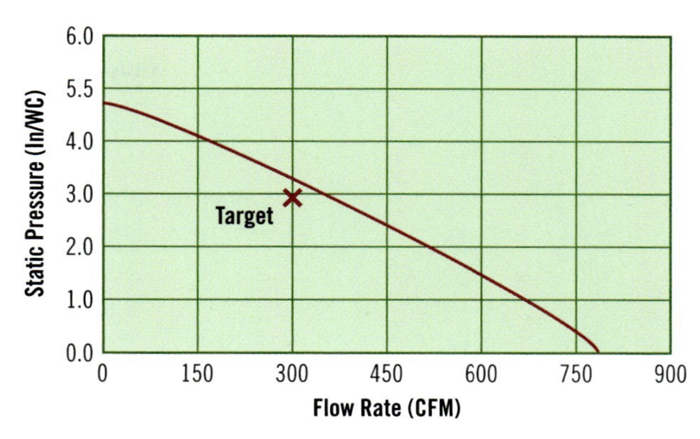

Figure 9.25. A theoretical fan blower curve that can provide approximately 300 CFM at around 2.9 inches WC.

control options come with a hefty price tag and often require the purchase of the whole ASP system or for you to hire a controls engineer to custom-design something for you. Green Mountain Technologies, however, sells a control system and temperature probes separately from their ASP. Their MACS system provides remote control whereby an operator can monitor temperatures and make adjustments from a mobile phone. It can also operate blowers based upon a temperature set point, running fans to cool a pile as needed.[20]

At the other end of the spectrum, a simple and low-cost option is to use a timer to control the fan on/off times. You could argue the benefits of each. We'll run through the basic options. Keep in mind that these can theoretically be used in combination with one another and with valves and VFDs (variable frequency drives, or fan speed controllers).

Cycle Timers

The type of clock timer that you might use to control a lamp is not well suited for controlling an ASP fan for a number of reasons that I won't get into. What you want is a cycling timer. Cycling timers let you set the *on* and *off durations* without regard for the time of day. You can set up the timer to turn the fan on and aerate for 5 minutes, then off for 15 minutes (5 minutes on per 20 minutes or 15 minutes on per hour), then easily adjust that rate up or down based on the pile's performance.

All composting systems require consistent monitoring of pile temperature, but with a timer-controlled ASP system, temperature monitoring is going to be your sole means of feedback to guide how much to aerate. Cycle timers can be combined with VFDs as well for a very low cost. The cycle timer in figure 9.26 costs under $100; the fan was wired to a VFD and plug so that the whole system could be plugged into an outlet.

Heat-Controlled Systems

Since the primary feedback for compost performance is temperature, directly controlling aeration based on pile temperature is very common in today's larger systems. I was told recently by a site manager that a heat-controlled VFD would be his ideal control system. This was at a larger site that has dealt with a lot of problems related to high temperatures and odor management in their ASP. The site has manually controlled cycle timers and VFDs that are adjusted based on temperature data from heat sensors in the piles. If the timers and fan speeds could be intelligently connected to the temperature sensors, a controller could replace the operator as the constant intermediary between temperature data and aeration rates. A really good system would log all of this information and allow the operator to optimize the algorithm that manages aeration based on this data.

In my experience the available heat-controlled systems are beyond the reach of most small-scale

Figure 9.26. An ART-DNe cycle timer and variable speed drive are a really simple and versatile control system. The static pressure gauge provides insight into system performance and is a good indicator that something is wrong if pressures are too high or too low.

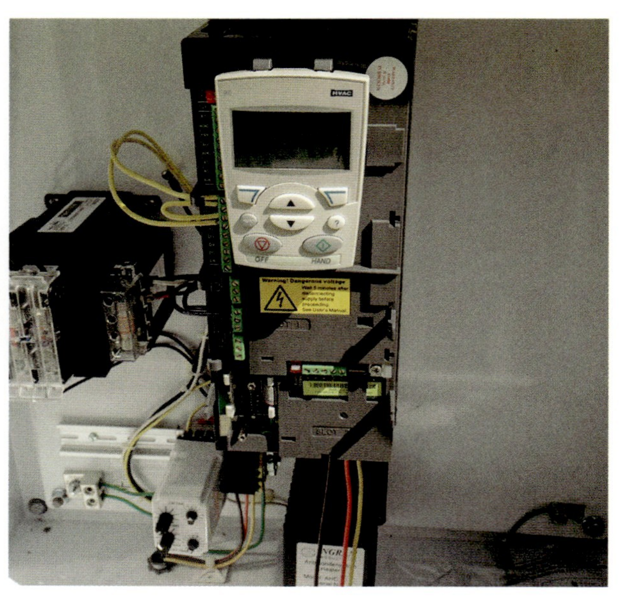

Figure 9.27. These controls provide essentially the same on/off and fan speed controls as in figure 9.26, but they could also be tied into a more advanced control system that has temperature feedback, for example.

Figure 9.28. The Gore-Tex ASP system distributed by Sustainable Generation is equipped with a mobile aeration system, which is contained in this box. It has a blower that can be controlled by a timer and/or with an oxygen sensor and also has a temperature data logger.

operations, but I hope that will not always be the case. As these technologies develop, the costs of prefabricated systems should come down, and it's possible that open-source designs will emerge.

Experience in the industry indicates that the best designs place heat sensors in the pile. The recommendation for windrows is to place them at about two-thirds the length of the pile away from the blower.[21] If you invest in a heat-controlled system that has multiple sensors and variables, expect a learning curve and learn as much as you can about how to work and fix the system as needed. One of my biggest concerns with these systems is that operators become dependent on outside engineers and vendors to keep a system operating over the long term.

Oxygen-Controlled Systems

Aerating the compost based on oxygen demand is perhaps the most logical approach, because temperature is in theory primarily an indicator of oxygen availability. Some oxygen-controlled ASPs allow composters to do just that, but designing these systems is no easy task. Like temperature-controlled systems, expensive prefabricated systems and custom systems designed by controls engineers are the only options for developing these systems right now, unless you are extremely talented with controls, sensors, and programming.

With oxygen sensor controls in place, it is still ideal to have a temperature override, because one of the risks with unlimited oxygen availability is overheating. The oxygen aeration settings will keep compost aerobic, but with a temperature set point, the fan can stay on if the compost is overheating (155°F or more, for example), until it cools off.

One of the really cool things about having oxygen sensors is that you can see the ASP in action. Some composters also buy a manual O_2 probe. These are quite expensive, but are the dream toy of every compost geek. One composter told me that they can't use their oxygen control, because it ends up overaerating their compost and drying it out, so they sometimes use it as a data logger while running the blower on a timer.

Designing Biofilters

One of the many touted benefits of ASP composting is the ability to capture and treat odors with filtration systems. There are numerous filtration techniques, but in all but the largest and nastiest of jobs, a biological filter, or *biofilter*, is a very appropriate technology.

The concept is simple: design the composting system so that there is a porous yet biologically active organic substrate between the fresh, odor-producing compost and the surrounding environment. The odorous compounds adsorb to the biofiltration substrate, then biodegrade, remediating the odor. The practice of *capping* fresh compost (as covered in *Capping for Odor Control* on page 268) is effectively a biofiltration technique. With turned windrow systems, the capping layer on the surface of the pile passively filters volatile organic compounds (VOCs) and other odors as they waft upward from the active core. The forced draft produced by the ASP actively moves the air and the compounds contained in it through the media, filtering it along the way.

In positive aeration the biofilter looks virtually identical to the capping layer in turned windrows systems, except that the depth and type of capping material will usually differ. With positive ASPs, capping with a thick (around 12 inches) compost biofilter layer is common (see figure 9.37, page 267). The compost should be mature enough that it doesn't produce any unpleasant odors of its own, but does not need to be fully cured or screened. One technique used by operations that have a screener is to double-screen their compost, then use the midsized screened particles as the biofiltration layer. For example, the compost might first go through a larger screen, such as 2 inch minus, to get large trash and particles out, then screen the 2 inch minus through their fine screen. The fine particles are cured while the midsized particles, of which a large percentage should be woody, are reused as the capping biofilter. Capping works best with semi-porous materials, otherwise it can act as a barrier for airflow.

With negative ASP systems, the compost should still be capped with a thin layer (around 6 inches) to contain odors and deter vectors, but the primary biofilter is external to the pile on the back end of the aeration system. Good external biofilters have a reputation for lasting five or more years with no need to replace the media and for being very effective at removing multiple sources of odors, including VOCs. It is important to remember that the media decomposes over time, which increases its density and hence creates resistance to airflow. A good manager will keep an eye on the biofilter media and monitor the static pressure within the system. Replace the biofilter media if it becomes too dense, to avoid blower inefficiency and the associated electricity costs.

External Biofilter Design

External biofilter configurations are categorized into *open-bed* and *closed-bed* systems, the key difference being whether the filter media is exposed to open air or contained with only an exhaust outlet.[22] Closed systems could be built into a permanent stationary vessel or designed to be mobile, but both open and closed systems should be easily accessible for replacing the filtration substrate when the time comes. Small systems are more conducive to being closed, because creating a vessel to contain a large volume of substrate would quickly become cost-prohibitive. But having an airtight closed-bed biofilter could also save you money, if you configure the system with the fan on the backside of the biofilter. This would allow you to pull air through the biofilter, removing condensate, corrosive gases (such as ammonia), and heat before these reach the fan. While I have not seen anyone build a system this way yet, it is similar to systems that place a heat exchanger or condensate trap before the fan (which I have seen), so in theory it should work as long as the biofilter is airtight and properly manifolded.

For permanent, stationary closed-bed systems, concrete or large wooden boxes are the most likely material, whereas with mobile systems, palletized boxes or roll-off containers are most likely. Prefabricated boxes like those used to store dry ice (figure 9.29) are one option. Pretty much any biofilter will

Figure 9.29. These insulated poly boxes used to store dry ice are a logical option for creating a very small biofilter bed.

require some custom design and fabrication, unless you're paying for a full ASP system from a vendor.

Open-bed systems look a lot like ASP systems, although they are typically short and wide. The best way to illustrate the design of an open-bed biofilter is to draw one. You will see that the design is similar, with one difference being that the media is not constantly being replaced, so in that way the design is much simpler. Pipes can be low-cost (PVC would be fine, but expect it to be brittle when you remove it) and above grade, but will need to be worked around or pulled when the substrate is replaced. The biofilter itself can be constructed above- or belowground. Aboveground is obviously less costly and easier to access when it is time to replace, but belowground systems will likely maintain their temperature better in cold climates and maintain moisture better to some degree as well. If you build a belowground system, be sure there is adequate drainage, because if the manifold clogs with water, the system will not perform.

Figure 9.30. A large biofilter bed at a biosolids compost facility is sunk into the ground.

Media for External Biofilters

The ideal media for external compost biofilters is high in carbon, porous, biologically active with lots of surface area for microbial habitat, and moist with an MC between 50 and 60 percent.[23] Biofilters perform best at mesophilic temperatures where microbial communities thrive but are more stable than under thermophilic conditions. The recommended range is between 70 and 104°F.[24] As we know, decomposers can tolerate a wide range of temperatures, but a mostly composted media substrate is going to be more naturally suited to that temperature range than fresh material with a lot of energy potential.

A wide range of media have been used for biofilters, from wood chips to soil. Williams and Miller report biofilters being constructed from uncommon materials such as peat, heather, and volcanic ash. They also report sand and compost as a common blend, with the sand providing the porosity and the compost providing the surface area and biological host sites.[25] In planning your biofilter, keep the relationship between media porosity and static pressure in mind. One of my go-to resources, *Biofilters* (Nicolai and Schmidt), recommends a recipe of finished compost and wood chips. They provide a weight-based recipe that, converted to volume, would be roughly 1 part mature compost to 6 to 8 parts wood chips.[26] High-carbon compost, such as composted yard debris or leaves, would be ideal for the compost component (15 to 30:1 C:N ratio).

The target porosity of your media is 40 to 60 percent. Wood chips are typically assumed to be 50 percent or more free air space (FAS), and mature unscreened compost is closer to 30 to 33 percent FAS, so the recipes in the previous paragraph are going to fall somewhere between 40 and 50 percent, which is very acceptable. If you are uncertain about the suitability of a particular media as a biofilter in terms of porosity, you can measure the percents in a lab (see *Porosity* in chapter 4, page 77).

Biofilter Media Volume and Exhaust Retention Time

The performance of a biofilter depends on a number of different variables in the media itself (porosity, moisture, temperature), as well as how evenly the manifold is able to distribute the exhaust air throughout the bed. Assuming these factors are operating effectively, the key variable is the retention time of exhaust in the media itself, which is determined by the total volume of the media in relation to the volume of compost exhaust.

Earlier in this chapter, steps 1 through 9 outlined the process of sizing an ASP system based on the volumetric air exchange method. That methodology will help us to determine the retention time in the biofilter as well. In step 5 we determined the target blower capacity, so we know the peak volume of exhaust air per minute the biofilter should be designed to accommodate. Assuming relatively even distribution of exhaust through the biofilter, the volume of media required for an average retention time of one minute is very easy to calculate. If the media has 50 percent pore space (air), then we need twice the media volume as the blower CFM. So for our fan operating at 300 CFM, we would need 600 cubic feet of biofilter media, or 22 cubic yards.

Recommended biofilter residence times seem to fall in the range of 30 to 60 seconds.[27] One minute is a safe retention time in typical cases; any longer would be a conservatively sized system. If you chose negative ASP as a means for odor control, chances are you don't have a lot of room for error. Plan your retention time based on the level of risk you're taking, should you encounter an odor issue.

If the biofilter serves more than one fan, that will need to be factored into the retention time and manifold designs. The given example assumes a simple configuration, with one biofilter manifold per ASP zone, although the biofilters would logically be stacked in one large bed.

Following the formula below, we can calculate a longer or shorter retention time for any volume of exhaust and calibrate it to the media pore space as well.

Step 1. Calculate Media Biofilter Volume and Exhaust Retention Time

Where:

BMV = Biofilter media volume (cubic feet)
PFAS = Percent FAS (pore space/voids—for example, 45 percent PFAS = 0.45)
TRT = Target retention time (minutes—for example, 2 minutes = 2)
CFM = Target aeration rate (for example, 300 CFM)*

The formula is:

BMV = CFM ÷ PFAS × TRT

Example

1,333 Cu. Ft. BMV = 300 CFM ÷ 0.45 PFAS × 2 TRT

A biofilter medium volume of 1,333 cubic feet, rounded up to 1,350 cubic feet or 50 cubic yards. This would be a very large biofilter with a long target residence time and high odor buffering capacity.

Biofilter and Manifold Design

After estimating the volume of biofilter media required to meet your target retention time, you can now calculate the corresponding footprint and depth of the system. There is a trade-off between media depth and static pressure loss. As the depth or height of the media increases, the footprint decreases, which is often beneficial in small spaces, but the static pressure increases non-linearly, even while the total volume and retention time remain the same. You can see this relationship clearly in table 9.9, page 259. It helps to keep that depth to a minimum, usually 3 to 4 feet in composting biofilters.[28]

* If the biofilter is serving more than one fan, make sure to account for the peak air volume it might be required to handle at any given time.

Open- or closed-bed biofilters could come in a few different geometric formations, but most likely it's going to be rectangular cuboid (a deep flat bed or box) or trapezoidal. If it's more of a windrow or a perhaps semi-cylindrical, use the appropriate math to solve for the biofilter's dimensions. As with ASP manifold designs, you want the distance from any manifold perforation to the pile's edge to be roughly the same (this includes the pile height).

Step 2. Calculate Dimensions of Biofilter

Where:

V = Pile volume (cubic feet)
H = Pile height (feet)
W = Width (feet)
L = Length (feet)

The simple rectangular pile dimensions formula is:

$$L = V \div H \div W$$

You can solve for pile height for a rectangular pile, if you assume a pile length and width. If you have a height in mind, you can swap height with length or width to solve for that instead.

For any edges that are not a vertical 90 degrees and are instead a triangular 45 (for a biofilter in a bay with three walls, for example), for the dimension with the non-vertical edge, divide the pile height by 2 and add that number to that length.

Example

V = 1,350 cubic feet
H = 3 feet
W = 18 feet
L = 25 feet (Add ½ height to length if one side of the biofilter is non-vertical: 25 feet + 1.5 feet = 26.5 feet length)

25′ L = 1,350 V ÷ 3′ H ÷ 18′ W

The manifold design will be based on the length and width of the pile. Following this example we'll keep all manifold perforations 3 feet from any edge (equal to the pile height) and space the pipes evenly across the width of the biofilter. So for an 18-foot-wide biofilter, with a 3-foot height, we'll subtract 3 feet from both edges to give us a 12-foot-wide manifold. The length of the perforated zone of the manifold will be determined the same way. Since the pile length is 26.5 feet, we'll subtract 3 feet from both edges to determine the perforated zone, but the actual pipe length will depend on whether the pipes stick out at the ends or not as well as the various connections (if they stick out, they will be easier to find in the future).

Next we have to decide how many distribution pipes to place in our manifold. The pipes must be evenly spaced across the manifold, so we're looking for factors of 12 feet for potential spacing intervals (for instance, 6 feet, 4 feet, 3 feet, and 2 feet). Since it's such a short pile, 3-foot and even 2-foot spacing would not be overkill, but for a standard manifold design you ideally have an even number of pipes, so rather than going to a single center pipe, the header that feeds the manifold falls in between two center pipes. According to flow dynamics, this encourages more even distribution across the manifold. Of these four choices, only 4-foot spacing provides an even number (pipes at 0 feet, 4 feet, 8 feet, and 12 feet).

Like other ASP design considerations, sometimes it's good to know more before making final decisions. In this case the pipe size and associated static pressure may be a determining factor in how many pipes you choose. You may even go back to the pile dimensions to adjust the width, so that you can fit a greater (and even) number of pipes. For example, if 3-foot spacing allows you to go a pipe size lower than 4-foot spacing, it's probably worth it in terms of cost, and will probably add distribution performance. On the other hand, smaller pipes could also cost the system in terms of static pressure loss, so I recommend checking multiple options.

In the example, the number and size of the distribution pipes were both very obvious after looking at the options. Not every case is straightforward, however, and while it might seem like you are splitting hairs over 1/4 inch of static pressure, I can assure you that this exercise is absolutely critical. Static pressure increase is non-linear, and designs that don't take this factor into account can become inefficient. I've seen people build systems with lots of 2-inch pipes, thinking they've covered their bases without having done the math. When I've tested the static pressure it's through the roof. Remember, the system may operate at both a lower and a higher flow rate at different times depending on the pile size, porosity, and moisture, so it's good to have some buffer.

See the *Manifold Design Rules* and *Air Channel Design Rules* sidebars (pages 231 and 234 respectively) if you need a refresher on designing for uniform pressure and airflow.

Step 3. Design and Size Biofilter Manifold

Here's what you need to know:

a. Target aeration rate (CFM from biofilter pipe header)
b. Length of the header ducts
c. Width of the manifold
d. Length of the distribution ducts

EXAMPLE

a. 300 CFM from header duct
b. 20 feet
c. 12 feet
d. 30 feet (this accounts for non-perforated sections as well)

Note that the header pipe that feeds the biofilter will be the same size as the header pipe for the ASP, unless more than one fan simultaneously feeds the biofilter, in which case the header pipe and all of the distribution pipes should be sized based on that cumulative airflow.

3.1 Calculate Static Pressure per Foot from Biofilter Header Duct

___″ WC per 100′ from Biofilter Header Duct ÷ 100 = ___″ WC/FT from Biofilter Header Duct

0.57″ ÷ 100 = 0.0057″

3.2 Design Length of Biofilter Header Duct

For sizing and estimating static pressure, consider the length of the biofilter header duct as being from the fan to the manifold tee. Make an educated guess if you don't know the exact length. If it's a long run, you might need to increase the pipe size to reduce pressure loss.

___′ Length of biofilter header duct

20′

3.3 Calculate Total Static Pressure Loss from Biofilter Header Duct

___′ Length of Biofilter Header Duct × ___″ WC/FT from Biofilter Header Duct = ___″ WC Static Pressure Loss from Header Duct

20′ × 0.0057″ = 0.114″

3.4 Determine Number of Distribution Ducts Minus One

Formula		Result	Example
___ Manifold Width ÷ ___ # Distribution Duct Spacing	=	___ # of Distribution Ducts minus 1	12′ ÷ 4′ = 3

3.4.1 Determine Number of Distribution Ducts

Formula		Result	Example
(___ # of Distribution Ducts minus 1) + 1	=	___ # of Distribution Ducts	3 + 1 = 4

3.5 Determine Airflow per Duct

Formula		Result	Example
___ CFM from Header Duct ÷ ___ # of Distribution Ducts	=	___ CFM/Distribution Duct	300 CFM ÷ 4 = 75 CFM

3.6 Size Distribution Ducts

Use table 9.6 to select possible pipe sizes based upon the estimated airflow rate in each distribution duct. Duct size should be based on the total length of pipe,* minimizing the cumulative static pressure in the biofilter.

Formula		Result	Example
___ CFM Through a ___″ Duct	=	___″ WC per 100′	~~75 CFM through 3″ = 1.35″~~ Not desirable due to the higher static pressure. 75 CFM through 4″ = 0.34″

3.7 Calculate Static Pressure per Foot from Biofilter Distribution Ducts

Formula		Result	Example
___″ WC per 100′ from Duct (A, B, etc.) ÷ 100	=	___″ WC/FT from Distribution Ducts/Air Channels	0.34″ ÷ 100 = 0.0034″

3.8 Calculate Total Length of Biofilter Distribution Ducts

Formula		Result	Example
___′ Length of Duct (A, B, etc.) × ___ # of Ducts	=	___′ Total Length of Ducts (A, B, etc.)	30′ × 4 = 120′

3.9 Calculate Total Static Pressure Loss from Biofilter Distribution Ducts

Formula		Result	Example
___′ Length of Ducts (A, B, etc.) × ___″ WC/FT	=	___″ WC Static Pressure Loss from Ducts (A, B, etc.)	120′ × 0.0034″ = 0.408″ Ducts A. 4″ distribution ducts (120′ at 0.34″ WC per 100′) = 0.408″ WC Ducts B. 6″ distribution ducts (12′ at 0.16″ WC per 100′) = 0.019″ WC (This section accounts for the manifold, where the header duct splits, running to the four distribution ducts.)

Repeat steps 3.4 and 3.9 for each unique duct section (any duct with a variable diameter or airflow).

3.10 Determine Total Static Pressure Loss in Biofilter Manifold

Formula		Result	Example
___″ WC Static Pressure Loss from Header Duct + ___″ WC Static Pressure Loss from Ducts (Ducts A) + ___″ WC Static Pressure Loss from Ducts (Ducts B) + ___″ WC Static Pressure Loss from Ducts (Ducts C)	=	___″ WC Total Static Pressure Loss in Biofilter Manifold	0.114″ + 0.408″ + 0.019″ + n/a″ = 0.541″

* Including non-perforated zones.

Static Pressure from Biofilter Media

Along with the manifold, the biofilter media places added static pressure on the blower and needs to be taken into account in terms of blower selection. In order to estimate the static pressure, you'll need a rough estimation of the biofilter's height, as well as the estimated retention time of exhaust in the biofilter (see *Step 1. Calculate Media Biofilter Volume and Exhaust Retention Time* and *Step 2. Calculate Dimensions of Biofilter*). The target retention time (or TRT) is used to calculate the velocity of the exhaust through the biofilter, which in turn is used to estimate static pressure loss.

Step 4. Estimate Static Pressure from Biofilter Media

Use table 9.9 to estimate the added resistance created by flow through the biofiltration material.

INPUTS

___ seconds = target retention time or TRT (step 1, convert minutes to seconds)
___ feet = biofilter height (step 2)

ANSWER

___ inch WC = estimated resistance from biofilter media

EXAMPLE

120 seconds
3 feet

0.003 inch

This is a very low static pressure from the biofilter media. Because it has such a long retention time in the biofilter, the velocity through the material is very slow, which equates to very little friction.

Total Static Pressure Loss from Biofilter Manifold and Media

The final steps are to sum the total loss from the biofilter and then to add that number to the total ASP system pressure loss, before selecting a blower (see the previous *8.8 (Optional) Add Static Pressure from Biofilter*, page 247). Remember to select a blower that can handle the noxious operating conditions associated with negative aeration.

Step 5. Calculate Total Static Pressure Loss from Biofilter Manifold and Media

___″ WC Total Static Pressure Loss in Biofilter Manifold
+ ___″ WC = Estimated Resistance from Biofilter Media
= ___″ WC Static Pressure from Biofilter

EXAMPLE

0.541″ + 0.003″ = 0.544″

Table 9.9. Resistance from Biofilter Media

RETENTION TIME (SECONDS)	15	30	45	60	75	90	105	120
Biofilter Height (Feet)	Inches WC	Inches WC	Inches WC	Inches WC	Inches WC	Inches WC	Inches WC	Inches WC
2	0.034	0.010	0.005	0.003	0.002	0.002	0.001	0.001
3	0.107	0.032	0.016	0.010	0.007	0.005	0.004	0.003
4	0.242	0.072	0.036	0.022	0.015	0.011	0.008	0.006
5	0.453	0.136	0.067	0.041	0.028	0.020	0.015	0.012

Note: These static pressures were calculated using equation 15.24 and airflow resistance coefficients (table 15.1) for wood chips described in Roger Tim Haug, *The Practical Handbook of Compost Engineering* (Boca Raton, FL: Lewis Publishers, 1993), 525–26.

Biofilter Location

The location of external ASP biofilters is an important consideration in terms of long-term functionality as well as potential impacts on the surrounding environment. There may be limited options for siting biofilters on many sites, especially on small and urban sites and where the ASP is an add-on to a site that is already laid out and built. Ideally, the location of a biofilter is not an afterthought in your design, as it may ultimately be a determining factor in how other elements of the site are designed.

Treat the biofilter as a component of your primary composting phase. In choosing the location of one or more biofilters, aim to meet as many of the following criteria as possible:

- Minimize the distance from the ASP to the biofilter to gain maximum operating efficiency. This is especially true for compost heat recovery systems, where heat is lost in transit.
- Maximize setbacks from neighbors and avoid air drainages that flow to neighbors.
- Plan appropriate access to the biofilter for periodic maintenance and media replacement. You may need to moisturize the material periodically, so proximity to water is of benefit.
- Shade and windbreaks will reduce evaporation and overdrying, and also act to filter and buffer airflow.
- Avoid locating the biofilter where site and roof runoff could saturate the media. Also keep in mind that excessive snow (from roofs, plowing, and so on) will hinder airflow.

Condensate Traps

One recommendation with negative aeration is to place a condensate trap in the exhaust line prior to the fan. A condensate trap is simply a point where the exhaust can cool, discharging moisture. The idea is that this reduces the wear and tear on the fan. The *On-Farm Composting Handbook* describes using a 55-gallon drum for this purpose, connecting exhaust pipes in and out of the drum. The drum slows the airflow and pools moisture enough to cool the exhaust, acting as a point of collection. It has an air-locking drain that passively releases condensate as it accumulates and overflows.[29]

Building the ASP and Biofilter Manifold

If you construct your own ASP manifold, it will help if you or someone assisting with the project has some experience with installing irrigation or plumbing. Even experienced builders sometimes hire contractors to install more complicated systems. Obviously it depends on your experience, tolerance for minor imperfections, desire to learn, and budget, among other things.

The good news is you're dealing with low pressures and air, so unless your system is negatively aerated and operating in a confined space (in which case be really careful), you should have enough wiggle room to learn as you go. Also, you can quickly access videos on how to do just about anything just by picking up your smartphone. The following section is by no means an exhaustive how-to for ASP manifold construction, but it covers a few of the more important lessons I've learned in building and seeing ASPs get built.

Cutting and Connecting the Pipes

For cutting PVC and HDPE pipes, use a circular chop saw (wear goggles always). It makes a very precise, clean cut. Make sure that if you're making cuts on the compost site, you have a way to collect the debris, such as a tarp. Even better, do it in your shop, where there is little chance of contaminating the compost (keep any clean sawdust separate).

Pipe materials are all slightly different in terms of how they are fit together, and many manifolds have more than one material, which will require coupling. One useful coupling is a Fernco or similar rubber fitting, which is virtually universally compatible. Fernco makes couplings as well as both acentric and eccentric reducers that are rubber on both sides and tighten with pipe clamps.

Due to the low pressures, some people working with PVC dry-fit all or parts of the manifold. I have used petroleum jelly (such as Vaseline) on PVC because it creates a seal while still allowing me to take the system apart and easily reuse the parts, avoiding use of noxious primers and glues. If you are dry-fitting the pipe (or using petroleum jelly), the design of the system will need to hold itself together somewhat. Self-tapping screws and a couple of straps in strategic places can usually do the trick.

Under negative pressure a glued system is recommended to avoid leaking compost exhaust. This is especially important where the manifold and exhaust pipes are indoors (in a chase or greenhouse, for example). Seal all of the pipes that come after the fan on a negatively aerated system, because these will have positive pressure and can leak exhaust. Glue is also advisable on the leachate collection system.

Two other common materials are galvanized duct and HDPE. Galvi duct is typically fastened together with self-tapping metal screws, then sealed with tape. HVAC installers use glues, but tape is enough for the low ASP pressures. Aluminum tape is stronger and looks cleaner than duct tape in my opinion.

HDPE is typically heat-fit, which takes skill and equipment, so if you are doing a lot of HDPE, best leave it to someone who knows what they are doing. HDPE, which is the preferred material for aboveground air channels, can be easily connected with Fernco fittings. In fact, when in doubt, use a Fernco as a general rule.

Slope

In the design and construction of the entire ASP manifold and leachate collection system, give careful consideration to how moisture will flow within the system. Condensate is especially a concern for negatively aerated systems, because of the moist and corrosive exhaust vapors, but it comes into play in positively aerated situations more than you might anticipate.

The manifold designs should account for condensate by sloping ductwork (around 2 percent) toward condensate traps located at low points in the system. Keep a close eye on slopes during construction, using levels or "snapping a line" as needed to ensure smooth flow. Any points where a duct transitions from hot to cold (such as inside to outside, or compost to ambient) or vice versa are a likely point for condensation.

Forming Concrete and Concrete Air Channels

Unlike other parts of the system, most people hire contractors to do concrete work. Experienced concrete installers will do the job well, but chances are good that they have never built a composting system before. Don't expect them to read your mind: Make

Figure 9.31. The simple wooden form above was used to create subsurface air channels in a concrete ASP slab. The system was designed as part of an experimental compost heat recovery system, which is why the channels are surrounded by 1-inch PEX and blue board insulation. Interestingly, the blue board and concrete were roughly the same price by volume, so we opted for very thick blue board in between the air channels. Courtesy of Highfields Center for Composting and Vermont Sustainable Jobs Fund.

absolutely sure that they understand the importance of things like pad and air channel slope.

In my experience concrete contractors usually work fast because they have an order of concrete coming and limited time to get the forms in place. This means that there is little time to make modifications once a project gets started. Be ready for them and talk through how the channel forms and pipe connections will be set ahead of time, so that there aren't any last-minute "oh shit" moments. Unlike other parts of the system, concrete is very hard to undo.

The formed concrete ASP designs in figure 9.20 were constructed using a really simple air channel form as seen in figure 9.31. The contractor placed a small spacer in between two boards that formed the side walls of the air channels, then poured concrete around it. Once the concrete had set, he cut the spacer with a Sawzall (reciprocating saw) and was then able to easily knock out the forms.

Testing ASP Design Performance

Once the ASP system is constructed and before building your first pile, test the manifold to ensure its basic function. You're testing for leaks, preferential or uneven flow, and excessive static pressure. This will give you the baseline for airflow and static pressure without the compost. You will also want to test the flow and static pressure with the ASP loaded. Comparing the system with and without compost will give you an indication of how the compost mix is performing in terms of its permeability and how well you've formed the pile, which is important for how well the aeration is distributed. Some of the common tests for ASP function and performance are described in more detail below.

Leak Testing

Turn the fan on and check for leaks; you should be able to hear or feel them. With positive ASP systems you can also pull smoke through the blower and into the system, which will allow leaks to be seen. Fix leaks with glue, duct tape, or, my favorite, aluminum-foil tape, which maintains its "stick" well under temperature extremes. Keep an eye (and ear) out for leaks in the future. Keeping the system pressurized is key to maintaining an evenly distributed, efficient system.

Airflow and Distribution

For systems with multiple branches and air channels, testing the distribution of air through the manifold is especially important and can be easily tested with the right tools. Pulling smoke into the system is a quick way to visually assess the distribution throughout the ASP, both laterally and lengthwise across the manifold.

To measure actual flow rate (CFMs), as well as flow distribution, the most common method is to use an *anemometer*. I have used a thermal anemometer (hot wire), which uses a temperature gradient to estimate air velocity. Based on the velocity and size of the duct, the CFMs can be calculated using tables or an app (I use an iOS app called Duct Sizer).

Carefully drill a small hole in any duct you wish to test. (Be sure to plug this with a rubber stopper or tape after you've taken the measurements.) You may want to measure flow in the header pipe to get the total flow rate in the system, as well as the flow in individual air channels, which will show you the distribution. Take your measurements in a straight section of pipe before it reaches the perforated section. The air velocities between air channels should be close to equal (the tool is not entirely precise).

A third way to estimate total airflow is to measure the static pressure in the manifold, then cross-check that measurement against your fan's blower curve. For example, if the fan in figure 9.25, page 249, is operating at 2.2 inches WC, then the fan will move approximately 470 CFM. Static pressure will give you an estimate of total flow, but it is not a method for measuring flow distribution between channels. More on measuring static pressure is described in the next section.

If fan speed is controlled with a variable speed drive, I recommend measuring the flow and static pressure at different speeds on the dial, in essence calibrating the controller. You can then use this information to make more informed decisions about speed and timer settings.

Static Pressure

Static pressure is one of the primary assessment tools for the ASP designer and operator. Much of this chapter is devoted to design methods for estimating and minimizing static pressure, because ultimately, the choice of fan is based upon the fan providing your target aeration rate at the estimated static pressure.

Once the system is built, I recommend testing the static pressure in the system and ideally checking it multiple times throughout the course of operation. You can leave a static pressure gauge (manometer) plugged in and make checking it a part of your monitoring routine. Static pressure is an indicator of both operating efficiency and total flow.

As a starting place, you can measure and compare the actual ASP static pressure with both the estimated design static pressure and the blower performance curve. Typically, you want your system to be operating somewhere in the middle of the curve, and certainly not above the rated static pressure. Flow starts to drop off at a certain static pressure (non-linearly), so operating above that static pressure uses more energy but doesn't really provide more flow. At a certain static pressure fans are basically operating at 0 CFM.

Static pressure will be a good indicator when there is no flow to the compost. Really low pressure might indicate a large leak somewhere, whereas high pressure is an indicator that there is a blockage in the system.

By testing the manifold prior to constructing the first pile, you'll know how much of the pressure is from the manifold versus the compost. If there is a problem with static pressure in the future, it is probably a compost porosity issue or possibly leachate, condensate, ice, or organic material clogging the manifold. Porosity issues can be dealt with by increasing the ratio of bulking agent (say, wood chips) in the recipe (see *Recipe Development for ASP* on page 265).

To manually measure static pressure is a simple process using a *manometer* or static pressure gauge. Simply connect the gauge using a piece of ⅛-inch rubber tubing connected to an NPT brass nipple, which you will have to install in the duct (the threaded side connects to the duct; the barbed side connects to the rubber tube, as seen in figure 9.32).* Theoretically, you can measure static pressure anywhere in the system, so choose a place that you'll frequently visit, like near the fan controls. Systems that have more sophisticated controls may have a digital manometer,

* Note there is a different connection point on the manometer for positive and negative pressures.

Figure 9.32. A manometer connected at the first tee coming from the fan header duct. Some brass NPT nipples have little valves, which are convenient for closing off the pressure if you are checking multiple points.

Figure 9.34. Hot wire anemometers (*left*) have a sensor on the end of a probe that uses a temperature gradient to estimate air velocity. Velocity and duct size can be used to calculate CFMs. On the right is a Koster moisture tester.

Figure 9.33. A homemade static pressure gauge (manometer) literally pushes water—in this case it looks like antifreeze—up a column (tube).

which can be monitored and logged as part of the controls system. It may be useful to record manually collected data in a log, so that you can track trends. Static pressure will probably increase slightly over the course of a pile's life as the compost settles and shrinks. Keeping an eye on this will help you make more informed decisions about fan speed and timing, or turning the pile to break up dense patches and preferential channeling, which is undesirable.

ASP Compost Pile Construction

The process of blending and forming a compost pile for ASP has some slight differences from other methods that we'll briefly describe in this section. The five basic steps are:

1. Develop a recipe for ASP
2. Blend feedstocks
3. Create a wood chip base or plenum
4. Construct the pile
5. Cap the pile for odor control

Considerations for performing these steps are described below.

ASP Performance Testing Tools

The two main tools for testing ASP performance are the anemometer (*ana-MOM-iter*) and manometer (*MAN-om-iter*).

Anemometer

Anemometers are not cheap ($500 or more) and—as described in *Airflow and Distribution*—they aren't entirely necessary for most people, as you can make a reasonable estimate of airflow by cross-referencing the system static pressure with the fan performance curve. However, an anemometer is a very good tool for larger operations with multiple ASPs and a higher likelihood of system imbalances. They are also useful for designers and technical service providers, who need to assess a system's performance with more precision.

There are multiple types of anemometers, many used to measure wind speed outdoors, and some that are installed permanently in a duct. For ASPs, you want one that can be temporarily inserted in a duct, so that multiple ducts can be quickly measured. I have purchased a used thermal or hot wire anemometer on eBay for around $300 (see figure 9.34). These have a sensor on the end of a "probe" that uses a temperature gradient to estimate air velocity. Based on the velocity and size of the duct, the CFMs can be calculated using tables or an app (I use an iOS app called Duct Sizer). Carefully drill a small hole in any duct you wish to test. (Plug these holes with a rubber stopper or tape after you've taken the measurements.)

The airflow in a duct is not uniform. The sides create friction and turbulence, so the flow at the center is faster than at the edges of the duct. This means you'll get a different reading depending upon where you hold the sensor. For smaller ducts (6 inches or less), I typically take just one measurement in the center of the pipe, whereas with larger ducts (8 inches or more) I take measurements in both the center and halfway between the center and edge of the pipe (half the radius), then use the average of those two measurements.

Manometer

The static pressure gauge commonly used in ASP performance testing and monitoring is called a manometer, and the unit is inches of water column. Quite literally, this unit refers to the height a column of water can be lifted (pushed) under pressure. For this reason, a static pressure gauge can be cheaply made using a pipe and water—or antifreeze if it might freeze (see figure 9.33).

There are several brands of manual (analog) manometers. I am familiar with Magnehelic dial manometers and have purchased inexpensive ones on eBay in new condition. Manometers come in all different ranges, so choose a range that fits your estimated static pressure (most will want the 1 to 10 inches of WC range).

Magnehelics connect to the duct system with ⅛-inch barbed brass fitting to which you'll connect ⅛-inch rubber tubing. You'll need to install a ⅛-inch NPT to ⅛-inch barbed nipple in the duct (the threaded side connects to the duct; the barbed side connects to the rubber tube, as seen in figure 9.32). Choose a place to install it that you'll frequently visit, like near the fan controls.

Recipe Development for ASP

There are a few unique pile characteristics that should be taken into account when developing a recipe for ASP composting. Creating a high-porosity mix is the number one recommendation. Porosity allows the air to flow more evenly throughout the pile, mitigating preferential flow, which also decreases the work of the fan.

Of course, porosity is important for all compost, so what constitutes high porosity? Some recommendations are as high as 40 to 50 percent pore space. I would approach it in terms of the ratio of coarse, highly porous materials, such as wood chips, to other feedstocks. My typical recipe recommendation for a turned pile is 5 to 10 percent chips or bark by volume, so for an ASP, I would recommend a 15 to 20 percent blend of chips or bark by volume. Use the recipe development techniques described in *Compost Recipe Development* in chapter 4 (page 83), as a starting place, then adjust the recipe based on your observations.

Some ASP composters screen out the wood chips after composting and reuse them as their bulking agent. As a feedstock, recycled wood chips have very different characteristics than do fresh chips and will provide little available carbon to the mix, which should be factored into recipe calculations. Some composters leave them out of their calculations altogether, but I might test them and then use those analyses in my calculations at one-quarter to one-fifth of their actual volume, because from what I've seen, while chips break down very slowly, they are anything but inert.

Moisture content (MC) is another recipe characteristic that needs to be considered slightly differently in terms of the ASP process. Compost has a tendency to become excessively dry with ASP, which can short-circuit the process. By starting at a slightly higher MC, there is the potential to reduce the dryness of the compost on the back end of the ASP process. The normal target I recommend is 60 percent MC, but the prime range is between 55 and 60 percent. With ASP you should try not to start below 60 percent MC and could experiment with starting a percentage point or two higher and observing the mix's density, the blower static pressure, the overall homogeneity of the compost that comes out, and of course odors. To counteract the odor risk of composting at higher moisture, use a thicker biofilter capping layer as described in *Capping for Odor Control* on page 268.

Blending Feedstocks

In order for uniform aeration to take place, a homogeneous compost mix is critical, which requires thoroughly blending the raw feedstock. Many ASP composters use a feed mixer, but aggressive blending with a tractor bucket, or a manure spreader (not recommended for food scraps), is a more common strategy for smaller composters. On a micro scale, blending by hand or with a small mixer is an option, but the typical layering method most commonly used for backyard systems is not ideal, as the denser layers will act as barriers driving airflow to the sides of the pile.

Creating the Wood Chip Base or Plenum

To increase air distribution throughout the pile, ASP composters cover the air channel, or channels, with a thick layer of wood chips, often called the *wood chip plenum*. The theory is that the air flowing in or out of the air channel will first hit the coarse wood chips, slowing and spreading the flow more evenly throughout the pile. The larger wood chip particles also reduce clogging in the air channel's perforations.

General recommendations are to form a wood chip bed either directly over each individual air channel (6 to 12 inches deep and a third the width of the pile) or a wide wood chip bed across the perforated section of all of the air channels (6 to 12 inches deep, stopping at least 3 feet from any wall—less for really small systems). Remember the weight of the compost will compress the wood chips, so a deep bed is better. There are a couple of benefits to covering individual channels versus making a wide bed: (1) You'll use less chip overall and chip can be expensive (or free, it just depends); and (2) as you form the pile, the loader's tires can straddle the channel, avoiding

Figure 9.35. Wood chips surrounding an HDPE air channel are meant to slow and distribute air coming out of the channel perforations.

compressing the chips. If you form a wood chip bed across the base of the pile, consider forming it as you build the pile to minimize driving on the chips.

Constructing the Pile

There are strategies for forming an aerated static pile so that it performs optimally. Take care not to drive on the base of the pile, which will compress it and limit airflow through portions of it (this is true of all piles, not just ASP). It can be challenging to stack the pile really high without driving on it a little bit, even with a large loader. For this reason, some composters have swapped out the traditional loader for a telehandler, which has a telescopic bucket and can dump into places that only an excavator could traditionally reach (without driving on the pile). These have a lot of advantages for material handling and management. The main drawback with telehandlers is the cost of maintenance. Along with not driving on the pile, try not to drive on the wood chip plenum when forming the pile, as described.

A less obvious consideration in pile construction is the importance of the pile's geometry in relation to the design of the ASP manifold. We've talked about the importance of uniformity throughout this chapter, but constructing the pile is where all of that design actually counts. Picture a radius around every perforated section of air channel. The pile's height is the radius, and the edges of the pile should be roughly that same radius. Of course the side of a pile is more triangular or parabolic than circular. Still, however imperfect, the goal is to have the distance from the air channel perforations to the top and bottom edges of the pile be close to equivalent (see figure 9.19, page 235). The air will take the shortest path through the compost, so this minimizes the tendency for portions of the pile to be short-circuited. It may be helpful to have markers in place to delineate where the edges of the piles should be.

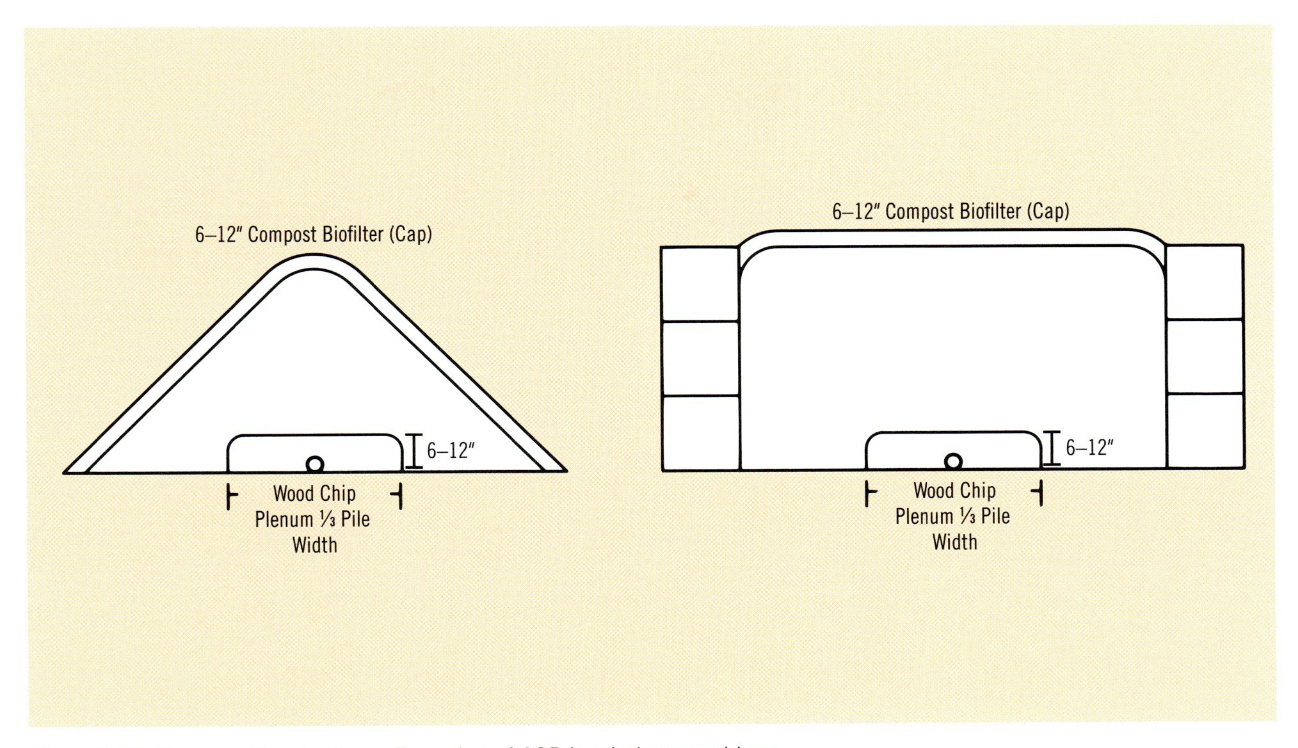

Figure 9.36. Cross-sectional pile configuration of ASP in windrows and bays.

Figure 9.37. Covering a newly formed pile with a Gore-Tex cover at Big Reuse.

Capping for Odor Control

To finish forming the pile, ASP composters use a thick blanket of compost or feedstock as capping on the pile, which acts as a biofilter for odors. Positively aerated piles typically have a 1-foot-deep cap. Negatively aerated piles have a thinner cap, around 6 to 9 inches, because they will usually have an external biofilter where the majority of the odor is treated.

The importance and usefulness of capping should not be underestimated in any composting method, but especially not in positive ASP where odors are forced out of the pile. The reality is that even perfect food scrap compost recipes generate some odor in the initial stages as proteins, fats, acids, and liquids are released. While recipe and other management issues should not be ignored, in my experience the vast majority of odor and vector issues in small systems can be addressed by good capping practices.

Depending on the shape and size of the pile, the method for capping can be slightly different. For most ASP operations, piles are closely stacked together, limiting access once the pile is constructed, so you'll need to cap the pile as you go. For wide piles, try to make the top of the pile as level as possible.

Choosing which material to use for capping has a big impact on the overall process flow of the operation. Capping with semi-mature compost is the preference of many operations and has a number of real advantages: (1) Compost is very effective—more so than other likely options, such as horse manure and leaves; (2) using "pre-heat-treated" compost on the outside of the pile remedies one of the criticisms of ASP, which is that the outside of the pile does not reach treatment temperatures; and (3) for operations whose main goal is to process the greatest possible volume of food scraps, recycling compost as capping material, rather than importing another feedstock, maximizes the ratio of food scraps to other imported feedstocks. Effectively, for every yard of compost produced, there will be a higher ratio of food scraps that went into that yard.

Every composter will have to weigh these advantages against the disadvantages of capping with compost, the most obvious being that you are handling the compost you've capped with more times to get it to a finished product. On average, most operations would need to devote about 15 to 20 percent of their output—after, say, three to six months—back in as a biofilter. From a handling perspective, this means moving it back to day 1. It is a pretty big con, given that for many composters, the rate at which finished compost is produced is critical.

Other capping material such as high-carbon animal manure/bedding or leaves can also function well as a biofilter in many scenarios. One middle ground would be to minimally process alternative biofilter feedstocks such as leaves, say for three weeks until they've been heat-treated, then cycle them into the mix as a biofilter.

Avoid using really wet and dense materials for capping, as these will inhibit airflow.

Managing ASP Composting

Experienced and nascent composters alike will need to learn what composting looks like using the ASP process. While the fundamental principles of composting apply, there *will* be a learning curve. As with other composting processes, temperature and odor are the two main indicators, and while the ASP process will ultimately offer you superior control over these conditions, it can also exacerbate them.

New adopters are often skeptical in the first days and months of ASP deployment. In my experience, as some time passes, former turned windrow composters in particular are able to do the math and recognize all of the time, space, and costs they are saving with ASP.

Diligent monitoring, observation, documentation, and trial and error are all critical. Operators who have managed ASP systems over time say that making small and calculated adjustments in response to monitoring results will mitigate ASP's potential for extreme swings in pile conditions. During the start-up phase, remind yourself frequently to take the long view. Diligent monitoring and documentation will speed up the learning period and build confidence in your ability to make more informed management decisions over time.

Compost Heat Recovery: Then and Now

by Gaelan Brown, Brian Jerose, and the rest of the Agrilab Technologies team

Imagine being able to heat water, buildings, and greenhouses using energy captured from composted biomass wastes while also creating compost products that are worth more per ton than coal. Compost aeration with heat recovery (CAHR) can do just that, while dramatically improving the economics of compost production.

Modern innovations around an old idea have made compost power a reality that a growing number of commercial compost operations and farms are putting to work around the world. In this sidebar you'll learn how modern compost heat recovery with aerated static pile (ASP) processing reduces material handling costs while capturing continuous and predictable heat from the process.

Aside from leveraging the natural biological energy of compost to work for sustainable agriculture and energy, CAHR systems often have the added benefit of reducing or eliminating odors and greenhouse gases like methane that are often otherwise issues with compost production.

Since the dawn of agriculture, farmers and foresters have understood that when biomass materials such as grass, food scraps, slaughterhouse waste, manure, or sawdust undergo the natural process of aerobic rotting, otherwise known as composting, heat is produced.

More than 2,000 years ago in northern China (and to this day in certain regions), peasant farmers commonly built raised "hot beds" filled with a 3-foot-deep layer of fresh manure underneath the topsoil, which produced enough heat to extend their growing seasons by a month or more in the autumn and spring.

The same is true in Northern Europe from the Middle Ages through the early part of the 20th century, with animal manure being used as a heat source under seedbeds. It was also common, particularly in Eastern Europe and Iceland, for animal manure to be piled up against the outside walls of homes, providing winter heat and insulation.

In the late 1970s a French farmer named Jean Pain invented a simple compost heat recovery process. Large piles of wet wood chips were piled up with waterlines buried in the material and a small anaerobic digester tank placed in the center. He was able to capture continuous streams of hot water for 12 to 18 months on each pile. The Jean Pain method ultimately was too prone to inconsistent results and required too much labor to be economically viable, but his work inspired the modern advances available today.

It might be seen as ironic or sad that the advancement of Western civilization also resulted in entire populations spreading their human/animal manure and other biomass wastes, causing runoff pollution into rivers, groundwater, and landfills instead of composting these materials as had been naturally done in the past. The good news is that advanced CAHR systems have relatively low up-front capital cost, and they dramatically flip the economics of composting with investments that can pay for themselves in just a few years for most farms and commercial/municipal compost operations.

Investments in expanded compost and soil amendment production around the world have sparked significant innovation in compost heat recovery. Innovation has been particularly strong in the past decade in Vermont and Canada, where several compost practitioners, engineers, and tinkerers have—independently of one another at first but eventually through collaboration—developed several economically viable methods of recovering predictable amounts of heat while reducing the costs of compost processing through computer-controlled compost aeration systems.

The most advanced and economical approach to CAHR has also been developed by Vermont-based Agrilab Technologies Inc. (AGT). The patented AGT approach uses negative aeration to keep the static piles well oxygenated, which reduces the need for the material to be mechanically tumbled and shortens the processing time required. Active aeration also reduces odors and methane emissions. The negative aeration system pulls air down through the compost into pipes or channels in the floor. The humid, 140 to 170°F compost vapor is run through a specialized heat exchanger by the same aeration fan, capturing continuous heat for use on-site. The exhaust from the aeration system can be vented into a biofilter to capture odors, or it can be vented back into another zone of the active compost to retain moisture and heat in the material. AGT systems in general pay for themselves with material handling savings and energy savings in two to five years.

In 2005–06 Agrilab Technologies installed its first compost heat recovery system using its patented technology at Diamond Hill Custom Heifers in Sheldon, Vermont. The system generated more than 200,000 Btu per hour continuously, and has been earning the farm tens of thousands of dollars annually since then. The first system was a success, but it took up a lot more space and cost a lot more money than AGT's current technology platform.

In 2011 Vermont-based CompostPower.org published a Modified Jean Pain Method Design Guide for do-it-yourselfers online while offering free advice and encouragement to people across the globe who wanted to do their own experiments. In 2012 they produced a short video demonstrating a Compost Power project producing water of 120°F or more in the middle of Vermont's winter, including a time-lapse segment showing every step of the construction process.

By 2013 CompostPower.org had assisted with dozens of projects around the world, and AGT had reorganized with new leadership and an expanded team of people from the Compost Power Network. Since then AGT has redesigned a standard platform of mobile "plug and play" computer-controlled aeration and heat recovery systems, including remote monitoring and controls, and units qualify for lease-to-own financing.

The resurrection of Jean Pain's methods, along with development of entirely new approaches like AGT's system, has created exciting new momentum. This momentum has also taken root across the globe with many people learning from the open-source experiments of others, finding ways to capture heat from compost.

As of this writing standardized CAHR systems are being sold with bank-approved financing, and continuous research is being undertaken at AGT project sites including the University of New Hampshire's Organic Research Dairy Farm. Compost power is now a market-ready and economically viable concept, and there is tremendous opportunity for innovative applications of this old/new opportunity.

Case Study: CAHR System

The Compost Hot Box 250R-D is a state-of-the-art, computer-controlled compost aeration system with heat recovery built into a 20-foot shipping container, developed by Agrilab Technologies and installed at Vermont Natural Ag Products (VNAP), a farm-based commercial compost company in Middlebury, Vermont.

The Hot Box 250R-D is "plug and play" with four-zone aerated static pile composting pads with up to 800 cubic yards in process at any given time. One of the four zones is used as a forced-hot-air positive ASP drying pad to dry finished compost products, while three zones are used for primary negative aeration processing.

Renewable thermal energy is used primarily for a forced-hot-air product drying system to dry finished compost to below 25 percent moisture content before it is screened and bagged. This significantly improves screener efficiency and eliminates the propane-drying system formerly used on-site, saving a lot of money on the

Figure 9.39. Leaving a plume of steam in its wake, the Komptech windrow turner at Vermont Natural Ag Products in Middlebury, Vermont, is integrated into the negative ASP and heat recovery system. Systems that allow for more frequent turning alleviate many of the challenges associated with static piles. Courtesy of Agrilab Technologies Inc

Figure 9.38. Insulated flex hoses connect the four zones of ASP with Compost Hot Box 250R-D. Courtesy of Agrilab Technologies Inc.

Figure 9.40. Agrilab Technologies Drum Dragon at Vern-Mont Farm in Vernon, Vermont. The system pulls warm air and vapor out of an in-vessel rotating drum, which is used by the farm to make bedding from their cow manure. The compost/bedding recovery system continuously processes fresh manures and bedding, making for targeted energy recovery reprocess. The fan pulls 140 to 155°F exhaust to a hydronic heat exchanger that can sustainably heat water to 120 to 140°F. Courtesy of Bob Spencer.

overall process. Heat captured is also used for radiant floor heating in the bagging facility.

Based on the first six months of operations, VNAP expects to see return on investment in under two years. Savings come from a 50 percent reduction in material handling costs and a 50 percent reduction in processing time, along with significant reduction in winter space heating costs.

This case study shows installation and operational details between November 2016 and August 2017. Phase 1 implementation was funded in part by Rural Energy for America Program (REAP), USDA Rural Development. The Phase 2 expansion in the summer of 2017 was supported by a grant from the Vermont Clean Energy Development Fund (CEDF).

CHAPTER TEN

In-Vessel Composting Systems

The term *in-vessel composting* typically refers to the production of compost within a contained vessel. For our purposes, the term will refer specifically to the fabrication of systems that are designed to manage the initial composting process, including aeration and sometimes agitation, within an enclosed system. There are a wide variety of commercially available in-vessel composters on the market, with varying degrees of capacity, mechanization, automation, and cost.*

This chapter begins with a discussion of the benefits, challenges, and common applications of in-vessel systems, which are most often deployed in urban and institutional settings, and then moves on to describe the basic styles of in-vessel systems. In 2016 I collaborated with the Institute for Local Self-Reliance (ILSR) to collect information about many of the in-vessel systems that are currently on the market. A small portion of this chapter is adapted from the guide that we created, and the system information collected is presented in appendix A, page 401. While most operations purchase prefabricated in-vessel composters, there are also examples of self-built systems in operation, which we'll discuss in this chapter as well.

* In-vessel-style composters are sometimes called *reactors*, but given that word's less environmentally sensitive counterparts, I tend to avoid using it.

respiration
(ˌre-spə-ˈrā-shən)

DEFINITION: The processes of oxygen uptake and carbon dioxide release that are generated by organisms during the metabolism of complex carbon compounds.

Common Applications for In-Vessel Composters

Currently in-vessel composting is one of the least common methods in use, accounting for just 4.5 percent of United States composters reporting in the 2017 *BioCycle* survey.[1] However small, commercially available in-vessel composters are primarily used in on-site institutional settings, and I would guess that a majority of the 1,800 composters reporting in the *BioCycle* survey were not on-site operations. In another survey, small in-vessel composter vendors reported that their systems were in use at schools, universities, jails, and other institutions.[2] For the most part these sites are processing their own material and using the compost themselves. I do occasionally see in-vessel systems at commercial food scrap composting facility, such as Rocky Hill Farm in Saugus, Massachusetts (figure 10.1), but large food scrap composting applications are currently few and far between.

Two of the most common applications for larger in-vessel systems are on farms for composting livestock mortalities and as a method for heat-treating manure press cake for reuse as animal bedding. They are also in use on the back end of small wastewater treatment plants for biosolids composting.

While in-vessel composting currently makes up a relatively small percentage of total composting capacity, I think the method has enormous growth potential. As more individuals and communities become interested in composting, user-friendly methods will become increasingly desirable. The main selling points of in-vessels, at least in theory, are lower maintenance requirements and fewer headaches. We're already starting to see micro in-vessels come on the market for home use (I have yet to see one that looks practical, but I still view this as progress).

Outside of home use, two applications for in-vessel systems really stick out to me: compost heat recovery (CHR) and low-cost micro systems.

The technology already exists to pair CHR with in-vessels, as Agrilab Technologies has demonstrated with their Drum Dragon (figure 9.40, page 272), or as Conan Eaton has done with his containerized ASP (figure 10.10, page 285). Both in-vessel and CHR are targeted at the most active primary phase of the composting process, when high heat generation corresponds with high management demands. The in-vessel system contains the heat during peak energy generation, so it can then be captured. It thus takes up the burden of managing material during the period when management demands are the highest. The captured thermal energy can help offset the higher installation cost of in-vessel systems. For a site that is already considering in-vessel, CHR is a very logical pairing, assuming there is justifiable end use for the heat.

Micro systems are the second underutilized application for in-vessels I want to mention. At a scale of, say, 100 to 2,000 pounds per week of food scraps, there are very few options that do the job without a sizable investment. The micro-systems market includes almost every small to medium-size school or cafeteria, community garden, and neighborhood in America. Having tinkered with self-made in-vessel designs, I know how challenging it is to produce a low-cost system—it requires a lot of research and development—but my hope is that with a combination of open source designs and entrepreneurship, lower-cost systems will become widely available, meeting the demands of those in the micro-system markets who wish to either build or purchase an in-vessel system.

Figure 10.1. Rocky Hill Farm designed and built two large rotating drum composters.

Benefits, Challenges, and Considerations with In-Vessel

The specific advantages and drawbacks of in-vessel composting vary from system to system, and I have consciously avoided going into the specifics of any particular vendor or model in this book. A significant amount of research is needed when considering vendor-supplied systems. Here I discuss the broader pros and cons of various in-vessel styles and some of the questions and considerations to take into account when assessing system options.

Potential Benefits of In-Vessel

The three greatest benefits associated with most in-vessel systems are effective containment of food scraps and their associated nuisances, low management requirements during the primary phase, and reduced processing times. Most systems can easily reach 131°F for the three-day heat treatment that is the standard target for pathogen reduction in in-vessels. Some systems provide or enable fast mixing and agitation, while others incorporate forced aeration. Temperature monitoring, data logging, and process control features are also sometimes included, along with add-on capabilities such as grinding, watering, and screening of product as it exits the system. Reduced labor and nuisance potential seem to be driving factors for operations that choose in-vessel systems.

Potential Challenges of In-Vessel

While conducting research for ILSR's 2018 micro-composting systems guide,[3] I was able to talk to and visit many in-vessel system operators. I have also researched in-vessel system options for clients, a couple of whom designed and built systems themselves. Still, I must admit that I have worked with in-vessels less frequently than other systems, probably because the up-front investment costs are prohibitive for my typical clients (farmers, schools, commercial composters). In-vessels generally have much greater up-front costs compared with lower-tech options such as bins, turned windrows, and even ASP.

For those who use in-vessel composters, the most common challenges reported are similar to those of other systems. There is a learning curve during start-up, often associated with non-ideal recipe conditions or a systems operator's lack of composting experience. There also seems to be a higher rate of technical problems when in-vessel models are first released, which is to be expected. If you are looking at a recently developed or beta-type system, it is important to have a technical support agreement with the vendor and to anticipate that you will likely deal with technical hiccups.

In-Vessel System Considerations

While in-vessel systems often have unique capacities, they are more similar to other compost methods than they are different, and should be considered based upon the same basic criteria:

- System capacity
  - Recipe
  - Volume
  - Retention time
- Ease of mixing
- Reliable temperature treatment
- Up-front cost
- Operational/maintenance costs
- Footprint
- External requirements

These major factors are discussed here, but I would also recommend reviewing ILSR's 2018 micro-composting systems guide,[4] which addresses many considerations specific to in-vessel systems.

System Capacity and Retention Time

Most in-vessel systems manage the active stage of composting, which typically includes the primary phase and at least some of the secondary phase as

I define them in *The Phases of Composting* section, page 58. The *primary phase* is when heat treatment occurs. The *secondary phase* is when degradation of attractive food sources and odorous compounds reaches completion.

In-vessels are designed to accommodate and facilitate a balance of recipe, heat, consistent oxygen, and frequent mixing during these active early stages, but the composting process can vary widely in how long it takes for raw material to become stable. The retention time and rate of composting depend on the system, its throughput, and how it's managed. Faster processing is a desirable feature in most cases, because it means less material is in active management at any given time, which equates to cost and space savings. Just how much the compost process can be sped up is a question that provokes skepticism from many in the composting world, however. Manufacturer claims often run counter to common knowledge about how long the process takes.

As you consider in-vessel options, you will need to estimate how much additional time and capacity will be required following in-vessel processing in order to create a stable end product. As a general recommendation, having extra capacity and planning for longer processing time provides a valuable buffer and encourages a more mature end product. At minimum a system should take your raw material through the primary phase; then in most cases you will need a plan for managing the secondary and final phases. Many compost sites use open piles or windrows to finish the composting process, assuming the material is already broken down enough that it won't cause issues in the open.

When selecting a system that can meet your capacity needs, it's important to differentiate between total throughput and food scrap throughput. Some systems assume a food scrap processing capacity based on recipes that assume the use of a very high-carbon, low-moisture-content feedstock such as sawdust pellets (the sort you see used as cat litter). You might need to recalibrate capacity after researching their recipe assumptions or piloting the system to see what it can realistically handle in a given time period.

To research in-vessel system options based on your capacity needs, refer to appendix A, page 401. I'll provide an example using the 10-ton-per-week food scrap processing capacity scenario found in the other compost method chapters. There are only a few in-vessel systems available at this scale, and none of them assume the 1:4 ratio of food scraps to additional feedstock that we used in those examples, so it's hard to make a direct comparison. But as an exercise, let's look at the largest-capacity model—the Green Mountain Technologies Earth Flow EF5010 (see figures 2.5, page 40, and 10.3, page 280, for pictures of a smaller Earth Flow model), which the manufacturer reports can handle just over 13 tons per week of food scraps. That capacity is based on a 1:2 food scraps to additional feedstock ratio and a 21-day processing time. An operation could possibly play with their recipe and create a workable blend within the manufacturer's parameters. It might force a shortened retention time, but it's a pretty safe bet as long as there is consistent access to a good carbon source such as sawdust, wood shavings, or well-bedded manure. An operation would be wise to contact others who are using the same system and find out more about their recipes and throughput as points of comparison. It would also be worth seeing if the vendor makes something slightly larger. I happen to know that Green Mountain Technologies does built-in-place Earth Flows, so it's likely that they size a model to an operation's precise specifications.

Mixing

The ability to efficiently turn the active compost is highly beneficial, and I see it as one of the main advantages of many in-vessel systems. Not all in-vessels have a fully automated mixing mechanism, though. Some facilitate manual mixing (typically by using a manually turned drum or auger), and some require other means of turning, along the lines of bins or ASP.

Temperature Treatment

Reliability in heating and meeting the pathogen reduction standards (131°F for three days for in-vessels) are important considerations for many food scrap composters. Most large systems are designed for this, but some struggle in cold weather without a heated or sheltered structure protecting them. Make sure that the system has been tested under weather conditions similar to your own.

Upfront Cost

In my experience up-front cost is the greatest barrier to wider adoption of in-vessel systems. To give you a sense, out of the 38 in-vessel models listed in appendix A (page 401), the base prices ranged from $7,845 to $163,336 per ton of maximum weekly food scrap processing capacity (not including external costs and add-ons). These costs are fairly reflective of the vendor-supplied in-vessel market as a whole. While the efficiencies associated with in-vessel systems may justify higher up-front investment, the sticker price is often just a non-starter, especially when compared with other options. That said, larger systems have an economy of scale, so they tend to look better on paper, and some of the smallest systems that lack automation have less intimidating price tags.

Operational/Maintenance Costs

In addition to up-front costs, you must consider the expenses involved with operating and maintaining a piece of equipment over time. Some vendors may be able to provide an estimate of these costs, and it would be worthwhile to try to understand what the assumptions are behind their numbers. An even better option is to ask an operator who manages the same systems about their costs.

As a labor reference point, I often refer to my experience operating a small food scrap operation using the turned windrow method. It took a 20-hour week to process around 4 to 5 tons of material, and to produce and sell 400 to 500 yards of finished product. This was at a clunky site in northern Vermont, so conditions were less than optimal. An in-vessel-based operation should require much less time than this to produce the same output (assuming you have a loader).

Footprint

While having adequate space to house your composting operation is a self-evident requirement, there are some space-related factors that are not always obvious. Vendors will provide the spatial requirements for their systems (see tables A.9–12), but the footprint of other external requirements, as discussed in the following section, should be considered as well.

External Requirements

As with other composting systems, in-vessel composters typically require additional facilities, space, and equipment to operate. These external requirements can vary greatly based on scale, existing infrastructure, and rate of composting/in-vessel retention time, and they can add significant cost and footprint. Common external requirements are listed in table 10.1. Planning for these system components is covered in much greater depth in chapter 6.

Common Styles of In-Vessel Composters

While in-vessel compost systems are generally lumped together as a single "method," in practice the different styles of in-vessel systems are each quite unique. There are three common styles that we see at a small scale: rotating drums, auger-turned systems, and containerized systems. Some sources categorize bin systems as in-vessel, and they can effectively serve as containerized composters, but for our purposes we are treating bin and bay composting systems separately from in-vessel systems. It's important to remember that there's a great deal of overlap among composting methods. It only makes sense that designers hybridize over time, borrowing and combining the most advantageous traits and mechanisms from each method.

I should note that there are other, less common in-vessel styles, besides the three mentioned above, that might prove useful at a small scale. For example, one in-vessel style that I have yet to see utilized for community-scale food scrap composting is the vertically fed silo.

Table 10.1. Common In-Vessel External Requirements

External Requirements	Notes
Secondary, finishing, curing, and storage areas	In-vessels typically require an additional composting period following initial processing. This can include anything following the heat treatment phase.
Feedstock storage areas	Planned space for feedstock material storage. Outdoor and indoor spaces, open piles, covered bins, bays, and improved pads are common.
Bucket loader	Used or new, most small sites can manage with a tractor that costs under $50,000, but prices vary widely depending on loader size and condition.
Hand tools	Shovels, rakes, brooms, wheelbarrows (invest in flat-free tires)—even sites with bucket loaders will require some hand tools; volunteer-managed sites will require lots of tools.
Improved composting pads/staging pads	Improved surfaces for working or setting the composting system on. Many large in-vessels are designed to be set on a level concrete pad.
Roads/access	Improved roads or access might be needed.
Receiving mixing pad or equipment	Specific dedicated site for receiving/mixing raw materials (especially food scraps). An improved pad is highly recommended, and sometimes a vertical auger TMR (total mix ration) mixer is used for in-vessel ASP systems.
Compost thermometers	Typically 3-foot stainless steel compost thermometers with handles, which are widely available from a number of sources.
Bins/block walls	Loader push walls, bay/bin walls for storing and organizing materials.
Covered space	Open or closed walled structures, superstructures, shed roofs, etc.
Compost screener	Purchased or self-built compost screeners.
Compost bagging equipment	For selling bagged product. Numerous options exist. On a small scale, composters often bag manually, then use a stitcher (less than $200) to sew the bag shut.
Electrical connection or solar power setup	Most large in-vessel systems require power.
Water supply	For wetting piles if need be, doing plant growth tests, drinking, etc.

Source: Adapted from James McSweeney and Brenda Platt, *Micro-Composting: A Guide to Small Scale and On-Site Food Scrap Composting Systems* (Washington, DC: Institute for Local Self-Reliance, forthcoming), courtesy of the Institute for Local Self-Reliance.

Rotating Drum Composters

Drum composters are essentially scaled-up versions of the backyard tumblers that most of us have seen or used. The primary purpose of drums is to provide mixing and containment. The mixing should, in turn, provide aeration. Large drums are typically set to automatically rotate using a timer-controlled motor, while small systems like the Jora must be turned by hand. Most commonly, drums are designed for continuous flow, fed at one end and emptied at the other, but they can also be designed as batch systems with two or more compartments or individual drums that can be both filled and emptied.

The main critique of drum composters that I hear is that they don't do an adequate job of ensuring 100 percent heat treatment. It's unclear to me how much of this is vendors trying to sell their own product, but I agree that inadequate mixing is likely to become an issue if the drum is loaded beyond half the diameter of the tube, which would prevent adequate heat treatment. It's not uncommon to see three-day processing times specified by manufacturers, which I, along with many others, find suspect. Undersized systems can easily lead to inadequate heat treatment, among other problems, and I think that inflated capacity claims have given drums a questionable reputation with some in the industry.

In general, though, I think that drum composters work as long as the system has the capacity to retain the target throughput for enough time and the cylinder is not more than 50 percent full, which prevents thorough mixing. In order for all of the material to hit 131°F for three days, the drum needs to hold enough volume to heat. To ensure that heat treatment standards are met, drums are often produced with diameters as large as 5 to 6 feet and insulated to retain heat. It is also vital that the composting process move slowly enough that it has at least three days in the hot zone. For batch systems this is simpler—you just need to have a slow enough batch rotation. With continuous systems you have both fresh and active material present, which creates a gradient: At the loading end the material might be 50°F, in the center it might be 140°F, and by the end it might be 120°F. For a drum to be effective, the material must be thoroughly mixed for at least three days while in the central hot zone. I always err on the side of longer retention times, especially for primary-phase systems, so I would say ideally you'd want around 21 days of retention in the system.

Auger-Turned Composters

Another common technique with in-vessel composters is to use one or more augers (rotating screw blades) to do the mixing. Some systems, such as the Rocket, have a large auger within a tilted cylinder. Other systems, including Green Mountain Technologies' Earth Flow and Earth Tub, utilize smaller augers that pass through

Figure 10.2. The Rocket is a continuous flow composter with a large internal auger.

Figure 10.3. Green Mountain Technologies' Earth Flow is a variation on the agitated bed method. It uses continuous flow: Material is loaded on one end of the bed and moves to the other end. The auger has the capacity to add moisture as it mixes.

Figure 10.4. The Earth Tub, also from Green Mountain Technologies, works as a batch system. Mixing is achieved by manually passing an auger through the compost.

Figure 10.5. A roll-off dumpster was retrofitted for use as an external biofilter at a biosolids composting facility. The filter media (wood chips) has reportedly remained effective for over five years.

a bed of compost from overhead. In most cases the mixing is automated and works on a continuous flow principle, but there are also batch systems that require manual operation of the auger. The Earth Tub is one of the smaller in-vessels on the market and it requires several people to manually maneuver the auger.

Containerized Composters

Compost systems that take the form of large box-type vessels are usually called *containerized* systems. These are essentially the fully enclosed equivalent of bays and are typically aerated batch systems. As in-vessel systems go, the containerized models that I'm aware of are geared toward larger-scale operations, and because they are batch systems, multiple units are recommended. Most containerized systems filter exhaust through an external biofilter.

Self-Made In-Vessel Systems

Creating an in-vessel composting system from scratch can range from easy and simple to very involved and complicated, depending on size and style. Although intriguing, if you're working above the home scale, the do-it-yourself route takes time, research, and expertise in composting, design, and mechanics. It also often requires significant trial and error. Vendors have done this research and development for you. Unless designing an in-vessel system is your goal, purchasing a prefabricated system will probably get you composting faster and possibly end up being cheaper than a self-made system.

That said, there are plenty of examples of successful self-built systems, even at relatively large scales. Some composters take great pride in the systems they have developed, and in many ways self-made systems are a

Figure 10.6. A containerized composter, also from Green Mountain Technologies, used for processing biosolids. The composter is aerated, and the exhaust goes to the external biofilter pictured in figure 10.5.

hotbed of innovation and learning. For operations that have an educational mission, as well as the desire and capacity to research, design, and adapt a system, there may be a lot to gain from a self-made approach. Novice composters who are interested in designing and building their own system would benefit significantly from working with a professional consultant or engineer.

One of the biggest challenges with self-made systems is that the real costs can be hard to quantify ahead of time. Budget cautiously. Cost savings over simply purchasing a system may depend largely on how the designer's and builder's time are valued. For example, if volunteers, rather than employees or contractors, provide the majority of the labor, huge cost savings and other value can be realized. Materials are often donated or salvaged and repurposed, too. When it comes to ambitious, advanced, and larger-scale systems, however, the risks you take doing your own research and development can be significant. I've seen cases where a composting method is recommended based on an elaborate self-made system that has yet to even be designed—this type of scenario should be avoided.

A good strategy is to pilot a new or experimental method on a small scale before investing on a larger scale. This will provide a hands-on discovery process, giving you a solid basis for larger-scale implementation. The key is to go into the project expecting a learning curve, not only for managing the composting process but also for designing and developing the system itself. Also, Murphy's law reminds you to be prepared for it to take longer than you initially anticipated to design and build your own system.

Self-made systems span the whole range of in-vessel styles, but the most common are drum and containerized systems. Some design factors for these styles are discussed here. The section *In-Vessel System Considerations*, page 275, is also relevant.

Self-Made Rotating Drums

Drum composters can be designed as either continuous flow systems or batch systems. Continuous flow systems are fed at one end and emptied at the other. To make a continuous drum system work, you need enough distance from one end to the other so that distinct pockets of fresh and heat-treated compost are separated by at least three days' worth of material in the hot zone (and ideally more than three). In theory this gradient of Phase 1 composting activity is functional as long as the material is retained in the system for long enough. The compost also needs to move from one end to the other, which is usually achieved by having a slight slope on the drum, from entry to exit. The self-made system at the North Country School (figures 10.8 and 10.9, page 284) originally had a slope of 4 inches over 20 feet, but they adjusted it to 2 inches over 20 feet or 0.8 percent, to optimize their retention time.[5] Angled paddles can help move the material down the drum as it turns. With batch-style drums, on the other hand, there is a filling period after which all of the material gets heat-treated simultaneously. Therefore it can be filled and emptied from one opening. For sites that have a constant flow of material, two or more distinct batches (or segments of drum) will be necessary so that you always have a place to put new material without corrupting the active material. For example, if you want a 21-day retention time in the system, and each drum segment can hold one week's worth of material, you would need at least three segments of drum in total. You would empty one segment and then refill it that same day.

Both continuous and batch-style drum systems have several shared design components and considerations:

Size. The dimensions of the drum have to be large enough that the material can heat adequately. The drum should only be filled halfway to avoid overloading the turning mechanism and to ensure good mixing. This means that you're only working with half of the volume of the vessel. Volume increases disproportionately to diameter, so a 6-foot-diameter drum holds almost four times the volume of a 3-foot-diameter drum (see table 10.2).

Materials. Drums can be made out of virtually any rugged cylinder, but the materials that I've seen used for self-made systems are steel and poly culverts. Steel corrodes, and stainless steel would be quite expensive and heavy, so I think that poly cylinders are the best choice in most small-scale cases.

Insulation. Both small and large drums benefit considerably from an outer layer of insulation. Most vendor-supplied systems are insulated—sometimes the drums are double-walled with insulation in between, while others have spray-foamed exteriors. Very small drums may benefit from an insulated box enclosing the system, in addition to a layer of insulation on the drum itself, particularly during the wintertime. If a system has trouble meeting temperature standards, adding insulation is a viable course of action.

Force/turning mechanism. In order to create good mixing action, the drum must be rotated to the point where the compost reaches its angle of repose (the angle at which the material starts to tumble down). The mechanism that turns the drum must have enough force to lift the weight of any compost that is not centered on the drum's vertical axis. Large systems utilize mechanical advantage such as gear reductions to achieve this. The compost density/porosity will have a large impact on the weight that must be lifted in terms of the angle of repose.

Paddles. The rotational action of the drum needs to lift the material up high enough that it tumbles down the face of the compost, mixing and covering material as it turns. Without paddles, the material may just slide down the inside of the drum.

Table 10.2. Volume of a Drum Composter (Half the Volume of a Cylinder) in Cubic Feet

Drum diameter	Length										
	3	4	5	6	8	10	12	14	16	18	20
3	11	14	18	21	28	35	42	49	57	64	71
4	19	25	31	38	50	63	75	88	100	113	126
5	29	39	49	59	79	98	118	137	157	177	196
6	42	57	71	85	113	141	170	198	226	254	283
7	58	77	96	115	154	192	231	269	308	346	385
8	75	100	126	151	201	251	301	352	402	452	502

Figure 10.7. This rotating drum was installed at the Lake Region Union High School in Barton, Vermont. The system was created by Richard Hudak (*left*) and his sons on the Hudak Farm in Swanton, Vermont. The drive wheels that turn the drum are powered manually using the gear drive of a boat lift (*right*). Courtesy of Highfields Center for Composting and Vermont Sustainable Jobs Fund.

Figure 10.8. North Country School's self-made drum composter. The open-source designs are available for free online. Courtesy of North Country School.

Figure 10.9. The discharge end of the North Country School's self-made drum composter is designed to screen as it turns, separating fines and overs. Courtesy of North Country School.

Self-Made Containerized Composters

Creating a containerized composter can be relatively simple compared with developing other in-vessel styles. Essentially these are fully enclosed bin or bay systems. They also frequently include forced aeration, making them containerized ASP systems. If this is the direction you are going, reference chapter 7 (on bin and bay systems) and chapter 9 (on aerated static pile systems) for detailed information about sizing, building materials, system flow, and aeration system design.

I should note that large containerized systems can be constructed with a number of materials not mentioned in those chapters. For example, containerized composters are sometimes made out of roll-off dumpsters or shipping containers. Surprisingly, building large containers using typical lumber construction techniques may be one of the cheapest options. With premade containers, in comparison, you get relatively little volume for what you're paying, and then there's the cost of customizations, such as adding insulation, wall height, or a roof. One thing that makes me nervous is operating loaders in large containerized systems, which can easily become filled with ammonia and other noxious gases. The ideal system should be vented or have a roof that opens up. Always let the container air out, and be careful!

Figure 10.10. Conan Eaton (*left*), compost heat recovery pioneer, built the 120-cubic-yard containerized ASP in the background on his dairy farm in Lunenburg, Vermont. The entire system, including concrete slab, aeration, and heat recovery, was built for around $20,000. Courtesy of Highfields Center for Composting and Vermont Sustainable Jobs Fund.

Commercially Available In-Vessel Systems

For those in the market to a purchase a pre-made in-vessel composting system, there are many systems available in the United States and internationally. This section presents information adapted from *Micro-Composting: A Guide to Small Scale and On-Site Food Scrap Composting Systems*.[6] In 2016, working in partnership with the Institute for Local Self-Reliance, we conducted a survey to gather information about small-scale composting systems currently on the market. We surveyed compost systems vendors, conducted phone interviews with compost systems operators, and made in-person visits to several small compost sites that are utilizing the surveyed technologies. The survey was sent to 24 compost systems vendors, of which 13 responded. The systems represented in these responses included 38 in-vessel models, 3 aerated static pile systems, and 1 vermicomposting system. They started at scales slightly above single residential (backyard composting) and capped off with systems designed to manage up to 5 tons of food scraps per week (though a handful of vendors provided information on models that process more than 5 tons per week). Combined, the vendors who shared information with us reported over 5,500 small-scale compost systems in operation worldwide. We have received independent confirmation that all of the systems have been successfully utilized to compost food scraps, which was the only criterion for inclusion in the guide.

We organized the compost system information from the survey into 20 tables, 12 of which were adapted for this book. Information is listed by model and sorted by maximum food scrap processing capacity, from smallest to largest. We asked vendors to provide information about using the same models at multiple scales to facilitate comparison among different systems of like capacities, which is why you will see Earth Flow, Rocket, and other brands listed multiple times.

When you're considering systems, I advise that you start by looking for models that meet your scale requirements (appendix A, page 401), and then move on to the other relevant factors. It should be noted that some vendors did not include all of their potential models, so if there are features in a system that are attractive, but the system is not at your scale, it would be worth contacting the vendor to see if they make a model that does fit your scale requirements. Keep in mind that choosing a system with a maximum capacity that exceeds your anticipated generation will give you more flexibility in terms of recipe and retention time.

The information in these tables was provided by vendors, and their claims could not be independently verified. These tables are intended to serve as a starting point and to help you narrow your options. Making a conclusive decision will require independently researching the systems and exercising due diligence, including checking system references.

CHAPTER ELEVEN

Composting with Animals

Feeding animals "garbage" or "swill"—as food scraps have historically been called—is a practice much older than managed composting as we know it today. It is even believed that the attraction of wild pigs to food scraps in early human settlements is what led to pigs' domestication around 8,000 years ago.[1] There are many people in the Boston area, where I live, who remember putting food scraps in a bucket outside as a kid; the pig farmer came and picked up their feed periodically. I've heard from several people who had a metal can buried outside their back door, which I assume held a small bucket of food scrap for the farmers to empty. A foot pedal was used to open the can's lid. Talk about a practical system: The ground temps kept the feed cool in the summer and warm in the winter.

No such system exists here today, although a 2012 report by Business for Social Responsibility (BSR) estimates that the majority of the estimated 21 million tons of food residuals diverted in the United States annually goes to animal feed (the majority of this comes from the food manufacturing sector, as we will discuss later in this chapter).[2]

Feeding food residuals directly to animals yields enormous energetic advantages as well as cost savings. Food scraps are a local and recycled resource and can replace imported grain, which is energy-intensive to produce and transport. An analysis of pork production in the European Union estimated that the land base used to grow hog feed could be reduced by one-fifth by reinstituting the reuse of swill on a large scale.[3]

upcycle
(ˈəp-ˌsī-kəl)

DEFINITION: To create a new product out of a waste material that is of higher value than its original form in terms of usefulness and/or environmental benefit.

The practice captures and makes use of a resource before it is technically "wasted." For this reason, most states consider feeding food scraps to animals (acceptable scenarios for which are covered later in the chapter) an "agricultural" rather than a "waste management" practice. The distinction is large when it comes to regulations. In most cases, being considered agricultural is beneficial in terms of the regulatory framework your operation falls under. Of the many options that divert organics from the landfill, the US EPA and many state governments consider feeding animals a higher use of discarded food than composting or anaerobic digestion.

When done effectively, animal management should only change slightly when scraps are used for feed, although this depends on scale, of course. Feeding your animals includes providing them with access to adequate compost (and supplementary feed as needed) to meet their dietary needs. Ideally this happens in a deliberate manner and in a thoughtfully designed location in order to manage refused (uneaten) feed. Refused feed includes unconsumed and rotting food scraps and potentially some trash, depending on how free the food scrap source is from contamination. As a mechanism to increase food source diversity, some operations create an active composting environment from which their animals forage, not unlike the nutritional benefits seen from deep litter systems (see the *Benefits of Deep Litter* sidebar, page 296).

By combining the management of food scraps as a feed with the composting process, farmers can effectively avoid the concerns that accompany the presence of rotting food in a livestock operation (pollution, odors, vectors). Effective techniques for composting with animals are the primary subject of this chapter. The focus is mainly on feeding laying hens, which is a practice that is, to the best of my knowledge as of this writing, legal in forty-seven states.[4] Feeding food scraps to pigs is also common, although this practice carries with it different considerations, both health-related and regulatory. Feeding hogs and other animals will be covered in this chapter, although in less depth than with egg laying operations. We will also look at incorporating animal feeding into the composting process and vice versa, a process we'll call *composting with animals*. Starting with understanding the potential nutritional value of mixed food scraps as a feed source and some of the accompanying unanswered questions, we'll then look at some of the successful strategies for feeding food scraps to animals, including mobile systems for pastured animals and stationary systems.

While feeding food scraps to animals is a very old practice, composting with animals in the ways being done today is relatively new, and there is still a lot that is unknown. Surprisingly, in some parts of the organics recycling industry, feeding animals non-food-processing streams, such as residential or commercial food scraps, is not a subject you hear about often in the larger discourse with regard to where the industry is headed. In many ways this is a good thing for small farms interested in composting their community's food scraps, because it means that the market is still open, although my hope is that as more people compost with animals, more success stories emerge and recognition of the practice as a viable option for communities becomes more widespread. For this reason, it is critical that early adopters of composting with animals are good stewards of this community resource.

Disclaimer

My expertise is composting, not food safety, livestock health, care, or nutrition. The goal of this chapter is to emphasize the ways that the traditional practice of feeding food residuals to livestock can be optimized through integration with modern composting methods. It is incumbent on individual operators to conduct due diligence to ensure the well-being of their animals and the safety of the products they are producing. To these ends, there remain large gaps in knowledge about feeding food residuals as an agronomic practice. While the anecdotal evidence is that the practices discussed in this chapter are safe, humane, and comparable to other sustainable farming methods, there is always a chance of both human and animal health problems in farming.

Common Applications for Composting with Animals

As any farmer who has integrated animals and composting in their operation can tell you, it's not for everyone. Raising livestock is called animal husbandry for a reason: Livestock require consistent care and should essentially be thought of as a separate business from composting altogether in terms of the skill sets and management involved.

The converse is also true: Livestock operations that are considering using food scraps as feed need to manage these unique feed sources appropriately, and the most likely route to do so is through managed composting. So composting with animals is a solution both for composters who want to farm and for farmers who want to compost. I say this because if you are inclined, as many entrepreneurial types are, to continuously add business components, consider the benefits, but also weigh the additional operational requirements and business prospects for doing so successfully.

For those considering food scraps as animal feed, if you're unfamiliar with the landscape, I would recommend stepping back for a minute to look at who's already doing what and why. The most important thing to understand is that not all upcycled feed sources or livestock species are equivalent or treated the same in the eyes of the law. To add a compounding factor, regulations vary dramatically from state to state, so depending upon what state you are operating in, regulations may limit your options. Essentially, the animals you want to raise need to be compatible with both the feed sources that are available to you and your regulatory environment.

Figure 11.1. A hog feeds on a mixture of de-packaged dairy products. High levels of protein and solids make this a high-quality feed. In addition, the farm feeds a wide variety of organic vegetables from a local processor and is implementing a distributed feeding and composting system that will be compatible with its pasture-based production methods.

In 2016 the Harvard Food Law and Policy Clinic published *Leftovers for Livestock: A Legal Guide for Using Excess Food as Animal Feed*, which outlines the laws from state to state.[5] The net effect of so much regulatory variability is similar variation in animal feeding practices across the country. From working in the organics recycling field, I know that many large generators of non-meat residuals (food processing/manufacturing generators) already divert food to hogs, and cows to a lesser degree, on a large scale. My experience is that large volumes of very consistent material from the food manufacturing sector are mostly spoken for as either hog or cow feed, and also as feedstock for anaerobic digesters. Anecdotally at least, medium- and large-scale chicken farms do not appear to be recycling food to the same degree, despite the fact that in many parts of the United States, food residuals are not required to be meat-free and/or pre-cooked.

Chicken feeding represents a real opportunity, especially on a small local scale and when combined with effective composting. Consider that there is high demand for local eggs in many regions; growing recognition of eggs as a low-carbon-footprint animal protein; and significant growth in markets for collection in the residential, commercial, and institutional food scrap generator sectors, which are more conducive to chickens than to hogs.

Utilizing material from the non-food-processing/manufacturing sectors—anything from plate scrapings to carrot tops—as animal feed is typically more challenging than food processing residuals. It's both less consistent and tends to be less fresh; therefore it can be nastier and more work to manage. There is a resurgence of interest by small diversified farms in composting with animals for a variety of reasons, however. We've already mentioned the feed replacement value and environmental benefits of replacing traditional feeds as two huge incentives. On top of these overarching benefits, one of the most common things I'm hearing is that vegetable and fruit farmers are becoming concerned with their reliance on exogenous (imported) nutrients for maintaining soil fertility and productivity. For example, small organic farmers who would like to displace their reliance on imported Chilean nitrates or semi-composted chicken manure from factory farms might diversify their operations by incorporating livestock, which has the added benefit of generating manure. Rather than importing commercial feed, utilizing local food scraps as feed more closely aligns with the ultimate goal of reducing imported fertility. While most food is technically imported, since the scraps typically come from local generators, they can be viewed as a local resource (people have to eat), in effect replacing imported fertility in the system. Ultimately, both farmers and composters would like to see more local food replace imported food to create a truly closed-loop food system.

Schools are often a source of both pre- and post-consumer food scraps. A school might partner up with a small chicken farm in a one-to-one generator-to-composter relationship. This would constitute a micro program in a community whereby the farm serves as food scrap recycler for one or several small generators. Often the school is the entry point for a program that might then grow to serve residential drop-offs, a small grocery, several schools, and so on, in that community. For example, 1 ton per week of food scraps could likely be captured from a 700-student elementary school, a 400-student middle school, and a residential drop-off in a town with a population of 10,000 people. In turn, 1 ton per week could feed approximately 140 laying hens.

Another common benefit of incorporating animals into a composting operation is potentially controversial in some circles of the organics recycling industry (and a no-brainer in others). By upcycling food residuals as an animal feed, you effectively transfer the recycling of "solid waste" into the agricultural domain. The language may vary from state to state, but logic puts this squarely in the realm of agricultural activity, and, as mentioned earlier, agricultural activities have far fewer regulatory requirements in general than do solid waste management activities and other non-agricultural commercial activities. It's a huge benefit, and while by no means do I advocate composters using this as a backdoor path to skirt regulations (say, 50 chickens at a compost facility that processes 20 tons per week of scraps), if you're passionate about both animal husbandry and composting, combining the two deserves your serious consideration. Unfortunately, during the writing of this book at least one very progressive state has moved to exclude this practice from their definition of agriculture, so those who utilize this practice would do well to band together as have others fighting for the right to restore and maintain the natural integrity of our food system.

When you compare composting with animals to a standard turned windrow or bin composting system, there is a significant reduction in scale that translates across the system to a reduction of footprint, labor, cost, feedstock requirements, and so on. A theoretical "composting with chickens" operation is sized later in this chapter (see *Sizing a System for Composting With Chickens* on page 308); it's comparable in capacity to systems covered in other chapters. For a chicken composting operation that can process 10 tons per week of food scraps, the composting portion of the operation requires only 10 to 50 percent of the composting footprint/capacity of a comparably sized composting operation, depending upon the volume that the chickens don't consume.

Essentially, composting with animals, like technologies such as ASP, windrow turners, and in-vessel systems, can increase the food scrap processing capacity of a farm without increasing the compost site footprint. Land can be used for raising animals and other food ventures rather than compost production. In rural areas and on the outskirts of cities are farms with barns, manure pits, silage bunks, greenhouses, and other existing infrastructure that can easily be converted for composting with animals, whereas these spaces are often harder to efficiently convert and permit for stand-alone composting operations.

Lastly, there are a host of other anecdotal benefits that I have heard reported by operators (largely chicken composters), including reduced fly populations, reduced exposed food on pile surface, increased feed value (biological and feed diversity), eggs for employees, and overall increases in employee job satisfaction. Farm animals provide a host of other valuable services, such as clearing, weeding, fertilizing, and otherwise preparing land for crops, to name just a few. All of these can be compatible with composting.

Safety and Regulation

There is currently a huge push both for greater food safety and reduction of the ecological footprint of our food system. With widespread recognition that diverting unconsumed human food to animals is preferable to all options other than reduction and redistribution, industry, researchers, regulators, and policy makers then have to ask themselves: When and where is this practice safe and appropriate?

The answer is animal-specific and might depend on whom you talk to. One of the potentially beneficial but also challenging elements of composting with animals is the overlap among the waste, organics recycling, and farming industries. Not everyone is on board with the practice, and there are a variety of interests and concerns at play. Feeding food scraps to hogs and cows is a well-established and regulated activity. Feeding to chickens is fairly widespread on a small scale, but it's not well researched and is sparsely regulated. The next three sections discuss some of the specifics related to safety and regulation in feeding hogs, chickens, and ruminants.

Feeding Swine

It is well known that pigs run a high risk of contracting human diseases from food scraps or from material that contains diseased animal parts. Feeding pigs raw garbage that might contain meat products is banned in the US as well as the EU, largely to protect the hog industry. In the US cooked feed must be heat-treated to 212°F for 30 minutes to be allowed under the Federal Swine Health Protection Act (although there is variation from state to state).[6] Diseases of concern include foot-and-mouth, hog cholera, swine fever, salmonella, campylobacteriosis, and toxoplasmosis.[7]

In 2001 an outbreak of foot-and-mouth in the U.K. was thought to have originated with the illegal feeding of uncooked food. The outbreak reportedly cost the U.K. economy £8 billion and led to an outright ban on feeding garbage across the EU the following year.[8] While many farms follow best practices, there are undoubtedly some who do not. Although it can be done, backyard and small-scale pig farmers are not in the best position to be cooking feed; there just isn't an economy of scale. In these cases the most logical strategy is obtaining material from meat-free sources, including breweries, dairies (whey), bakeries, and vegetable farms/processors. There are concerns in the organics recycling community that some small pig operations might not follow best practices when it comes to cooking feed.

Figure 11.2. These organic vegetable residuals are quite beautiful. They are one example of the diverse feeds that can be sourced to fill out an animal's diet. A video of the hogs on this farm coming to feed from the trailer is one of my daughter's favorites.

Feeding Chickens

Although I wish I could say that there was rigorous study and debate, there is in fact very little peer-reviewed information on feeding mixed food scraps to poultry as a commercial practice. (Food processing residuals are a whole different story and hard to compare for a number of reasons.) In the *Handbook of Poultry Feed from Waste: Processing and Waste*, second edition, the authors state that "Raw garbage is not suitable for poultry because it contains numerous harmful bacteria owing to its storage and subsequent fermentation in relation to high environmental temperatures; it cannot therefore be used immediately."[9] Earlier in the same chapter they reference swine and cattle contracting diseases from this practice, following its increase during World War II, which resulted in regulations that banned the raw feeding of certain scraps to these livestock in many parts of the world. However, the book offers no reference to actual issues or incidents with poultry, and to date I have never seen or heard of any sickness originating in the feeding of raw scraps.

One of the reasons feeding food scraps to chickens is different from feeding them to hogs is, as I understand it, that humans share fewer diseases with chickens (in other words, we are less likely to make them sick). Therefore, the concern has been more focused on people getting sick from the eggs or meat. Since raw food scraps are a widely used feed source for chickens, we can infer that at the very least, the practice is relatively low-risk for consumers (compared with, say, breathing or riding in a car). That is not to say that no one has, or ever will, get sick from eating an egg from poultry raised on food residuals. Remember, chickens are living in and around their own excrement, so chicken farms are not particularly sterile.

The FDA estimates that around 79,000 people annually get food poisoning from eating *Salmonella*-contaminated eggs, and 30 of them die.[10] Along with many others, I would like to better understand how feeding food scraps compares in risk with other practices overall. One question for researchers is whether there is any difference in disease risk between raising laying hens and broilers. I hope that rigorous testing of the practice will be a priority for the next generation of sustainable agricultural researchers.

Producing Organic Animal Products

While the national and state regulatory framework surrounding composting and animal feed can initially be somewhat confusing, how using food scraps as animal feed falls into the organics standards is very straightforward. Feed must be organic to meet National Organic Program requirements, so only residuals from entirely organic generators (such as an organic dairy or food manufacturer) meet that standard.

Feeding Ruminants

In the United States, feeding ruminants (cows, sheep, goats) food residuals is limited by the Ruminant Feed Ban Rule (FDA) to material that is free of protein derived from the tissue of mammals.[11] The primary concern is mad cow disease and other prions (spongiform encephalopathy diseases).

Animal Feed as Highest Use

The US EPA Food Recovery Hierarchy promotes a logical order for diverting food that might be otherwise be wasted. Near the top of the hierarchy, below "Source Reduction" and "Feed Hungry People" but above "Industrial Uses," "Composting," and "Landfill/Incineration," is "Feed Animals." When I talk about feeding animals, be it with farmers, with policy makers, or with regulators, the concept of highest use almost always comes up. I can't overemphasize the value of having not only the practice but also the priority of animal feeding validated by a federal agency.

However, what is ultimately needed is a more complete picture of the true social and environmental impacts of these food recovery options. In the case of animal feed, there is a dearth of comprehensive quantitative research. We know that grain is energetically intensive to produce (and to a lesser degree to transport). For example, something on the order of 69 percent of global greenhouse gas (GHG) emissions in egg production and 60 percent emissions from pork production come from the production of feed.[12] We know that commercially composting food scraps, as opposed to sending them to the landfill or incinerator, can offset significant GHG emissions, even when trucking and equipment operation are factored in.[13] For example, composting 10 tons of food scraps is the equivalent of *not* burning 974 gallons of gasoline (-9 metric tons of CO_2 equivalent, or $MTCO_2E$) as calculated by the EPA's Waste Reduction Model (WARM),[14] which quantifies GHG emissions reductions based on reportedly conservative assumptions. The landfill-offset portion of this, due to reduced methane, is about 758 gallons of gasoline (-7 $MTCO_2E$); the rest assumedly comes from soil carbon storage, according to their composting guidance document.[15]

As the hierarchy implies, I think we can be fairly confident that feeding animals in most cases will have a similar or greater emissions offset than composting or anaerobic digestion (AD). If 10 tons of food scraps per week can feed around 1,430 laying hens (at 2 pounds per bird per day), and assuming those birds have a conservative 75 percent laying rate, they would produce 625 dozen eggs a week. A 2014 study in *Poultry Science* on the GHG emissions of free-range eggs in England found that a dozen eggs produced 2.2 kg of CO_2E,[16] so 625 dozen eggs would generate 1.376 $MTCO_2E$. If food scraps replaced 100 percent of the conventional feed, the estimated emissions associated with producing 625 dozen eggs that are non-feed-related (31 percent global average as per the previous paragraph) would total 0.426 $MTCO_2E$.

The reduction in emissions alone is significant, but when you combine the remaining emissions (0.426 $MTCO_2E$) with the landfill diversion GHG offset (-7 $MTCO_2E$), you actually get a negative GHG scenario (-6.574 $MTCO_2E$). In other words, we should have a "carbon negative" food, where the organics recycling element pays for the other elements of production more than 10 times over in energetic terms. Based on these rough calculations, we're offsetting just over 10 kg of carbon emissions or 1.1 gallons of gasoline with each carton of eggs. I can just picture a C-EGG label on food-scrap-raised eggs.

Yet unlike other forms of organics recycling, there is little to no full life-cycle assessment that I can point to. My math is far from comprehensive, and frankly I'm not qualified to conduct a full life cycle assessment, but this is too important a point to ignore. There are viable arguments to be made that animal feeding should be considered from an emissions perspective and that the Food Recovery Hierarchy is rooted in sound and reasonable assumptions. The lack of comprehensive comparative studies is a problem, though, because communities and decision makers should not have to rely on loosely connected dots, in the way that I just did, to make the argument for prioritizing animal feed, food donation, or any other option for that matter. Luckily, the hierarchy does seem to stand on social principles and logic, despite the lack of quantitative research backing it up. It's a pivotal time for the development of food scrap recycling infrastructure, and sectors with significant industry support have louder advocates working on their behalf. I think we would all benefit from a clear and un-biased picture of the options communities have in front of them, one that informs highest use and is ultimately well supported by the hierarchy.

Nutritional Value of Food Scraps as Animal Feed

Through a combination of research and work with farmers, I have seen strong evidence that mixed food scraps contain valuable sustenance for livestock. While not all food residuals are equal, and not all animals need the same things, the typical mixed material coming from commercial, institutional, and residential settings seems to be well suited as feed, and in particular for chickens and hogs.

The *Handbook of Poultry Feed from Waste: Processing and Waste*, second edition, cites several researchers going back to 1945 and states that "this waste has a reasonable nutritive value for the nutrition of poultry, especially broilers and layers,"[17] with crude protein ranging from approximately 15 to 25 percent (dry weight) depending upon the study and feed source.[18]

Since I am neither an animal scientist nor nutritionist, I am in no position to analyze the details of these studies, but I do trust that the authors are correct when they posit that there is valuable nutrition present in food scraps, which farmers utilizing this feed confirm. Farmers need to work to understand their feed's qualities and variability. Coefficients of variability for different characteristics can be over 100 percent from day to day, though the average nutrient contents tend to be good.[19]

For hogs, a 1999 study by Westendorf et al. looked at the feed value of institutional food scraps and found it to be of excellent quality, although the high moisture content was a limiter of overall consumption.[20] Based on the findings in that study, Westendorf estimates hogs under 100 pounds consume around 8 to 10 pounds per day and hogs of approximately 250 pounds consume 20 pounds or more per day.[21] Later in the chapter, we'll look at methods for sizing chicken composting operations. Two pounds per bird per day is the recommended proportion of food scraps that I cite, but with little evidence other than anecdotal and some rough math.

Systems for Composting with Animals

The remainder of this chapter is dedicated to the design and management of systems for composting with laying hens. For those raising other animals, the strategies for composting with laying hens may very well inform the development of systems for those animals. As with everything in this book, I recommend taking what fits and leaving the rest.

Composting with Laying Hens

Feeding kitchen scraps to the backyard or homestead chicken flock is an age-old practice that is currently receiving increased attention from egg producers large and small. Due to rising costs of purchased feed and increasing public participation in composting/food scrap recycling programs, egg producers are exploring food scraps as a primary feed source for their laying flocks.

Chickens, like humans, are omnivorous, and farmers are showing that they can maintain condition and production on a diet primarily comprising post-consumer food scraps. What the hens don't eat can be composted along with their manure and bedding in a proper compost recipe, returning valuable nutrients and organic matter to the soil. When properly managed, composting with laying hens achieves the dual goals of affordable egg production and organics recycling. Ideally, these two activities overlap and are mutually beneficial.

The remainder of this chapter covers systems and approaches to integrating chickens into the composting process, operational design, sizing, basic operational considerations, and best management practices (BMPs) for implementing feeding/composting systems.

Developing a Feeding and Composting Strategy

Feeding food scraps to chickens and composting food scraps can be, and routinely are, performed independently. However, there are potential benefits to combining the two activities in space and time: composting the feed and feeding the compost. Amendments such as leaves, livestock manure/bedding, wood chips, and hay not only aid in creating a balanced compost recipe but also mitigate problems associated with exposed food scraps. Such nuisances include odors, rodents, scavenging animals, and unsanitary conditions for you and your flock.

Similarly, allowing chickens access to food scraps during the initial stages of composting can benefit the composting process. By consuming a large percentage of the food scraps, chickens reduce the volume of materials to be handled. As they forage, chickens help mix and aerate the outer layers of the pile and add nutrient-rich manure to the compost.

System designs for chicken composting vary greatly based on several case-specific variables, including site attributes, operational scale, infrastructure, equipment, seasonality, and operator preference. However, functional design should account for some common operational needs:

- Access for food scrap delivery, chickens, and material handling
- Mitigation of odors and vectors
- Defined and contained area for feeding food scraps
- Use of a compost recipe and sufficient feedstock supply
- Incorporation of an aerobic composting process (any of those described in the other chapters of this book)
- Appropriately scaled chicken flock and composting infrastructure

Taking these factors into consideration, it is critical to plan the physical processes for feeding and subsequent composting of residuals. In other words, how will the two processes interface to meet the operation's goals and limitations?

General strategies for integrating the chicken feeding and composting processes include but are not limited to the four options described in the next sections.

Frequent Cleanout of Refused Feed and Bedding

In this two-step process feeding is performed in a defined area, often on an improved surface such as a concrete pad. Residuals are then removed as often as necessary to mitigate vectors and odors (daily to

Figure 11.3. Food scraps are layered on a bed of carbon in the receiving/feeding area at Tinmouth Compost in Tinmouth, Vermont. The carbon and uneaten scraps get removed periodically and transported a short distance to compost. The operation is located in an old dairy barn that was previously used for storage on Wheaton Squire's family's land and was converted for the purpose of composting with chickens. The barn also houses two small aerated static pile composting bays, as well as space to finish the compost in turned windrows. Courtesy of Wheaton Squire.

weekly) to be managed in a separate composting system. Frequency of cleanout will vary, but this process's defining characteristic is that composting doesn't really become active until the residuals have been removed from the feed area and combined to form a pile elsewhere.

Although the food scraps are not composted in the feeding area, it is recommended that amendments and bedding be routinely added to the feed area to meet the goals of sanitary conditions and a balanced compost recipe. If space allows, the recipe can be entirely built in place before the feed area requires cleaning out, in order to initiate compost formulation. Transferring the residuals to the composting system then aids in blending the recipe.

Benefits of Deep Litter

The deep litter strategy is commonly used for manure management in small-scale poultry operations. Similar to bedded pack barns, it involves the buildup of bedding over time. Aside from being a low-maintenance, pleasant, and comfortable bedding system, there are in fact health benefits to having a deep litter composting system in place. Chickens receive extra nutrition from the high population of microbes as they scratch and peck at the material. Citing the United Nations FAO and research conducted at the Ohio Experiment Station, Harvey Ussery explains in *The Small-Scale Poultry Flock* how as chickens peck at the deep litter surface, they consume ". . . metabolites of the microbes—by-products of their life processes—[including] vitamins K and B_{12}, in addition to other immune-enhancing compounds."[22]

The main benefits of frequent cleanout are its simplicity, adaptability, and scalability. It's a really simple strategy to comprehend, and the frequency of cleanout is the main variable that changes with an increase in scale (other than flock size). Easy access for cleanout is key with all of these strategies, but especially vital here.

Layered Static Pile (Deep Litter Pack)

The *layered static pile* involves the alternate layering of carbon/bedding amendments and food scraps, similar to the deep litter or bedded pack methods of manure management. On top of a thick base of coarse and absorbent materials (such as a mixture of wood chips and sawdust), layer food scraps, then cap them thinly with a layer of additional amendments based on a target recipe. The chickens will scratch and forage in the pack to access the food. Subsequent feedings are layered lasagna-style, and the pile builds up over time. When removed, the partially composted residuals should be managed appropriately and may be actively composted in a secondary composting system.

Similar to a bedded pack barn, if it's sized correctly, a layered static pile requires less frequent removal. The clearest advantage of this method is that the removal of residuals is consolidated, performed either as space becomes limiting or based on your operational needs. Another potential benefit is that if the pile has adequate porosity and volume it will heat up, which may be useful in winter months to keep food scraps warm and accessible as feed. Additional feed resources are also thought to be available in the form of bacteria and fungi that thrive in the composting material (see the *Benefits of Deep Litter* sidebar). Imagine the chicken's digestive microbiome!

One of the potential challenges of the layered static pile is its vulnerability to rodent activity. Due to the presence of a steady food source and the lack of pile disturbance, the pile will become a very attractive rodent habitat. I've included the method in the book because it's something that I've seen deployed, but it's really only suitable for micro and backyard-scale systems as opposed to more substantial commercial operations, and it's not something I generally

Figure 11.4. The hens at Gould Farm had no problem cleaning up. As you can see, the feed wagon has nothing but a few lemon rinds. Courtesy of Mark Little.

recommend. For larger-scale operations a hybrid of the layered static pile and frequent cleanout might be a logical strategy, building up a pack for several days or weeks between cleanouts.

Mobile Feeding Systems

The *mobile feeding system* is the obvious choice for operations where the flock is kept in a remote location and access to a stationary area where food scraps can be received and fed to the birds would be challenging. The system allows food scraps to be effectively fed and managed with rotationally grazed layer flocks, for example. The strategy as I've seen and recommended involves using a trailer as a mobile feeding wagon (not to be confused with a feed mixer), which transports the food scraps to the flock. The wagon keeps the food scraps contained for convenient removal, avoiding a mess in the pasture or barn.

As with the other systems, food scraps and amendments are layered in the trailer according to a compost recipe. Chickens forage for food scraps within the feed wagon, simultaneously mixing, aerating, and fertilizing the contained material. Due to space limitations, this system typically requires frequent removal of material, but with smaller systems you can add several layers of feed and amendments to the wagon over time until it reaches capacity. Manure spreaders and dump trailers make ideal feeding wagons, because they have the added benefit of being able to discharge the residuals after feeding is done. Ordinary trailers will need to be cleaned out by hand (or with the assistance of a tractor bucket if possible), which will be easiest with more frequent cleanout. The residuals should be composted, as with the other systems, which can happen in a dedicated composting area; or, if you have room, piles can be formed, managed, and used in the pasture.

One disadvantage of this system is the added step of transporting the feed to the birds, rather than just receiving food in the feeding area and letting the birds have at it. I often recommend mobile feeding as a temporary system while a farm is first piloting food scraps as feed. If possible, a permanent stationary location is usually more labor-efficient, but using an old trailer or manure spreader when piloting food scraps is a good intermediary step, prior to investing in permanent infrastructure. For farms that are dedicated to pastured birds, the labor trade-off is going to be a given no matter what, so a good piece of equipment is a wise investment. Larger operations may even need several feeding wagons.

In colder latitudes using a mobile feeding system in winter can potentially be a big challenge. While there are certainly possible workarounds, unless you can keep the feed warm, and feed it on a daily basis, the birds may have a hard time eating the frozen food. Cleanup will be really tough as well, because the scraps will get stuck in the wagon. Seasonal challenges are discussed later in this chapter, but many operations would do best to consider one of the other strategies during winter, when pastured birds are often more stationary.

One other piece of advice for mobile feeding systems is to cover the wagon with something that is easily removable to allow for loading (say, a wooden-framed roof on a hinge). The main purpose of a roof is to deter scavenging birds, which is especially important for farms near roads and with neighbors, but it will also keep out precipitation.

Feeding Blended Compost

As the title implies, with this strategy food scraps and amendments are *blended* into a compost recipe; chickens then forage in the compost pile for food scraps. Although the food scraps and amendments have been mixed, chickens further blend and aerate the outer materials as they forage. Portions of the compost blend can be fed out in order to expose fresh feed resources, or the birds can simply be given access to the pile, depending on whether the birds are finding enough food without assistance. As food scraps decompose and the fresh compost reaches high temperatures, chickens typically lose interest, focusing instead on the fresher compost.

Benefits of feeding blended compost include streamlined flow, increased odor control and sanitation, and rapid initiation of the composting process. The interaction between the food scraps and amendments in the composting environment may also increase the nutritional value as a feed. Active compost is alive, teeming with bacteria, fungi, and other life. Little research has been conducted on feeding blended compost, but the benefits of fermenting food scraps as feed has been looked at in China, where different food scrap pretreatment methods were piloted in almost 100 cities starting in 2011. Researchers who looked at the fermentation practice saw an increase in crude protein of 11 percent (from 25.5 percent to 28.4 percent) in just one to two days of fermentation (both hot and apparently not hot), which they attributed to the formation of "unicellular protein."[23] While the practices in this study were different from composting in a number of ways, both composting and fermentation are marked by an increase in microbial populations, and hence a higher concentration of living protein.

A potential challenge in this system is managing pile surface area to provide the flock access to all the food scrap resources within the compost.

Clearly, systems for composting with chickens can take many different forms. All of these general process options can be modified or adapted for various types and sizes of layer operations. These concepts often overlap or are used seasonally to serve different functions. Renewed interest in food scraps as feed will no doubt lead to improvements in systems for composting with chickens and other animals, as on-farm ingenuity, adaptation, and design play out.

In addition to these four direct feeding strategies, the following section discusses three other systems that are becoming more common, which involve pre-processing of food scraps prior to feeding.

Feeding Community Food Scraps to Laying Hens in an Active Composting System

by Tom Gilbert, farm owner, Black Dirt Farm

Black Dirt Farm sits at 1,700 feet on Stannard Mountain in the Northeast Kingdom of Vermont in a USDA Zone 3b climate. We undertake a variety of farm enterprises and homestead functions, including feeding laying hens on a food-scrap-based compost mix. As homesteaders we had utilized this practice for over 11 years with success, and now we have been using the approach for 15 years total. Upon establishing Black Dirt Farm in 2014, we were eager to bring this model to scale. Integrating food scraps and compost into their diet mirrors hens' natural ecosystem and provides benefits for both the hens and the other integrated processes on our farm.

From the 30,000-foot perspective, our process looks like this:

Collect food scraps each week → Tip these at the farm → Blend with co-composting materials daily for one week → Feed hens daily → Empty feeding bin after one month → Make compost → Extract heat for greenhouse (under construction) → Make worm castings from heat-treated compost → Grow crops → Sell eggs, compost, worm castings, and crops.

The compost-based feeding system is an attempt to realize the nutritional value from regionally available food scraps while also mimicking the ecological system of the forest floor (a decomposer system) and creating a year-round pasture environment for hens to forage. The practice fits our operation for a variety of reasons. Black Dirt's farm model is designed to increase value within the farm system through process integration that will mitigate input costs, such as hen feed or heat for the greenhouse, reducing pressure to increase revenues. By stacking functions and dovetailing value from one enterprise into the next, we aim to keep the scale of each enterprise as contained as possible, limiting the need for growth. In the case of feeding laying hens, food scraps arrive at the farm at a net profit, reducing our need for purchased feed, which lowers operating costs and the scale at which we must operate to be profitable.

Food Scrap Collection

We currently collect roughly 250 containers of food scraps (48 gallons each) per week from 65 customers, mostly within a 30-mile radius of the farm. We deliver roughly 10 tons a week to Tamarlane Farm and bring 15 tons to our own farm each week. We prefer containers made with a somewhat flexible plastic (to absorb impact), wide wheelbase, large (6-inch) wheels, lids, and the capacity to be stacked. We use Toter International 48-gallon totes. The vehicle used to collect materials must correspond to the type of container, as well as the total volume and weight of material being collected. We operate an F-350 Super Duty diesel but would likely upgrade to a 550 in the future. We pull a 10-yard Down Easter dump trailer with the truck.

To avoid contamination, we train all of the staff and students at every generator we collect from, provide refresher trainings, provide immediate feedback to generators when contamination is identified, and bill generators in cases of excess or routine contamination. Participants must genuinely believe in the value of the resource they are stewarding and understand the larger system of which they are a part. We also maintain clearly stated protocols and penalties for handling contamination when it arises. At Black Dirt Farm, if a generator is found to have contamination we try to work with them to prevent further contamination in a short time frame. If the problem reoccurs, the entity is "fined" at a rate of 100 percent of the cost of each contaminated container (two times the usual container price).

Feed Ration

To account for the material that hens will consume and reject, we leave the food scrap portion of the recipe high, at roughly 40 percent, in the initial mix. We estimate each hen eats 1.2 to 2.5 pounds of food scraps per day, although we supply considerably more than that to ensure enough food diversity and pile mass for building and maintaining biological activity. By the time we empty the feeding bin, the food scrap portion of the mix is roughly 20 percent and the blend is complete.

Operators need to be thoughtful about the material they are exposing for the hens as they blend. For instance, certain materials we collect add value to our overall composting blend but are either inedible (such as niter from filtering maple syrup) or undesirable to the hens (say, fermented juniper berries from gin distilling). Ideally we like to leave the hens with a good mix of whole foods that contain vegetables (especially leafy vegetables), fruits (chickens can peck through apple and watermelon skins but not orange rinds), fresh meat, and grains. Maximizing diversity of feed sources will provide benefits to the hens. We provide hens with either pasture or second-cut hay during most of the year. Other products, such as okara (the pulp that results from making soy milk) or whey, would be good additions to our mix.

Along with food scraps, we feed the hens modest amounts of grain to supplement their protein, energy, and calcium intake. While these could potentially be provided through materials such as whey, second-cut hay, and other feeds, we decided to use purchased grain. We currently feed 0.15 pound of grain per hen per day in the winter (60 percent of a standard ration), and 0.025 pound of grain per bird per day in the summer (10 percent of standard ration). It is important to ensure all hens can eat at the same time so that some hens do not consume a full ration while others eat none at all. Having enough feeding trough space for every hen to stand at simultaneously is therefore critical. We allow 6 inches per bird of linear trough space. Troughs can be accessed from both sides.

Feed System

We have developed a food scrap receiving, blending, and feeding area that is cross-functional. While there may be advantages to an open pile (access to all sides of the material, greater surface area for foraging, improved passive aeration), we like the bin system because it provides containment and enables us to exclude the birds from the bin during tipping and cleanout.

You want a feeding bin to be both contained and accessible for tipping, feeding, and cleanout. Our bin consists of a concrete slab, walls made of 2 × 2 × 6-foot locking concrete blocks, and a roof, which excludes moisture and encourages the hens to forage even in inclement weather. The block wall on one side provides the "dock" from which we tip our loads of incoming food scraps. This is elevated by about 2 feet so that the operator does not need to move the trailer forward in mid-tip to empty it completely. The bin has three access points: the tipping dock, tractor door, and chicken access door.

Our bin is currently roughly 50 feet from the barn. The hens foraging compost therefore need to travel through the elements to get to the bin and work harder to earn their breakfast. In our northern climate this can be a real challenge for the hens. Constructing feeding bins or piles contiguous with hen housing would eliminate this issue.

Managing Compost

There are several ways to optimize the compost for the hens and beyond. The pile must be agitated on a daily basis to support hen access to fresh feed. We also inoculate our mix with active compost throughout the week to improve microbial growth and expose hens to nutrients from the microbial communities of different stages of composting. In addition, we have played with applying worm castings tea to the mix to stimulate activity.

After the grazing period, the residuals move to a secondary composting pad. While we encourage operators to maximize the duration of the hens' access to the compost mix, to meet organic standards and possibly state laws, you are likely to want to remove the material from chicken access to a second site during and after your PFRP (Process for the Further Reduction of Pathogens) phase.

Challenges and Lessons Learned

While experimenting with this system for two years, we have encountered challenges. In many cases we have also developed strategies to make the process easier and more beneficial for both us and the hens:

Consistent nutrition. The hens' uptake of food, and therefore nutrition, is somewhat unpredictable. Individual hens may have different food preferences and consume differing levels of nutrients. By contrast, in a grain system the ration is calibrated and consistent.

Moisture management. The mix can become wet and your birds will spend time in a damp environment. This can result in dirtier birds and eggs. In addition to keeping the feed mix as dry as possible, it is important to set your bin up to drain liquids into a treatment system. We also bed heavily in the coop so that the birds can clean themselves frequently. It is good to make sure your coop layout provides a good area for the birds to clean off before they reach the nesting boxes.

Rodents. This feeding system is at prime risk for rodents if controls are not in place at the outset. Managing rodents, especially rats, can be very challenging once populations are established. Other operators have experienced challenges with crows.

Material handling and bin system. With only one bin with tractor access from one end, we are limited in our ability to agitate material in the back of the bin. To keep the back of the bin active, we often fork the material over by hand. This is relatively inefficient.

Northeast Sustainable Agriculture Research and Education (NESARE) Research

In 2015 and 2016 we evaluated the laying hens composting system via a grant from NESARE. This project assessed the opportunities and risks, such as nutritional value and pathogenic risks, associated with food scraps as a feed, as well as the economic viability of this practice

Figure 11.5. The receiving bay at Black Dirt Farm is prepped with hay and other carbon sources. The back of the bay is filled in to keep food scraps from getting stuck in the corners where they are hard to blend and access (by both the chickens and the loader). Courtesy of Black Dirt Farm.

for small-scale commercial production using 50 to 2,500 hens. We compared a group of 50 hens fed only grain with a group of 50 fed only food scrap compost.

For our research we collected results related to egg quality, labor, water, and expenses. Egg results were promising: No samples showed evidence of *Salmonella enteritidis*. Egg production was largely even between the groups, with the grain group producing roughly 2 percent more eggs over the year, though during certain times of the year the compost group produced as much as 10 percent more. In terms of nutrients, leucine, an essential amino acid, was higher in the compost group. Egg weight was slightly lower in the compost group, and the compost group's eggs were also slightly more fragile. Lastly, yolk color for the compost group was significantly darker than for the grain group, at a dark orange compared with a moderate yellow.

The grain group required only 57 percent the amount of management time over one month as the compost group did. However, labor in the compost group is more scalable (labor for 300 birds is the same as for 50 birds), unlike the grain-fed option, in which labor and grain costs per bird will parallel flock growth for the most part.

The grain group consumed nearly two times the amount of water that the compost group did. Still, the salt content of the food scraps

Figure 11.6. Food scraps are tipped into the Black Dirt Farm feeding and compost blending area. Courtesy of Black Dirt Farm.

means that access to clean water is equally important for compost-fed chickens.

The cost savings on feed and the secondary, value-added compost product in the compost group considerably improved gross and net revenues, despite increased labor costs. The grain group operated at a loss with the cost of feed representing over 87 percent of the value of the egg sales. The cost of setting up for feeding hens on compost is greater than that of grain feeding. Both systems require the same basic housing, pasture, watering, and grain feeding systems, but the compost group also required the setup of a feeding bin for receiving, managing, and feeding food scraps. The cost for this system, which has since been able to increase the number of birds it feeds by six times, was roughly $3,000. Additionally, this system requires a bucket loader to manage the material.

Moving forward, we will continue to look into several elements. We will conduct further testing of *Salmonella enteritidis* and an expanded list of foodborne pathogens, make more productivity comparisons, evaluate differences in egg nutrition, and answer questions about feed rations: What is the role of microbial value in the ration? What other locally available by-products could be used to supplement a compost ration? What is the baseline grain requirement, if any, required to meet production and cost goals?

Growing Soldier Fly and Other Larvae

One strategy that many practitioners are enthusiastic about is using food scraps and other organic matter (including manure) as feed to raise insect larvae, which can then be fed to chickens, fish, or other livestock. There are many motivations for people to go to the effort of essentially raising "maggots," the irony of which is not lost on those of us who go to significant efforts to prevent maggots (hopefully no one hates me for saying this; I mean it in the nicest of ways).

Soldier fly (*Hermetia illucens*) larvae are one of the more established feed insects. On day 14 of larval development, one study found them to have a crude protein content of 39 percent and crude fat of 28 percent;[24] both levels are significantly higher than commercial grain or mixed food scraps. The flies can consume manure and other wastes and turn it into large amounts of fat and protein with very little input other than labor, infrastructure, and heat in cold climates. These are admirable traits to have in a food system that relies heavily on fuel, mechanization, and a plethora of other inputs. A pound of dried soldier fly larvae goes for $11 in some places on the internet, and the larvae are widely available. There are an abundance of do-it-yourself-type videos and blogs on the subject, as well as several books. I myself have never raised larvae for animal feed, but I find it really interesting. If I was considering raising livestock, I would seriously consider this as an option.

Food Scrap Dehydrators

Preparing food scraps for feed by heat-treating, dehydrating, and pelletizing is another option that is worthy of consideration. There are numerous technologies on the market that perform some version of this, and although I am aware of only one brand that advertises itself as a feed production system (called Sustainable Alternative Feed Enterprises), other dehydrator vendors have at times plugged the feed value of their end products to me. Typically these systems are set up on-site at a large food scrap generator, such as a grocery store or resort. The economics of on-site systems generally appear to work for the generators, because they often claim to reduce the volume of the material by as much as 90 percent, which reduces their hauling costs (or they use it on-site). It is not typical to see composters use these as a pre-processing system for food scraps coming onto the site. However, a farm that's getting paid a tipping fee to receive food scraps while offsetting feed costs might justify the cost of installing and operating one of these systems through increased efficiency. In states where heat-treating feed is a requirement, a dehydrator system that meets the heat treatment standards might be an option, and would alleviate some of the unknowns about feeding food scraps directly to chickens.

Dehydrators produce a material that has a consistency similar to pelletized organic fertilizers. Think couscous, but denser, and with a slightly fermented smell. It's dry, typically somewhere in the range of 10 percent moisture content, and has a much higher nutrient content than raw food scraps. Because the moisture content of food scraps is a limiting factor in consumption by animals, this could be a benefit.

From an operational perspective, handling a smaller volume of drier material would bypass most of the potential issues, labor, and infrastructure required with feeding food scraps directly. Feeding the dehydrated material would really be more of a nutritional adjustment than a major operational adjustment in the way that feeding raw material would be (although removal of refused feed for composting would still be necessary).

There are also several potential drawbacks to using dehydrators to pre-process food scraps as feed. First, if you are not dehydrating on the farm, the economics of getting that material may change. Since the material has a theoretically higher value, there *may* not be a tipping fee, or there may even be a cost for the material, hauling or otherwise. The benefits of the processed material to the farm may outweigh the loss of a tipping fee, and may even justify paying a little for the material. These systems change the economics for the generator, which could shift the economics for the farm, too.

The largest concern that I have about dehydrating food scraps at the point of generation, then feeding this to animals, is the presence of contamination. Since many of these systems grind as well as dehydrate, plastics that might have made it into the system might get pelletized along with everything else. Obviously we don't want to feed plastic to animals, but we also don't want uneaten plastic particles to contaminate the manure and compost. Also, the purist in me worries about concentrating any chemicals (pesticides, heavy metals, and the like) that were present in the food scraps into the pellets and feed.

For those pursuing this option, it is worth researching practices for making animal feed in Asia, where the combination of dehydration and fermentation for feed purposes appears to be further along than in the United States. More discussion of dehydrators can be found in the *Food Scrap Dehydrators* section in chapter 2, page 45.

Bokashi Food Scrap Fermentation

The use of select microbes as a pretreatment or additive for compost feedstocks is popular in some segments of the composting world. Bokashi is one such treatment, and it is sometimes used to prepare feed for animals as well. Adding microbes to compost is not a practice that I generally subscribe to except in select circumstances, and cultivating organisms that have beneficial feed value is one of those applications. I have little direct experience with this, other than making bokashi at the Effective Microorganisms (EM) lab in Tucson, Arizona, in college, and working with some folks in Hawaii who used bokashi (I think I even tasted a bit at one point; I could see the appeal if I were a chicken). There is a wide array of information available on the practice.

Feed Diversity

Criteria for assessing food scraps as chicken feed should include *diversity*. Certain processing wastes might be extremely high in protein (such as meat and seafood) or homogeneous in other ways. One study noted that restaurant and residential residual was more balanced than that from homogeneous sources like fruits and vegetables, meats, or fish. Restaurant and residential scraps had crude protein of 27.5 and 16.3, respectively (dry weight), compared with 57 percent from fish wastes.[25] Mixed food scraps are more likely to provide a diverse and balanced diet for the laying flock and can include food processing residuals in the total mix. It is recommended to supply laying hens with the same supplements that would be used in a grain-based diet, such as free-choice grit, minerals, oyster/crab shell.

Basic Laying Operation Components

Apart from the food scrap feeding and composting aspects of production, the farm will look just like any other. There is an abundance of great information easily available to help you adequately plan for space and capacity in terms of nesting boxes, perches, pasture, water, and so on.

As a longtime student of permaculture, I love to see the creative ways that farmers and horticulturalists incorporate chickens and other animals into their workflow. Farms are working chickens into their crop rotations, using their natural instincts to reduce weeds and pests and prepare the ground by scratching and fertilizing. In Hawaii we used chickens to prepare ground for a food forest by growing cover crops of buckwheat and cowpea, and then letting the chickens do the mowing and tilling.

With the emergence of great systems such as solar-powered electric fences, mobile coops (also known as chicken tractors), light-sensor-controlled coop doors, and low-cost hoop houses that make great winter housing, to name only a few of the more recent innovations, poultry farming has an abundance of options for getting off the ground quickly and easily. Although food scrap feeding and composting is not yet as mainstream as, say, chicken tractors, there are effective methods for integrating this practice into laying operations no matter how small or big. In order to be successful, set up the feeding and composting side of the operation so that it is compatible and streamlined with all of the other vital components of the operation. I realize that this seems obvious, but doing it requires thought and often some experimentation. An efficient system really comes down to the proximity of the food scrap receiving and feeding area to the rest of the operation, having good access, and, of course, having enough capacity to manage the added material flow.

Operational Considerations (Best Management Practices)

Compared with a grain-based diet, food scraps present unique logistical considerations for chicken farmers. Scraps are typically wet and heavy, and have the potential to be odorous and attract unwanted visitors.

The systems described in this chapter are designed to minimize the challenges and maximize the benefits of utilizing food scraps as feed. Each system is designed to promote and be used in combination with best management practices (BMPs) for optimal function. Chicken composting BMPs have a lot of overlap with composting BMPs, but at the risk of being overly redundant, it felt important to touch on them in the unique context of chickens. Farmers will develop their own individual systems for feeding over time; that's the nature of both farming and composting. These BMPs are some common tools to mitigate issues or address them as soon as they arise.

Cleanliness

Minimize exposed feed and keep it contained to the designated feeding areas. As I often tell my three-year-old, small problems become big problems really quickly if left unaddressed (not in those exact words). General cleanliness is your primary means of preventing issues before they start.

Carbon Trough

Before receiving or feeding food scraps, lay down a base of absorbent carbon materials to contain the scraps and to balance excess moisture. A good way to describe this base is as a shallow trough made of carbon materials. The *carbon trough* keeps the receiving and feeding areas clean by keeping the food scraps contained and off the ground. Think of it almost like greasing a pan before baking: These carbon materials minimize the extent to which the scraps contact and make a mess of your receiving/feeding areas and equipment. All of the systems described here involve the use of a carbon trough as the base.

Capping

The primary tool for minimizing exposed food scraps in the designated receiving and feeding areas is covering the scraps with organic matter, which in turn minimizes unwanted visitors (vectors, flies, birds, and so on). The organic layer is often called *capping* or a *compost blanket*. Capping is a simple and efficient way to prevent the most common problems associated with composting. The organic matter creates a physical barrier that contains and absorbs odor and deters pests. It also can serve the dual function of diversifying the chickens' diet by adding plant fiber, fungi, or seeds, for example. Some degree of capping is inherent in all of the systems described, but it's also a variable that can be increased or decreased based on the level of need. Just 3 inches of capping will have a decent effect, and 6 inches will be very effective. If you are worried about adding too much carbon to the mix, use a neutral material in terms of C:N, or semi-mature compost, which will affect the end recipe minimally. The chickens will easily scratch through this layer to access the food underneath.

Cleanout

The receiving and feeding areas serve specific temporary functions in your overall management system and are not an ideal place to manage the breakdown of food scraps over time. Stockpiling scraps or leaving them unmanaged for longer than intended will increase the risk of problems, odors, rodents, and flies in particular. Clean out the material and manage it through active composting before it becomes a nuisance.

Composting

The final step in composting with chickens is, of course, composting the residuals. By feeding the food scraps to the chickens, you've reduced the overall volume you will manage considerably, but there will no doubt be some residual that gets cleaned out and

Figure 11.7. Food-cicles—frozen blocks of food resembling enlarged Popsicles—mixed in with loose food in the Highfields receiving and blending area. Courtesy of Highfields Center for Composting and Vermont Sustainable Jobs Fund.

will need to be managed. Since the rest of the book is all about composting systems and BMPs, we don't need to get into those details here, but I will say this: *Composting is not stacking a bunch of rotting food into a pile and waiting for it to break down (also known as feeding the birds, coyotes, and skunks).*

I say this because I (and other more authoritative types) fear that in some cases there is a higher likelihood of poor composting practices happening where people are composting with animals, because the compost outputs are a secondary priority to the livestock. In fact, I believe in most cases the opposite is true: By adding the efficiency and motivational factor of animal feeding to the operation, the system itself is more conducive to good composting practices than composting food scraps alone would be. Regardless, if the farm doesn't have the resources to manage food scraps well, abandoning ship is a better option than giving this practice, or composting in general, a bad name.

Seasonal Challenges

In cold climates operators who compost with chickens face major challenges due to severe winter weather, including frozen materials, snow, and cold environmental conditions for the laying flock. When temperatures stay below freezing for multiple days, food scrap containers stored outside can freeze solid, making them difficult to tip and inaccessible to chickens to eat. In order to solve this problem, composters use various methods depending on scale and equipment.

- Shooting pressurized water (heated is preferable) at the inside perimeter of a frozen container usually helps the food scraps to loosen enough for tipping. This process adds extra moisture to the material, which needs to be managed.
- If space allows, frozen food containers can be thawed in an above-freezing area prior to tipping.

- If hot, active feedstocks are available, initial blending of the food scraps prior to feeding will help them thaw. Having the ability to manage feeding as part of an active composting process is one of the best strategies for management throughout the winter months.

Sizing a System for Composting with Chickens

There isn't a textbook covering the precise methodology for sizing a chicken composting system. I recently spent several hours with Tom Gilbert (my technical mentor and owner-operator of Black Dirt Farm) discussing methodologies for the sizing of his chicken feeding and composting operations. Over the years, between the two of us, we have literally sized hundreds of composting systems. While our approaches were similar and included the same variables, I was reminded how big an impact small variables can have on a design. This is especially challenging when volume reduction is in such a rapid state of flux, as it is during feeding.

The approach in this chapter is to first size the hen flock based on the available feed (or vice versa). Then, based on the ratio of birds to available feed, we will estimate the volume of residuals that needs to be composted following feeding. Next, recipe gets factored in and used in sizing both the receiving area and the feeding areas, if these are different. The volume reduction that happens during the feeding process is then factored into the batch sizing calculations for residuals going to the farm's composting system, following the methodologies in the other chapters. Finally, we'll look at the receiving and feeding areas in terms of their sizing requirements.

Each chicken composting system is nuanced and has numerous embedded assumptions. As with any of the methods in this book, adapt the calculations to reflect your scenario as best you can and leave yourself flexibility in the design to adapt to on-the-ground conditions once the operation is deployed. We'll start with flock size and feed requirements.

Flock Sizing and Feed Requirements

When composting with chickens, I have traditionally used a very simple rule-of-thumb ratio of food scraps to birds as the starting place for determining the ideal scale of the operation. Two pounds of food scraps per bird per day is the ratio that I was taught is necessary to adequately meet the nutritional requirements of a laying hen, though there is currently a lack of scientific research on the subject, and I have never been able to locate the original source of that recommendation.

In researching this chapter and looking at the nutrient contents of food scraps from a feed value perspective, I noticed that on average, 2 pounds of food scraps is not that dissimilar to a grain ration in terms of protein. Consider that:

1. The feed value of food scraps on a dry-weight basis is fairly similar to or higher than a laying hen grain ration, which has a protein content of around 16 percent (analyses of food scraps show a crude protein range of 15 to 28 percent[26]).
2. Food scraps have a much higher average moisture content than grain feed at 75 to 90 percent, and water is heavy.
3. If you take 2 pounds (32 ounces) of food scraps and remove 80 to 90 percent of the moisture, you'd be getting close to the typical 3 to 4 ounces per day per bird of layer ration; the dry-weight protein content remains the same, because the percentage has already accounted for the removal of the moisture.

Obviously, this is rough math, but food scraps are so variable, the only strategy short of individually doing a lot of feed analysis is to get it in the ballpark. The character of food scraps from both a composting and a feeding perspective is all about averages.

It took me a while, but I have come to recognize that to accommodate for food scraps' variability and to make the process work logistically, we need to overfeed. The 2 pounds per bird per day ratio should therefore be viewed as a minimum, and it is system-dependent. One reason for this relates to the geometry of 2 pounds of food scraps. At a bulk density of 1,000 pounds per cubic yard, 3 pounds of scraps occupy about 1 square foot at a depth of 1 inch. Two pounds only occupy a 12 × 8 × 1-inch space. Now consider the footprint of a chicken while it's eating. Maybe 0.66 square feet per bird (12 × 8 inches) of feeding space would be adequate for a small flock using the frequent cleanout method, but scale that up and it would quickly feel crowded. That geometry just doesn't scale well, unless the scraps are spread out over a large space, probably at a ratio of 1 square foot or more per bird in terms of the feeding area. To achieve this in an active compost pile, where several feet or more of material is accumulated, and where there is still adequate feed on the surface, requires much more feed per bird.

Black Dirt Farm's 2017 SARE study report illustrates this well (see the *Feeding Community Food Scraps to Laying Hens in an Active Composting System* sidebar on page 299). The farm estimated that their laying hens consumed an average of around 1.2 to 1.5 pounds per bird per day of food scraps. However, they recommended budgeting a minimum of 3 pounds per bird per day and noted that feeding in an active composting environment means providing much more than that in practical terms (as much as 15 to 20 pounds per bird per day).[27]

Providing food scrap throughput in excess of what a flock can consume is acceptable and even appears necessary. It does require the capacity to effectively manage and compost the excess volume. Overfeeding means being extra diligent in managing the residuals or else running the risk of creating really big problems. It also means more feedstock demands to balance out the refused feed and more material to clean up and compost.

Inadequate food scraps, on the other hand, will present health and production problems for your flock, and steps should be taken to secure enough food scraps or supplemental feed to maintain flock health. The benefit of maintaining a lower food scrap to bird ratio (assuming additional grain or other feed) is less cleanup and less composting capacity.

When transitioning grain-fed chickens to a food-scrap-based diet, I recommend gradually weaning the flock from grain in order to minimize stress and provide a reasonable learning curve for both chickens and operators.

To put the 2 pounds per bird per day ratio into perspective, the portion of scraps for 1,000 birds is around 1 ton per day. A small layer flock of 25 birds could handle about 350 pounds per week of scraps, which is the volume generated by a 310-student elementary school.

Step 1 in planning a chicken composting is calculating flock size.

Step 1. Calculate Flock Size

A laying hen flock size can be approached in one of two ways based upon the ratio of food scraps to birds:

1.1 (A) Calculating Flock Size Based on Food Scrap Throughput

___ Food Scraps (Pounds per Day)
÷ ___ Pounds per Bird per Day = ___ # of Laying Hens

or

1.1 (B) Calculating Feed Requirements Based on Flock Size

___ # of Laying Hens
× ___ Pounds per Bird per Day = ___ Food Scraps (Pounds per Day)

Note that throughout most of this book we work with weekly throughput, not daily throughput. Make sure you convert your numbers to make the time periods uniform.

EXAMPLE

10 tons of food scraps per week or 2,857 pounds per day
1 receiving area
Feeding areas use frequent cleanout

Step 1.1 (A)
2,857 Pounds ÷ 2 Pounds
= 1,429

Step 1.1 (B)
1,429 × 2 Pounds
= 2,858 Pounds

Estimating Refused Feed for Composting

The next step is estimating the percentage of food scraps that the birds will consume. If you have hit the 2 pounds per bird per day ratio, you would expect 60 to 80 percent consumption of the scraps, and 20 to 40 percent residual feed that needs to be collected and composted. If you're feeding a ton of scraps a day, that would leave 400 to 800 pounds per day of refused feed, which is less than 1/5 to 2/5 of a ton of refused feed. That's a really good reduction in volume, right off the bat.

If the flock is undersized in relation to the ratio, adjust the consumption rate based on the ratio. Subtracting 60 to 80 percent of the feed calculated based on the ratio from the total amount you plan on receiving will give you a decent estimate of what scraps will remain and need to be composted. Farms that operate on the other end of the spectrum, for example providing 1 pound per bird per day, will see near-complete food scrap consumption; 95 percent would be a reasonable estimate, because with typical food scraps there are still going to be a few broccoli stems and avocado seeds that the chickens can't eat.

In addition, depending upon the feeding/composting strategy, there may be significant volume reduction during mixing or decomposition. We can adjust for this when we calculate the batch size for composting.

Step 2. Estimate Refused Feed

Subtract the consumed feed based on the ratio from the total amount you plan on receiving. This will give you a decent estimate of what scraps will remain and need to be composted. In turn, this will also provide an estimate of the additional feedstock needed to balance out the recipe for the remaining feed.

2.1 Estimate Feed Based on Ratio

___ # of Laying Hens × 2 (Pounds) = ___ Pounds Food Scraps Based on Ratio (Pounds per Day)

EXAMPLE

1,429 × 2 Pounds
= 2,858 Pounds

2.2 Estimate Consumed Feed

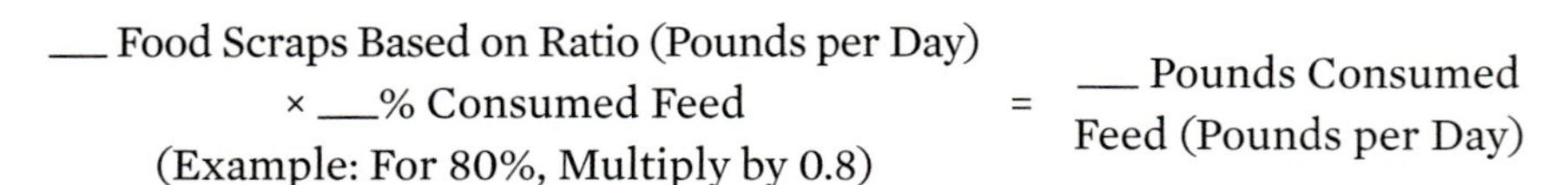

___ Food Scraps Based on Ratio (Pounds per Day) × ___% Consumed Feed (Example: For 80%, Multiply by 0.8) = ___ Pounds Consumed Feed (Pounds per Day)

2,858 Pounds × 0.8
= 2,286 Pounds

2.3 Estimate Refused Feed

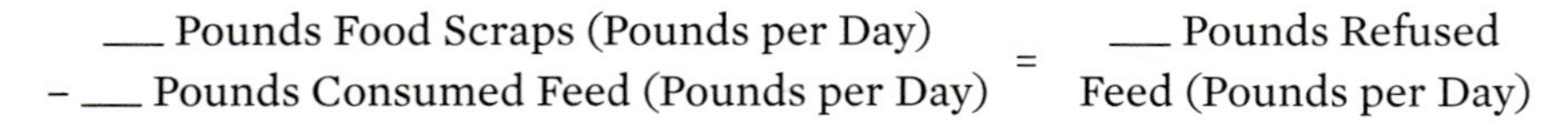

___ Pounds Food Scraps (Pounds per Day) – ___ Pounds Consumed Feed (Pounds per Day) = ___ Pounds Refused Feed (Pounds per Day)

2,858 Pounds – 2,286 Pounds
= 572 Pounds

Note that actual food scraps might be the same or higher than the food scraps based on ratio. Also note that your result for refused feed will be used in step 3.2. You will need to convert your numbers to make the time periods uniform.

Accounting for Compost Recipe in Feeding

When planning the receiving, feeding, and residuals compost areas, you need to take composting recipe into account. Creating the composting recipe you'll use for managing food scrap refusals (residuals) during and following feeding requires a nuanced approach. The amounts of feedstocks used should be based on both sanitary maintenance/bedding of the feeding area and the desired end result of the composting process. Ideally the same means achieves both ends.

For planning purposes, additional feedstocks will be proportional to the estimated volume of refusals. These estimates will be used to formulate the capacity of the receiving and feeding areas, as well as the residuals composting system. The discussion of compost recipe for estimating batch size (see *Compost Site Processing Capacity and System Scale*, page 114) will probably be the most relevant to this process. A much more in-depth look at recipe can be found in chapter 4.

Since feed consumption rates (that is, the volume of refusals) are challenging to accurately predict, once you begin feeding scraps, I recommend testing the residuals to develop a solid recipe around the material that's actually coming out of the feeding area. Based on this number, the farm can adjust the rates of bedding/feedstock going into the blend, prior to or immediately following the feeding process.

In addition, most sites compost the chickens' bedding and manure. These volumes should be accounted for in the recipe and batch size as well. Manure production figures for laying hens are presented in table 4.4, page 100.

Sizing the Feeding and Receiving Areas

Receiving and feeding area designs are unique to every farm, but most commonly these two infrastructure components are integrated in some way. They both need to have the capacity to hold your target volume of food scraps until those are either consumed or moved on to composting in another location. Nuance comes into play largely in terms of the retention time and pile height with the feeding strategy you intend to deploy. A third factor is that the feeding area needs to have enough space for the hens to congregate and feed. Anecdotally, 1 square foot per bird appears to be adequate in terms of the actual footprint of the feed, but this assumes there is additional space for the birds' range.

Depending upon the layout of the farm and the strategy you deploy, the receiving and feeding area may be one and the same, directly adjacent to each other, or in separate locations. Although minor adaptations to the strategy can be made later on, before creating detailed plans for these site components, you should have a sense of the flow of feed and other materials through the farm.

Food scrap receiving areas are common to most compost sites. Topics such as materials, slope, cover, push walls, tipping ramps and docks, and more are discussed in greater depth in *Food Scrap Receiving and Blending Areas, and Equipment*, page 132. These design considerations will apply to food scrap feeding pads as well.

In most cases food scraps that are being fed upon will have a shorter pile height and wider pile width—in other words, a larger surface to volume ratio—while food scraps that are being stored for subsequent feeding will be more consolidated. In addition, the system will require access and work spaces for bucket loader operation.

The following steps are similar to the batch and receiving area sizing approaches used with the other composting methods, except for a few adaptations, which are accounted for as we go through each step. The main difference is that we are accounting for less additional feedstock in the initial mix due to the volume of food scraps subtracted as feed.

Step 3. Calculate Volume in Receiving and Feeding Areas

Building upon steps 1 and 2, we'll now calculate the total material volume in the receiving and feeding areas. In order to achieve this, we'll need to have identified:

a. The overall feeding strategy (see *Developing a Feeding and Composting Strategy*, page 294)
b. Whether the receiving and feeding areas are combined or separate
c. The frequency with which material comes into and leaves the receiving and feeding areas
d. The ideal pile height in the receiving and feeding areas

If the receiving and feeding areas are combined, factor that into the assumptions. If they are separate, repeat step 3.5, adjusting your assumptions to account for material flow to the feeding areas.

*3.1 Convert Weight of Food Scrap to Volume**

___ Pounds Food Scraps per ___ Time Period ÷ ___ Food Scrap Bulk Density (Pounds per Cu. Yd.)	=	___ Total Cu. Yd. Food Scraps per ___ Time Period

*3.2 Convert Weight of Refused Feed to Volume**

___ Pounds Refused Feed per ___ Time Period ÷ ___ Food Scrap Bulk Density (Pounds per Cu. Yd.)	=	___ Total Cu. Yd. Refused Feed per ___ Time Period

3.3 Estimate Volume of Additional Feedstocks Based on Refused Feed

___ Total Cu. Yd. Refused Feed per ___ Time Period × ___ Ratio Additional Feedstock to Primary Feedstock (Typically 3–5:1)	=	___ Cu. Yd. Additional Feedstock per ___ Time Period

3.4 Estimate Total Volume of Feed and Additional Feedstocks

___ Total Cu. Yd. Food Scraps per ___ Time Period + ___ Cu. Yd. Additional Feedstock per ___ Time Period	=	___ Total Cu. Yd. Feed and Feedstock per ___ Time Period

This is the volume the receiving bay/feeding area needs to handle.

* Convert from pounds per day to whatever time period of throughput the infrastructure will be sized to accommodate. See *Converting Food Scraps to Tons and Cubic Yards*, page 88.

EXAMPLE

a. Food scraps come to the receiving area, then a ration is transported to two concrete feeding pads every day.
b. Separate.
c. Food scraps are delivered to the receiving area weekly. Food scraps are transported to pasture for feeding daily (feeding areas need to accommodate ⅐ the volume of the receiving area).
d. Receiving area is 5-foot pile height after consolidation. During feeding, food scraps will be spread out, leaving 1 square foot per bird for space, which equates to a 1-foot or less depth of total material.

Step 3.1
20,000 Pounds per 1 Week
÷ 1,000 Pounds
= 20 Total Cu. Yd. per 1 Week

Step 3.2
4,004 Pounds per 1 Week
÷ 1,000 Pounds
= ~4 Total Cu. Yd. per 1 Week

Step 3.3
4 Total Cu. Yd. per 1 Week
× 4
= 16 Cu. Yd. per 1 Week

Step 3.4
20 Total Cu. Yd. per 1 Week
+ 16 Cu. Yd. per 1 Week
= 36 Total Cu. Yd. per 1 Week

3.5 Calculate Receiving Bay and Feeding Area Footprints

Based on the volume calculated in step 3.4, refer to *Step 1. Size Receiving and Blending Area (Pad)* in chapter 6, page 139, to find the receiving bay/feeding area footprint in table 6.1 (page 140). Make sure to use the total "Yards in Batch" column, rather than the "Tons" or "Yards Food Scraps" column.

If the farm has separate food scrap feeding area/s, repeat step 3.5 for those areas, making sure to adjust the volumes in each feeding area according to the feeding strategy. Based on anecdotal evidence, I recommend that feeding areas provide at least 1 square foot per bird, assuming that there is additional space for the birds to range while not feeding.

Step 3.5

Receiving Area: 300 square feet based upon a 5-foot pile height. Assume a 15 × 20-foot pad, with an additional 20 × 20-foot access way.

Feeding Areas: At 1 square foot per bird, the two feeding areas will occupy around 1,500 square feet. Two 25 × 30-foot pads. The two pads will have no additional work space other than an access road. Ideally these pads will be covered.

Estimating the Volume of Residuals for Composting

The final step in this chapter is to estimate the volume of material coming out of the feeding process and into the residuals composting system. The volume calculated in this step replaces the batch size calculations in chapter 5 (see *Compost Site Processing Capacity and System Scale*, page 114), which are a precursor to all of the other system sizing methods described throughout the systems chapters of this book.

The main difference between this step and more traditional batch sizing is that it accounts for the volume reduction of the food scraps going into the composting system, based on the estimated percentage of food the chickens will consume. In this step we'll also account for other reduction factors such as blending and degradation and convert from cubic yards to cubic feet.

In addition, most farms will compost the chickens' bedding and manure in with their other materials. Chicken manure should be added to the recipe following feeding so as not to expose the hens to their own manures unnecessarily. These volumes should be accounted for in the recipe and batch size as well. Manure production figures for laying hens are provided in table 4.4, page 100.

Step 4. Estimate Volume of Residual for Composting

Combining chicken manure into your overall recipe can be accounted for in steps 4.1 and 4.2. If not, skip to step 4.3.

4.1 (Optional) Add Chicken Manure

Since manure generation estimates do not include bedding, a portion of the additional feedstock volume will already have been added as bedding. The remainder can be added during blending. Manure production figures for laying hens are presented in table 4.4, page 100.

___ Cu. Yd. Chicken Manure per ___ Time Period
× ___ Ratio Additional Feedstock to Manure (Typically 3–5:1)
= ___ Cu. Yd. Additional Feedstock per ___ Time Period

EXAMPLE

~ 0.75 Cu. Yd. per 1 Week
× 4
= 3 Cu. Yd. per 1 Week

Assumes a manure generation rate of 0.001 cubic yards per bird per week (table 4.4), capturing an average of half the manure and cleaning out weekly.

4.2 (Optional) Estimate Manure and Additional Feedstock Volumes

___ Cu. Yd. Chicken Manure per ___ Time Period + ___ Cu. Yd. Additional Feedstock per ___ Time Period	=	___ Cu. Yd. Chicken Manure, Bedding, and Feedstock per ___ Time Period	0.75 Cu. Yd. per 1 Week + 3 Cu. Yd. per 1 Week = 3.75 (Rounded to 4) Cu. Yd. per 1 Week

Add this volume to the total cubic yards feed and feedstock per ___ time period in step 4.3.*

4.3 Estimate Volume of Blended Residuals

___ Total Cu. Yd. Feed and Feedstock per ___ Time Period × ___ % Blending Shrink Factor (Typically 20%, or Multiply by 0.8) × ___% Additional Shrink Factor (Optional)† × 27 (Cu. Ft. per Cu. Yd.)	=	___ Cu. Ft. Blended Compost per ___ Time Period	40 Total Cu. Yd. per 1 Week × 0.8 × n/a × 27 = 864 Cu. Ft. per 1 Week

4.4 Estimate Total Blended Volume/Batch

___ Cu. Ft. Blended Compost per ___ Time Period × ___ # of Time Periods per Batch	=	___ Cu. Ft. Blended Compost per Batch	864 Cu. Ft. per 1 Week × 2 Weeks = 1,728 Cu. Ft.

This is your standard "batch size" and can be used with any of the compost system sizing methodologies. For larger systems and especially for both positive and negative ASP systems, use of a biofilter as a capping material is a best management practice and highly recommended. If you are unsure about whether or not you should use a biofilter, see chapter 9. The estimated volume of capping materials for ASPs is different depending upon whether the system uses positive or negative aeration. If you don't know, positively aerated windrows are the conservative assumption.

4.5 (Optional) Estimate Total Volume per Batch with Biofilter Cap

___ Cu. Ft. Blended Compost per Batch × ___ Biofilter Capping Volume Factor (see table 5.1)	=	___ Cu. Ft. Blended Compost per Batch with Biofilter	1,728 Cu. Ft. × 1.15 = 1,987 Cu. Ft.

This volume will be used for sizing your residuals composting system.

* Depending upon the manure management strategy, the chicken manure may not be available all the time, in which case you can set this volume aside from your weekly food scrap batches and account for it separately in your compost system sizing.

† Regarding additional shrink factor, for feed that has a longer residence period in the receiving and feeding areas, additional volume reduction may be accounted for. Keep in mind that reduction due to bird consumption and blending shrink factor have already been accounted for. An additional shrink factor of 10 to 20 percent (multiply by 0.9 or 0.8, respectively) over one to two weeks might be realistic.

CHAPTER TWELVE

Food Scrap Generation and Collection

A growing number of innovative food scrap collection businesses are sprouting up around the country, with models at every scale. Some transport food scraps from people's homes, others from businesses and institutions, and many from both. The range of examples is large, from a school sending food scraps on their school bus to a local herb farm, to haulers with truck bodies capable of multi-stream collection. Many composters also function as hauling businesses, while others work with independent or municipal haulers.

In this chapter we'll first discuss the pros and cons that come with vertically integrating composting and collection, and then move on to route-planning considerations, such as route density, service area, and stop time. Integral to route planning are methods for estimating the scale of collection from different sources, including individual residences, larger generators, and whole communities or regions. If you are strategizing about how to estimate the volume of food scraps you're planning to process, reading this section should be one of your first steps. Lastly this chapter will tackle some of the basics in terms of collection fundamentals, scenarios, equipment, logistical considerations, and practical solutions that composters and haulers have developed.

As with other parts of this book, sections of this chapter were adapted from my prior work with the Highfields Center for Composting and the Institute for Local Self-Reliance.[1]

detritivore

(di-ˈtrī-tə-ˌvȯr)

DEFINITION: An organism that uses organic waste as a food source.

Vertical Integration: Collection and Composting

Models that handle food scraps generated off-site, by definition, must involve both a collection and a recycling component. It's a marriage of necessity, and in some regions where there aren't potential partners to play the corresponding role, composters have no choice but to be their own hauler, or vice versa. Even where there are potential partners, vertical integration offers a number of potential benefits.

First and foremost, since collection services act as the middleperson between the generator and the organics recycler, they have a large influence on the availability and quality of organics in a region. Food scrap composters are sometimes left high and dry without material when a hauler finds another outlet, even if they have been integral to the hauler's growth. Likewise, a hauler might adjust routes, leaving a composter with different or lower-quality feedstock. Even with contracts in place, the organics recycler is ultimately at the mercy of the hauler. With a vertically integrated hauler/composter model, the composter's interests are the hauler's and vice versa. This provides an advantage over the partnership models, where there is natural tension between individual interests, particularly as scale and competition grow.

Another benefit to the hauler/composter model is that, at least in the current organics processing economy, food scrap collection appears to be more profitable than composting on its own. I have heard this again and again from composters who use the vertically integrated model. Ideally the composting and hauling components are set up as distinct business entities, so that one can be closed or sold if it fails to provide adequate revenue, as was the case with Virginia's Black Bear Composting, whose collection business was turning a profit, but whose compost site was not.[2]

On paper the two business entities would be at odds with each other, with the hauling entity paying the composting entity a tipping fee, but with dual ownership you are privy to both businesses' finances and get to set the tipping fee as well as the rates charged to the generator. Ownership and management can thus be set up in way that will allow both businesses to succeed.

There are also drawbacks to vertical integration. Collection and composting have some overlap, but they are inherently distinct businesses, and running two distinct businesses well is much harder than running one business well. It's not an insurmountable challenge by any means—just a matter of capacity and skill sets. Another potential downside for a composter is that managing collection may create competition with other collection partners. Maybe that competition is actually an advantage in negotiating with other haulers, or maybe it's not an issue at all. It just depends on the scenario. Either way, it's worth considering how being a hauler might negatively impact the composting side of your business, along with the advantages it may yield.

Collection Service Area and Scale

In order to plan your collection route and composting system requirements, you'll want to have a fair degree of confidence in the estimated volume of material that you will handle over time. This section of the chapter discusses methods and factors involved in estimating food scrap capture from common food scrap generator (FSG) sources, including residences, businesses, and institutions, both at the individual entity scale and in whole communities or regions.

Businesses collecting and composting material from multiple sources often consider a range of factors, including service area, generation by the sectors you intend to target (residential, commercial/institutional, and food processing), capture rates by sector over time, stop time per generator, and route density.

In areas where there is a large population and little ongoing composting activity, conducting a detailed assessment of what's out there might not be absolutely necessary at first, although it certainly couldn't hurt, especially when trying to gauge route density (how far you'll need to go to collect a target volume of material). Still, jumping in and testing the waters will ultimately be the best way to figure out how open the market is to a new collection service. In dense untapped regions, the volume that's actually captured will more likely relate to the capacity of collection apparatus and hauling density, rather than to total volume of capturable material.

In small cities, towns, and rural regions, or where there is significant competition, a more detailed estimate of generation, participation/capture rates, and route density will be needed to assess feasibility, plan routes, and plan system scale.

Food Scrap Generator Sectors (Sources)

Although it's been touched on throughout the book, it seemed important to briefly reiterate the different sectors of food scrap generators as they are classified here. In planning a local composting strategy, it's critical to have the ability to differentiate between FSG sources when anticipating food scrap capture. For example, residential food scrap generation is one of the simplest sectors to predict. Population data is widely available, and a per capita generation rate is a reasonable predictor. Generally speaking, there are three different outlets for residential food scraps:

- Backyard composting
- Residential food scrap drop-offs
- Curbside collection

You can estimate what the participation rate in any one of these outlets might be, then use that information to estimate route density and plan for both collection and processing capacity.

The other two sectors that are discussed here are *commercial/institutional* and *industrial/food processing*. You'll often see these two lumped together in the waste industry as industrial/commercial/institutional or ICI, but at least in my experience, I find it useful to differentiate between them. The reason for this is that the industrial/food processing (we'll just refer to it as food processing) output is extremely hard to predict, while output from other commercial and institutional generators such as restaurants and schools can be estimated with better accuracy. Saying *mixed food scraps* differentiates commercial, institutional, and residential food scrap sources from *food processing residuals*, which refers to what is typically a more uniform product that comes from a food manufacturer. Potato peels from a potato chip factory would be an example of food processing residuals.

Route Density

Route density is one of the biggest factors in collection feasibility. This term refers to the amount of capturable material in a given space and time, taking into account proximity between destinations. In both economic and environmental terms, hauling efficiency can be gauged by the time it takes and the distance traveled in filling up a load and delivering it.

What's feasible in terms of route density is relative to the generator sector, the participation/capture rate, and, of course, the collection method. Policies such as organics diversion mandates and single-hauler municipal collection services (franchising or consolidated collection) also have a huge impact. These factors combined will ultimately determine what strategy, if any, would be appropriate to service a given region.

There is surprisingly little published data to pull from in assessing route density. Ultimately, for planning purposes you will be looking to estimate factors such as:

- Stops per hour
- Weight per stop
- Revenue per stop
- Operating costs

I know from anecdotal experience that trucks (with enough capacity) can collect and tip 8 to 10 tons in 8 to 10 hours from the commercial/institutional/food processing sectors. These numbers come from towns and small cities with 10,000 to 30,000 people, so more urban areas and routes with large generators could certainly have higher route density. Well-established routes may hit this level without organics mandates, but I've also seen routes hit this rate in under a year with mandates in place.

Residential collection rates were reported to run as high as 1,000 households per route, and to average around 750 households per route in Boulder, Colorado, where the cost of organics collection is embedded in mandatory trash collection fees. Although composting is not mandatory, the organics diversion rate was 16 percent.[3] That might be viewed as the extreme end of the route density spectrum, given that the cost of participation is built into the waste fee structure. At the other end of the spectrum, bicycle-based collection routes (without electric assist) appear to average about 200 to 300 pounds per route, and about 20 to 30 households over an average of about one to three hours.[4] These are private customers who are not mandated to compost or pay for services, so the route densities are not at all comparable in terms of participation rates.

Stop Time

Another critical factor related to route density and collection efficiency is *stop time*, which is the average amount of time it takes to service an account. One of the key targets when considering collection equipment is minimizing stop time while still providing excellent service.

Estimating Generation, Capture, Participation, and Diversion Rates

Before we move on to estimating generation, we need to discuss some of the terminology and assumptions that accompany data related to food scraps. When researching food scrap produced and available to be collected in a given region, you may encounter several terms, including *generation*, *capture*, *diversion*, *participation*, and *disposal*. Good sources will clearly define the terms that they use, but I find that this is not always the case, so I will define these terms as I use them, and as I understand their meaning in common use.

Even when a generator is composting, not all of the food scraps that are produced necessarily make it into the compost bucket. (Hopefully that will change as more young composters grow old!) *Generation rates* technically refer to the total food scraps that go unused by a given entity. These rates are typically given as a weight produced in time, such as pounds or tons per week.

Diversion and *capture* refer to the total amount of organics that actually gets recycled. These are typically given as a weight processed in time, or as a percentage of generation where there is a good estimate of generation (through waste audits or waste data). Although *diversion* and *capture* are, for all intents and purposes, interchangeable, *diversion* is more often used in reference to a population, while *capture* more often refers to a specific entity. What remains uncaptured is called *disposal*, and this is what still ends up in the landfill or incinerator. With good systems and education in place, disposal decreases and capture/diversion increases.

Participation rates, as I understand the term, refers to the percentage of a given generator sector or group that composts. Examples include the percentage of households that join a collection or drop-off program, or the percentage of students in a school that scrape their plates into the right bucket. So participation relates to generators, not generation.

I apologize for subjecting you to these semantic details, but it's important to define terminology, because there is too much extrapolation required in this work to begin with. For example, one challenge that you are likely to encounter is that published methodologies for estimating the volume of food scraps produced by particular entities are often vague about whether the data is an estimate of generation or diversion/capture, and the two are often used in the same data sets interchangeably. The data that I've compiled here is unfortunately no exception.

Ultimately, what matters to composters is what is captured, not what is generated, but the practical differences are so small that it's probably not worth sweating.

Planning for both collection and its corresponding composting infrastructure involves developing a reasonably accurate estimate of total capture. Once capture has been estimated, route density and stop times can also be estimated in relation to the particular collection strategy and equipment.

The next sections look at some of the strategies and metrics used for estimating generation and capture from the different FSG sectors.

Estimating Food Scrap Generation for an Entire Community

There are two different general approaches you might take when estimating food scrap generation from across a whole community, as opposed to from individual entities. One method is simply to take a region's population and use broad-brush metrics to calculate all of the material in that region. For example, based on an estimate of 61 million tons per year total generation in the United States,[5] and a population of 323 million, domestic per capita generation is right around 7.26 pounds per person per week. Alternatively, states publish waste characterization studies that estimate the volumes of organics that are being disposed of, based on sampling of loads coming into landfills. These are usually easy to find online. I am skeptical of some of the results that I've seen from the statewide studies, but as a ballpark metric, they can still be valuable.

Using any of these methods, you'll need to account for how the published metrics were calculated. Questions to consider include:

- Which generator sectors were included?
- Which materials were included? Yard debris? Paper?
- Are they reporting generation or disposal? (Disposal will not include what's already being diverted.)

Once total generation has been determined using these methods, you can very roughly estimate the breakdown among sectors. In the research that I've done, the split has been roughly 30 percent residential, 40 percent commercial and institutional, and 30 percent food processing. You can use per capita residential generation estimates as a cross check (see *Residential Sector* on page 320).

These simple methodologies are at the very least a way to familiarize yourself with the demographics of an area, and for dense untapped regions, they may provide you with all the information that's needed to just jump in and go. As I understand it, however, population-based estimates are not precise at a granular level for anything other than the residential sector, nor are these rough estimates useful when it comes to route planning or targeting generators. For rural or suburban regions, especially, using metrics based on statewide or national data can be highly inaccurate, because many areas within these regions are purely residential, or have much lower rates of commercial or food processing generation. Due to these inaccuracies, planners and policy makers are working on more refined tools and methodologies for estimation, at least for the commercial and institutional sectors. Sector-specific strategies for estimating food scrap generation and capture are discussed in the following sections.

Commercial and Institutional Sector

There are a growing number of resources that can be used to estimate generation by commercial and institutional food scrap generators. Foremost are online databases, often with GIS mapping capabilities, which use simple algorithms and publicly available data to estimate generation for different generator types. These databases compile publicly available information, such as metrics from state departments of health and education, apply an algorithm to estimate generation, and then map the data in space.

These databases also typically provide generator names, estimated generation rates, the assumptions used to make those estimates, and even generator contact information. This generator-specific data can be extremely valuable, as it can be used to map out a collection strategy, target large and likely generators, and track outreach.

Connecticut and Vermont are two states that currently have websites with GIS viewers and downloadable databases. It is important to look into any existing sources of data about food scrap generation in your region and investigate existing composting activities that might impact your own capture rates or provide partnership opportunities.

As with the estimation methods discussed earlier, know your data source's strengths and weaknesses. For example, are certain sectors or generator types missing or based on poor assumptions? Look to see if food processing data is included. This data is likely to be highly inaccurate and should therefore be removed (keep the generator's name if it's a potential client, but don't assume that the generation data is even remotely accurate—see *Food Processing and Manufacturing Sector* on page 320). The basis of this data influences the estimates significantly, and you may want to adjust your estimates after learning more about any weaknesses in your source. Whenever possible use more than one data source as a cross-reference to compensate for these weaknesses. I have spent countless hours with different iterations of Vermont's generator database, and even compared the estimates for specific generators with the actual capture by haulers who were willing to share their collection data.[6] My comparisons are noted, alongside some example algorithms, in table 12.2 (page 325).

I imagine that over time, state databases will become more refined. Yet even now, despite their imperfections, they appear to be far more useful than population data for estimating generation in the commercial and institutional sector.

Table 12.1. Commercial and Institutional Food Scrap Generators

Generator Type	Measurable Metric	Generation Factor to Get to Tons per Week	Notes
Elementary schools	# of students	(# of Students) × (1.13 Lbs. per Week) ÷ (2,000 Lbs. per Ton)	Very accurate.[a]
Middle schools	# of students	(# of Students) × (0.73 Lbs. per Week) ÷ (2,000 Lbs. per Ton)	Very accurate.[a]
High schools	# of students	(# of Students) × (0.35 Lbs. per Week) ÷ (2,000 Lbs. per Ton)	Very accurate.[a]
Colleges/ universities	# of students	(# of Students) × (1.13 Lbs. per Week) ÷ (2,000 Lbs. per Ton)	Have not validated.[a]
	# of meals served per week	(# of Meals Served per Week) × (0.35 Lb. per Meal) ÷ (2,000 Lbs. per Ton)	Have not validated.[b]
	# of students on-campus and # of students off-campus	(# of Students On-Campus) × (2.73 Lbs. per Week) ÷ (2,000 Lbs. per Ton) + (# of Students Off-Campus) × (0.73 Lbs. per Week) ÷ (2,000 Lbs. per Ton)	Have not validated.[b]
Correctional facilities	# of beds	(# of Beds) × (0.5 Lbs. per Meal) × (3 Meals per Day) × (7 Days per Week) ÷ (2,000 Lbs. per Ton)	Based on 2 actual generators, this rate was 33% low.[a]
	# of inmates	(# of Inmates) × (1 Lbs. per Day) × (7 Days per Week) ÷ (2,000 Lbs. per Ton)	Have not validated.[b]
Hospitals	# of beds	(# of Beds) × (0.5 Lbs. per Meal) × (3 Meals per Day) × (7 Days per Week) ÷ (2,000 Lbs. per Ton)	Based on 3 actual generators, this rate was 13% low.[a]
	# of beds	(# of Beds) × (3.42 Lbs. per Bed per Day) × (7 Days per Week) ÷ (2,000 Lbs. per Ton)	Have not validated.[b]
	# of meals served per week	(# of Meals Served per Week) × (0.6 Lb. per Meal) ÷ (2,000 Lbs. per Ton)	Have not validated.[b]
Food stores	# of employees	(# of Employees) × (57.7 Lbs. per Employee) ÷ (2,000 Lbs. per Ton)	Have not validated. Generation rate.[c]
Large hotels	# of full-time employees	(# of Employees) × (27.3 Lbs. per Employee) ÷ (2,000 Lbs. per Ton)	Have not validated. Generation rate.[c]

Residential Sector

Within the residential sector, on the other hand, population data remains a useful tool for estimating food scrap generation. While per capita generation and capture rates vary, a fairly accurate estimate can be made within a range. I recommend using the residential generation rates in table 12.1 in combination with local population data. Note, however, that if your program restricts meat and dairy, then diversion/capture rates are going to be lower. Although I was not able to find domestic data, one estimate of meat and dairy in food scraps in England was 10 to 15 percent.[7]

Food Processing and Manufacturing Sector

Little public data is available about food processing and manufacturing generators, and the data that is out there can be highly inaccurate, so this is a sector that requires some detective work. Generator

Table 12.1. *continued*

Generator Type	Measurable Metric	Generation Factor to Get to Tons per Week	Notes
Nursing homes	# of beds	(# of Beds) × (0.5 Lbs. per Meal) × (3 Meals per Day) × (7 Days per Week) ÷ (2,000 Lbs. per Ton)	Based on 3 actual generators, this rate was 25% low.[a]
	# of beds	(# of Beds) × (1.8 Lbs. per Bed per Day) × (7 Days per Week)÷ (2,000 Lbs. per Ton)	This appears to be a more accurate metric.[b]
Restaurants and cafeterias	# of seats	(# of Seats) × (1 Lb. per Meal) × (3 Meals per Day) × (7 Days per Week) ÷ (2,000 Lbs. per Ton)	Based on 13 actual generators, this rate was 61% low.[a]
	# of meals served per week	(# of Meals Served per Week) × (0.5 Lbs. per Meal) ÷ (2,000 Lbs. per Ton)	Have not validated.[b]
	# of full-time employees	(# of Full-Time Employees) × (28.8 Lbs. per Employee per Week) ÷ (2,000 Lbs. per Ton)	Have not validated.[b]
Bars, delis	# of seats	(# of Seats) × (0.5 Lbs. per Meal) × (3 Meals per Day) × (7 Days per Week) ÷ (2,000 Lbs. per Ton)	Have not validated.[a]
Grocery stores	# of full-time employees	(# of Full-Time Employees) × (57.7 Lbs. per Employee per Week) ÷ (2,000 Lbs. per Ton)	Have not validated.[c]
Venues and events	# of seats	(# of Seats) × (0.6 Lbs. per Seat per Day) × (7 Days per Week) ÷ (2,000 Lbs. per Ton)	Have not validated.[b]
	# of meals served per week	(# of Meals Served per Week) × (1 Lb. per Meal) ÷ (2,000 Lbs. per Ton)	Have not validated.[b]
	# of visitors per week	(# of Visitors per Week) × (0.45 Lbs. per Visitor) ÷ (2,000 Lbs. per Ton)	Have not validated.[b]

Source: Adapted from James McSweeney and Brenda Platt, *Micro-Composting: A Guide to Small Scale and On-Site Food Scrap Composting Systems* (Washington, DC: Institute for Local Self-Reliance, forthcoming), courtesy of the Institute for Local Self-Reliance.

[a] "Food Scrap Generator Database Calculations," Vermont Agency of Natural Resources, Department of Environmental Conservation, Solid Waste Program, 2014, http://www.anr.state.vt.us/dec/wastediv/solid/documents/FSGCalculations-Final.pdf

[b] "Food Waste Estimation Guide," RecyclingWorks, Massachusetts. Last modified February, 23, 2016, http://www.recyclingworksma.com/food-waste-estimation-guide/#Jump01

[c] "Targeted Statewide Waste Characterization Study: Waste Disposal and Diversion Findings for Selected Industry Groups," California Environmental Protection Agency: Integrated Waste Management Board, 2006, 22, 34, http://www.calrecycle.ca.gov/Publications/Documents/Disposal/34106006.pdf

databases (discussed in the *Commercial and Institutional Sector* section on page 319) may be a good source of entity names, but don't trust the data they provide without independently verifying it.

Another factor to be aware of in the food processing and manufacturing sector is that many large generators have already found an outlet, often anaerobic digestion, animal feed, or creating some form of value-added product. They also may have large amounts of liquid waste, which would be easier for a digester to handle or for a farmer to feed to animals. For these reasons, this sector may not contribute a large percentage to your business.

On the other hand, one large food processor could serve as an excellent anchor generator for your business (see *Anchor Generators* on page 331). Large processors are often eager to find an alternative to disposal if they haven't already, or if it's possible that your service will be competitive and preferable to existing outlets. My advice is to not write these generators off altogether, but to also be aware that processing residuals is a totally different animal than mixed scraps.

Applying Capture/Diversion Rates

Once you have a solid estimate of the total food residuals that are generated in your target sectors (commercial/institutional, residential, food processing/manufacturing), the next step is to estimate how much of each sector you actually anticipate collecting. Capture/diversion rates are impacted by how user-friendly the program is for generators, local economic and policy factors, the effectiveness of marketing, outreach, and education, and local competition, among other things.

The rate of capture in each generator sector is likely to be very different. While I was unable to locate good citable data on commercial and institutional generators, rough estimates for new programs in Vermont fall in the 3 to 5 percent range in the first year or two, and 10 and 15 percent, possibly even higher, appears to be quite possible over time. Where there are multiple haulers, you will need to factor in market share.

Figure 12.1. Miles Kolonoski (age eight) with a curbside collection bin during the second year of Cambridge, Massachusetts's pilot residential food scrap program. When cities take on the collection role, private enterprise often get phased out, unless there is a public-private partnership. However, competition from the public sector is fairly unique to residential curbside collection.

When estimating collection volume from the commercial/institutional sector, one way to cross check capture rate estimates is to actually go through generator lists and make educated guesses about who is likely to compost with you. You might make preliminary calls to determine interest, start building a relationship, and find out anything you can about the generator's current practices, including gauging volume, pricing, and other factors.

As we discussed earlier in this chapter, diversion in the residential sector usually takes three forms: curbside collection, residential drop-offs, and backyard composting. Diversion estimates include a participation rate—the percentage of households that compost—as well as a capture rate among participating households, which is the percentage of generation that is composted rather than disposed. In a study of five curbside residential programs, participation rates ranged from 45 to 90 percent (averaging 66 percent), and capture rates ranged widely from as low as 10 percent to as high as 70 percent in Hamilton, Massachusetts (averaging 30 to 40 percent).[8]

For residential food scrap drop-offs, on the other hand, there is less published data. I conducted an informal survey of rates in Vermont before the state began requiring all transfer stations to offer food scrap drop-offs. Although I found no direct information on the number of families that used the program, most transfer stations had some data on weight or volume collected on a weekly basis. Using data from both rural and urban sites scattered throughout the state, I developed the ranges in table 12.2. These capture rates could be extrapolated to roughly determine participation based on average household size.

When planning for infrastructure, anticipate growth in participation rates over time so that the program has room to freely expand, but for business planning, estimate low participation rates so that

Figure 12.2. The pickup in action. Courtesy of Compost Now.

Figure 12.3. A residential drop-off at Big Reuse in Queens, New York City.

Figure 12.4. Thea Gilbert showing off their Black Dirt Farm–style residential drop-off in Stannard, Vermont. Courtesy of Highfields Center for Composting and Vermont Sustainable Jobs Fund.

Figure 12.5. A former newspaper designer, Debbie Ullman, repurposed newspaper boxes around New York City as part of a guerrilla food scrap drop-off campaign. Courtesy of Laura Rosenshine.

your financial model is based on conservative revenue streams. Testing the waters using a pilot program on a small scale is almost always a good idea and may allow you to better calibrate capture rates at full scale.

In the end, estimating food scrap capture requires you to make an educated guess, from which you can set realistic and achievable goals. For example, you may set an objective to collect food scraps and coffee grounds from 200 local residences, three coffee shops, one school, and five restaurants. You may set an objective to build program capacity to capture and recycle 30 percent of your community's commercial/institutional sector and 20 percent of the residential sector. While it's not an exact science, it is a useful step in the early planning of both collection services and composting infrastructure. Time to bust out the spreadsheet . . .

Estimating Food Scrap Generation/ Capture for Individual Generators

The following section deals with metrics used for estimating food scraps from individual commercial and institutional generators. These might be useful for entities that are considering on-site composting, or for collection services looking to refine estimates from specific generators or entity types. Some entities that are planning to compost on-site will already have a solid sense of the volume of food scraps they capture. For example, an elementary school already composting off-site may know it captures about 12 buckets (5 gallons each) of scraps each week. If this is the case, it may be as simple as weighing the buckets or converting from volume to weight.

The two primary food scrap estimation methods for estimating food scraps from individual generators are:

1. Conducting a food scrap or full waste audit
2. Using a metric from which food scrap capture can be estimated, such as number of meals served, number of employees, or number of students

The most accurate method is for a generator to do a physical audit of themselves in which organic material is sorted and quantified. Audits have the additional benefit of allowing the generator to see their waste with their own eyes and might cause them to think more broadly about waste reduction. (See the *Conducting a Food Scrap Audit* sidebar for further information.)

Short of an audit, using entity-specific metrics to determine average food scrap capture can give a reasonable estimate. Researchers and practitioners have captured numerous data points across various FSG types, and from these have developed rubrics and formulas. Individual generators including schools and restaurants can use basic metrics such as number of students or number of meals served to get rough estimates of food scrap generation. In table 12.2 formulas from several sources have been compiled.*

The Act of Collection

Conceptually, collection involves consolidating material that is in a diffuse state so that it can be collectively managed in another location. In order to facilitate efficient food scrap collection, it's helpful to break the process down in terms of the physical steps involved.

First the scraps need to be sorted at the generator into a distinct vessel that can be collected.† Typically this is a bucket or a cart. The vessel is then lifted into the collection unit. In some cases it's dumped into a separate vessel in the collection apparatus; in others each original vessel is transported individually. The generator still needs a bucket or cart to fill again, so either the same vessel is left there or a new one is

* Note: In many sources on food scrap estimation, it is not clearly defined whether the figure refers to generation or capture. It is assumed that 100 percent of the food from most generators is not always captured, but it is unclear which of the "generation" factors already account for this. Overall, if you are doing a good job at educating participants on how to compost, the generation figures in the tables should be relatively close.

† To get clean material, generator education is needed prior to collection actually taking place.

Table 12.2. Residential Food Scrap Capture

Program Type	Measurable Metric	Capture Factor to Get to Tons per Week	Notes
Municipal curbside collection—voluntary[a]	# of households	(# of Households) × (6.6 Lbs. per Household per Week) ÷ (2,000 Lbs. per Ton)	Actual generation in the study was measured at 7.4 Lbs. per Household per Week, but there was a slightly lower capture rate. This was a voluntary pilot. There may have been some weight loss due to evaporation prior to collection.
Curbside collection—nationwide averages[b]	# of households	(# of Households) × (7–9 Lbs. per Household per Week) ÷ (2,000 Lbs. per Ton)	Reported from a national survey of collection services, including a mix of private and contracted commercial haulers, as well as some municipal haulers.
Curbside collection—well established[b]	# of households	(# of Households) × (Up to 12 Lbs. per Household per Week) ÷ (2,000 Lbs. per Ton)	Reported from well-established residential collection programs that continued to invest in significant outreach and education.
Bicycle collection[c]	# of households	(# of Households) × (8.5 Lbs. per Household per Week) ÷ (2,000 Lbs. per Ton)	Program in Austin, Texas, includes collection of all organics, but does not include meat or dairy (26% according to one study—Wrap, 2016).
Food scrap drop-off[b]	# of households	(# of Households) × (3–4 Lbs. per Household per Week) ÷ (2,000 Lbs. per Ton)	Note this is based on the number of households participating in the program, not the number of households in the community. Higher rates are likely possible over time.
Food scrap drop-off (low participation)[d]	Participating region population	(Population) × (0.01–0.05 Lbs. per Person per Week) ÷ (2,000 Lbs. per Ton)	Assuming generation of 8 Lbs. per Household, this corresponds to a ≤ 1% capture rate. Representative of start-up period.
Food scrap drop-off (medium participation)[d]	Participating region population	(Population) × (0.050–0.099 PLbs. per Person per Week) ÷ (2,000 Lbs. per Ton)	Assuming generation of 8 Lbs. per Household, this corresponds to a 1–3% capture rate. Representative of start-up to first couple of years.
Food scrap drop-off (high participation)[d]	Participating region population	(Population) × (0.100–0.199 Lbs. per Person per Week) ÷ (2,000 Lbs. per Ton)	Assuming generation of 8 Lbs. per Household, this corresponds to a 3–6% capture rate. Representative of an established and successful program.
Food scrap drop-off (very high participation)[d]	Participating region population	(Population) × (0.20–0.28 Lbs. per Person per Week) ÷ (2,000 Lbs. per Ton)	Assuming generation of 8 Lbs. per Household, this corresponds to a 6–8% capture rate. Representative of an established and very successful program. Higher rates may be possible with additional resources and/or policies.

Source: Adapted from James McSweeney and Brenda Platt, *Micro-Composting: A Guide to Small Scale and On-Site Food Scrap Composting Systems* (Washington, DC: Institute for Local Self-Reliance, forthcoming), courtesy of the Institute for Local Self-Reliance.

[a] Randy Mail and Everett Hoffman, "Curbside Organics Collection From Residents: Phase 2 Report," City of Cambridge: Department of Public Works, 2015, 21, https://www.cambridgema.gov/~/media/Files/publicworksdepartment/recyclingandrubbish/PDFs/2015%20Cambridge%20Curbside%20Organics%20Phase%202%20Report.ashx?la=en

[b] Juri Freeman and Lisa Skumatz, "Best Management Practices in Food Scraps Programs," Econservation Institute for US EPA Region 5, 6, 25, http://www.foodscrapsrecovery.com/EPA_FoodWasteReport_EI_Region5_v11_Final.pdf

[c] Dustin Fedako, email to author, 2016.

[d] Compost Technical Services LLC., unpublished survey, 2017.

delivered. Ideally a clean vessel is left at the generator, as we'll discuss below. These steps are repeated over and over at multiple generators. Once the collection unit is full or the route is complete, the material is transported to the processor and deposited.

Conducting a Food Scrap Audit

Conducting a food scrap audit can provide a clear picture of the potential food scrap generation available to be captured. The basic goal is to identify the components of an entity's waste (food, paper, plastic), and then calculate each material type as a percentage of the total waste generated. This information can be used to identify areas to reduce waste and divert resources. If an entity contracts trash and/or recycling collection, records from the contractor on waste/recycling volumes over time would be useful to gather.

The audit consists of collecting and opening trash bags from a normal day of operation; sorting their contents into recyclables, compostables, garbage, and miscellaneous articles that can be repurposed; and then recording their respective weights. You can target the audit by choosing to collect only from certain places—for example, just collecting trash from the school cafeteria—as opposed to the school's entire waste stream. Or you could look just at pre-consumer waste by auditing the school kitchen. There are a number of waste audit guides. The Environmental Protection Agency has a waste assessment with clear directions, helpful setup tips, log templates, and guidance on analysis.[9]

So to summarize, the act of collection involves the following basic steps:

1. Travel to generator.
2. Lift vessel/material into collection unit.
3. Return clean vessel to generator.
4. Repeat until route is complete.
5. Transport material to processor.
6. Deposit material at processor.

All of the collection systems and equipment discussed involve these same steps. While simple, there are a surprising number of nuances involved in completing these tasks efficiently, without injury (food scraps are heavy!), without making a mess, and so on.

In general we can break collection strategies into those that tip into the collection unit and those that don't. We'll refer to these as *tipping* and *swapping*, respectively. There are also haulers who wash their client's carts and buckets and those who don't. These four variables are key to understanding collection equipment and methods.

Tipping, Swapping, and Washing

As a general rule, collection equipment that can tip food scraps into the unit has an advantage over equipment that can't. In other words, tipping is typically preferable to swapping. Swapping involves the hauler taking the full cart or bucket, lifting it onto the unit, then leaving an empty, and typically clean, one behind. Haulers who swap need to transport clean replacement vessels around with them, and they also need to tip the carts individually at the processor, which is typically where the carts are cleaned. It's a lot of shuffling of buckets or carts.

If you don't have a watertight and tippable vessel on your collection vehicle to contain food scraps and washwater, then you are working with the swapping method. On the larger end of tote swapping options, box trucks are the most common. (Box trucks are your common delivery truck.) I've also seen people use large, low-to-the-ground trailers, horse trailers,

Figure 12.6. Ted from the Central Vermont Solid Waste Management District washing a compost cart on the back of the truck's carry can. Courtesy of Highfields Center for Composting and Vermont Sustainable Jobs Fund.

and other vehicles. On the smaller end of the scale, folks have tried just about everything under the sun. Food scraps are heavy, so either (1) buckets/totes need to be light (that is, small); and/or (2) trucks/trailers need a lifting mechanism (such as a lift gate or hoist); and/or (3) the hauling equipment needs to be low to the ground.*

Folks have also developed tools to make tipping of totes more efficient and ergonomic. When I was at the Highfields Center for Composting, we designed and fabricated steel tipping bars that were used as fulcrums to balance the weight of the totes as they were tipped. Designs for tipping bars are available in *Cart Tipping Equipment*, page 345.

Equipment with tipping capacity is essentially a waterproof vessel that facilitates the tipping of buckets and totes into the unit. Once the route is complete, the equipment can typically dump all of the consolidated material at once. There is no transporting empty carts or buckets around, other than a few spares. Carts and buckets are handled once on-site and that's it. The catch with tipping carts at the generator is that in order to provide a clean cart or bucket, the collection unit needs to have a washer. This is not something that most trash trucks have, so a custom-made unit is typically used. (Washing on-site would be very hard for bicycle collection. I'm unaware of anyone who does it, though there may be other strategies, like spraying bokashi.)

Trucks with tipping capacity can usually handle about 10 tons. Trucks that swap tend to do about 1 ton, from what I've seen, but could theoretically handle more, though the more vessels you're managing, the less efficient things tend to get. In terms of capacity, tipping has an advantage over swapping, because more volume means fewer trips to the processor per unit collected, which means less transportation time.

Understanding these key differences is critical to the discussion of collection equipment later in the chapter.

* Being wicked burly is obviously a prerequisite, but it is not a long-term strategy.

Figure 12.7. Black Dirt Farm carries a water tank and hot-water pressure washer in the back of their pickup truck. The system allows them to wash carts at the generator. Courtesy of Black Dirt Farm.

Providing Clean Carts and Buckets

Many food scrap haulers clean collection carts as a service to their clients. It might seem like a small thing, but it's actually a big deal. Food scraps get nasty, and clients would prefer not to clean containers (and often won't). Regularly cleaning carts reduces odor and flies, which in turn enhances the overall experience of composting for the generator. However, the ability to clean carts has a cost, and cleaning efficiency has a significant effect on the bottom line.

There are three primary methods of keeping carts clean: cart liners, swapping and cleaning off-site, and cleaning on-site.

Cart and Bucket Liners

Simply lining the cart or compost bucket vastly reduces buildup of odor and material caked to the walls. Compostable plastic liners are the main liner used, although brown paper bags are also common, especially for residential cases. The liner will of course have to be acceptable to the composter. Most will accept clean paper, but many do not accept compostable plastics for various reasons (see *Compostable Plastics and Service Ware* in chapter 4, page 102). For a long time, while living in a rented apartment, my family used a residential drop-off and we used plastic liners in our countertop collection bucket, the contents of which were then transferred into a 5-gallon bucket with a lid that we kept outside. This worked well and reduced the frequency with which we needed to clean the 5-gallon bucket, but it still got nasty over time. Juices seep out, bags break, and so forth. In other words, liners do not replace the need to clean periodically, but they do make the cleaning process easier and less frequent. Now that my family is back to composting in our backyard, I no longer use bags, but I do line the bottoms of our buckets with dry carbon material, which makes cleaning much easier. Leaves work; shredded straw works; and the pine needles from my Christmas tree work incredibly well (they don't clump, and they actually repel water).

Keep cart lining in mind as an option if you frequently face challenges with cleaning and freezing.

Other Collection Fundamentals

Food scrap collection is a large topic, and admittedly this chapter barely scratches the surface. There are several topics that are fundamental to collection operations that I feel the need to at least touch on. The following short sections jump around among some of these topics.

What Motivates Generators to Compost?

As the intermediary between the generator and the composter, collection services are often the public face of composting. They bring generators on board and train them in source separation. The level of service they provide will at least in part determine whether or not a generator continues to participate. It is critical that haulers understand the different factors that motivate generators to start composting and to continue.

Do-Gooders, Hippies, Hipsters, and Environmentalists

There is a population that is motivated to compost for what might be characterized as "virtuous" reasons. Be careful not to lump all of these folks together; they span a spectrum of interests. Some might be hoping to stave off global warming, while others are just preparing for the apocalypse. There are no doubt true patriots who find it unimaginable that someone would throw a piece of the United States of America in the trash. And of course there are the hordes of hippies and hipsters who just need that next composting fix. Oh, and I almost forgot about the guilt-ridden baby boomers!

These folks, and other nuanced segments of the virtuous composter population, are going to compost practically no matter what. They are the low-hanging fruit, so to speak, but it's still important to get into the habit of listening to them and working to understand their motivations—for a few reasons.

First, people pick up on others' ability to appreciate their interests and will often respond in kind. Second, community-minded composters need to find and cultivate informed consumers who align with them in mission and vision. Inevitably there will be other composting or recycling operations that approach the client, so you want them to appreciate the things that set you apart. There are whole hosts of people, for example, who will support a business that directly feeds into the local food system, if they understand that the alternative feeds a non-local wastewater treatment plant.

Lastly, the percentage of generators who start composting for truly virtuous reasons is fairly low—I'd guess under 5 percent of the total population in all but the most progressive of places—but as generators start composting for other reasons, such as cost savings, convenience, or regulation, I believe that more of them will see and ultimately be motivated by composting's virtues. We need to stay in tune with that and not assume that the market is homogeneous.

Pricing and Cost Savings

Composting is unfortunately married to waste disposal, at least in economic terms. Cost will almost certainly factor into a generator's decision about whether or not to compost. In places where disposal is cheap, composting may not be able to compete on price; in such cases other motivations need to be the key drivers. In areas where disposal is more expensive, composting is often comparable to disposal—sometimes a little more, sometimes a little less.

When making a sales call, collection services often try to get a sense of the current costs (either of disposal or of other composting options). Once generators begin composting, they typically continue to produce some waste. In order to offset disposal costs with composting, they will need to renegotiate trash fees; otherwise they are paying for composting on top of their existing trash bill. Renegotiating fees is much easier where trash fees are weight-based, but this is rare—most trash disposal is paid on the basis of volume. In order to see a cost reduction, generators need to reduce the volume they are paying for (by switching to a smaller dumpster or less frequent pickup, for example). For businesses that generate only a small volume of compost, this might not be possible, but many entities do succeed at reducing their trash costs and overall disposal costs, even after adding the expense of composting. As a composting service, you can empower the generator to save as much money as possible by understanding trash fee structures and negotiations. Although counterintuitive, it may actually be beneficial to have a partner trash hauler whom you can refer them to if their current hauler is not willing to work with them.

Green Credentials

There is a more recent wave of generators for whom composting aligns with what we could call green business practices, the triple bottom line, or social entrepreneurship. Honestly, this is sort of a subclass of the virtuous composter category, but there is an assumed win-win in terms of both the environment and business. Without a doubt, this is something to familiarize yourself with and capitalize on. Large generators such as grocery stores and hotels are likely to fall into this category.

Convenience

While disposal is often viewed as easier than composting, once in place, composting with a collection service is really no more work than conventional disposal, and it can actually have some unexpected benefits, which are selling points. Generators like the fact that food scraps can be separated from trash, which makes the trash less smelly, less heavy, and less attractive to rodents. That said, providing good service is essential, because if composting becomes an inconvenience, it is unlikely to last.

For businesses that offer residential collection, the convenience compared to backyard composting can be a big selling point. I know people who composted in their backyard for years, but when given the option of collection, they took it because it was more convenient.

Another incentivizing factor is the diversity of what can be collected. For example, most generators would prefer to compost all food, including meat, and some generators want to be able to compost service ware, paper products, yard debris, and so on. While you should only accept the material that you are equipped to handle (see *Feedstocks* in chapter 4, page 90), being able to accept a greater diversity of material provides an important incentive for customers.

Young Composters Make Old Composters

At risk of sounding idealistic, I believe that in many if not most places, there will come a time in the not too distant future when composting will be an inherent part of everyday life. There are a lot of areas where schools compost, and as these youngsters grow up, they won't need training and they won't need a sales pitch. If composting is available, they'll just know what to do.

Any community that is serious about composting would be smart to start with schools. It's an investment that will pay off for generations. Look at how Apple focused on, and even subsidized, getting their computers into schools early on. (My dad was a technology teacher who was hired after his elementary school received a grant that supplied an Apple computer for every kid in the school.) It worked. Over the last 20 years, if you've observed the brand of laptops in coffee shops, Macs have slowly overtaken PCs. Now they are totally dominant, especially among young people. The same logic applies to composting—the more ubiquitous composting becomes in schools, the more it will spread to restaurants, homes, and grocery stores. Kids are the perfect composting proponents, too. After all, if a five-year-old can compost, anyone can compost, right?

Policy and Waste Bans

State and local policies are undoubtedly a large factor in the success of community-wide composting. Although relatively new in the United States, laws that require organics diversion, often called *waste* or *organics bans*, are believed to be one of the strongest policy options. Even though these laws rarely have enforced penalties for generators who do not comply, they still provide a strong incentive. Most businesses, institutions, and even residents want to be in compliance with the law, but, as state employees, policy makers, and facility operators will tell you, bans are far from a silver bullet—funding, support for outreach, education, and technical assistance are equally critical to educating the public and initiating true social change.[10]

Policies such as *pay-as-you-throw* and *disposal surcharges* financially incentivize composting by raising the cost of putting waste in the landfill. Fees generated by these policies may then be reinvested in composting activities.

As a food scrap hauler, it is completely within reason (and actually highly necessary) to go to policy makers and share with them how they can support you in being successful. Maybe it's a waste ban, maybe it's giving grants to schools to initiate composting and pay for training, maybe it's mapping food scrap generators as discussed earlier. Connect with other stakeholders, food scrap generators, other haulers, and composters and discuss your shared interests. These are very likely to also be the shared interests of at least some in state and local governments. It is truly in composters' best interest to ally with one another, to create formal associations or trade groups, and to partner with nonprofit groups that can play outreach, education and advocacy roles. A more in-depth range of policy options can be explored by bringing stakeholders together, and the policies that are enacted will be more targeted and realistic.

It's also important to recognize that there are policies that can put some or all forms of organics recyclers at a disadvantage. The organics recycling and larger waste industry is far from a level playing field. The policies and systems in place in your region may actually disincentivize composting or make the specific option you provide less appealing. Privately operated composting and collection entities may be in direct competition with municipal operations that are publicly funded, for example, although there is always an important role for both private and public entities to play, which is why open lines of communication are so vital. I would advocate for a "first, do no harm" philosophy when it comes to policy.

Anchor Generators

One of the ways that haulers build route density is to target what are sometimes called anchor generators. An *anchor generator* is typically a large generator that would make up a large percentage of a route, at least compared with the average generator. Grocery stores, colleges, large hotels, food manufacturers—these are all examples of likely anchor generators. In some cases one good anchor generator will allow a hauler to justify a new route, because once they have a large client in the area it only makes sense to also serve the smaller stops.

Collection Fees and Fee Structures

Typically compost collection services charge the generator a fee. While the food scraps have a value, the cost to collect the material and to then make it useful far outweighs that value, at least in the current organics recycling marketplace. This fact is not always obvious to new composters, who assume that black gold is the next Bitcoin; nor is it obvious to generators who might view their material as a valuable donation to the cause. The tricky thing to convey is that "of course it's valuable, it's just not valuable in economic terms, and you'll actually have to pay me to make it valuable."

One of the big questions haulers ask is: How much should I charge? To which the answer is, of course: It depends, but you need to charge enough to make it worth your while. Since it's a changing landscape in a multitude of ways, the best thing I can do is to provide some examples of collection fee structures.

For the most part collection fees seem to follow typical economic principles, including economies of scale and supply and demand. Large generators pay less on a per weight basis because the hauler is essentially doing less work per unit. Small generators, and generators located where collection and composting services are in short supply and high demand, will pay more on a per unit basis. In terms of the different generator sectors, this translates to residential generators paying the most per unit, while the food processing sector often pays the least. Currently most non-municipal residential services collect weekly and charge flat monthly rates of around $15 to $40 per month. On a fee-per-ton basis residential collection probably averages over $600 per ton, which is massive compared with larger generator sectors. Of course, the economics are totally different in the residential sector, with tipping fees making up less than 10 percent of the cost, while labor makes up a much higher percentage.

In the commercial and institutional sectors, pricing is typically based upon either a set rate per stop or a minimum volume per stop. Fee structures are designed to ensure that the time it takes to get there and make the stop is covered at a minimum. A per unit cost is then tacked on top of that minimum, typically by volume, but sometimes by weight. I recently did some research on commercial food scrap collection options for a collective of businesses, which provides a good example of pricing dynamics in the commercial sector. There were two main options that I won't name. They both use 64-gallon carts. One had a minimum stop requirement of two carts, which cost $40. For three carts it was $55, and for four carts it was $70. Use of the carts was included in the cost. The other option had a stop fee of $20, with an added fee price of $22.25 for two carts, and then charged by weight for anything over 400 pounds at a rate of $0.06 per pound (their tote tipper also functions as a scale, and a full 64-gallon tote averages about 300 pounds). They also charged for either rental ($10 per cart per month) or purchase of their carts. Fee structures often include a stop fee, and then either a flat or tiered per cart fee. A fee of $15 to $20 per cart for carts in the 48- to 64-gallon range seems to be the going rate, at least in the Northeast, though I'm sure there are variations well above and below that. On a price-per-ton basis that's somewhere in the range of $150 to $200 per ton, 15 to 30 percent of which could be going to tipping fees.

Really large generators are almost certainly paying much lower rates for collection per unit, though it's primarily a commodities-based world that I am not that familiar with. Often these entities are filling up whole dumpsters on a frequent basis, so one pickup could yield more material than an entire day of residential collection. It is my impression that collection fees for these larger clients can be all over the map, so while it might take some investigation, don't assume you can't be competitive.

Collection Frequency

In the industry, food scrap collection most commonly occurs on a weekly basis. In early consultations, a surprising number of people assume that they need to collect several times a week, or even daily, but this is just not realistic unless you have a very large generator that could max out your equipment in a single trip. During summer months, odors and flies can be an issue, so you should supply generators with sawdust or other high-carbon materials to keep food scraps capped. This layer is enough to keep maggots at bay, and while it adds a cost, it allows you to maintain the weekly frequency while keeping the generator happy—plus the composter gets the carbon back when it comes time for collection.

Food Scrap Bulk Density and Collection

We talk a lot about the *bulk density* of food scraps in this book, mostly from a composter's perspective, and mostly in terms of recipe and throughput capacity. However, it is also an important factor in terms of planning collection equipment, as well as understanding bucket and cart sizes, tipping fees, and so on.

Tipping Fees

Composters usually receive what is known as a *tipping fee* from a hauler when they accept and process food scraps. These fees typically come in the form of a dollar rate per ton. As a hauler, this is an important cost to be aware of. For more on tipping fees, see *Basic Composter Economics*, page 14, and for estimating tonnage based on volume (the number of carts), see *Converting Food Scraps to Tons and Cubic Yards*, page 88.

Collection Vessels (Buckets, Carts, Et Cetera)

A number of different vessels are commonly used to contain food scraps prior to collection. The basic equation is that on-site storage needs to accommodate the material generated in between servicing, which in most cases happens weekly.

In both residential and commercial/institutional settings, it is common to have smaller vessels in the kitchen, which are then consolidated into a larger bucket or cart stored elsewhere. The collection service typically collects directly from this larger vessel, and it's usually something that they provide to the client, although this varies. (Some services also provide kitchen or countertop buckets.)

Wheels are critical for large vessels or vessels that must be transported any distance. At a density of around 4.5 to 5 pounds per gallon, even a 5-gallon bucket can weigh over 20 pounds, which is heavy for a smaller person. Schools might use garden carts or the like to transport buckets from the cafeteria to collection carts, for example.

Obviously, there are all kinds of countertop containers out there. It's worth looking into some options in order to advise residential clients on the styles that you like. Good lids are key. Some companies make screw-on lids for 5-gallon buckets that completely contain odors. These are an awesome option for larger kitchens or for a residential consolidation point. (My family of five fills about one of these per week; then I blend it all at once and put it into our compost bin.)

Moving up in capacity, common wheeled carts come in all different sizes, colors, and designs. Bought in bulk, these can be stamped with your logo. The common sizes used for composting are 13, 21, 32, 48, and 64 gallons. Some companies in the Boston area use 64-gallon carts, but several haulers whom I've worked with find 48 gallons to be the largest practical option. Try to avoid using colors that are already in use for trash and recycling in your region. Green and brown seem to be the most common compost signifiers, but this is not uniform nationally yet. Typically containers say something like COMPOST ONLY in as many places as possible.

Figure 12.8. Totes on a tipping dock. Courtesy of Highfields Center for Composting and Vermont Sustainable Jobs Fund.

Figure 12.9. A custom-made cart tipping bar. Courtesy of Highfields Center for Composting and Vermont Sustainable Jobs Fund.

Figure 12.10. At the Hudak Farm's compost site in Swanton, Vermont, they added these triangular stoppers to their tipping bars. It acts to stop the cart at the right angle for tipping and washing, reducing the energy expended during repetitive tipping. Courtesy of Highfields Center for Composting and Vermont Sustainable Jobs Fund.

Figure 12.11. Richard Hudak showing the tipping bar and wooden stopper in action. Courtesy of Highfields Center for Composting and Vermont Sustainable Jobs Fund.

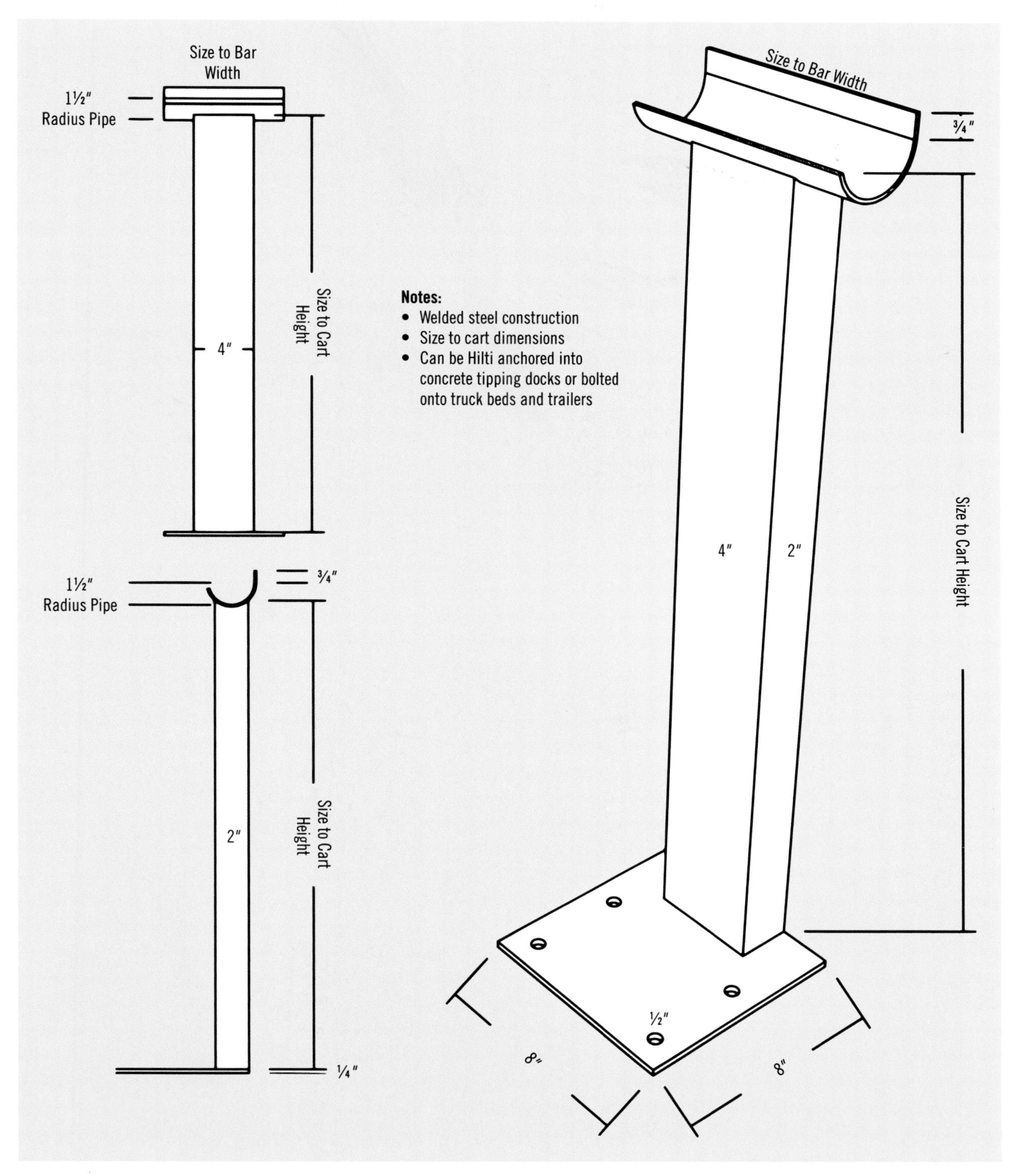

Figure 12.12. Tipping bar designs. Original designs created by Yuri and Richard Hudak. Adapted from Highfields Center for Composting, courtesy of Vermont Sustainable Jobs Fund.

Larger carts are designed to be compatible with cart tippers. There are connection points that fit into the cart and allow it to be tipped without falling into the collection vehicle. Some carts are even designed to be tipped by tractors.

Compost carts are getting more sophisticated every day. There are carts designed for residential drop-offs that have locks with digital key codes. This allows all-hour access, but only to program participants, which prevents random contamination and allows for targeted education. You can even buy carts that are designed to be raccoon-proof. Although I have not encountered problems with raccoons, I understand that they are quite common, and troublesome, in some areas.

There are also options above the compost cart scale, including dumpsters, roll-offs, and compactors. Some haulers in regions that aren't close to the compost site will consolidate scraps into a roll-off, or something along those lines, so that the larger vehicle can transport a higher volume less frequently. (This may only be legal at a permitted transfer station, depending upon state regulations.) It's an option that could be relevant to small haulers, too, especially bicycle collectors. If a town allowed lots of consolidation points, bikes could make all of the individual stops, and then larger-capacity equipment could pick up from those consolidation points.

Providing the right collection vessel to generators can take detective work and some trial and error. Good containers are not cheap. (Yes, some vessels are free, but 55-gallon drums are cumbersome and certainly not scalable.) Optimizing for the right amount of storage capacity—not too much and not too little—will save you money, but be prepared to add capacity if need be and to replace broken containers periodically.

As a side note, buckets and carts are likely candidates to be funded by grants or partner entities. These are not things that a hauler needs to technically own, and they are recoupable. Explore all options. Maybe a town or solid waste entity will allow you to use their buckets and carts for $1 a year, for example.

Figure 12.13. A battery-powered cart tipper (Toter brand) at NYCCP hosted by Big Reuse.

Collection Equipment

Food scrap collection equipment is a relatively new phenomenon that has largely been borrowed and adapted from other hauling sectors. As discussed in the introductory chapters, loads are relatively small out of necessity. Even the largest loads fall within a range that could fit the scale and principles of a community model.

The next sections are an introduction to some of the options for lifting, tipping, and transportation. We'll move up the ladder in terms of scale from smallest to largest, starting with micro collection.

Micro Collection

Micro collection is the counterpart to the micro routes mentioned in chapter 1. There is no precise line where micro collection ends and larger collection begins, but a typical micro-collection operation relies on existing transportation, rather than dedicated collection

Figure 12.15. Mary Ryther of Compost With Me in Falmouth, Massachusetts, demonstrating the lift gate she uses to facilitate the lifting of heavy carts and buckets into her truck bed. Like so many community-scale composting businesses, hers started on a micro scale and has grown rapidly.

Figure 12.14. Micro collection often takes advantage of existing transportation options, such as bus routes and other commutes.

Figure 12.16. Many residential collection routes use vans to swap buckets. Bootstrap Compost services the Greater Boston region.

Figure 12.17. A modified dump bed used in Cambridge, Massachusetts's residential curbside collection program. The small size makes it great on the side streets that would otherwise be hard to traverse.

vehicles. This could look like a pig or chicken farmer grabbing discards from the bagel shop and throwing them in his truck bed, but often it's slightly more involved. If the collection vessels are small enough to be manually lifted and stacked in small spaces (such as 5-gallon buckets), a truck bed, van, or car trunk can suffice. At a certain point, though, the volume involved makes minor vehicular adaptations highly beneficial. To collect larger vessels, small lift gates or small crane arms are sometimes attached to the back of a truck. Trailers are another useful option—I've literally seen 48-gallon carts being hauled around by sedans. The key to making micro-collection routes work is keeping initial investment low, because revenue streams are low by design. The equipment options I have just mentioned can all be put into place for well under $1,000.

Figure 12.18. Compost Pedallers of Austin, Texas, carrying two large carts of compost on a cargo bike. These carts accommodate a bucket tipping system. Over time Compost Pedallers has phased out this box bike (Dutch *bakfiets*) style in favor of the electric assist bike and cargo trailer model. Courtesy of Compost Pedallers.

Bike collection might fall into this category as well, although bike haulers typically operate well above what I would define as a micro route.

Bicycle Collection

In recent years we have experienced a boom in pedal-powered food scrap collection routes. According to one of the most popular bicycle collection equipment makers, Bikes at Work, a person in decent physical condition can pull 300 pounds at 10 mph, 600 pounds at 8 mph, or 1,000 pounds at 6 mph, assuming no wind and level ground. On moderately hilly terrain, a 300-pound load could still be reasonably managed by someone with a well-geared bike.[11] Electric assist bikes provide a boost to capacity, especially in hilly terrain. Zero to Go, a collection company based in Beacon, New York, reports that their Cycles Maximus electric assist tricycle can pull up to 1,000 pounds with a custom trailer.[12]

Since many compost sites are located miles away from the food scrap generators that they could potentially service, the distance traveled to unload limits efficient collection. To get around this bike

Figure 12.19. A tricycle transporting bins of food scraps in Denver, Colorado. Courtesy of Scraps.

Figure 12.20. Compost Now in Durham, North Carolina, swaps their members' bins. They load up the trikes with the correct number of bins per route and then swap members' full ones with empty ones. Back at headquarters, they aggregate everything into large community dumpsters, which their commercial composter collects to process. Courtesy of Compost Now.

Figure 12.21. Jennifer Mastalerz of Philly Compost sporting a tricycle with a cargo box in the front and a trailer in the back. Courtesy of ILSR.

collection services often act as consolidators, moving material from hundreds of homes into one central location where it is picked up by a truck and delivered to the compost site. There are exceptions to this model, however, such as Compost Pedallers in Austin, Texas, which has created a successful bicycle collection enterprise using a dispersed network of small compost sites, located in close proximity to their routes.[13]

Box Trucks

For small-scale collection, *box trucks* are some of the most common vehicles seen in use. These are the type of delivery trucks you see everywhere, all the time, with large boxes mounted on the back. Box trucks are most often used with a cart or bucket swapping system, because they don't typically have a place where material can be dumped. They are filled with clean vessels that are then swapped out for full ones on the route.

Most box trucks used for compost collection have a lift gate on the back, which is useful for moving heavy carts and buckets up and down from the truck.

Figure 12.22. Clean carts being unloaded from a box truck using the lift gate.

When collection is complete, carts and buckets will need to be manually dumped at the compost site. I have seen tipping at the site facilitated in several different ways. Some systems involve the use of a tipping bar (see *Cart Tipping Equipment*, page 345), which is a custom-fabricated steel bar that acts as a fulcrum for the compost cart. A tipping bar can be fitted directly onto a lift gate for tipping off the back of the truck. Another solution is to have a tipping dock at the compost site itself (figure 12.10, page 334). This frees up space within the box truck to shuffle, clean, and store carts, because if the box truck is full, there's very little room to maneuver. A third option is the type of stand-alone tote tipper that they use at NYCCP at Big Reuse (figure 12.13, page 336).

The main benefits of collection with a box truck: (1) They are relatively low-cost; (2) they are versatile and can be used for many transportation needs; and (3) there are no moving parts other than the lift gate, and no major fabrication needs, so maintenance can be kept to a minimum. The drawback is that box trucks must use the swapping method, as opposed to the tipping and washing method, which severely limits capacity. Box trucks loads typically max out at around 1 to 2 tons, although this is obviously highly variable. One thing to always keep in mind as a hauler is that renting a box truck is an easy fallback if your primary equipment goes down.

Modified Pickup Dump Body

A recent innovation in collection equipment is the small dump-body cart-tipper combo (figure 12.17, page 338). These are modular vessels that can fit in the back of a pickup truck and mechanically dump food scraps: The cart tippers dump in and the vessel tips out. Perkins sells models that hold from 3 to 6 cubic yards, and E-ride makes an electric collection vehicle with a 2-cubic-yard dumper.

Some of these devices are now being marketed specifically for food scrap collection. They might be particularly useful for small residential markets or for combination residential/commercial/institutional markets. Customization could even allow for washing on-site at the point of generation, because the washwater could then be tipped into the truck body. Weight is probably the greatest limiting factor, as opposed to volume, when it comes to what this type of equipment can hold, so make sure you know a particular device's capacity and don't overload it. Otherwise you'll be shoveling by hand.

Trailers

Over the years I have seen all kinds of trailers used for small-scale collection, from horse trailers to goosenecks—you name it. Most have only been equipped for the swapping method, so in many ways they were similar to box trucks; all of the same limitations and factors apply. The one exception is height: Low trailers are in some ways easier to work with than tall box trucks.

In 2013, with funding from both the state of Vermont and USDA Rural Development, the Highfields Center for Composting developed and piloted a dump trailer that was equipped for both tipping and washing. This customized system can carry around 4 or 5 tons per load, which is midway between a box truck and a full-size roll-off or packer truck in terms of capacity. The scale and low cost of this equipment make it quite relevant to small-scale collection enterprises. The truck and fully customized trailer together might run $50,000 to $75,000 new, which is less than half of what you would pay for a full-size rig. When Highfields closed in 2014, the route was fully under way, but the equipment still had a lot of kinks. Black Dirt Farm purchased the equipment and took over the collection route later that year. Since then Black Dirt has worked out many of the initial problems, and the system now collects somewhere between 15 and 25 tons per week. It's capable of doing two loads a day if you're working long days, but as I understand it, keeping it to one run a day during the wintertime is more manageable.

To my knowledge, this is the only collection rig of this style, although I'd be surprised if there weren't similar systems in use or development elsewhere. There are two case studies that provide more details on the system, one from Highfields[14] and the other from Black Dirt Farm.[15] The second case study also has an interesting cost comparison of different types of collection equipment.

Figure 12.23. A few tons of compost rolling down the road on the inaugural run of Highfields Center for Composting's trailer collection pilot. The trailer still functions with some modifications and is utilized by Black Dirt Farm. Courtesy of Highfields Center for Composting and Vermont Sustainable Jobs Fund.

Figure 12.24. A full trailer load. Note the candy cane cart tipper. Courtesy of Highfields Center for Composting and Vermont Sustainable Jobs Fund.

Modified Roll-Off and Packer Trucks

The largest class of organics collection equipment includes large trucks—typically roll-off dumpsters and packer trucks—that are customized to handle food scraps. These trucks can typically handle 8 to 10 tons of food scraps in a route, though they are sometimes built in smaller sizes as well.

Packer trucks are what most of us think of as trash trucks. They have a covered container body with a low open back, which is called a loading sill. A compactor unit pushes the material into the container body. Small buckets and carts can be loaded by hand, as you've likely seen garbagemen doing, or a cart tipper can be used to accommodate larger vessels.

Roll-off trucks carry large, removable or "roll-off" dumpsters, and can be customized with tote tippers. (You may have seen these types of dumpsters situated outside construction sites.) Some companies use roll-off trucks that can tip small dumpsters of food scraps, like your typical small trash dumpster, but these are used for large generators and aren't particularly relevant at a small scale.

In my experience modified roll-offs or similar styles of equipment are more common than packers for use in food scrap collection. Roll-offs are more contained, while packer bodies are not well sealed at the compactor blade, which results in leakage and makes them harder to clean.[16] This also means that washwater from cart washing will build up there, so they are not as effective for on-site tote washing. For bagged material, packers seem to work decently, and for dense residential collection routes they have the advantage of a low tipping sill for fast-paced manual loading.

Roll-off trucks are often modified to contain liquids by adding seals around the dump gates. Because

Figure 12.25. A roll-off modified for food scrap collection at TAM Organics in Bennington, Vermont. Note the cart tipper, water tank, and pressure washer. Courtesy of Highfields Center for Composting and Vermont Sustainable Jobs Fund.

Figure 12.26. A smaller roll-off truck with a triple cart tipper.

the containers have high walls, they require cart-tipping mechanisms with high lifting capacity (see *Cart Tipping Equipment*, page 345). The cart washing apparatus is a custom add-on, as discussed in *Washing Systems*, page 346.

The main advantage that comes with these larger rigs is capacity, and greater capacity means fewer trips to the processor. Where there is material available, collection services can collect 8-plus tons per route using larger rigs, which is a solid day's work. However, that is probably only possible with commercial, institutional, and food processing accounts, unless residential compost collection is being provided as a municipal service in a densely populated area. Full-size vehicles can cost $150,000 or more, so they need to be full at the end of the day in order to justify the high capital investment. Until recently, most of these vehicles were custom-designed and -fabricated, but companies such as Brown Industrial have begun manufacturing standard models.

Figure 12.27. Residential food scraps in compostable bags collected in a packer truck by Save That Stuff (Cambridge, Massachusetts).

Multistream Collection

While most compost collection services are *single stream*, meaning they only collect one type of material, greater efficiency and revenue can in theory be gained by also collecting recycling and trash. *Multi-*, *dual-*, or *triple-stream collection* can be achieved through the use of a split-body rig, which has the ability to load and tip from separate vessels.

Another strategy for multistream collection is to use brightly colored and heavy-duty trash bags for organics, which are thrown in with the trash or recycling and later separated out manually. I have heard this method talked about for about five years, but have never actually seen it practiced, which makes me question how practical it might be to implement. The collection side is easier, certainly, but the separation side would require picking out heavy bags and cutting them open. I could see it working for extremely small routes, but would imagine it being highly impractical at larger scales.

Cart Tipping Equipment

Food scraps are heavy, and the repetitive motion of dumping a full cart or bucket is awkward. Vessels above 25 pounds (5 to 6 gallons) become dangerous to dump manually. It can of course be done, but even with two people it's just inviting a back injury. I've always been a proponent of the "work smart not hard" philosophy. Tools that make tipping more efficient ensure that you can safely manage this essential step in the collection process.

There is a surprisingly large array of equipment designed for the purpose of tipping carts. It's not a subject that I've studied deeply, but what's interesting to me is that thinking back on all of the different composting operations I've seen over the years, few if any used the exact same system. Either cart tipping equipment is going to be used for tipping into the collection vehicle at the site of the generator (what we're calling the *tipping* method), or it's going to be used for tipping carts back at the compost or transfer facility (what we're calling the *swapping* method).

Tipping carts into trucks is usually achieved using cart tippers. Cart tippers do three things: (1) lift; (2) rotate to dump; and (3) hold on to the cart so that the food comes out, but the cart doesn't fall. There are a variety of cart tippers on the market. Some, such as *candy cane tippers*, lift higher. As you can see in figure 12.24, the tipper lifts over the side of the high trailer wall.

Larger trucks with high sides typically use a rear hydraulic lifting arm to lift either carts or a *carry can*. Many of the more recent systems I've seen allow the operator to load two to three carts onto a platform at the rear of the truck; then the arm lifts and tips all of the carts simultaneously. Another variation is attaching a tote tipper to a carry can, which is just a small vessel at ground level. The cart is first tipped into the carry can using a small cart tipper, and then the lifting arm dumps the carry can into the truck body.

Cart tippers themselves come in many forms, with different lifting heights. They can be powered hydraulically or pneumatically. As previously mentioned, candy cane tippers can lift the highest. The simplest tippers rotate on a single fulcrum and have very little lifting clearance. These are what you see on the back of carry cans, packer trucks, and pickup truck dump bodies.

The most advanced collection systems now use automated side loaders, which are robotic cart tipping arms that can be operated from inside the truck cab. We see these in trash disposal and recycling, but at least one company (Roto Pac) is marketing this option for compost collection. Other companies will undoubtedly begin producing similar equipment, if they haven't already.

Mechanical cart tippers can also be used for tipping on-site, and there are several prefabricated systems available for just this purpose, including both stationary systems and mobile systems that are battery-powered. Another solution for facilitating tipping at the processor is the use of tipping bars, as we touched on earlier in this chapter. At the Highfields Center for Composting, we would have tipping bars custom-fabricated from time to time for haulers who were using the cart swapping method. Richard Hudak of Hudak Farm, a composter in Swanton, Vermont, designed the tipping bars shown in figure 12.12, page 335. The wooden frames in figures 12.10 and 12.11 (page 334) stop the cart mid-rotation and hold it at a good angle during cart washing.

Facilitating Even Weight Loads

Tipping mechanisms need to be designed so that all of the weight does not end up on one side of the vehicle. Food scraps do tend to ooze and level out somewhat, but this is less likely to occur in cold weather, so you need to make sure the weight ends up in the center of the truck body to ensure safe operation.

Washing Systems

Pressure washers are the most efficient tools for cleaning carts, and hot-water pressure washers are particularly effective.

In order to wash carts at the generator, the collection apparatus will need, at minimum, a water tank and a pressure washer. Some systems heat the water off the truck's engine, while others use a hot-water pressure washer. Black Dirt Farm's trailer-based collection system contains a washing system in the back of their pickup trucks, which is made up of a water tank and hot-water pressure washer. They have also enclosed the back and vented the gas-powered pressure washer motor out the side.

Washing systems on-site at the processor could involve similar components, though tanks would not be necessary for sites that have a water source easily available.

Managing Water and Food-Cicles

Managing collection routes in cold climates can be extremely challenging. Food scraps and water are prone to freezing, so the entire process takes longer

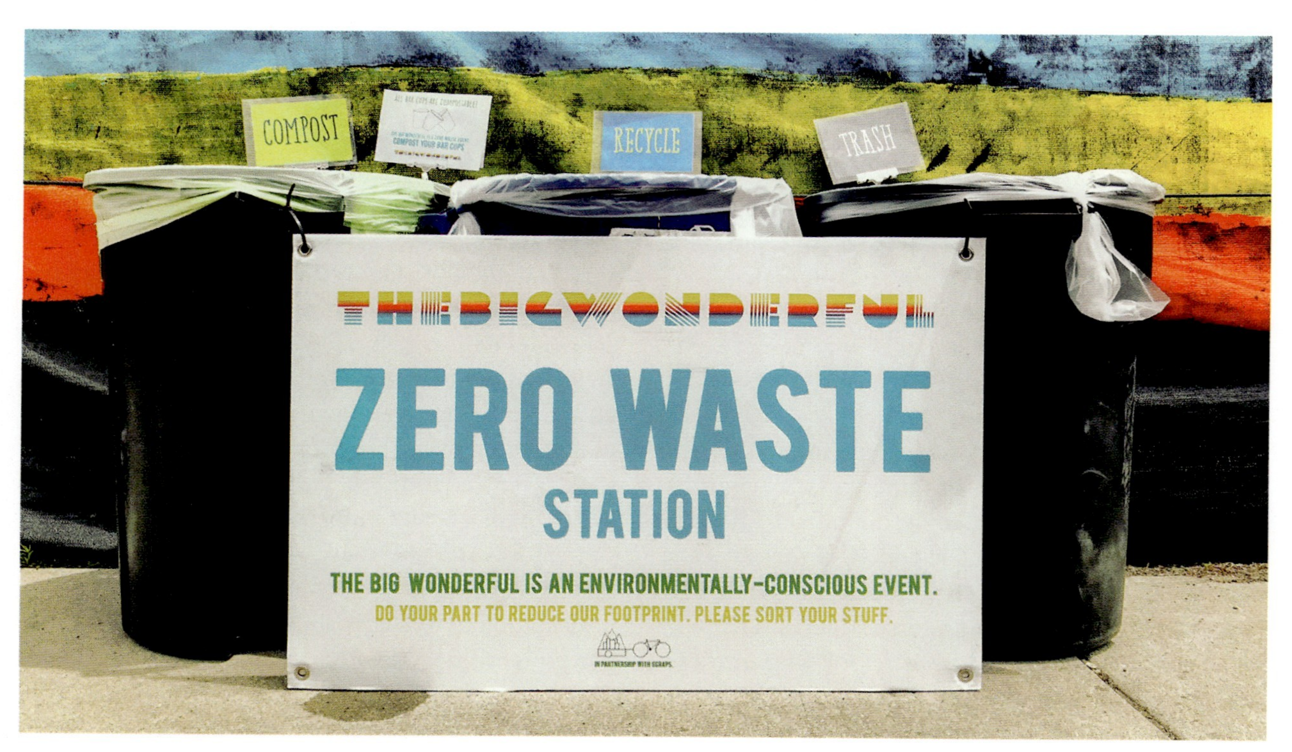

Figure 12.28. A zero-waste station provided by Scraps at the Big Wonderful, a music and beer festival for artisans in Denver, Colorado. Courtesy of Scraps.

and requires extra care. For those in cold climates, collection systems need to be designed with these challenges in mind.

Food-cicles (frozen food scraps, also called tote-cicles) get stuck in carts and buckets and can be extremely challenging to remove. Compostable bags and other liners help with this (see *Cart and Bucket Liners* on page 328), as does having a hot-water pressure washer to thaw the edges of the vessel. Sometimes the cart tipper itself can give it enough jerk to loosen things. Another tool to consider is the *dead blow hammer*, which is a sand-filled mallet that reduces the direct impact of the blow to the cart or bucket and hopefully prevents it from breaking. Strong carts and buckets help as well.

Generator Training and Education

Collection services serve as gatekeepers between the public and the compost facility. As the most public-facing component of any community-wide composting program, the hauler plays the role of bringing generators on board, showing them the ropes, and teaching them best practices, such as proper separation of materials. Effective training can increase capture and reduce contamination of unaccepted materials.

While there are collection services that provide little to no education to generators, this requires a composter who is willing to tolerate highly contaminated material, and most community-scale composters do not fall into this category. Furthermore, generator training is an opportunity to engage the public as more than just business clients. People want to feel like their efforts are contributing to their community, and sometimes even to the planet as a whole. Training is an opportunity to engage people as active participants, community members, and stakeholders. In this respect collection services play a critical role in moving forward the waste-to-resource

Figure 12.29. An educational refrigerator magnet designed to guide residential generators on proper source separation. Each program needs to customize their own training materials in relation to what the composter will accept.

paradigm shift. They can also move it backward, however, if they leave a bad taste in people's mouths by providing poor service.

An abundance of resources for generator education and engagement are available, including videos, posters, and talking points. I am partial to the videos on the Highfields Center for Composting's Vimeo page, which feature some very cute kindergartners. Although generator engagement is not an area where I have been particularly focused, I have been around enough of it to know that it takes work, both at the beginning of the relationship and over time.

CHAPTER THIRTEEN

Compost Site Management

Anyone who's had the experience can tell you how good it feels the first time your compost pile hits upward of 130 or 140°F. No matter how well I understand the science behind composting, it still feels like I'm on the very edge of some great mystery, a new discovery taking place right in front of my eyes. Maybe this is why composting has taken on an almost religious following—dirt worshipers at the altar of the compost thermometer, witnesses and stewards to the cycle of life.

Even for those who have been composting for years, it never becomes entirely static or formulaic.

Figure 13.1. Logging pile temperature at the Vermont Compost Operator Certification Training.

bioturbation
(ˌbī-ō-tər-ˈbā-shən)
DEFINITION: The movement of soil and other particles in the environment by animals and plants.

There is ample opportunity for lifetimes of learning and discovery. Nevertheless, most who continue composting successfully develop a set of practices to guide them in the process. That set of practices will be slightly unique to each operator, acquired through countless thermometer probings, sniff tests, recipe adjustments, and pile turnings.

Chapter 13 is designed to support site operators and managers in building that set of practices. We will cover the core management techniques used in the composting trade, including sourcing feedstocks; materials blending and pile formation; compost pile monitoring; pile turning; mitigating and remediating nuisances; and testing. As an introduction to these topics, I will discuss operator training, management plans, and other strategies for institutionalizing best practices. Creating buy-in, continuity, and agility in site management is often one of the greatest challenges, especially for micro and on-site operations, but it is absolutely critical to the sustainability of any composting operation.

Compost Operator Training

I would be remiss if I did not start this chapter by emphasizing the value of the many composter training programs already available across the United States. You can learn a lot by reading this book and other related literature, but it will not provide you with same experience as attending programs such as the Maine Compost School or the US Compost Council's operator trainings. Both are invaluable, hands-on, weeklong trainings taught by leading compost experts.

There are also shorter trainings in many areas, and some states have good Master Composter programs, although these are not really designed for commercial operations. For compost sites that experience frequent turnover, finding a good training program and building training and continuing education into the operational roles is critical, as knowledge inevitably gets lost in the transition between operators. (I'm thinking of student-managed on-site systems at colleges in particular.)

Institutionalizing Composting and Compost Best Management Practices

There are hundreds if not thousands of well-documented composting success stories, but you rarely see documentation about why compost sites fail. Yet the potential problems that can arise from undermanaged food scraps composting are very well known: odors, birds, rodents, and flies. Really, any of the nastiness you associate with food scraps putrefying unchecked could become a reality. All the worst-case outcomes and nuisances can arise due to poor management, and the risks increase the larger the site is.

Yet there are far more cases where composting doesn't cause problems, so why do some sites fail, while others succeed?

Nuisances such as those I have just discussed do not have to be part of the composting process. At successful sites, if a problem arises, the operation responds quickly and the problem is dealt with. When compost sites fail it is usually the result of systemic issues preventing responsive site management. We see the symptoms play out as stereotypical nuisances, but there are typically other issues underlying them. In my experience it often comes down to an entity's failure to institutionalize compost best management practices (BMPs). This is especially true with entities for whom composting is not the primary mission. For all composters, but especially for non-composting-specific entities such as schools and colleges, having institutionalized supports, roles, procedures, and BMPs in place before problems arise is critical.

Institutionalization of composting practices at many levels of the entity, and ideally in the community at large, is a way to ensure the longevity and resiliency of composting sites and programs. I use *institutionalization* for lack of a better word, but what I really mean is "continuous support system."

At a school, for example, multilevel institutionalization of composting might begin with support from the local municipality and/or solid waste management system. At the school board level, a composting and recycling mandate will ensure continuity beyond any one principal, and at the staff level the role of compost coordinator as part of an employee's job description will ensure continuity beyond any single person. At the student level, institutionalization might include the involvement of a school's sustainability group to roll out source separation training. The group might then study and document the program and educate fellow students about its impacts on the sustainability of their local food system and the broader environment. All this external support outside of the composting system itself creates a culture of composting over time that can override the tendency of compost projects to falter when resolvable issues arise.

Finally, institutionalization of composting means ensuring that best management practices get implemented over the long term. With a small amount of training and practice, anyone can compost effectively, but most operators don't start with much hands-on experience composting. Even experienced composters can run into trouble managing a new system or following a recipe with unfamiliar materials.

Here are some examples of ways that sites institutionalize compost site best management practices:

- Providing new operators with technical training and experienced operators with continuing education
- Creating and regularly updating a site management plan
- Creating and utilizing a system for recording composting activities and metrics such as materials received, recipes used, and pile monitoring
- Posting signs and reminders of protocols at sites with multiple operators
- Regularly meeting with and collaborating with other composters to develop a "community of practice"
- Identifying and utilizing qualified technical mentors, consultants, and service providers

Figure 13.2. Every year mountains of leaves are shuffled around. In most urban and suburban communities, they are not hard to capture.

For larger sites, most of the things on this list are a given, but at smaller sites they sometimes get overlooked. Find ways to implement best management practices that work for your operation. There is a tendency to overextend an operation, which is why it's so important to start small and build protocols as you grow. BMPs are the foundation of any operation, and not even the most advanced composting systems can work well without them. Take this to heart. A yard of prevention is worth a trailer load of cure.

Sourcing and Capturing Raw Materials

In many places an abundance of raw organic materials, or *feedstocks*, remain uncaptured. In others sourcing materials is competitive and costly. In either case it can require significant effort to find and acquire feedstocks. Chapter 4 covers the qualities of different materials in significant depth. Here I've broken down the topic of sourcing and capturing feedstocks in terms of logistical and economic considerations.

Feedstock Logistics

Identifying sources and assessing the quality of potential feedstocks can take considerable effort. Often the availability and quality from a specific source can change without warning, and site operators quickly learn that maintaining consistency requires regular communication with their sources. One solid feedstock source is worth 10 intermittent sources.

Remember that generators are looking for reliable outlets as well. With food scrap generators and haulers collecting compost, the importance of reliability is obvious, but with other materials it's easy for operators to get caught up in the day-to-day details on-site and lose track of the external relationships that keep the site running fluidly. Particularly at larger sites, this can be a large part of the site manager/operator's job.

Ideally sites should be located in close proximity to consistent feedstock sources (say, on a farm that generates quality manure), but where this is not possible, the next best thing is to set up a feedstock delivery system that basically manages itself. If you are doing the hauling, then set a schedule with the generator and stick to it. If you are contracting the hauling to a local trucking company, set up a consistent delivery schedule and then put the company in direct communication with the generator so that you manage the relationship without having to manage every individual delivery. In all cases, make sure that both the source and hauler have a clear understanding of what you're looking for in the material (for instance, fresh manure free from contamination versus nine-month-old manure).

Creating a *backhauling* setup is another ideal way to manage feedstock transportation. This means that a delivery truck that is traveling one way with material (such as finished compost) then fills up with feedstock and brings it to you on its way back; the truck never travels empty. It is often easier said than done, but looking for opportunities for backhauling, particularly for longer trips, makes a lot of sense.

The system that Stop & Shop recently deployed in New England is a great example of backhauling. The supermarket chain built a methane digester at their distribution facility in Freetown, Massachusetts, and trucks now travel to their stores full of food and return with food scraps to be digested. A community-scale version of this system would be a farm that travels to a farmers market with produce and returns to the farm with food scraps or other feedstocks. In these cases, and with feedstock hauling in general, it's important to prevent cross-contamination of equipment in either direction by compartmentalizing the equipment or cleaning it as needed. Avoid hauling feedstock in truck bodies that also carry trash.

Loading the truck or collection equipment also requires consideration. Sometimes the source will be able to load the truck if they have a bucket loader on-site. Other times it will have to be taken care of in another way. Micro operations may be loading the back of a pickup or a trailer by hand. Larger sites might need to drive their loader to the source, or contract someone nearby to do the work.

Many composters have found creative ways to add efficiency to their feedstock capture process. Maybe you have a small trailer, carts, bin, or even a roll-off dumpster that you leave at the source. Efficient loading and unloading, as well as other logistical factors, are all relevant to the economy of a given feedstock source.

Feedstock Economics

At the end of the day, for composting to be an economically sustainable enterprise, the income generated from your outputs and services (compost, organics recycling), needs to be higher than the cost of your inputs (feedstocks, labor, equipment). Feedstocks are a key factor in that equation.

However, the economic value of any one feedstock is often hard to quantify with specificity. For example, most composters get paid to take food scraps, which should be a profit area, right? Yet food scraps are mostly water, so they require a lot of dry matter. They stink, so they require carbon and management. They shrink, so they don't produce much actual finished product. They are highly regulated, so permitting and infrastructure costs are often higher than with other materials.

Recently I've heard composters say that unless they are getting paid $55 to $65 a ton as a tip fee, they are losing money on food scraps, and that's not a tip fee that could be easily passed back to food scrap generators in many places. Food scraps are an extreme example, but I use it to stress the fact that unless you parse the economic underpinnings of a given feedstock, you are operating in the dark.

In another specific example, a composter whom I worked with recently was having issues with several high-moisture feedstocks and had started purchasing wood chips as a source of dry matter. Together we did a quick recipe comparison of wood chips versus purchased sawdust in terms of effectiveness in reducing moisture content. Here is a redacted excerpt from that analysis:

> *While all of the wet feedstocks received by the site pay a tipping fee (along with the yard waste), dry matter and high-carbon bulking agents often need to be purchased. Cost of these supplementary materials is therefore a primary consideration in feedstock choice. Sawdust may be a more cost-effective choice for the farm in terms of managing high MC feedstocks than wood chips (assuming both are purchased). Table 13.1 illustrates this point.*
>
> *While wood chips are valuable as a bulking agent to increase pile porosity, they would need to be pre-processed (to increase the surface area) and very dry in order to be cost-competitive with sawdust in terms of reducing MC. Even then, they are not as readily absorbent as sawdust, so the calculated MC in table 13.1 may be misleading. It would cost approximately twice as much and require four times as much volume to use wood chips assuming they have a 50% MC. Wood chips do not break down as completely as sawdust, so a large percentage of the chip would need to get screened out, reducing overall yield per yard processed and adding to the cost. The chip could be reused if the leftovers are clean enough, but it would no longer act as a dry matter source. (Source: Compost Technical Services)*

Table 13.1. Comparison of Dry Matter in Sawdust and Wood Chip Recipes

	Sawdust Recipe	Wood Chip Recipe
Wet feedstock 1 (87% MC)[a]	1	1
Wet feedstock 2 (70% MC)[a]	1	1
Sawdust[a]	2.5	0
Wood chips[a]	0	10
Cost per cubic yard of carbon	$17	$10
Total cost	$43	$100
Moisture content	61%	62%

Note: Not actual recipes; for use as dry matter comparison only.

Source: Adapted from and courtesy of Compost Technical Services LLC.

[a] Units in cubic yards.

Here are some of the factors an operation could consider when assessing the economics of specific feedstocks:

Material cost. Does the material cost money to acquire in and of itself?

Transportation. Are you paying for hauling?

Loading. Is there a cost to load the material?

Shrinkage. How much does a given material shrink during composting?

Composting rate. Will the material speed up or slow down the composting process? Will it be broken down by screening?

Volume. How much is available and how frequently? (It is typically cheaper on a per yard basis to transport 15 yards than 5.)

C:N ratio, moisture content, and bulk density. How effective is the feedstock at getting your recipe where you need it to be? If you were to compare two feedstocks that are both 100:1 C:N and 50 percent MC, but one had a BD of 400 pounds per cubic yard and the other had a BD of 1,200 pounds per cubic yard, it will take far less of the denser feedstock to have the same effect on C:N and MC.

Handling and other pre-processing. Is it efficient to handle? Consider a bale of hay, for example—it's laborious to break up and may require a bale shredder.

Physical contamination. Increased costs in hand-picking, screening, and disposal.

Chemical contamination. Herbicides can cause damage, and damage mitigation can be extremely costly. Chemical contamination may ruin compost and/or hurt your brand.

If you can't get paid for taking feedstocks, and you can't get enough material for free, how much should you be willing to pay for compost feedstocks? To answer that question, I suggest you look at it with the specifics of your operation and business plan in mind.

The reality is that feedstock costs run the gamut, and I think a lot of operations don't actually know what their specific costs are because they don't do an accurate job accounting for their own labor. If I had to throw out some numbers based on the past five years (2013–18), I would say that $10 or more per yard is high, approximately $5 per yard is okay, and $3 or less is good. Of course some $13-per-yard feedstocks are far more valuable than their $3-per-yard alternatives; it just depends.

In the next section we will talk about feedstock tracking. You should track both individual feedstock costs and average feedstock cost. Whether you're looking at the larger business plan, examining your annual budget, or comparing one feedstock source with another, it's incredibly valuable to have this information on hand.

Feedstock Tracking and Inventory

Carefully tracking what comes into your operation on an ongoing basis is a critical practice in feedstock management. While the piles of slips and sticky notes might get you through, diffuse paper trails are neither an efficient nor a useful source of feedback about how an operation is running.

Make a feedstock tracking spreadsheet, track relevant information such as feedstock costs, and update the spreadsheet at least once a month. This information will be useful for a number of reasons:

Production feedback and projections. It's extremely useful to know the throughput of individual feedstocks. From that, you can get a realistic sense of your processing capacity and project outputs, and understand labor rates and other factors in your cost of production.

Permit reporting. Permitted compost sites typically need to report to the state on the volumes/weights of materials processed.

Regulatory compliance. Permitted compost sites have a limit on how much they can process and therefore may need to show that they are under that limit. Conversely, if a site is exempt from permitting because it falls under a certain volume threshold, it would help to have documentation showing that.

Problem feedstocks. If it is learned that a particular feedstock has a continuation or other issue, records on that material may be useful.

Compost Recipe Development

For the specific steps involved in creating a compost recipe, an essential best management practice and a precursor to the next section on blending and forming your compost piles, refer to *Compost Recipe Development* on page 83.

Food Scrap Receiving and Blending

So you've developed your site, acquired equipment, identified and secured feedstocks, and created a recipe. Now comes the moment you've been waiting for. Drumroll please . . . T-p-t-p-t-p-t-p-t-p-t-p-t-p-t-p-t. You're ready to make compost!

Going from raw material to cooking pile is straightforward; that said, it's during the initial steps of receiving and blending food scraps that operations often run into the most problems. During the receiving and blending process, food scraps are at their rawest, and typically they are temporarily exposed, so expedience and care are essential. The steps of receiving, blending, and constructing the compost pile can differ slightly by site, method, and scale, and there are ways that work and ways that don't.

The quality of blending and pile formation can also have an immediate effect on pile performance and the overall efficiency of your operation, so it's worth taking some time to evaluate your approach. Once you've blended and built a few piles, you'll undoubtedly find ways to make the process more efficient for yourself.

Receiving Food Scraps

With the exception of small volumes of food scraps that are going directly into a vessel, food scraps are generally going to be tipped into a designated receiving area where they can be covered until blending. Even with an impervious receiving pad (for example, concrete), I highly recommend preparing a bed of dry carbon materials beforehand into which the food scraps can be delivered. I refer to this as the *carbon trough* method. Essentially, you form the carbon feedstocks into a low bed or trough with edges 8 to 12 inches high. Picture a kiddie pool made out of a combination of wood chips, leaves, horse manure, and sawdust that is wide enough to retain all of the food scraps received.

The carbon trough achieves a few things:

- It assists in the blending process by spreading the food scraps thinly, layered bottom and top with your added feedstock.
- It contains both food scraps and liquids in a confined area.
- The dry carbon materials immediately begin to absorb and balance out the moisture, nitrogen, and volatile compounds that cause odors.
- The receiving area is protected from the residues and moisture that accompany food scraps, which make it less likely to harbor odors and attract pests.

Figure 13.3. It broke my heart to see this pile of food scraps tipped about 4 feet away from the concrete receiving bay. As you can see, organic matter buildup is becoming a real problem on this site—there is gravel on that access road, but you wouldn't know it. Courtesy of Highfields Center for Composting and Vermont Sustainable Jobs Fund.

- It buys you a little bit of time if you need it before blending.

In addition, receiving food in a carbon trough puts your operation on track to keep food scraps contained right from the start. Visitors, regulators, and neighbors will see that you are putting thought into your work.

Once the food scraps have arrived, you can layer additional materials on top to contain them (following your recipe). Ideally this should happen immediately, and some states have specific requirements about how long food scraps can remain exposed. I can't adequately emphasize how important this is from day one of operation. Birds in particular will plan their vacation around your food scrap delivery schedule if they think there is an opportunity for an all-they-can-eat buffet. Use your judgment and balance what's efficient with what keeps your site free of vectors.

Blending Techniques

The timing and thoroughness with which food scraps and other highly putrescible materials are blended can have a large impact on the operation as a whole. Decomposition won't wait, so it's happening in either a managed or an unmanaged environment—in the presence of carbon and oxygen or not, with balanced moisture or without.

We don't need to review all of the benefits of a managed process here, but I do want to emphasize that by increasing the distribution of carbon, nitrogen, moisture, and free air space, good blending supports just about every other aspect of a well-tuned operation. Every day that material sits on-site without being processed optimally is wasted space and capacity. Not to put this in strictly reductionist terms, but if our goal is to reduce waste, building efficiency through blending is entirely worth the effort involved.

Figure 13.4. In contrast with figure 13.3, these food scraps in Hudak Farm's receiving bay were delivered by the Northwest Solid Waste Management District directly into a trough of carbon feedstocks. Easy to blend and easy to clean up. Courtesy of Highfields Center for Composting and Vermont Sustainable Jobs Fund.

Setting aside in-vessel composters that do the mixing for you, I am familiar with four basic methods to get a blended compost recipe (five if you count using animals, but that's covered in chapter 11). These methods are discussed below in order of blending thoroughness, and in terms of their management and operation. (For mixing infrastructure and equipment see *Additional Infrastructure*, and *Compost Site Equipment*, page 159).

Total Mix Ration Mixers

The standard mixing equipment that composters use are livestock feed mixers, or total mix ration (TMR) mixers. Specifically, most sites use *vertical auger mixers*. These are powerful tools that allow composters to take immediate advantage of the dry matter and carbon in a mix. According to experts a thorough mix actually decreases odors by bringing them into contact with oxygen and carbon on particle surfaces. (Remember that odor-causing compounds tend to be nitrogenous.)

For those using an aerated static pile (ASP) system, achieving a uniform and highly porous mix is especially important, because it mitigates preferential airflow, which can impede pile performance, reduce quality, and increase the need for turning. Mixers are a standard recommendation for ASPs in the industry, although not everyone uses them.

Feedstocks processed with mixers need to be handled a little bit differently than they are with a bucket or hand-blended system. Composters typically make multiple recipe batches in mixers, which are then combined into a single compost pile. Feedstocks are loaded into the mixer following a recipe, blended, and then discharged. Your recipe will therefore need to be compatible with your mixer size in terms of proportionate feedstock volumes. If you're using an 8-yard mixer, then your recipe needs to be calibrated to 8 to

Figure 13.5. The Jaylor mixer at the NYCCP at Big Reuse in Queens, New York City, moves blended material to the discharge point.

10 total yards (remember, blending reduces volume by approximately 20 percent) in order to build out the recipe completely while using your time efficiently.

If you are using mixers, the carbon trough method may not make sense, because it would be hard to keep your recipe consistent unless you pre-blended the materials well. At Big Reuse in Queens, New York City, they tip totes directly into a Jaylor mixer using a tote tipper. Using a mixer requires a method such as this that keeps the receiving and mixing site free from exposed food scraps. If you receive raw food scraps in an open pile, this may just involve fast mixing, meticulous cleanup, and having sawdust or wood chips on hand to absorb any free moisture that is lost from the scraps.

Mixing with Manure Spreaders

While manure spreaders can be used to effectively mix compost, I am hesitant to recommend them as blending tools for food scrap composters. Large particles have a habit of falling out of the pile, especially when flung, so you are likely to end up with more of a mess around the base of the pile than you are with other methods. Spreaders are also ill equipped to handle frozen materials.

Caveats aside, if you are using a manure spreader for blending materials, the methodology is similar to that of a mixer. Materials need to be layered vertically, like lasagna, because the mixing happens in horizontal succession from back to front. You will want to place lighter and drier materials on the bottom (wood chips, sawdust) to keep the machine from getting clogged. Hay and straw can bind, so they are best put on top. Once the materials are layered, you discharge the spreader, which will make your blend. With many spreaders you will then need to restack the material, because the piles they make are small.

For windrows, re-form the pile as needed so that it is uniform. Make sure that the spreader is cleaned out well, especially in the wintertime to avoid it freezing up, and if you are using it for food scraps, keep it free of food to avoid vectors. Adding carbon back to the spreader immediately will help reduce any odors that might be lingering.

Mixing with a Loader or by Hand

I have the most hands-on experience blending with a tractor bucket. In many ways, blending with hand tools is similar. With this method, I highly recommend using a carbon trough for receiving food scraps or other wet materials.

If there is enough space in your blending area, the entire recipe can be built at once. At sites where space is more limited in the receiving area, operators may need to build and blend mini batches in order to get a good mix, but ideally the receiving area should have enough space to process it all in one batch. In either case feedstocks are spread evenly over the pile following the recipe. Account for any material already in the carbon trough or cover.

Be sure to evenly distribute materials as they are added, feathering each bucket out thinly across the trough. Alternate materials frequently. I recommend getting your driest materials directly into contact with your wettest materials. Picture creating evenly layered lasagna, where each stratum of material is uniformly distributed across the pile.

The larger the batch, the more challenging it is to get a proportional mix using this method. As the mound gets thicker (every 18 to 24 inches), do some mixing so that the new materials come into contact with the food scraps.

Figure 13.6. Manual mixing.

The actual blending process is as simple as lifting and dumping materials, working to get as much mixing action as possible. Lift the materials high and feather them as you dump. (Think of feathering as a gentle waterfall rather than a crashing wave).

Blend intermittently as you work and then do some really thorough blending at the end prior to forming the pile. With piles that are small relative to your loader or human power, you can blend everything at once. When dealing with larger piles, blend and remove sections as you go to open up access to unblended material at the center. One strategy that can save time is to add some of the recipe directly into the pile as you form it. Assuming the pile is large enough, a single bucket or shovelful of material will disperse itself very thinly as it rolls down the sides, so you won't have to worry about unblended layers. Find a pattern that works well for your site. Blending in tight spaces is more time consuming, which is why I recommend leaving access on at least two or three sides of the blending area and planning loader work space accordingly.

Layering

For those working at a micro scale, layering compost material is a very common method. In bin systems, open piles, and windrows, layering is simple and fast. The pile is formed like lasagna, with different strata of feedstock stacked in succession following the recipe.

Although it might put me at odds with layering proponents, I would advocate against layering in most cases. I just haven't seen the evidence that, in practice, layering works as well as mixing. So given the option, I will always recommend mixing. Especially at a micro scale, it takes very little time, and can happen in wheelbarrow or a very small mixing pad. But regardless of the method you choose, I do strongly recommend adding a layer of compost cover to keep odors in and vectors out. It's often beneficial to have a small layer of carbon as a base as well. In between these layers, blended material is the ideal.

Tracking and Recipe

Track what materials went into each mix and where it went:

- Date of blend
- Recipe—volumes of each material
- Feedstock sources—where each material came from
- Batch name—which pile or batch the blend went into

Pile Formation

Once your recipe is blended, you will use that material to construct a compost pile. There are a few factors to keep in mind for optimal pile performance:

Distinct batches. Make sure you are creating or adding to a distinct batch of compost that is of like age. I don't recommend adding new material to compost that is more than six weeks old. Ideally you are forming a new batch every one to three weeks to ensure that the newer material has time to catch up to the older material.

Minimum volume. Plan your batches so that you have enough material to heat (1 cubic yard or more of fresh material is a safe assumption).

Pile architecture. The shape of the pile makes a huge difference in how the compost will perform. Height in particular matters; piles shrink as they settle, release moisture, and break down. Slumping decreases passive aeration. Form the pile as steeply as possible, aiming for at least 45-degree angles on the side walls. There is no such thing as stacking at too steep an angle, because the angle will decrease very quickly. Bay and bin walls help with this, particularly with smaller batches that would have trouble achieving much height on their own. For small volumes, creating a cylinder within a bin that's being filled slowly will increase pile activity. I use cardboard for this in my own backyard bins. With ASP systems, be sure the pile is centered on the air channel perforations.

Avoid compaction. Take care to form the pile loosely, using your tractor bucket to blend and aerate the material as it dumps. Don't drive on the edge of the pile with your loader. The base of the pile is especially important in allowing both convective and forced air to distribute oxygen to the rest of the pile.

Food scraps contained. You aren't done forming the pile until food scraps are covered with a blanket of compost or carbon materials (4 to 12 inches depending upon your pile size and odor/vector pressures). Sometimes compost tarps are used to contain piles. I recommend using these in addition to capping with compost. Similarly, if you are using bins, you will want to keep food scraps away from the edges and surface of the pile by covering and surrounding the raw materials with carbon to avoid nuisance issues.

Clean site. Both in blending and in pile formation, some food scraps and other bits of organic matter will inevitably escape. Taking the extra time to clean up these materials is key to avoiding vectors and odors. Keeping your site free from organic matter preserves the site and minimizes potential pollutants in runoff.

Compost Pile Monitoring and Management

Following pile construction, active management begins, the foundation of which is pile monitoring. There are several reasons to monitor your compost piles. Monitoring is primarily done to provide the operator with insights into activity in each compost pile. These insights guide management choices regarding the specific piles you are monitoring, and also help you determine how you make and manage compost on an ongoing basis. Monitoring provides you with a feedback loop for maintaining optimal composting conditions and producing a quality product. Temperature monitoring, for example, can be very useful in determining when a pile should be turned to sustain optimal microbial activity.

Figure 13.7. Tom Gilbert demonstrating a moisture squeeze test. Where are the gloves, Tom? Courtesy of Highfields Center for Composting and the Vermont Sustainable Jobs Fund.

The monitoring practices I describe will apply to composters processing food scraps and manures, as well as other materials. Pile monitoring is a requirement for many composters who operate within a regulatory framework. Most regulatory requirements for pile monitoring aim to ensure that a heat treatment called the Process to Further Reduce Pathogens, or PFRP (or similar), is achieved. In addition, consistent reduction in pile temperatures can be an indicator of compost maturity.

When to Monitor

To paraphrase an old Chinese proverb, the best fertilizer is a farmer's footsteps. Likewise, frequent attention is the best ingredient for making good compost. Consistent monitoring will help you correct small problems before they become big problems and allow you to refine your practices over time.

Monitoring temperature two times weekly is usually adequate for most turned windrow composters, but more regular monitoring can be beneficial, especially if you're a new composter. Some permits may require you to monitor temperatures at a specified frequency during the temperature treatment period and to keep records of this. Even following the treatment period, however, continuous monitoring is critical to producing quality compost, especially during the first two to three months of pile activity. Those using aerated static pile and in-vessel methods benefit from keeping a daily monitoring regime, since temperature swings can be dramatic with these systems.

How to Monitor Your Compost

When you're producing and managing compost, you are working to ensure that you have created a suitable

Capping Piles

The use of organic matter as a cover for fresh and potentially odorous compost is a best practice in the industry, and in my experience one of the simplest and most underutilized tools in the composter's tool kit. The practice is called by many names, such as creating a *compost blanket* or *skinning*. My preferred term is *capping*. The idea is to use either semi-mature compost or high-carbon feedstocks to encapsulate fresh compost, containing odors in the pile, deterring vectors, and insulating the fresher compost on the edges of the pile so that it is able to heat and break down.

Slightly different strategies for capping are utilized, depending upon the composting stage. It's most often used when food scraps are received and blended, as well as in the final step in windrow formation.

Capping Incoming Food Scraps

As soon as possible upon arrival, cover incoming food scraps with additional feedstocks based on your composting recipe. Consolidating food scraps into a larger pile will reduce the surface area to volume ratio, making capping the materials more efficient. If possible, blending the food scraps with part of the composting recipe prior to capping the pile will help to absorb moisture and facilitate aerobic composting conditions. You will blend capping materials into the compost before you form the windrow.

Figure 13.8. Capping with semi-mature compost at the NYCCP at Red Hook Community Farm, Brooklyn, New York City.

Figure 13.9. Note the difference between the raw compost layer and the stable capping layer as it is being added.

Capping Windrows

Cover windrows evenly with a thick layer of carbon materials or compost (4 to 12 inches is typical depending on the likelihood of nuisances). Over time capping materials will be blended into the compost, but this contains and filters odors during the rawest stage of the process. Materials for capping are ideally somewhat porous, so as not to suffocate the pile.

Figure 13.10. The final phase in compost pile formation is cleaning up the edges. This prevents nuisances and pollution, prolongs the life of the compost pad, and creates an aesthetically pleasing and well-managed atmosphere.

habitat for aerobic decomposer organisms. As a best management practice, pile monitoring is designed to assess the ongoing health of the pile. There are four primary pile conditions to assess when employing monitoring practices:

1. Temperature
2. Moisture content
3. Structure
4. Odor

The cause of a problem can often be determined by cross-checking among these different conditions. For instance, if a pile is generating an ammonia smell, you may be able to infer that it is probably a result of pile dryness, using moisture monitoring to verify. While individual monitoring measurements can provide valuable information, only the results of the combined monitoring techniques collectively can tell the whole story. In addition to pile monitoring practices, consistent observation of the site may alert you to other issues.

Techniques for monitoring these four factors are covered in the rest of this section, including a short description of why each monitoring practice is used, the tools required, how the monitoring is performed, and general recommendations for responding to some non-ideal conditions you might observe.*

Generally, each batch is given its own monitoring log, so that batch histories can be easily reviewed. Sample *Compost Pile Monitoring Logs* can be found in appendix B, page 414.

Monitoring Pile Temperature

Metabolic heat generated by microbial activity is the primary factor that will influence the temperature of a compost pile. Pile temperatures can also be affected by the physical characteristics of the materials being composted (more versus less insulating), as well as chemical reactions (at high temperatures) and external environmental variables.

Pile temperatures are an imperfect but useful indication of microbial activity. Newly formed piles commonly reach or exceed 130°F within several days to several weeks from pile construction. Piles constructed during extremely cold weather or with frozen feedstocks will take longer. If you are trying to ensure weed seed and pathogen destruction, all of the raw material will need to reach a temperature of at least 131°F for three or more days (see *Temperature Treatment Standards and the Process to Further Reduce Pathogens* on page 56).

Field Tools

- Compost temperature probes. For turned windrow operations, use a 3-foot probe with a 5/16-inch stem. For aerated static pile systems, probes 4 to 6 feet long are often beneficial. Multiple temperature probes with quick-response stems can be useful for efficient monitoring.†
- Compost Pile Monitoring Logs (appendix B, page 414).

* Monitoring pile oxygen is also a useful practice, although it is fairly uncommon and not discussed here.

† Larger sites can monitor really efficiently with two pairs of 12- and 36-inch thermometers. Work in between the two by setting and reading each pair in a constant rotation.

Figure 13.11. A bad habit that some composters get into is leaving their probes in the piles. Moisture tends to accumulate in the dial, and the probe shaft tends to bow over time. Removing the probe improperly or leaning on it can also cause it to bend. A probe handle extends the device's life dramatically, and is well worth the extra $30 or so.

How to Measure

Pile temperatures should be taken roughly every 10 to 20 feet along the pile, depending on the total pile length and your regulatory status. Five locations across a 50 to 100-foot windrow will give you a reasonably representative sample of the pile conditions. Temperatures should be taken at depths of 12 inches and 36 inches at each location. The probe should be left in place for at least one minute or until the dial stops moving. Record pile temperatures in a monitoring log, and keep these logs on hand.

In smaller piles such as bins, I recommend taking at least four temperatures per bin at varying depths.

One school of thought suggests that you should always monitor a pile in the exact same locations to provide you with a picture of how specific areas of the pile perform over time. Some sites leaves flags at those locations. This practice will allow you to show that you've achieved temperature of at least 131°F for three days in specific zones of the pile. Another approach is to find the hot spots in the pile and record those. The hot spots move around, and if the pile temperature declines in one area and is highly active somewhere else, why not let that area have its process? I lean more toward this second strategy, knowing that monitoring is only giving me a small snapshot of what's happening in the pile. In terms of hitting 131°F or greater for at least three days, I assume that through mixing, material will heat and cool slightly more than once. Most of the time, piles will reach well above 131°F, so the time/temperature relationship will be reduced.

General Response

Temperature will impact your decision whether or not to turn a pile, and may indicate that factors in your pile recipe need to be adjusted. There are a number of possible explanations for depressed temperatures, such as a carbon to nitrogen ratio that is too low or too high, imbalanced moisture content, compaction in the pile, or excessive density.

If you determine through monitoring that low temperatures correlate with a high or low moisture content, the problem can generally be addressed by adjusting moisture content. If you experience low pile temperatures and moisture is not the issue, carbon to nitrogen ratio and pile density are the next factors to explore. If everything in your recipe seems fine, try turning the pile once to mix and aerate it.

If your pile is heating, your temperature monitoring will help you determine when to turn the pile (or, in the case of aerated static pile composting, what sort of aeration schedule to maintain).

You will want to turn your pile after its initial heating has peaked and is beginning to decrease, or if your pile temperatures at the 12-inch height are consistently more than 20°F different from those at a 36-inch height. Additionally, if your pile is heating too much and your temperatures have surpassed 160°F, you should consider turning the pile to cool it down and prevent excessive loss of nitrogen and microbial diversity.

See *Pile Turning*, page 367, for more information about how and when to turn in response to monitoring results.

Monitoring Pile Moisture

Moisture in a pile is a critical factor in microbial activity and the decomposition process as a whole. If you have too much or too little moisture, microbes cannot function effectively. Typically, operations target a moisture content of roughly 60 percent in the early stages, and then the pile dries out over time.

Figure 13.12. This 10-cubic-yard test pile at a recent operator training I was leading hit 180°F. It had some high-energy feedstock (see figure 4.17, page 101), but the bigger factor was likely that it was very dry (in other words, very insulating).

Pile moisture contents of 50 to 65 percent are okay, but material on the boundaries of these parameters must be closely monitored, and if the moisture content moves beyond 50 to 65 percent it should be addressed. Moisture surrounding the pile can also adversely affect the composting process by inhibiting oxygen intake through its base. Standing water around the piles will result in saturation, creating undesirable, anaerobic conditions and stagnation.

Field Tools

- Hands
- Eyes
- Shovel
- Latex or vinyl medical gloves
- Compost Pile Monitoring Logs (appendix B)

How to Measure

Dig at least 1 foot into the pile with your shovel. Wearing a glove, take a small handful of compost in one hand, remove excessively large particles, and squeeze the material hard. If the contents in your hand begin to drip moisture between your fingers, the moisture content is likely above 65 percent. If there is no dripping, but the moisture glistens between your fingers, the MC is roughly 60 percent, which is considered ideal. In cases where there is no visible moisture, open your hand so that the contents sit on your palm. If the contents remain in a ball, depending on how tightly they maintain their form, your moisture content is 50 to 60 percent. If the contents fall apart, your moisture level is below 50 percent. It may be beneficial to assess moisture in

Figure 13.13. Moisture content below 50 percent. Note: The material *cannot* maintain its conformation in a ball.

Figure 13.14. Moisture content of 50 to 60 percent. Note: The material *can* maintain its conformation in a ball, but no moisture can be squeezed out of it.

Figure 13.15. Moisture content of roughly 60 percent. Note: A very small amount of moisture can be squeezed out of the material to create glistening between the fingers. That's what we're shooting for!

Figure 13.16. Moisture content over 60 percent, and likely 65 percent or greater. Note: There is dripping between the fingers.

several locations in the pile if MC does not appear to be uniform.

A visual inspection of the pile and the surrounding site will also provide you with feedback regarding moisture. Site moisture and pile moisture may be connected or not, and therefore clarifying where the moisture is originating from—the pile or the site (including water coming onto the site from the surrounding environment)—is important.

Record pile moisture content in a monitoring log, and keep these logs on hand.

General Response

See *Adjusting Moisture*, page 370.

Assessing Pile Structure

The term *pile structure* refers to the "architecture" of the pile and the character of the blend in terms of density, particle size, and porosity. The pile's ability to maintain its parabolic shape and porosity over time impacts how well the pile can be aerated, both actively and passively. The structure of your pile is the result of how it was constructed, the recipe and the thoroughness of blending, and the integrity of the particles. Some mixes become overly dense, while others maintain some coarseness of form until the mix relaxes on its own accord. Poor structure causes overly dense compost and slumping piles, and ultimately leads to anaerobic conditions.

Field Tools

- Eyes
- Hands
- Shovel
- Latex or vinyl medical gloves
- Compost Pile Monitoring Logs (appendix B, page 414)

How to Measure

When you're monitoring and working on-site, observe the compost pile's overall shape. Piles that are slumping or unable to maintain a horizontal conformation (parabolic or triangular shape) likely have poor structure. While performing the moisture content squeeze test, observe the compost's density and diversity of particle sizes, looking for a range of visible particles 1 inch or smaller within the mix. Woody particles in particular provide excellent structure.

Observe whether there is crusting on the pile's surfaces, which will reduce air exchange in the pile.

More involved field and lab tests are usually not needed unless there is a consistent problem. Information about field and lab testing of bulk density and porosity can be found in chapter 4 (see *Measuring Bulk Density* and *Measuring Porosity*, pages 76 and 78 respectively).

While you're monitoring, record your observations about pile structure in a monitoring log, and keep these monitoring logs on hand.

General Response

See *Improving Pile Structure* on page 371.

Monitoring Pile Odor

Smells coming from a compost pile provide important indicators of the pile's internal dynamics and may direct management choices. The odors from

Figure 13.17. David DiDomenico giving the old sniff test at Vermont's first Compost Operator Certification Training in 2011. Courtesy of the Highfields Center for Composting and Vermont Sustainable Jobs Fund.

compost piles and composting feedstocks are commonly associated with the release of volatile organic compounds (VOCs) and other chemical compounds such as ammonia. Smell is a natural by-product of microbial decomposition, but when you're composting food scraps, there is a high nuisance potential, and strong smells are an indication that the compost needs attention.

Field Tools

- Nose
- Compost Pile Monitoring Logs (appendix B, page 414)

How to Measure

Take note of the smell of both the individual piles and the site as a whole by consciously breathing in through your nose while working around the piles, including during monitoring and turning. You may be able to discern a particular character to an odor or track the odor to a certain location, pile, or portion of a pile.

Record your descriptions of pile odors in a monitoring log, and keep these logs on hand.

General Response

See *Managing Odors*, page 371.

Pile Monitoring Logs

Compost Pile Monitoring Logs are used to track the life of each individual batch of compost. Examples of logs for different pile types are shown in appendix B, page 414. You could adapt these to go directly into a spreadsheet accessible on a tablet or smartphone (they make screen protectors that reduce glare), or you could stick with the traditional paper route.

Pile Turning

For many composters, turning piles is one of the most enjoyable parts of the process. Thankfully microbes do most of the work of composting for us,* diligently consuming away while we're busy with less important human stuff. Yet there's something satisfying about getting in there and actively managing the material from time to time.

Personal benefit aside, turning a pile supports both its microbial life and the quality of the end product. Turning, whether by hand or mechanically, releases stagnant gases and exposes the material to fresh oxygen. I think of turning a pile like giving it a breath of fresh air after a long exhale. Turning also mixes up materials, moving them from the less active portions of the pile into the active zones. The fresh oxygen and food sources reactivate the pile, and if everything else is in balance, you'll typically see an increase in pile activity in the form of rising pile temperature.

Pile turning is an important component of meeting temperature treatment standards. Typically the goal is for all of the material to reach at least 131°F for at least three days, which is achieved by turning the pile and thus rearranging the location of material within it. Meeting temperature treatment standards is discussed in much greater depth in the *Temperature Treatment Standards and the Process to Further Reduce Pathogens* section, page 56.

One of the most common questions I get is, "How many times should I turn my pile?" My best answer is that if you are asking that question, then you probably could be turning it more. If you are turning your pile enough, you will know it.

Pile conditions as observed through monitoring are your main indicator of when to turn the pile. Here are a few recommendations.

I WOULD RECOMMEND TURNING THE PILE WHEN:

- You see pile temperatures trending downward
- You need to turn the pile to meet temperature treatment standards and the pile has achieved 131°F or more for at least three days
- Pile temperatures reach or exceed 155 to 160°F (in order to cool the pile)
- A fresh pile has had stagnant temperatures for more than three weeks (If you turn a fresh pile and expose raw food, consider odor concerns and the fact that you may need to re-cap the pile.)
- Pile moisture is non-uniform

* Or do we work for the microbes?

Digital and Remote Compost Temperature Probes

Wouldn't it be nice to know what's been happening with your pile's temperature in the interim between your monitoring visits? Well, there is a way, and it's called a temperature data logger. There are several suppliers of this equipment, including REOTEMP and Green Mountain Technologies (GMT). GMT's equipment also integrates with their ASP controls. The cost for a small site to set up a digital logging system is somewhere in the $10,000 range, depending upon the site's specific needs.

Over the past couple of years, I have also started to experiment with data loggers that are not specifically made for the composting industry with some success, although it's nothing that I'm ready to publish just yet. That said, I will be very surprised if lower-cost solutions are not developed or hacked in the coming decade.

At one school that I worked with, a tech-savvy teacher figured out a way to have their compost bin temperature continuously broadcast on the internet. Another simple monitoring solution for small-scale operations is the use of small digital high-low thermometers with probes on wires. These will give a quick read of both the current temperatures and the range of fluctuation. Some systems might give high-low external temperatures as well. Schools in particular will likely find high-tech data collection methods useful in terms of both engagement and educational value.

Figure 13.18. A temperature data logger taking readings at 12 and 36 inches in a windrow at Kingdom View Compost in Lyndonville, Vermont.

I WOULD *NOT* RECOMMEND TURNING THE PILE WHEN:

- Pile temperatures are rising and have not reached 131°F for at least three days
- Temperatures are rising and no immediate action is required to meet temperature treatment standards
- The pile contains food scraps and is less than a week old (assuming nuisances are a concern)

ADDITIONAL GENERAL TURNING TIPS:

- You probably do not need to turn the pile more than two times per week, and less frequently is okay and even beneficial by some measures.
- Concentrate your turning efforts in the first two to three months (the primary and secondary phases). An extra turning in month 1 is worth four turnings in month 4.
- Continue turning as the pile matures to ensure uniform and timely completion (one time a month is adequate once temperatures are consistently below 120°F).
- Moving material counts as turning it, but it is not the most efficient way to turn it. *See below.*
- For hand-turned windrows, look at how Red Hook Community Farm manages their piles (page 191). They are the masters!
- With loader bucket windrows, it is most efficient to turn them from the sides as opposed to the ends. You do not need to turn the entire pile over each time. You can *roll* one-third to one-half of the width of the pile on each turning, assuming you turn fairly frequently. Here are the steps that I follow:

1. Lift the bottom edge of the pile up to the top or over the windrow, so that you can get closer with your loader.
2. Push the top of the windrow down the back side.
3. Push the upper part of the windrow down and over the back side, and use the power of the loader to loosen/turn this upper section over and down. This was essentially the core hot zone in the previous iteration of the windrow.
4. Turn the side of the pile closest to you, moving one to three more bucket widths from that edge up/over the pile, feathering the material as it's dumped.
5. Loosen and open up the edge of the pile closest to you and leave it as smooth and uniform as possible.
6. Optional: Add new capping material if needed to prevent nuisances. Slightly older compost would work well in this instance.
7. Clean up organic matter from the edge of the pile and alleyways with the tractor bucket.

See figure 8.5 (page 197) for an illustration of these steps.

Material Flow and Consolidation

Compost material management should first and foremost be quality-based. But to keep everything moving smoothly, composting also requires a lot of time-consuming labor. I know I'm not the first one to realize and even find some humor in the fact that a large part of composting, farming, and so many things in life is literally just moving stuff from one place to another.

Water goes here. Wood chips go here. Odors stay here. Birds stay away. Recipe, monitoring, and turning all follow sourcing, organizing, and moving substantial volumes of materials. Both good site design and good site management come down to making the movement of materials as smooth and efficient as possible, while avoiding common pitfalls.

The design side of material flow is covered in the compost infrastructure and compost-method-specific chapters. Once a compost site is operational, its designs will be put to the test. Since it may not always be obvious what the designer was thinking when planning the site, it takes due diligence on the site manager's part to understand the flow and capacity assumptions as they were designed. If the designs were well thought out, the flow of materials through the site should be efficient, flexible in response to fluctuations, and scale-appropriate. Of course this is not always the case, and while site modifications may be necessary, operating well also requires that you adapt to the realities of the site.

Managing material flow well requires planning not only day-to-day, but week-to-week, month-to-month, and year-to-year. Think of the site as a water pipe, and the site manager as the flow control valve. If there is a blockage anywhere in the system, then flow into the pipe also stops, which becomes a big problem if the pipe's job is to empty a pond. Eventually that pond will overflow. The job of the site manager is essentially to keep the flow unblocked, but it's more complex than that, because there's not always a place to put the water coming out of the pipe. The system also has to have some storage capacity—storage pools that can be used as needed.

Every site is unique and complex when it comes to material flow, so I'm going to throw out some concepts that I find useful when considering how to efficiently manage the material on a site:

Become proficient in thinking in space and time. I talk about this more in chapter 5, but the basic concept is that every unit of space is also a unit of time. For example, a windrow might be two weeks' worth of throughput at 10 tons per week food scraps processed. A compost pad that has room for four windrows would then represent eight weeks' capacity. Map out your site like this both on paper and in your mind, and understand the assumptions behind your thinking.

Consolidate materials of like age. One large pile takes up less space than two small piles. To use space efficiently, maximize pile size while staying within the constraints of BMPs. Materials shrink dramatically in the primary and secondary phases of composting, so materials of like age can be combined to make space. For example, an operation might combine a pile that is six weeks old with one that is eight weeks old once they have both met temperature treatment targets and lost at least 40 percent of their volume.

Manage more, move less. Moving materials from place to place on a site is necessary, but it is also a time sink. Consider the previous concept, but also consider whether it is always necessary. Why consolidate two piles when there is no reason to? If there's a way to move material less distance to create necessary room for new material, go for it, but if there's plenty of room, why touch the stuff? Management practices that support moving less material include:

- Targeted recipes, turning, and frequent aeration all lead to faster degradation, which facilitates less movement of material and less volume per movement.
- If possible, manage a batch from start to finish in one location. This might not be possible, but where space is less of a constraint it's definitely something to consider. As long as material and leachate from pretreated compost is not contaminating the treated compost, there is no reason to move it.
- Cure the compost where it will be used. On a farm, there's no reason why compost that is finished (and has met any applicable maturity standards) can't be cured and stored off-site, saving you from labor of moving it twice, first to curing and storage, then to the field.

Compost Management Issues and Nuisances

Composting is rewarding, but it involves managing numerous challenges, often simultaneously. When problems arise, early intervention can keep them contained. Luckily, there aren't many issues that you will be the first composter to encounter, and an abundance of resources and training opportunities are available. One of the key points you should take away from this chapter is that making an investment in monitoring can keep a small problem from becoming a large one. Best management practices can keep a problem contained within a pile, or at least within a site, so that it does not impact neighbors or involve regulators. Let's turn now to prevention and remediation strategies for some of the basic challenges composters are likely to face in everyday operation.

Adjusting Moisture

If your moisture content is high (above 60 to 65 percent) and either creating an odor or failing to heat, you may need to dry out your mix in order to mitigate issues and produce a quality product. If the mix is not significantly above 60 percent moisture content, simply turning the pile may achieve the desired drying effect. Turning, as well as general exposure to dry climatic conditions, can reduce pile moisture over time. In many cases, multiple turnings over several dry days will be sufficient. Another strategy to consider is opening up the top of the pile with the tractor bucket to create more surface area from which the air and wind can wick away moisture.

If your mix has moisture content significantly over 65 percent, or during particularly wet times of year, you will need to add dry matter. Do this by opening the top of the pile with the bucket, forming a trough, adding some dry feedstocks, and then rolling or otherwise turning the pile to incorporate the new material. Windrow turners are particularly effective for drying the pile mechanically or incorporating new dry matter into a recipe.

In dry climates or weather conditions, operators may mix to a higher moisture content than is traditional to offset the climatic drying effects. If your moisture content is below 50 percent, the addition of moisture is probably required, though impending rain may sufficiently wet the pile in some cases. Irrigating may be the most effective method, and adding water does not impact the carbon to nitrogen ratio. This can be a good use for leachate or dirty storm water collected from the site if the pile is still actively achieving thermophilic temperatures (to ensure pathogen destruction). If other indicators of pile health are good and your moisture content is on the low side but within the acceptable range (50 to 55 percent), minimizing pile agitation will help to retain as much moisture as possible until the pile is naturally moistened by rain. Turning or flattening the pile during rain events will increase the volume of moisture that the compost is able to absorb.

When you are adjusting pile moisture up or down you need to be careful not to adversely impact the pile recipe in other ways, such as carbon to nitrogen ratios. If you are bringing the moisture content down, the use of ingredients with a neutral carbon to nitrogen ratio (around 25 to 30:1) with low moisture content will help. Ingredients such as dry, heavily bedded horse manure, hay, or small ruminant bedding often meet these criteria.

Improving Pile Structure

Poor pile structure is an indication that your compost recipe needs adjustment or that materials were not adequately homogenized during blending. If a pile is too dense or slumping, it will diminish the availability of oxygen and create stagnant conditions. Slumping piles should be restacked, and if the slumping continues, you can remediate the mix by thoroughly blending in more bulking material. Typically adding wood chips or bark so that they make up 5 to 10 percent of the mix by volume is adequate for food scrap compost, assuming another 65 to 75 percent of appropriate feedstock is in the mix to achieve a well-balanced recipe.

Lastly, if the piles were large to begin with (8 or more feet tall), then consider reducing pile size. Note, however, that compost piles will naturally reduce in size over time. This is not an indication of pile slumping but rather of volume reduction. If you observe surface crusting, turn the pile to incorporate the crusted materials and reopen the surface to airflow. Efforts should be made when constructing and turning piles to limit compaction.

Managing Odors

The general character, intensity, and offensiveness of odors are all indicators of whether a particular smell is a problem. Subtle odors from a pile may indicate potential problems or areas to improve upon in the next batch of compost, but they do not necessarily need to be addressed immediately. Likewise, some insubstantial odors may arise when raw feedstocks are combined or when fresh compost piles are first turned. These odors should be noted but may not require an operator's response.

If odors are distinct and strong when the pile has not been agitated, this usually indicates a problem in the compost pile and should be investigated and dealt with.

Most odors are indicative of one of the following:

- Low carbon
- Low oxygen/high moisture
- High density
- Inadequate containment

Since the first three factors are inherently connected, it could be any combination of them that is causing the odor. Carbonaceous materials and compost capping materials act as a biological scrubber for odors, retaining them in the pile until they can be metabolized by microbes. In most cases containing odors with a compost cap until they can be biodegraded is all that is needed to manage the inevitable low-level odors that come with composting. If capping is not enough, it's an indicator that something else is out of balance.

If a pile is low in carbon, you should take steps to incorporate additional carbon-rich material into the mix. Often, addressing deficiencies in carbon will also address low oxygen and high moisture conditions. For additional suggestions to reduce pile moisture see *Adjusting Moisture*, page 370.

If a pile is too dense, the best response is to incorporate a bulking material, something with a large enough particle size to allow more airflow in the pile, such as wood chips or bark (5 to 10 percent by volume). This can be done in a similar manner to adding carbon or dry matter. If such a material is not immediately available, several successive turnings may suffice to elevate pile oxygen sufficiently.

Vector Mitigation and Controls

Vector is the term commonly used to describe any animal that is attracted to compost and that could potentially introduce pathogens from the site to the outside. In addition to being a potential public health concern, these unwanted visitors are also a nuisance and a source of stigma for composting in general.

Vector controls are methods used to mitigate and manage pests. The most effective way to mitigate or deal with a vector issue is to follow compost BMPs to contain and break down raw food scraps and their associated odors while rapidly achieving hot temperatures. Some BMPs that are most applicable to vectors include:

Prevent habituation. Animals form habits. Extra diligence is needed to break those habits and to avoid future habituation. Watch for signs of vectors and act quickly.

Contain raw food scraps.

- Cap piles with carbon materials or compost.
- Keep raw food scraps off of the ground surrounding piles and in receiving areas.
- Create physical barriers. Use in-vessel systems or bins secured with hardware cloth.

Manage pile health. Follow recipe, monitoring, and pile-turning protocols to achieve the rapid breakdown of food scraps and their associated odors.

Mitigate standing leachate. Manage pile moisture and absorb leachate on the site with dry carbonaceous materials (such as leaves and wood chips).

Disrupt vector reproductive cycles and habitat. Temperature treatment and regular turning of windrows effectively disrupts reproductive cycles and habitat.

In addition to these general practices, there are a variety of methods specific to certain vectors. Rats and seagulls (and other avian scavengers) are the most challenging to deal with, because they are practically ubiquitous wherever people are found, and they are well adapted to survive despite our best efforts to suppress them. Below I will discuss what I've learned about dealing with these two types of pests specifically.

Rats

There are a number of methods used to reduce rat populations, but most come with a downside. Poison is the simplest and is fairly effective. While there are pet-safe poison lures, once a rat has ingested poison, it is a threat to any animal that eats it. Traps are safer for other animals than poisons, but they require more work and may not do the job if you are dealing with a large population.

Some sites employ cats and dogs as rat control. Cats tend to be better rat hunters than dogs, but they can also carry a parasite called *Toxoplasma gondii* that is ironically contracted from rats and transmitted in their feces. Sites with cats will need to keep them away from post-heat-treated piles in order to prevent contaminating finished product.

Some sites use buckets of water to trap and drown rats. This is a practice that I have never tried, but I know there is an abundance of info about it online. Recently I learned of a humane and effective method that employs dry ice to asphyxiate the rats in their nest. Dry ice is the solid form of carbon dioxide, which returns to its gaseous form as it warms. The method involves locating the rats' holes, stuffing dry ice down the holes, and then burying all of the holes to trap the gas and the rats in the holes, where they will be become oxygen-deprived. If the holes are still covered afterward, the job is done.

For sites that have a rat problem, assume that no one method is going to do the trick alone. Rotating between a variety of controls on top of your best management practices will be the most effective strategy.

Seagulls and Other Avian Scavengers

Diligently covering piles and keeping the site clear of attractive and accessible food sources should be enough to mitigate scavenging birds. Many sites are virtually free from avian activity, and this is not for a lack of hungry birds set on finding food sources wherever they are available. Because of this, I believe that simply employing best management practices can prevent problems with birds.

When birds do become an issue, increasing BMPs may solve the problem, but if it doesn't, and if birds find that brief window when food sources are available, there are a few methods that sites use for abatement.

One of the simplest means is to cover receiving areas. It's a deterrent that can sometime reduce access enough to get the problem under control. Another method is to cover the active management areas with a matrix of wires or fishing line. Birds really don't like to fly where there are invisible obstacles.

When other options fail, sites sometimes employ more aggressive tactics, including drones, noise machines, pyrotechnics, and hunting. While it's not necessarily legal in all places without a permit, and it's not a replacement for BMPs, I have heard that placing a dead bird on a site gets the message across fairly effectively to some scavenging birds like seagulls, which are highly intelligent.

Leachate

The loss of *free moisture* from the compost pile is typically called *leachate*. Free moisture is essentially pile liquid that doesn't have dry matter to absorb it. When it exits the pile we call it leachate, which is the technical term for liquid compost discharge. Leachate is high-odor and high-nutrient, and it could potentially carry pathogens, so it should be treated with the

Figure 13.19. Leachate forming at the base of a pile can lead to odor issues and is prime habitat for insects.

same precautions as the raw feedstocks from which it was derived.

Infrastructure for leachate control is discussed in the section *Managing Moisture On-Site* (page 152) as well as in chapter 9. Beyond the site design elements for managing leachate, there are best practices for dealing with leachate where and when it occurs. Leachate should be prevented to the greatest extent possible by controlling moisture content. When it exits a pile, leachate should be absorbed with carbon-rich dry matter such as sawdust. This can then be mixed into the next batch of compost. Absorbing leachate reduces odor, prevents insect habitat, and mitigates dirtying runoff from the site.

Pad Maintenance

Many composters fail to keep compost pads level and free from debris until it's too late. With concrete and asphalt, grading is pretty much covered, except for filling potholes or sealing cracks. The main task with concrete and asphalt pads is keeping the pad clear of organic matter in order to reduce contamination of storm water. (Remember the foundational tenet of stormwater management: "Keep clean water clean.")

Unlike concrete and asphalt, packed gravel pads have a tendency to rut, pool, and collect organic matter. The best preventive strategy is to regrade and relevel when low spots form. Grading and leveling minimize standing water and collection of organic matter, which—unchecked—lead to more rutting and ponding in an increasingly destructive loop. If needed, fresh gravel should be added to low spots when the pad is dry. Adequate grading can usually be achieved by back-dragging with a loader bucket.

While you're grading packed gravel pads, remove as much organic matter as possible. Organic matter has a tendency to make its way into the gravel surface and hold moisture, which eventually creates a muddy and non-uniform surface. Maintenance will increase the life of the pad dramatically, but eventually it will need to be resurfaced or replaced altogether. The life span of a well-maintained gravel pad seems to be in the 5- to 10-year range, but with heavy wear and tear it can be much shorter than that.

Quality Control Standards

People use and sell compost products in various stages of readiness and quality. But I am a firm believer that regardless of scale, or even whether a composter intends to sell the product, they need metrics in place by which to gauge (1) when the composting process is complete; (2) whether a compost is suitable for the intended end use; and, conversely, (3) whether the end use is suitable for the compost.

Regulations compel some composters to track their product through the process and to test periodically for maturity and safety. As I've discussed in other parts of the book, it is incumbent upon every operation to conduct due diligence on the regulations that may apply to you. In terms of end product, there may be specific state and federal regulations that apply. Some composters will choose to meet the criteria for producing compost suitable for use on organic farms (see *Producing Compost Suitable for Use on Organic Farms*, page 128). At a minimum, a

Figure 13.20. The compost that was delivered to this customer was unfinished; it was hot to the touch and had a strong ammonia smell. It also had a large percentage of both uncomposted plant parts and non-volatile solids (sand).

site's quality control standards will need to cover any applicable regulatory requirements, but often these will only be a start, if any apply at all.

Baseline Quality Control Standards

For many in this book's audience who are unregulated or who would like to take further steps to produce the highest-quality products, I suggest implementing voluntary internal standards. At a minimum, I recommend adopting the baseline quality control standards outlined in table 13.2, and both monitoring and documenting every batch created on your site based on these parameters.

Additional Quality Control Standards

For commercial operations and those supplying large volumes of compost to the public, implementing a program of routine lab analysis and plant growth trials is highly recommended. The frequency of testing these parameters varies depending upon scale. Testing 10 to 25 percent of the batches should yield good accuracy, with frequency of testing increasing with facility scale and if variability in results raises a concern. (Plant growth trials are typically more frequent—as often as every batch.)

The goals of these tests include:

- Quality assurance
- High product consistency and performance
- Process optimization

Based on the results of your tests and plant trials, you can make adjustments to your recipe and other management activities.

It's worth noting that the US Compost Council has created a Seal of Testing Assurance (STA) program that supplies composters with third-party testing and labeling for product quality. Their testing regime contains pretty much all of the parameters listed in table 13.3, although C:N is not listed and that's one

Table 13.2. Baseline Quality Control Standards

Quality Control Parameter	Target or Minimum Criteria
Individual batches tracked	Pile ID. Start date. Completion date (no new material added after this date).
Pile recipe	Record pile inputs by volume. Track feedstock sources.
Temperature treatment target	All compost has reached 131°F for a minimum of three days. Monitor and record temperature for every batch twice a week at a minimum. Record the date the temperature treatment target was met.
Maturity	Temperature is less than 90°F or ambient.
Visual quality	Compost looks finished, dark.
Smell	Compost smells finished, like forest soil.
Contamination	Compost is free of known physical (trash) and chemical contaminants.

Figure 13.21. Rooftop eggplants grown in Vermont Compost Company's (VCC) Fort Light versus those grown in another local compost (corner planter). VCC potting mixes are notoriously productive and reliable. They are therefore the benchmark against which I and many in New England compare other products' performance.

of the main things that I recommend targeting. The STA program also includes testing for pathogens and heavy metals, which adds to the cost considerably. STA might be a practical option for larger sites that see marketing value in an independent seal of quality on their label. For small sites that follow best management practices, my opinion is that pathogen testing is unnecessary. It's one thing if you have reason to believe your process may not be adequate to reduce pathogens, but heat treatment and the maturation process are very effective with respect to pathogen reductions. Likewise, heavy metals are very rarely of concern for food scrap or even yard debris composters. This does come up with biosolids composters, who are essentially managing anything that goes down the drain in a municipality. Even if there was a heavy metal issue at a community-scale compost site, unless it was consistent, the odds that it would be caught with infrequent testing are very low. Rather, I recommend that community-scale composters focus on ensuring best management practices and use other compost quality indicators where they are useful.

Tables 13.3 and 13.4 outline the key standards and targets most composters need to look at. Of course these are just targets and there are always outliers, so adopt and adapt these as you see fit.

Table 13.3. Quality Control Standards (Lab/Field Testing)

Quality Control Parameter	Target or Minimum Criteria
Maturity	Solvita Compost Maturity Index (CMI) > 6 or practical maturity[a]
Carbon : nitrogen ratio	Target of 12–15:1 for standard compost; 15–40:1 for high-carbon compost
Salts (Conductivity)	≤ 5 mmhos/cm
pH	6.5–7.5
Organic matter	≥ 50% dry weight
Moisture content (if screening)	≤ 50%

[a] *Guide for Indexing Compost Maturity*, Tech Memo 0317-6 (Mt. Vernon, ME: Solvita, 2017). https://solvita.com/wp-content/uploads/2017/06/Solvita-TechMemo6.pdf.

Table 13.4. Quality Control Standards (Plant Growth Tests and Bioassays)

Quality Control Parameter	Target or Minimum Criteria
Plant vigor tests	Plant growth ≥ control or target product (see Plant Growth Trials)
Persistent herbicide bioassay	See Washington State University protocol[a]
Weed seed germination testing	See *Weed Seed Germination Testing*, page 380

[a] Washington Department of Ecology (WA-DOE), *Bioassay Test for Herbicide Residues in Compost: Protocol for Gardeners and Researchers in Washington State*, draft (Washington State University and Washington State Department of Ecology, 2002). https://s3.wp.wsu.edu/uploads/sites/411/2014/12/PDF_Clopyralid_Bioassay.pdf.

Plant Growth Trials (Bioassays)

Well-made compost enhances plant growth. However, phytotoxic (plant-toxic) compounds are often present in raw material and/or generated during the composting process.[1] Plant growth trials, or *bioassays* as they are often called, can be a reliable way to test for the adequate degradation of these compounds, as well as to improve and ensure product performance. Some labs that specialize in compost analysis also offer bioassays as a service.

Bioassays most commonly fall into four different categories: compost maturity, phytotoxins, persistent herbicides, and of course plant vigor. Plant growth trials are typically more frequent than lab testing; some composters literally test every batch.

After reviewing the research on bioassays, I was surprised to find that the most widely used methods for compost maturity bioassays have been called into serious question. Dr. Will Brinton is highly critical of the use of cucumber as the STA's chosen bioassay species, although he does appear to see some value in the use of watercress following the Organization for Economic Cooperation and Development (OECD) seedling emergence and growth protocol.[2] There

Pursuing Biological Indicators

by Lynn Fang, MS

Standard compost testing can be an important window into many of the properties of compost, but these tests are poor indicators of its qualities as a living, breathing ecosystem. Compost's microbiology is the source of the many beneficial functions compost provides; assessing microbial health and vitality is a good start to ensuring the creation of functionally vigorous compost. Because the microbial ecosystem is complex, integrating a variety of organisms, metabolic functions, and environmental variables, it can be easy to get lost in the details.

Currently the biological indicators that are most commercially available are fungi to bacteria ratio (F:B) and overall microbial activity (as assessed through CO_2 respiration). Some labs can also assess nematode community composition, which can provide more integrated information about the ecosystem. Bacteria and fungi are first-order organisms that feed directly on decaying organic matter. Nematodes are second-order organisms that feed on bacteria, fungi, and other nematodes, and provide indicators of ecosystem maturity and stability. Improving indicators of F:B ratios as well as overall microbial activity will help to ensure a more effective soil amendment that provides a potent biological inoculum and, subsequently, the release of more plant-available nutrients.

One of the most common biological indicators is the fungi to bacteria ratio (F:B). Generally speaking, as the compost ecosystem matures, the fungi will increase. Vermicompost tends to have greater abundance and diversity of fungi than standard compost. Fungi are important for improving and strengthening soil structure, help suppress plant pathogens, and are also more efficient in sequestering carbon. Direct F:B ratios are recommended to be in the range of 1.0 to 2.0. Nematode analyses can also generate a F:B ratio, which is typically calculated based on the number of fungal-feeding nematodes to the total number of fungal-feeding and bacterial-feeding nematodes. For the nematode F:B ratio, mature compost generally falls in the range of 3.5 to 6.0. Most labs provide suggestions based on the results, as well as consultation services for more detailed interpretations and recommendations.

A good principle to follow is that microbial diversity is beneficial, as it helps with suppressing plant pathogens and increasing plant nutrient availability. Diversity breeds diversity, so ensuring diverse feedstock materials can help increase compost microbial diversity. Microbial diversity can best be assessed using high-throughput genetic sequencing. Feedstocks and processing methods all contribute to shifts in microbial community composition. Higher-carbon feedstocks (such as woody materials) and more static piles favor fungal abundance. Annually cultivated horticultural and agricultural plants may prefer more bacterially dominant soils, so F:B ratios can be manipulated based on your end purpose uses.

Keeping tabs on standard compost testing will show you whether incorporating specific feedstocks or amendments would help provide the extra nutrients your soil and crops need. Testing for biological indicators can provide a broader window into the overall biological activity, F:B ratios, and microbial diversity and abundance, which are the most relevant pieces of information that can be assessed using currently available methods.

Available Biological Testing

Biological assays. Generally conducted using the direct microscopy counting methods popularized by Elaine Ingham, PhD, these assays are the gold standard for determining actual populations of organisms in soil and compost. They look at both active and total fungi and bacteria. Other laboratory methods are used for assessment of protozoan and nematode communities.

Labs

- Earthfort: www.earthfort.com
- Soil Test Farm Consultants: www.soiltestlab.com
- Soil Food Web New York: www.soilfoodwebnewyork.com

High-throughput genetic sequencing. A new technology that is becoming more available and commercialized for soil and compost microbial ecological analyses, this offers detailed identification of which microbial groups are present. Sequencing is considered to be more robust at species identification than laboratory culture or direct microscopy counting.

Labs

- Trace Genomics: tracegenomics.com/#/

Pathogen testing. Typically targets fecal coliform and *Salmonella*, available from standard compost testing labs.

Future research may show that enzyme analysis provides even more significant understanding of compost maturity, stability, and overall health. Enzymes are secreted by the microbial community and are a direct indicator of the microbial metabolic functions currently active in the compost ecosystem. Enzymes integrate information about nutrient availability, microbial community composition, and overall metabolic function.

About the Author

Lynn Fang, MS, works on developing integrated ecological compost farming systems in the urban environment around Los Angeles, California, with community organizations Huerta del Valle, Integrative Development Initiative, and LA Compost. She received her master's on the use of compost in suppressing soilborne plant pathogens from the University of Vermont in 2015. She applies her background education in permaculture and biochemistry to composting and actively teaches and speaks on soil and compost ecology.

does appear to be more acceptance of the use of certain species in terms of their sensitivity to particular phytotoxins as well as for assessing the presence of persistent herbicides, and of course there is enormous inherent value in being methodical when testing for positive responses from compost.

Maturity and Bioassays

Bioassays have been a widely accepted indicator of compost maturity for some time; however, the research as to the accuracy of these tests appears to be mixed, and there remains a gap in standardized and easily replicable protocols. Garden cress (*Lepidium sativum* L.) is the most commonly used species, whose use dates back to studies in the early 1980s. However, several more recent studies have demonstrated that cress shows little sensitivity to compost immaturity,[3] although it is sensitive to salts.[4] A Woods End report recommends using 25 percent to 50 percent compost by volume mixed with peat, then measuring the biomass of the compost treatment compared with a commercial potting mix control.[5]

In a test of 14 different plant species, including cress, lettuce, carrot, and Chinese cabbage, Amino and Warman found *Amaranthus tricolor* to be the most sensitive in terms of compost maturity, and therefore the most promising.[6] While other indicators of maturity are probably more reliable measures (temperature, Solvita), some states require multiple maturity tests, and bioassays have the benefit of being DIY, albeit potentially inaccurate. Operators might make the best of their efforts by coupling them with other more valuable indicators such as vigor and persistent herbicides. Despite the lack of clarity in the scientific community, I stand by my belief that it is always better to grow plants than to not.

Phytotoxin Bioassays

Some plant species appear to be more sensitive to specific compounds that are potentially plant-toxic, including salts, ammonia, and volatile organic compounds. Effects can include delayed/reduced germination as well as general inhibition. A 2011 review of the use of bioassays in assessing phytotoxins found that barley appears sensitive to organic acids, lettuce and tomatoes to phenolic acids, and garden cress to salinity.[7]

Persistent Herbicide Bioassays

Herbicides that remain phytotoxic even after composting have come to be known as *persistent herbicides* (see *The Threat of Persistent Herbicides*, page 110). Plant tests have become the preferred method of operators for assessing the presence and potentially detrimental effects of these compounds in a particular compost. Lab testing (chemical) is expensive, and there is a lack of availability in testing with certain chemicals.

The herbicide that is currently the most widespread in compost is clopyralid. Trusted bioassays for this chemical have been around for some time. The best known is probably a protocol created for gardeners and researchers by Washington State University in 2002 using garden peas, which uses a ratio of 2 parts compost to 1 part peat as a test treatment.[8] A 2002 study found that pinto beans have a no observable effects level (NOEL) of 5.0 to 0.5 ppb for clopyralid and 0.5 to 0.05 for picloram, another herbicide of concern, meaning that these herbicides are damaging to certain species at very low levels. Legumes are the most susceptible, whereas tomatoes were found to have a NOEL of 5 to 50 ppb for both compounds. While this is much higher, soils amended with contaminated compost tested as high as 37 ppb picloram. Also of interest: The study noted that peas show damage at two weeks, while pinto beans show damage at three weeks.[9]

Unfortunately, clopyralid and picloram are not the only herbicides of concern and to my knowledge there is less published work on bioassays for two notable herbicides, aminopyralid and aminocyclopyrachlor. Anecdotally, bioassays with legumes including clover and fava beans have been found effective with aminopyralid, although there is little published research to date on the specifics of compost bioassays for the herbicide.

It's also worth noting that Woods End Laboratories, Inc., offers herbicide bioassays. They are a well-respected resource for composters in this area and one that I would recommend to any composter, especially if there is a known concern.

Plant Vigor Trials

I hope I have not overemphasized the need for phytotoxin testing at the expense of testing for positive attributes. Composters simply need to use their products. Taking a methodical approach, similar to the other assays discussed, will lead to more refined and consistent end products. If you have a particular end user in mind, do trials with that crop, both in-house and/or directly with that grower.

Testing for positive results also provides several opportunities that the toxin assays do not. You'll be able to see how your products perform with multiple crops and at multiple dilutions. Simultaneously, if you grow crops that you know have certain sensitivities to phytotoxins as discussed above, that could serve both purposes. You can also see how your product performs when compared with other products or even commercial fertilizers. Particularly if your goal is to create great added-value products such as seeding mixes, both lab and plant tests should help you emulate the best products on the market. It's at least good to know how your product ranks against similar, lesser, and potentially better products, and to know how it compares with completely different options such as commercial fertilizers.

A methodological approach to testing for positive response will almost certainly involve some qualitative observation, but it will also be beneficial to have more quantitative documentation: height, tissue weight, days to germination, leaf number, root density. Growers look for plant health not only after two weeks, but also at four weeks, six weeks, and so on. The longevity of a product's performance is highly valuable. There are any number of traits you could monitor and document. I recommend picking a few, sticking to them, and photo documenting.

Weed Seed Germination Testing

As a best management practice, compost that you intend to sell to growers, landscapers, or greenhouse producers should be tested for weed seed viability. A product that contains a significant level of viable weed seeds will cause problems for the purchaser and damage your reputation as a source of quality compost product.

Finished compost that has undergone an appropriate hot-composting process should contain few, if any, viable weed seeds. Weed seeds can contaminate finished compost during curing and storage, when no further heating takes place. Finished compost must be protected from weed seeds that can be brought in with raw materials, wind, birds, equipment, and similar factors.

Weed seed viability should be tested frequently (again I recommend testing 10 to 25 percent). Use the following procedure* to test for weed seeds:

1. Take a sample of at least 4 liters of compost from a curing pile, following randomized sampling methodology described in the *Compost Feedstock Sampling Protocol* sidebar, page 90.
2. Moisten the compost sample if necessary to 60 percent moisture.
3. Fill a seeding tray, preferably two, with the sample to a depth of about 1 to 1.5 inches and record the total liters used, along with identification of the compost batch and date.
4. Place the seed tray in a warm location with decent light, where the temperature is maintained at or above 70°F.
5. Maintain soil moisture. Before sprouting begins, moistened cloth or paper towels placed on the compost surface help maintain moisture level, as will a transparent seed tray cover.
6. Once sprouting begins, place the trays in full sun or under lights if necessary.
7. Maintain sprouting conditions for at least three weeks.
8. Count the total sprouts found and divide by the liters of compost used.
9. Record the results along with any observations, such as types of weeds that germinate.

As a control, follow the same protocol with native soil (skipping the counting part). The control should grow weeds, assuming you are keeping the soil moist and warm enough.

Weed Seed Germination Standards

Internationally, weed content tolerance ranges from 0.8 to 5.0 seeds per liter,[10] although any high-quality compost should have a policy of zero tolerance. Some European countries have a legal requirement to test commercial compost products for weed seed germination. Denmark's voluntary standard includes three content levels, which provide a good benchmark for the industry at large.[11]

Very low. Less than 0.5 seed and plant part per liter.
Noticeable content. Up to 2.0 seeds and plant parts per liter.
Large (high) content. More than 2.0 seeds and plant parts per liter.

* Adapted from the Highfields Center for Composting with permission from the Vermont Sustainable Jobs Fund.

CHAPTER FOURTEEN

Compost End Uses and Markets

actinomycete
(ˌaktənōˈmīˌsēt)
DEFINITION: A group of bacteria that form long, whitish, filamentous chains that can be seen by the naked eye and that are responsible for the rich, earthy aroma associated with mature compost and rich soils.

It's appropriate that the final chapter in the book is devoted to compost's end uses. Composters' literacy in their material's diverse applications is essential to composting's growth across the board and can increase sales and encourage targeted uses. While composters need partners in advancing the use of compost, we also need to be confident advocates of its value, because nobody knows the products better. Luckily, for most people knowing the value of something makes it easy to promote, and compost is a product that has value.

Rarely in my work do I get to focus on the product side of the operations that I work with, so in writing this chapter I took the opportunity to increase my own literacy. I've seen plenty of compost in action, and when it came to compost's unique ability to repair and fortify soil, I drank the proverbial Kool-Aid a long time ago. Honestly, though, my own knowledge lacked detail, so at the outset I realized that I needed to ask a fundamental question: Is compost really as great as I think it is?

Looking back, maybe I should have started the book by answering that question, but after researching and writing this chapter I can say definitively: Yes, compost is truly awesome. But it's important for you to decide that for yourself, as well as to recognize that we still have a lot to learn about compost's benefits and how to increase its adoption in virtually every potential sector of end user. You'll be better both at making your product and at getting it out there if you see, believe in, and understand its efficacy. Some readers will make compost for direct use, while others will support their operation at least in part with compost sales. Regardless of the value that is yielded from the operation, I strongly urge you to become intimately familiar with your products and with compost's many useful applications.

The first part of this chapter is devoted to discussing compost's applications and benefits (that is, to answering the question posed above). The second part is geared toward those who will be marketing their compost. Community-scale composters are uniquely positioned within the compost market, with the ideals of sustainable, small scale, high quality, local, green, community sovereignty, and resiliency all creating a strong wind at the movement's back. Sustainable business principles tell us that economic profitability and living one's ideals are not at odds. These are actually key to making a strong and sustainable business case. Still, successfully building a composting business while being a living model for these ideals is no small feat.

Compost Applications and Benefits

End uses are built into many community-scale models from the start. With some farms, for example, composting is not simply a cost benefit or market analysis. It's just a given that the soil needs care in which composting plays a foundational role. A farmer feeding food scraps to 300 chickens and farming 2 acres of organic vegetables is mainly concerned about the sale of eggs and vegetables, regardless of whether or not they are composting. The small volume of compost produced on the farm may be sold, but it is just as often viewed as too valuable an amendment to send off the farm. If the farm uses significant amounts of potting mix, they may develop their own (one of their most expensive soil products). Offsetting real costs such as animal feed and potting mix may ultimately be the most valuable benefit.

Small farms, schools, community farms and gardens, particularly in urban regions where good soil and other inputs like manure are scarce: All of these scenarios can obtain value from composting without sales. On the whole, this is a benefit of community-scale models: Using compost where it's produced reduces transportation and time required for marketing and prevents a flood of compost on the market, which is a real concern especially for independent composters whose livelihoods depend on marketing their compost commercially.

In light of the many different motivations for composting, economic and otherwise, this final chapter looks at the benefits yielded from community-scale composting operations from a non-reductionist perspective, with cost offsets and compost sales revenues in mind, but also with an acknowledgment that compost's role on farms and in communities is extremely hard to fully quantify.

How Compost Works

Although no two compost products are quite the same, the underlying concept behind their use is essentially similar. Good compost is a mix of stable and complex organic matter, microbes and other bugs, macronutrients (NPK) and micronutrients/minerals, and water. All of these are also present in healthy soil, but can be diminished by disturbances such as tilling, overgrazing, fire, chemical applications, compaction, and removal of topsoil. By adding compost, these beneficial properties are bestowed on that soil. Compost can improve soil health and productivity practically anywhere humans are working with and disturbing the land.

Compost is different from most other soil amendments and fertilizers in two distinct ways. First, compost is abundant in beneficial soil life, which is a precursor to healthy soil and plants. Second, compost is packed with organic matter, which provides food and improves habitat for soil life and plant roots. A large number of other benefits stem from these two traits, many of which have yet to even be recognized or quantified. Other soil amendments and remediation techniques provide some of the same benefits, but compost is truly unique in terms of its diversity, speed, and effectiveness in improving soil quality.

That said, building healthy soil takes time no matter what the methodology, and the same holds true with compost applications. A great deal of the literature on compost use shows that the benefits from compost are not always immediate, and most show that the benefits increase over time. Even with a single application, the effects can be measurable well into the future. In one trial as much as half of the original organic matter applied as compost was still present five years after application; in another, 18 percent of the carbon applied as compost remained after seven years.[1]

In the following sections we'll look at some of the research on compost's applications and discuss some of the benefits, which include everything from reductions in disease and fertilizer use to increased water infiltration and retention.

Compost and Crop Productivity

Research on compost's effects on plant productivity usually compares composts to synthetic fertilizers in terms of cost per yield. Compost rarely competes with fertilizers on this point. Some yields

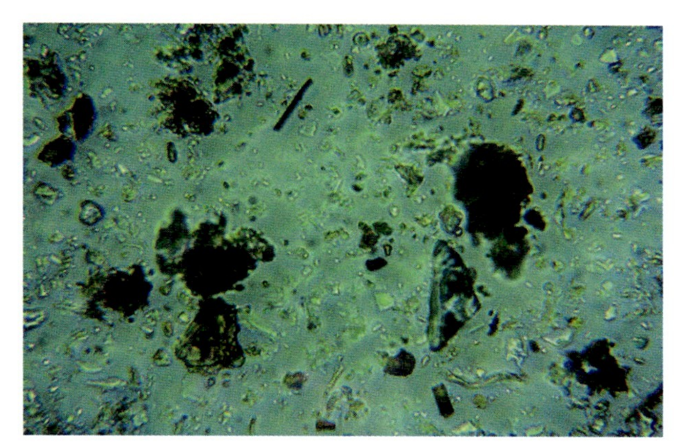

Figure 14.1. A control greenhouse soil (*left*) and a greenhouse soil inoculated with compost tea (*right*). Compost tea has many of the benefits of compost itself, particularly as a biological inoculant. Courtesy of Monique Bosch.

are comparable, and results of course depend on the cost of the compost and the time frame of the study. Given that compost has other qualities that fertilizers don't, comparable productivity is actually a net win for compost.

One approach to compost research that opens the door directly to wider adoption is on-farm trials. One such study in Ontario showed that while compost use appeared to have clear agronomic benefits, including increased crop yields in some cases, many farmers were unsure about the economics and therefore were most likely to use compost in targeted applications, where they trusted they would see the greatest return on investment. These included applying compost to the most denuded soils, as well as sandy soils, where the benefits from the compost were clearer than in other types, particularly in terms of its water-holding capacity.[2] Other studies have found similarly clear benefits on sandy soils. Even a relatively low application rate of 7.5 tons per hectare per year produced an increase in yield of greenhouse tomatoes while also improving soil quality (increased C, N, decreased soil bulk density).[3]

In another on-farm trial of fruit production in British Columbia, four years of poultry compost applications (5 to 10 tons per acre) based on N demands yielded similarly if not slightly higher than synthetic fertilizers applied at the same rate of N. Other benefits included increases in soil organic matter, beneficial nematodes, and a reduction of root knot nematodes.[4] Beet and carrot yields were higher with compost and fertilizer applications than with only fertilizer application in all three years of a study, at rates as low as 10 tons per acre of leaf compost.[5]

Numerous examples exist of farms that maintain productivity with compost or a combination of compost and both organic and non-organic soil management techniques. In addition, studies that also look at other aspects of soil and plant health often find benefits that are absent with straight fertilizer applications, which is an important takeaway. One of the biggest challenges, however, is to get farms to take the first step of trying compost.

Compost, Nitrogen Release, and Decreased Fertilizer Use

One well-documented characteristic of compost is that it releases nutrients slowly over time. Nitrogen in particular is slow to become plant-available. Below a C:N ratio of 15:1, decreasing C:N ratios correlate with an increase in plant-available nitrogen.[6] Remember that microbes temporarily tie up nitrogen in soil when there is excess carbon to metabolize in relation to the available nitrogen. Various studies show that the plant-available nitrogen (PAN) in composts varies widely, but generally PAN falls in the range of 1 to 9 percent of total nitrogen in the first year, again with lower C:N composts having higher PAN.[7] The remaining N will be mineralized slowly over time.

From a long-term perspective, having slow-release N in the proverbial soil bank is usually viewed as a net positive. But balancing available nutrients with crop needs is also a challenge in terms of promoting wider adoption: Many farms expect to see immediate yields from their investments, and it's hard to predict the optimal rate for application of both compost and supplemental amendments over time. Both PAN and phosphorus loading play a role in this (see *The Phosphorus/Nitrogen Relationship*, page 389). Despite the nuances, those who try compost and particularly those who invest in compost year after year should expect to use less amendments and fertilizers.

For example, one Ontario farmer who grows both fruits and grains reports that he has completely replaced fertilizers in his cherry and apple crops with a light annual application of compost. Other cash crop farmers using compost have found that by using soil testing, they have been able to target fertilizer applications based on demand, in some years reducing fertilizer use by 50 to 100 percent.[8] A study of tomato production in Connecticut using leaf compost showed similar results, with similar yields from a compost-only application and a 10–10–10 fertilizer application. The highest yields came with both compost and fertilizer, although a 50 percent application of fertilizer yielded the equivalent of the full fertilizer application.[9]

Compost and Disease Suppression

Another area where compost plays a potentially important role in the development of ecologically sound agriculture practices is in the suppression of pests and diseases. The general concept with disease suppression is what I think of as the "diversity promotes diversity" principle. Although the complexities of both the living and once living components of compost are practically unfathomable to me, the logic stands that incorporating that complex richness into the soil adds a natural resiliency that would otherwise not be present to the same degree. Although there is clearly a lot more work to be done in this area, the significant body of existing research bears out the benefits that farmers have seen for centuries regarding the connection between healthy soil and disease suppression.

The mechanisms of suppression can generally be understood as falling into one or more of several categories. These include:

- Improved plant health (that is, decreased susceptibility)
- Production of antibiotics and other compounds
- Activation of disease-suppressive genes
- Direct predation and parasitism
- Competition for food and other resources by beneficial organisms[10]

In a 2012 review of over 100 research papers relating to suppressive composts, Hader and Papadopoulou wrote, "Indisputably, suppression of soilborne plant pathogens by compost is a widespread and ubiquitous phenomenon."[11] They cite another review of over 250 studies of suppression of fungal pathogens by compost and other organic matter amendments. In this study 45 percent of cases were found to be suppressive, and 50 percent with compost applications.[12] Another study of peat-based potting mixes amended 20 percent by volume with 18 different composts found significant disease suppression in 54 percent of tests, which looked at seven plant pathogens including *Verticillium*, *Rhizoctonia*, *Phytophthora*, *Cylindrocladium*, and *Fusarium*.[13] In the nursery industry, composts are recognized as being as effective as fungicides at preventing root rots.[14]

A study comparing compost applications to the use of fumigation to control root lesion nematodes in new cherry tree plantings had interesting results. Fumigation was more effective in the first year, but by the second year, the treatments with 25 tons per acre of compost derived from manure and fruit pomace had fewer nematode infections, suggesting possible long-term advantages. If the trend lasts, it could lead to less frequent replanting. The cost of the compost was about 70 percent of fumigation.[15]

While the general relationship between compost use and disease suppression is well accepted, specific outcomes remain hard to predict. Many studies looking at other aspects of compost note reduced disease incidence as a side benefit. As with the other aspects of compost use, other than in certain specific applications, compost's disease-suppressive qualities are best viewed holistically.

Compost and Water Infiltration and Retention

Healthy soils are well recognized for their ability to infiltrate and retain, rather than shed, water. Organic matter and soil structure play a key role in these important traits. The science on the topic has been clear for several generations: The addition of soil carbon decreases bulk density (compaction) and runoff, while increasing aggregation and water-holding capacity.[16]

The impacts of organic matter content are highly variable by soil type, yet literature reviews of research on the subject from both 1981[17] and 2017[18] confirm a positive relationship between the use of organic matter applications and increased water-holding capacity.

One study of plots amended with 22 and 44 tons per acre of compost and continuously monitored for two years with moisture meters found maximum differences of 22 to 28 percent soil moisture in the higher treatment and 10 to 20 percent in the lesser treatment when compared with no compost treatment.[19] In a survey of 49 farms in western Washington State, 68 percent of respondents saw increased yields following compost applications, and 55 percent noted increased water retention during a drought in 2015.[20] In engineered soils, such as those used in stormwater bioretention basins, there are linear relationships among compost use, organic matter, and moisture retention.[21]

Plants obviously respond well to increases in water availability, particularly during dry spells. Building healthy soils will be a key adaptation strategy as extreme weather events increase in scale and frequency, a trend that is predicted to escalate with climate change. One study of tilled-in compost applications in compacted park soils found increased infiltration during short high-intensity rains to be 17 percent compared with the control, and 33 percent compared with tillage alone. Ksat values were 1.8 to 5.6 times those of the control.[22] During dry times, soil's ability to store water is extremely valuable, and during wet times our soils need the capacity to effectively infiltrate moisture.

Other Benefits

Crop yields, reduced fertilizer use, increased water infiltration and retention, disease suppression: These are surely not the only measurable and tangible benefits from compost use. Other studied and reported advantages include maintaining and sequestering soil carbon,[23] long-term mineralization (reduced loss and pollution from fertilizer),[24] and the introduction and increase of beneficial species such nematodes.[25]

While compost applications often yield visible and measurable results, many of the benefits are less tangible. One farmer whom I worked with in my early days as a compost technical service provider once told me that he wasn't sure he'd be able to afford to use much of the compost that he made. He knew that he could sell the compost for a good price, but didn't believe that people would pay more for a carrot grown in compost-amended soil. A year or two later, after he ended up using most of the compost he was making, he described to me how the land and carrots had a vibrancy that they'd never had before. The reality is that many people who work with soil understand that it needs care. They see the effects of neglect and are left with the choice to either abandon the land or rebuild the soil.

The direct benefits, and those stemming from them, are far too numerous to cover adequately in this chapter; nor do I claim that the research I've done here is even close to representing the full scope of the science. My goal is to provide a broad cross section of the research so that as composters, and compost advocates, we can have confidence in our products. Of course the value of our products is extremely hard to represent on paper, so my best advice is to grow, and get others to grow, because seeing is believing.

Many Applications for Compost

Continuously improving your understanding of the benefits of compost in specific applications will make you, as a compost ambassador and purveyor, more effective at communicating with and/or selling to potential end users. The most common applications for community-scale composters are:

Food production. Gardeners, farmers, community gardens, and farms
Horticulture. Nurseries, landscaping, cannabis cultivation, turf/lawns
Green infrastructure. Low-impact development (LID), rain gardens, green roofs, bio-swales, local soil remediation

Other common applications that are less relevant to readers here include animal bedding and landfill

Figure 14.3. Compost used as mulch on a well-kept landscape, rich in both biology and nutrients. Elegant looking compared with many mulches, which are often dyed and rarely come with the assurance of being weed- and disease-free.

Figure 14.2. A highly denuded soil in mid-repair. This Kauai soil has endured millions of years in one of the wettest places on earth, then sugarcane, then pineapple, and finally was managed as turf. Compost, cover cropping, mulching, and ground covers/alley cropping with N-fixing legumes were all important tools in restoring this soil. It is now a food forest with a view of one of the most beautiful coastlines on the planet.

cover (for very low-quality material). Creating products targeted for these applications is covered in *Common Compost Markets*, page 392.

People often need help understanding how to use compost products effectively. Sales is education. The next section looks at application rates and techniques.

When to Apply Compost

The best time to apply compost was yesterday, and the next best time is today. What I mean is that it takes time to build good soil, so the compost that you applied in the past is still benefiting you today. The compost you use today will benefit your soil from this point forward. The lag is real and well documented.

Fresh compost has more available food sources for bugs, so it is good to target compost applications with plantings, which usually means applying seasonally in spring or with crop cycles. For crops that you plant in fall, such as garlic, or that are already in the ground in spring, such as fruit trees, fall applications will be available as the plants first emerge the following spring.

Preserving and Feeding the Life in Compost

One of the most important things to remember when applying compost is that it's alive and will be most effective if that life is preserved. Don't drown it or suffocate it, which will cause it to go anaerobic. Avoid solar exposure by applying in cool weather (spring and fall), and incorporate it into the soil quickly if tilling is your intention. Prior to using, don't disturb good friable compost unnecessarily if you don't have to; fungi and other bugs have set up networks, a biological infrastructure if you will, and as a general rule it's good to minimize disturbing this. However, if the compost is wet and dense, it will probably help it to

Figure 14.4. Rain gardens, used to infiltrate, buffer, and purify stormwater, are an increasingly common form of green infrastructure. Healthy soil characteristics such as high organic matter and low bulk density are key to their functionality. In this rain garden water flows from the street through the stormwater grate on the right. It then passes through the vegetation and soil. Any overflow exits through grates at the ends of the garden.

get in and open up (turn) the compost before using it, because it's really hard to use when it's mucky and clumpy. Last, remember that while compost contains fertility in and of itself, think of its living component more as the intestines of the soil than the actual food source for plants. Once the compost is added, if you stimulate the bugs by giving them some food, such as a complete organic fertilizer, kelp, or green manure treatment, you compound the effectiveness of both the compost and the amendment. Crops really respond to this.

Compost Application Rates

There are some farmers and gardeners who grow in straight compost—I've encountered this with raised beds in greenhouses, for example—but this is uncommon, has some potential drawbacks, and is potentially wasteful (see the following section). The ideal volume of compost to apply requires calibration based on a number of factors.

First, you have to weigh what's beneficial against what's realistic. In situations where the available compost or budget for compost is low compared with the land base it will be applied to, any compost is better than none. Where the availability of compost is not the limiting factor, the application rate depends upon the soils. Denuded soils in need of intensive remediation could use as much as 2 to 4 percent by volume. (Ten tons per acre is about 1 percent by volume or a 1:100 ratio of compost to soil in terms of dilution.[26]) Soil tests will often tell you what nutrients are deficient, and you can apply based on meeting those deficiencies if you have an analysis of the compost. If nutrient overloading is a problem, apply based on the nutrient of concern, and amend with other organic supplements to fill the gap. Leguminous cover crops could supply N in a crop rotation, but the rate of N fixation is related to N depletion, so applying compost beforehand might not be as effective as applying it after as the crop gets incorporated.

Sometimes basing application rates on soil analysis just isn't practical, in which case calibration must be based on what you know of the soil. If it's healthy and has had multiple applications of compost in the past, a spreading depth as low as ⅛ inch can have a positive impact. One study of established raised bed applications showed little to no additional benefit at rates above 9 tons per acre,[27] supporting the observation that relatively small doses of compost are effective at maintaining productivity of already healthy soils. Common compost spreading rates for farms with decent soils are in the 10 to 15 tons per acre range, which—as you can see from table 14.1—equates to an approximate depth between ⅛ and ¼ inch of compost. If it's a new plot, highly disturbed soil, or the first time it's had a compost application, you can go heavy: ¾ to 3 inches, depending upon how you intend to prepare the soil (more if you will

Table 14.1. Compost Application Rates

Spread Depth	Application Rate (Tons per Acre)	Yards per Area								
		50′	100′	500′	1,000′	5,000′	10,000′	20,000′	½ Acre	1 Acre
⅛″	8.40	0.02	0.04	0.19	0.39	1.93	3.86	7.72	8.40	16.81
¼″	16.81	0.04	0.08	0.39	0.77	3.86	7.72	15.43	16.81	33.61
½″	33.61	0.08	0.15	0.77	1.54	7.72	15.43	30.86	33.61	67.22
¾″	50.42	0.12	0.23	1.16	2.31	11.57	23.15	46.30	50.42	100.83
1″	67.22	0.15	0.31	1.54	3.09	15.43	30.86	61.73	67.22	134.44
1½″	100.83	0.23	0.46	2.31	4.63	23.15	46.30	92.59	100.83	201.67
2″	134.44	0.31	0.62	3.09	6.17	30.86	61.73	123.46	134.44	268.89
3″	201.67	0.46	0.93	4.63	9.26	46.30	92.59	185.19	201.67	403.33

blend it or till it in). Out of necessity, farms will typically go much lower due to the large volumes and cost, but gardeners might not blink an eye at heavy applications. In many cases, home gardeners will just buy based on the units you sell (say, 1 yard). It keeps well and is pretty hard to waste!

Remember, too, that in gardens the square footage of the garden *beds* may only be half of the footprint (due to paths). End users may want to calculate the square footage of the beds if they have the ability to target the application, especially if it's a large area.

What Happens When You Apply Too Much Compost?

The obvious problems of nutrient overloading, imbalance, and conductivity related to salt accumulation are topics that are fairly well understood and covered elsewhere in the book (see *The Phosphorus/Nitrogen Relationship* and *Salts*, this page and page 82 respectively). There is much less clarity about what happens when you have too much organic matter. Organic matter contents between 3 and 5 percent are widely considered ideal. Anecdotally, I have been told that ranges above 5 to 8 percent can cause problems, although I was unable to find any convincing evidence to back up these claims, or to explain why too much organic matter might be a problem, other than the nutrient and salt issues already mentioned.

The story of ideal soil organic matter (OM) content is much more nuanced than a defined threshold, and its impacts and benefits seem to vary largely across soil types. In sandy soils, for instance, 2 percent OM is considered very good, while in clay 2 percent is considered greatly diminished. To form water-stable soil aggregates, 6 percent organic matter would be required in a soil that's 50 percent clay, versus only 2 percent in a soil that's 16 percent clay.[28] In certain applications higher percentages of organic matter are common. For example, some practitioners have advocated for the benefit of organic matter content as high as 50 percent in green roof media,[29] although that much OM would be considered very high, even for straight compost.

Nutrient Management Planning

One tool farmers use for managing soil fertility and health is nutrient management planning. Nutrient management plans are essentially an accounting tool used to monitor and maintain optimal soil fertility on farms. Classes and nutrient management planning support are often available through agricultural extension or National Resource Conservation Service (NRCS) offices.

The Phosphorus/Nitrogen Relationship

As is typically the case with manures, the natural ratio between phosphorus and nitrogen in compost is typically high in terms of the rate of crop plant uptake. While the plant available N varies between 1 and 9 percent immediately following application,[30] available P and K is in the range of 70 to 100 percent.[31] Therefore, applying compost to meet nitrogen demands overloads phosphorus,[32] while applying compost based on phosphorus demands leaves plants limited in nitrogen. Phosphorus overloading is a major pollution concern, primarily due to runoff. In freshwater and riparian ecosystems (rivers, lakes), phosphorus is limiting, and adding phosphorus creates algal blooms that lead to apoxic dead zones.[33] When P loading is an issue, apply compost based on phosphorus demand, then supplement with other sources of organic nitrogen that have low P concentrations; these include cottonseed meal, blood meal, feather meal, fish emulsion, crab meal, soybean meal, and Chilean nitrate.[34]

How to Apply Compost

Application methods for compost depend largely on the current state of the soil. Getting compost deep into existing soil requires disturbance. But typically, compost is used to remediate prior disturbance. So the question is: How disturbed is the soil already? Really denuded and neglected soils will benefit from intensive remediation. If not, minimizing further disturbance and topdressing might be a preferable approach.

Intensive Tilling and Mixing

In the words of Hendrikus Schraven, organic landscaping and farming guru, sometimes "you have to go in with a knife" to rehabilitate poor, denuded, and compacted soils. Originally I viewed deep digging, tilling, and mixing of compost into the soil as contradictory to the permaculture and no-till philosophy, but now I look at this practice as extremely complementary. What I learned from Schraven is that it's better to redisturb a soil one more time and get compost and other amendments as deep down as possible, in order to minimize the need for continuous disturbance in the future. The deeper the soil can be remediated, the better, because it benefits the plants for their roots to grow deep where they are able to access subsoil water and nutrients. In highly disturbed soils, plants and especially food crops need all the help they can get, because the soil needs time to repair itself, even with heavy intervention.

Many sites, such as community or school gardens, may have the ability to mix compost deeply into raised beds during construction or a deep soil remediation. Adding as much as 25 to 50 percent compost by volume into a poor soil and getting it really well mixed in is not unheard of in garden and raised bed applications. Farms might take a slightly slower approach depending upon the available equipment, combining compost and green manure application with crop rotations. With compacted soils, it may be beneficial to cultivate as deeply as possible initially.

Topdressing

With soils that have good structure, tilth, and organic matter, I'm a big advocate for topdressing with

Figure 14.5. Green roof media being applied to the Burnham Building (formerly Filene's) in downtown Boston, Massachusetts. The medium was conveyed using a high-powered soil blower for an extensive sedum roof that was planted by Recover Green Roofs. Courtesy of Recover Green Roofs.

compost and using no-till methods to minimize soil disturbance. Compost can be used like a mulch to reduce evaporation and weeds. Plant roots prefer to grow near the soil surface, so it's actually a very effective point of application.[35] Even a sprinkle of compost around a plant and/or mixed into the hole where a transplant is going can go a long way.

Topdressed compost benefits the soil, although it may take longer to notice the effects than with tilled-in compost. Somehow the organic matter, organisms, and nutrients have a way of making their way into the soil where the plants can access them. One of the mechanisms by which this is likely to occur is known as *bioturbation*, essentially the movement of soil and other particles in the environment by animals and plants. (So technically, composters are bioturbators, human animals facilitating the movement of organic matter from place to place.) Another bioturbator, the earthworm, literally pulls organic matter into its burrows.

Topdressing can be done by hand on a small scale. On a farm scale, manure spreaders are the common tool. There is also an array of applicators used mainly by landscaping professionals for turf applications. For landscaping and green roof applications, there are soil blower trucks that move an astonishing volume of pre-mixed medium through long tubes. These can also be used for hydro-mulch planting mixes, mulch, or straight-up compost.

When you're planting woody perennials, my recommendations in this section apply to soils in need of deep remediation. It's not unreasonable to dig a deep, wide hole, mix compost with soil, and then refill it. I also would encourage applying a layer of compost or a blend of compost and topsoil. Topdress to a depth of 1 to 4 inches around the drip zone of the tree for optimal long-term growth. As the plants mature, applications can be well outside of the drip zone of the tree, as tree roots sometimes make their way one or even two times the radius of the drip line.[36] If you plan to mulch, make sure that the compost is in direct contact with the plant's root zone, with the mulch on top. You want to inoculate that beneficial biology as close to the root zone as possible.

Compost Marketing and Sales

To repeat what I said at the start of the chapter, sales and marketing is not my specialty. However, I've worked with enough compost sites from start through deployment to know that compost distribution can make or break a compost site early on. It involves a unique set of skills and targeted interaction with potential end users. Having both fielded sales calls and loaded trucks, I also know how awesome it feels to spread that rich goodness around in your community.

The following sections discuss some of the key considerations in developing a successful marketing and sales strategy.

Developing a Marketing Plan

For operations where selling compost is the key to economic solvency, a marketing plan built on the foundation of the good work you are doing will be the cornerstone of your business. Experts in the compost industry generally recommend that start-up companies set aside 20 to 30 percent of their budget in the first year for marketing and sales. After becoming established, the marketing budget will be expected to decrease to around 5 to 10 percent.[37] If you plan to add value to your compost—through bagged products, for example—product development will add significantly to start-up costs. Sales are not everyone's strong point, but procrastinating with the marketing end of the business inevitably results in stagnant business growth. A good plan may need to include market and business development support if distribution is not something that is naturally built into your operation.

Expanding the Pie

If you look at the compost made and/or sold in a given region and think of that current market as a pie, the different compost products sold there constitute different-size slices of that pie. One of the concepts that Ron Alexander, compost marketing consultant and author of *The Practical Guide to Compost*

Marketing and Sales, talks about is how composters need to work on growing their own slice of the pie by growing the whole pie, not simply by competing for another composter's portion. Of course some competition is both healthy and inevitable, but with the rate of new compost coming onto the market in some areas, industry growth really can't be sustained without expanding the market as a whole. While the US Compost Council and other local organizations and trade groups have a role to play in promoting markets, composters need to be at the forefront. At the same time, we need to be pushing partners (such as states) to put their money where their mouth is, because compost adoption is going to make or break increased diversion goals.

Pricing and Competition

In the northeastern United States, the price of high-quality compost has risen consistently since 2008, when I first started to really pay attention. It's a trend that I expect holds true in most areas of the country and that I would expect to continue, as long as costs continue to rise. It's important to set your pricing realistically, but also not to undersell your compost. Know your *cost of production* and use that as a gauge. Research the pricing of compost in your area and also look at how composts of similar quality are priced elsewhere.

Recognize that not all products are alike and therefore they are not equally subject to the same forces of supply and demand. If you live in an area where most compost is low-grade and underpriced, don't try to compete on price; differentiate your product by selling higher and educating consumers about what makes your compost high-quality. Sure, major garden retailers may be selling mass-produced bagged products for dirt cheap, but high-quality products are selling for two, three, and even four times what major-branded bagged products are. By the yard, pricing of the highest-quality products can be equally as high in comparison with, say, municipal yard debris composts, which can be priced as low as free. Competition from large industrial and municipal composters can be intimidating to small producers, but I assure you, people want and will pay more for high-quality composts produced on a small scale in their community.

Common Compost Markets

Presently, compost markets for community-scale producers can be divided up into a few spaces: wholesale, bulk retail, and value-added products. Here I'll discuss a few of the considerations associated with these markets. Note that seasonality and pricing are approximations based on my observations in current markets. There are certainly exceptions, and in the spirit of expanding the market, composters should always look for opportunities to expand beyond the trends, disrupt markets (to use a very overused term), and adapt to the unique needs of their communities.

Bulk Wholesale Markets

SEASONALITY

- Around 80 percent early to mid-spring and about 20 percent summer and fall to garden centers
- Around 80 percent early to mid-spring and about 20 percent fall to farms
- Around 50 percent spring and about 50 percent summer and fall for landscaping and construction

PRICING

- 10 to 50 percent less than bulk retail

Selling bulk wholesale compost typically involves shipping large trucks (15 to 100 cubic yards) of screened compost to retailers, such as gardening centers, or to users who need quantity, such as farms, landscapers, and construction projects. Selling wholesale means dropping the price considerably, sometimes as much as 25 to 50 percent, depending upon the quantity and who's buying it. Retailers typically expect the largest markdown. The pricing of bulk wholesale markets can be a stretch for small producers, so while the ability to move a lot of compost all at once is intriguing, it won't be practical unless your costs are very low.

Bulk Retail

SEASONALITY

- Around 80 percent spring and about 20 percent summer and fall to home gardeners and small farms
- Around 50 percent spring and about 50 percent summer and fall for landscaping and construction

PRICING

- Research market and price competitively based on product quality. Use as a basis for pricing in other market sectors.

Bulk retail, or "by-the-yard," sales is one of the most common avenues for small composters, especially on-farm operations. For sites that have a loader and enough compost to sell, it's a really simple way to move compost out the door at a reasonable price for both the end user and the composter. The way it generally works is that customers stop by with a truck, into whose bed you will very carefully dump a yard or tractor bucket load of compost. It's also helpful to have a dump truck for making deliveries available. Customers will pay for delivery and will sometimes buy more compost if they are paying for delivery, to help spread out the cost of delivery. Make sure to build some service cost into your delivery price for scheduling and coordination.

A Note on Compost Brokers

Some composters move all or most of their compost through a broker or middleperson, who takes their products and redistributes them. It's an unlikely avenue for this book's main audience, but something that's good to be aware of. Creating localized aggregation and distribution networks for small-scale high-quality composts is a potential business opportunity, and could really support the movement at large. Think food hubs for compost. This might involve a truck, a mobile screening and bagging unit, and a marketing and distribution apparatus. These are all things that many small composters have a hard time investing in and sustaining themselves, so there is definitely a potential business niche here.

Value-Added Products

A range of composts and compost blends can be broadly classified as *value-added products*. In some cases these are products that are simply packaged in such a way that the compost sells for more. In others the product is enhanced in some way. Both types of value-added products are discussed here.

Common Value-Added Products

Common value-added products include potting mixes, amended topsoil and raised bed mixes, and vermicompost, among others.

POTTING MIX

Mixes for container growing are both high in demand and high in value. These include blends specially designed for seed starting as well as for larger pots, greenhouses, nurseries, and all manner of specialty applications. Both from direct experience and from conversations with growers, it is quite apparent that all potting mixes are not even remotely the same. I have literally purchased seed mixes in beautiful bags that have killed my seedlings within days of emergence. Growers rightly place huge value on their potting mixes, because so much of what happens in the field depends upon providing a good start to the plant's life. This is why many growers choose to buy their potting mix rather than producing it themselves.

Creating your own mix requires the ability to produce consistency, as well as the ability to experiment and test products both up front and on an ongoing basis. Of course, to make a compost-based mix also requires access to a mature and high-quality product. Compare growth tests of your mix with other products and ideally with the best other products.

Set benchmarks. See *Quality Control Standards* and *Plant Growth Trials*, pages 374 and 376 respectively.

I am no expert in creating potting mixes, but from the published recipes I have found for seeding and potting mixes, general compositions fall in the range of 20 to 33 percent compost, 40 to 80 percent peat, 0 to 25 percent perlite (some substitute sand), and 0 to 50 percent vermiculite.[38] Seeding mixes need good water retention and a fine texture, and seedlings are especially sensitive to salts, so low electrical conductivity (EC) and balanced nutrients are also critical.[39] Traditional soil-based mixes have included equal parts topsoil, sand, and peat.[40] These might be used for transplants. Some or all of the peat in potting mixes can be replaced with compost.[41] Some mixes also contain bark fines.

Most mixes also contain amendments. Mixes that include peat will need a buffer to adjust its low pH. Eliot Coleman's recipes employ a ratio of 240:1 brown peat to lime by volume, although Coleman recommends adjusting the recipe based on pH testing, because different products vary. He also lists several additional amendments including green sand, colloidal phosphate, and blood meal.[42] Other amendments/inoculants might include vermicompost, mycorrhizae, and trichoderma. One organism that I wish more products contained was fungus gnat nematodes, which effectively feed on the pesky little fungus gnats that sometimes take up residence in composts, vermicomposts, and manures. In theory, these could be inoculated into a mix or directly into the mature compost. They are especially a pest in indoor applications.

In the right balance many of the amendments used in organic farming could be used to boost the

Figure 14.6. These vegetables from the Boston Medical Center's rooftop farm feed patients at the hospital and are donated to local food kitchens.

effectiveness of potting mixes, so there is a lot of room for custom product development.

One of the questions producers of potting mix need to consider is what to do if any components require sterilization. Conventionally, soil additives were sterilized, although today most potting mixes are soil-free. Sterilizing compost or a mix containing compost would obviously be sacrilegious. Peat is considered acidic enough to be self-sterilizing. The main constituents I would be cautious about are plant-based substrates such as leaf mold or bark. These would ideally be heat-treated to compost standards to avoid having to cook them.

AMENDED TOPSOIL AND RAISED BED MIXES

There is a huge market for premixed topsoils and garden blends. These products could be composed of a lot of things, but typically they are a mix of what people here in Massachusetts call loooom (translation: silty loam, the soil particle sized between sand and clay, but more like a mix of sand and silty loam) and compost. Some blends might have some added sand; others are amended or inoculated in other ways. I've heard of landscapers growing specific plants in their mix before installation to encourage the presence of plant-specific mycorrhizal fungi.[43]

The basic calculation with producing these products is that by adding compost to straight-up soil, you increase the value of that soil dramatically. Soil can often be purchased relatively inexpensively, but after it's been blended it can go for close to the cost of straight compost. Mixes are typically around 25 percent compost by volume. It can be challenging to source weed- and seed-free soil, but it's important to try, which means that you're actually looking for a loamy subsoil (as opposed to a topsoil).

AMENDED COMPOST

Some composters make a "super" compost, a product that's been amended to balance out deficiencies or to give more of an immediate bang. These products are often used more as a fertilizer or amendment for top-dressing than as a base medium as a typical potting mix would be used. Amended compost can sell for more than straight compost (50-plus percent more) and shouldn't add a huge amount of cost, at least in terms of added materials.

VERMICOMPOST (WORM COMPOST) AND WORM CASTINGS

Worm composting is a cool (warm rather than hot) process that utilizes worms' decomposition capabilities to produce a product that is largely composed of their *castings* (excrement). Concentrated castings are marketed as such, whereas products with a higher diversity of other compounds are considered worm compost. The official names for worm composting's products are *vermicompost* and *vermicastings*. Since many laypeople don't know what that is, from a marketing perspective using the term *worm castings* or *worm compost* may appeal to a wider market. The prefix *vermi-* also sounds like *vermin*, which doesn't have the best associations.

Regardless of name, worm composts are widely regarded as superior to standard composts in a number of ways, and are often sold by the yard for 8 to 10 times the cost. With bagged worm composts the price per volume could be 20 times higher or more.

Like other composts, vermicompost promotes the growth of beneficial soil bacteria, increases soil aeration, improves texture, and so on. One study showed that plants grown in vermicompost as opposed to standard fertilizer and manure application showed all-around healthier features (longer roots, more branches, more flowers and pods, and so on), and even increased both free-living and legume-associated nitrogen-fixing bacteria.[44]

A worm farmer friend told me that currently the main vermicompost market is what he called the three G's: grapes (wine), grass (turf, golf courses), and ganja (*Cannabis*). All of these are high-return situations where growers can justify paying considerably more for the insurance that they are using the best product available.

HIGH-CARBON COMPOSTS

While most of this book focuses on creating and using high-nutrient composts (food-scrap-based), there is

Figure 14.7. A compost made primarily of wood grindings will be carbonaceous and highly porous, perfect for applications such as filter socks. At 136°F, though, this compost isn't quite ready for distribution.

also a growing need for low-nutrient, high-carbon composts for certain applications. In recent years regulations have been building upon successful use of compost in green infrastructure and low-impact development applications. The state of Washington's model program Soils for Salmon was one of the early pioneers in this area. It was initiated by the Washington Organic Recycling Council (WORC) and brought together stakeholders such as the state's DOT with the construction and landscaping industries.[45]

These policies promise to increase the market for compost and, in particular, this distinct grade of high-quality compost. High-carbon composts have the ability to filter and absorb nutrients, pollutants, and bacteria in stormwater runoff. They should be porous enough to slow but not stop the flow of water, allowing sediment to settle and filtration/treatment to take place. Common applications for these composts include compost filter socks (Filtrexx and similar products), bank stabilization, rain gardens (vegetated stormwater treatment areas), bio-swales, and vegetated buffers. All of these systems are also common for managing leachate and stormwater treatment on compost sites.

In systems where stormwater flows directly to drainpipes and waterways, high-carbon composts are preferable to higher-nitrogen composts. Companies that create and install these products specify standards for specific applications, including particle size distribution, pH, salts, and physical contaminant tolerance. Interestingly, carbon to nitrogen ratio and phosphorus are currently not inherent in these specs. As C:N ratios decrease, nutrient availability increases (see *Compost, Nitrogen Release, and Decreased Fertilizer Use*, page 383), so it is safe to assume that higher C:N ratio composts will perform better in terms of reducing pollution.

GREEN ROOF MEDIA

Particularly in urban areas, installation of living roofs, or *green roofs*, has been expanding rapidly in recent years. When soil and plants are added to roofs, stormwater runoff is reduced and filtered, cooling costs are reduced, and local biological diversity is increased. They are also beautiful, and I think that they play a vital role psychologically, reestablishing the connection between our built and natural environments.

Green roofs come in four main forms:

Extensive roofs. Two to 3 inches of soil media, typically planted with the hardy alpine plant sedum.

Intensive roofs. Much deeper soil media, planters, a wide varieties of plants including woody shrubs and trees, landscaping similar to that typically found at ground level.

Rooftop farms. Raised beds, containers, crates, greenhouses, a mix of media types including potting mix, typical green roof media, and standard composts for amendments. The main target with a green roof media is to make it lightweight, drainable, and stable in terms of nutrient loss (say, to pollution).

Blue roofs are designed to temporarily store rainwater on roofs as a buffer for stormwater runoff. These can also include plants and soil media, but not always.

Extensive and intensive roof media typically are composed of puffed shale, compost, and sand.

Producing media for this distinct market requires some industry savvy, but it may be something that's worthy of exploration. Typically green roofs are sold as a system, and a "certified blender" produces the media to a spec for the system. A composter might become certified as a blender or might sell bulk compost to a blender. I can tell you from firsthand experience that good clean media is highly valuable, and poor media is costly.

Rooftop farms are somewhat different in that they are not typically sold as a system in the same way as other roofs. Therefore the farmer/installer has some leeway in terms of what soils to use. Still, a lightweight blended product is highly desirable. Due to its potency and high levels of biology, I view worm compost as a likely candidate for green roofs. In general, the less volume and weight that needs to be transported onto a roof over time, the better.

COMPOST-AMENDED BIOCHAR

Biochar is a high-carbon charcoal product produced through a process known as pyrolysis. It has a variety of desirable traits as a long-term soil amendment. Like compost, biochar is produced from biomass, but unlike compost it's produced at temperatures that make it biologically sterile. For this reason, it is often suggested that composts could be used as an efficient way to reintroduce microbial life to biochar products.

I have come across a number of compost-amended biochar products in stores in recent years. Since these are relatively new products, it is unclear how big the market is or what a widely marketable price

Figure 14.8. Farming in a more typical compost-amended green roof medium at the Brooklyn Grange Rooftop Farm. Note the expanded shale.

Figure 14.9. A vermicompost-inoculated biochar.

point is. It is safe to assume that as with other unique compost-related products, significant research and education will be required on end use.

Bagged Products

Whether we like it or not, we can get a lot more money for compost when we put it in a little bag. The bag allows us to brand and communicate a message about the product, which in itself adds value. Also, as with most products, there's a price per volume increase inherent in selling smaller quantities. In table 14.2 you can see a few examples of bagged compost pricing from some unnamed brands available in the northeastern United States in 2017. These prices are easily four times what the composts would retail for by the yard, and in many cases they come in even smaller bags that are marked up even more.

Typically composts are bagged in either woven poly or polyethylene (PE) film bags. Both can be purchased with printed labels on them. Smaller quantities are sometimes packed in plastic-lined paper bags as well. When you're choosing a bag, the moisture content of the compost product you'll be bagging is a big consideration. Woven poly bags are in some ways preferable because they breathe, which is usually considered an attribute, even when it comes to very mature compost. But moisture also wicks through woven poly, so wet composts can stain and saturate the bags, which is unflattering on a store shelf. This is one of the main reasons why people choose PE bags. They are pretty watertight, which may not be as great for the compost, but these preserve the branding, which is extremely valuable from a retail perspective.

There are a wide array of bagging systems available, with a large spectrum of efficiencies and automation capabilities. Many small sites start by hand bagging to test the market. This involves nothing more than a large funnel or small hopper, a measuring implement, bags, and a stitcher or heat sealer. (Woven bags are stitched, while PE bags are heat-sealed.) Even small baggers are not cheap, but I've seen them on the market for as low as $5,000 (Weaverline); these claim you can bag 100 bags per hour with one person, and more with two.

SUPER SACKS (SLING BAGS)

Some composters transport their product in large woven poly bags called *super sacks* or *slings*. These are designed to be moved on pallets and can also be strapped and lifted from above with the right equipment. Slings are most commonly used for value-added products, but can be used for straight-up compost or any product that would be hard to transport otherwise. Super sacks typically hold 1 to 2 cubic yards and are commonly used for transporting compost products to greenhouses, green roofs, and other places where handling loose material is impractical.

FORTIFIED COMPOSTS

Composts that have been inoculated with specific strains of beneficial microbes (biological controls) have been called *fortified composts*. This has yet to become a common product category, but it is something that has been talked and dreamed about for

Table 14.2. 2017 Bagged Compost Pricing

Product	Organic Use	Bag Size	Bags per Yard	Bag Retail Price	Retail Price per Yard
Local worm castings	Yes	20 quarts	34.7	$29.99	$1,040.65
Local amended compost	Yes	20 quarts	34.7	$17.99	$624.25
Local composted manure	Yes	20 quarts	34.7	$12.00	$416.40
Local composted cow manure	Yes	1 cu. ft.	27.0	$5.99	$161.73
Seafood/berry compost and peat	Yes	1 cu. ft.	27.0	$10.99	$296.73
Nationally branded compost	Yes	1 cu. ft.	27.0	$8.59	$231.93

Source: Adapted from and used courtesy of Compost Technical Services, LLC.

Figure 14.10. Recover Green Roofs cranes super sacks of green roof media onto the Whole Foods rooftop farm in Lynnfield, Massachusetts, in spring for use as a soil amendment.

Figure 14.11. Mary McSweeney, age three, displays a worm nest we dug up in our yard. It's essentially a moist tunnel of leaves. Bioturbation at work.

some time. Hader and Papadopoulou reported on the development of composts fortified with strains of *Bacillus subtilis, Trichoderma hamatum, Verticillium biguttatum,* and *Flavobacterium balustinum,* as well as other cyanobacterial/bacterial cultures. Some of these products have been found to have broad as well as specific disease-suppressive effects.[46] In one study *Bacillus subtilis,* which is effective in the control of *Rhizoctonia* spp., was cultured in composted grass clippings. The procedure they describe entailed a very hot stage at 176°F, which is essentially a microbe-mediated sterilization phase, followed by inoculation and a three-day mesophilic stage, followed by an extended thermophilic phase at more ideal temps for maintaining high microbial populations.[47]

To my knowledge these types of fortified products have yet to be mass-produced or marketed, although I'd be very surprised if nothing like this existed on the market. The concept does at least appear to hold water, so who knows? In 50 years we could be asking ourselves, "Remember when compost was just compost?"

I would like to salute
the ashes of American flags
and all the falling leaves
filling up shopping bags.

—"Ashes of American Flags,"
sung by Wilco

APPENDIX A

Compost Systems Overview, Capacity, and Requirement Tables

Source: Adapted from James McSweeney and Brenda Platt, *Micro-Composting: A Guide to Small Scale and On-Site Food Scrap Composting Systems* (Washington, DC: Institute for Local Self-Reliance, forthcoming), courtesy of the Institute for Local Self-Reliance.

Table A.1. Compost Systems Overview (< 1,000 Pounds per Week)

	Company Name	Model Name/Number	Composting Method	Batch / Continuous Flow
≤ 500 Pounds per Week	Jora Compost Canada	NE 401	In-vessel, rotating drum	Batch
	Sustainable Ag Technologies Inc.	Model 5x8	Vermicomposter	Continuous flow
	Green Mountain Technologies	The Earth Cube	In-vessel, containerized/enclosed	Batch
	Food Waste Experts	Ridan Medium	In-vessel, auger-turned	Continuous flow
500–1000 Pounds per Week	Food Waste Experts	A500 Rocket Composter	In-vessel, auger-turned	Continuous flow
	Food Waste Experts	Ridan Large	In-vessel, auger-turned	Continuous flow
	Green Mountain Technologies	Earth Tub	In-vessel, auger-turned	Batch
	EC ALL Ltd.	Big Hanna T60	In-vessel, rotating drum	Continuous flow
	Vertal Inc.	CityPod model S	In-vessel, rotating drum	Continuous flow
	Jora Compost Canada	NE 20T	In-vessel, auger-turned	Batch, continuous flow
	Food Waste Experts	A700 Rocket Composter	In-vessel, auger-turned	Continuous flow

Table A.2. Compost Systems Overview (1,000–4,000 Pounds per Week)

	Company Name	Model Name/Number	Composting Method	Batch / Continuous Flow
1,000–2,000 Pounds per Week	EC ALL Ltd.	Big Hanna T120	In-vessel, rotating drum	Continuous flow
	O2Compost	Micro-Bin Compost System	Aerated static pile (ASP), bins/bays	Batch
	Vertal Inc.	CityPod model M	In-vessel, rotating drum	Continuous flow
1–2 Tons per Week	Food Waste Experts	A900 Rocket Composter	In-vessel, auger-turned	Continuous flow
	FOR Solutions	Model 500	In-vessel, rotating drum	Batch, continuous flow
	EC ALL Ltd.	Big Hanna T240	In-vessel, rotating drum, containerized/enclosed	Continuous flow
	Eco Value Technology Inc.	EVT-412B	In-vessel, rotating drum	Batch
	Eco Value Technology Inc.	EVT-412	In-vessel, rotating drum	Continuous flow
	Global Composting Solutions Ltd.	HotRot 1206	In-vessel	Continuous flow

At Maximum Capacity				External System Requirements
Maximum Food Scrap Processing Capacity (Pounds per Week)	System Retention Time (Days)	Additional Time Required to Finish Compost (Days)	Base Price per Unit (2016)	
88	56	0	$530.00 (varies with exchange rate)	Stable level surface
120	90	0	$5,185.00	Electrical hookup, cover (e.g., roof), concrete pad, heated space, water
245	21	60–90	$2,995.00	Stable level surface
260	14–21	28–90	$5,900.00	Stable level surface
400	14	14–21	$21,500.00	Electrical hookup, stable level surface, cover (e.g., roof)
525	14–21	28–90	$7,900.00	Stable level surface
525	21	30–60	$9,975.00	Electrical hookup, stable level surface
550	42–70	≥ 0	$45,000.00	Electrical hookup, stable level surface, cover (e.g., roof)
700	42	7	$35,000.00	Electrical hookup, stable level surface, cover (e.g., roof), water, ventilation piping
700	28	0	$41,000.00 (varies with exchange rate)	Electrical hookup, heated space (e.g., heated building)
900	14	14–21	$31,500.00	Electrical hookup, stable level surface, cover (e.g., roof)

At Maximum Capacity				External System Requirements
Maximum Food Scrap Processing Capacity (Pounds per Week)	System Retention Time (Days)	Additional Time Required to Finish Compost (Days)	Base Price per Unit (2016)	
1,100	42–70	≥ 0	$55,000.00	Electrical hookup, stable level surface, cover (e.g., roof)
1,200	21–30	30–60	$1,500.00	Electrical hookup, stable level surface, separate mixing pad or area, water
1,500	42	7	$45,000.00	Electrical hookup, stable level surface, cover (e.g., roof), water, ventilation
2,300	14	14–21	$46,500.00	Electrical hookup, stable level surface, cover (e.g., roof)
2,500	5	0–21	$137,500.00	Electrical hookup, concrete pad, cover (e.g., roof)
2,600	42–70	≥ 0	$84,000.00	Electrical hookup, stable level surface, cover (e.g., roof)
3,400	4–7	20–40	$15,000	Electrical hookup, stable level surface
3,400	4–7	20–40	$40,739	Electrical hookup, stable level surface
3,800	10	30–42	$100,000.00	Electrical hookup, stable level surface, separate mixing pad or area, water, biofilter, bin lifter

Table A.3. Compost Systems Overview (2–5 Tons per Week)

	Company Name	Model Name/Number	Composting Method	Batch / Continuous Flow
2–5 Tons per Week	Food Waste Experts	A1200 Rocket Composter	In-vessel, auger-turned	Continuous flow
	FOR Solutions	Model 1000	In-vessel, rotating drum	Batch, continuous flow
	EC ALL Ltd.	Big Hanna T480	In-vessel, rotating drum	Continuous flow
	Eco Value Technology Inc.	EVT-424	In-vessel, rotating drum	Continuous flow
	Green Mountain Technologies	Earth Flow—EF16	In-vessel, agitated bed or vessel	Batch, continuous flow
	Green Mountain Technologies	Earth Flow—EF20 Intermodel	In-vessel, agitated bed or vessel	Batch, continuous flow
	Green Mountain Technologies	Earth Flow—EF20	In-vessel, agitated bed or vessel	Batch, continuous flow
	Eco Value Technology Inc.	EVT-432	In-vessel, rotating drum	Continuous flow
	Green Mountain Technologies	Earth Flow—EF24	In-vessel, agitated bed or vessel	Batch, continuous flow
	FOR Solutions	Model 2000	In-vessel, rotating drum	Batch, continuous flow
	Sustainable Generation	SG Mini System	Aerated static pile (ASP)	Batch

Table A.4. Compost Systems Overview (> 5 Tons per Week)

	Company Name	Model Name/Number	Composting Method	Batch / Continuous Flow
> 5 Tons per Week	Eco Value Technology Inc.	EVT-440	In-vessel, rotating drum	Continuous flow
	Green Mountain Technologies	Earth Flow—EF30	In-vessel, agitated bed or vessel	Batch, continuous flow
	Eco Value Technology Inc.	EVT-532	In-vessel, rotating drum	Continuous flow
	Green Mountain Technologies	Earth Flow—EF40 Intermodel	In-vessel, agitated bed or vessel	Batch, continuous flow
	Green Mountain Technologies	Earth Flow—EF40	In-vessel, agitated bed or vessel	Batch, continuous flow
	Eco Value Technology Inc.	EVT-540	In-vessel, rotating drum	Continuous flow
	Global Composting Solutions	HotRot 1811	In-vessel	Continuous flow
	FOR Solutions	Model 4000	In-vessel, rotating drum	Batch, continuous flow
	Green Mountain Technologies	Earth Flow—EF53 Intermodel	In-vessel, agitated bed or vessel	Batch, continuous flow
	Green Mountain Technologies	Earth Flow—EF5010	In-vessel, agitated bed or vessel	Batch, continuous flow
	Agrilab Technologies Inc.	Hot Skid 250R	Aerated static pile (ASP)	Batch

At Maximum Capacity				External System Requirements
Maximum Food Scrap Processing Capacity (Pounds per Week)	System Retention Time (Days)	Additional Time Required to Finish Compost (Days)	Base Price per Unit (2016)	
4,625	14	14–21	$95,500.00	Electrical hookup, stable level surface, cover (e.g., roof)
5,000	5	0–21	$200,000.00	Electrical hookup, concrete pad, cover (e.g., roof)
5,300	42–70	≥ 0	$154,000.00	Electrical hookup, stable level surface, cover (e.g., roof)
6,800	4–7	20–40	$47,760.00	Electrical hookup, stable level surface
7,000	21	30–60	$59,980.00	Electrical hookup, stable level surface
7,000	21	30–60	$59,960.00	Electrical hookup, stable level surface
8,400	21	30–60	$69,950.00	Electrical hookup, stable level surface
9,307	4–7	20–40	$52,441.00	Electrical hookup, stable level surface
9,800	21	30–60	$79,895.00	Electrical hookup, stable level surface
10,000	5	0–21	$250,000.00	Electrical hookup, concrete pad
10,000	28–56	30–60	$9,000.00	Electrical hookup, stable level surface, separate mixing pad or area

At Maximum Capacity				External System Requirements
Maximum Food Scrap Processing Capacity (Pounds per Week)	System Retention Time (Days)	Additional Time Required to Finish Compost (Days)	Base Price per Unit (2016)	
11,815	4–7	20–40	$57,121.00	Electrical hookup, stable level surface
12,600	21	30–60	$99,895.00	Electrical hookup, stable level surface
14,875	4–7	20–40	$66,478.00	Electrical hookup, stable level surface
15,400	21	30–60	$74,975.00	Electrical hookup, stable level surface
16,800	21	30–60	$119,900.00	Electrical hookup, stable level surface
18,530	4–7	20–40	$72,758.00	Electrical hookup, stable level surface
19,000	10	30–42	$186,000.00	Electrical hookup, stable level surface, separate mixing pad or area, water, biofilter, bin lifter
20,000	5	0–21	$375,000.00	Electrical hookup, concrete pad
22,400	21	30–60	$87,875.00	Electrical hookup, stable level surface
26,600	21	30–60	$144,985.00	Electrical hookup, stable level surface
50,000	21	60–270	$55,000.00	Electrical hookup, stable level surface, separate mixing pad or area, bucket loader

Table A.5. Compost Systems Capacity Information (< 1,000 Pounds per Week)

	Company Name	Model Name/Number	Maximum Food Scrap Processing Capacity (Pounds per Week)	Minimum Food Scrap Processing Capacity (Pounds per Week)
≤ 500 Pounds per Week	Jora Compost Canada	NE 401	88	88
	Sustainable Ag Technologies Inc.	Model 5x8	120	40
	Green Mountain Technologies	The Earth Cube	245	0
	Food Waste Experts	Ridan Medium	260	65
500–1,000 Pounds per Week	Food Waste Experts	A500 Rocket Composter	400	100
	Food Waste Experts	Ridan Large	525	100
	Green Mountain Technologies	Earth Tub	525	70
	EC ALL Ltd.	Big Hanna T60	550	330
	Vertal Inc.	CityPod model S	700	400
	Jora Compost Canada	NE 20T	700	700
	Food Waste Experts	A700 Rocket Composter	900	225

Table A.6. Compost Systems Capacity Information (1,000–4,000 Pounds per Week)

	Company Name	Model Name/Number	Maximum Food Scrap Processing Capacity (Pounds per Week)	Minimum Food Scrap Processing Capacity (Pounds per Week)
1,000–2,000 Pounds per Week	EC ALL Ltd.	Big Hanna T120	1,100	660
	O2Compost	Micro-Bin Compost System	1,200	n/a
	Vertal Inc.	CityPod model M	1,500	700
1–2 Tons per Week	Food Waste Experts	A900 Rocket Composter	2,300	575
	FOR Solutions	Model 500	2,500	1,000
	EC ALL Ltd.	Big Hanna T240	2,600	900
	Eco Value Technology Inc.	EVT-412B	3,400	0
	Eco Value Technology Inc.	EVT-412	3,400	0
	Global Composting Solutions Ltd.	HotRot 1206	3,800	1,500

Volume Ratio of Food Scraps : High-Carbon Feedstocks	System Retention Time (Days)	Additional Time Required to Finish Compost (Days)	# of Units Required	System Hits ≥ 131°F Reliably Year-Round	Estimated Yards Compost Produced (@ Max Capacity)
7:1–9:1	56	0	Each unit manages 2 batches	Yes	~ 40% of inputs
1:4	90	0	n/a	No	12
1:1	21	60–90	≥ 1	Yes	9–12
1:1–6:1	14–21	28–90	n/a	Yes	12
1:1	14	14–21	n/a	Yes	21
1:1–6:1	14–21	28–90	n/a	Yes	24
1:1	21	30–60	1-2	Yes	18
5:1–16:1	42–70	≥ 0	n/a	Yes	6
1:6	42	7	n/a	Yes	20% of input
7:1–9:1	28	0	Each unit manages 2 batches	Indoors	~ 40% of inputs
1:1	14	14–21	n/a	Yes	46

Volume Ratio of Food Scraps : High-Carbon Feedstocks	System Retention Time (Days)	Additional Time Required to Finish Compost (Days)	# of Units Required	System Hits ≥ 131°F Reliably Year-Round	Estimated Yards Compost Produced (@ Max Capacity)
5:1–16:1	42–70	0	N/A	Yes	11
1:3	21–30	30–60	3–4 bins typically make up a single system	Yes	85–100
1:6	42	7	n/a	Yes	20% of input
1:1	14	14–21	n/a	Yes	119
4:1	5	0–21	n/a	Yes	130
5:1–16:1	42–70	≥ 0	n/a	Yes	26
1:1–1:4	4–7	20–40	1 unit	Yes	209–365
1:1–1:4	4–7	20–40	n/a	Yes	209–365
1:1	10	30–42	n/a	Yes	250–300

Table A.7. Compost Systems Capacity Information (2–5 Tons per Week)

	Company Name	Model Name/Number	Maximum Food Scrap Processing Capacity (Pounds per Week)	Minimum Food Scrap Processing Capacity (Pounds per Week)
2–5 Tons per Week	Food Waste Experts	A1200 Rocket Composter	4,625	1,150
	FOR Solutions	Model 1000	5,000	1,000
	EC ALL Ltd.	Big Hanna T480	5,300	1,750
	Eco Value Technology Inc.	EVT-424	6,800	0
	Green Mountain Technologies	Earth Flow—EF16	7,000	1,000
	Green Mountain Technologies	Earth Flow—EF20 Intermodel	7,000	1,000
	Green Mountain Technologies	Earth Flow—EF20	8,400	1,000
	Eco Value Technology Inc.	EVT-432	9,307	0
	Green Mountain Technologies	Earth Flow—EF24	9,800	1,000
	FOR Solutions	Model 2000	10,000	1,000
	Sustainable Generation	SG Mini System	10,000	0

Table A.8. Compost Systems Capacity Information (> 5 Tons per Week)

	Company Name	Model Name/Number	Maximum Food Scrap Processing Capacity (Pounds per Week)	Minimum Food Scrap Processing Capacity (Pounds per Week)
> 5 Tons per Week	Eco Value Technology Inc.	EVT-440	11,815	0
	Green Mountain Technologies	Earth Flow—EF30	12,600	1,000
	Eco Value Technology Inc.	EVT-532	14,875	0
	Green Mountain Technologies	Earth Flow—EF40 Intermodel	15,400	1,000
	Green Mountain Technologies	Earth Flow—EF40	16,800	1,000
	Eco Value Technology Inc.	EVT-540	18,530	0
	Global Composting Solutions	HotRot 1811	19,000	7,700
	FOR Solutions	Model 4000	20,000	1,000
	Green Mountain Technologies	Earth Flow—EF53 Intermodel	22,400	1,000

Volume Ratio of Food Scraps : High-Carbon Feedstocks	System Retention Time (Days)	Additional Time Required to Finish Compost (Days)	# of Units Required	System Hits ≥ 131°F Reliably Year-Round	Estimated Yards Compost Produced (@ Max Capacity)
1:1	14	14–21	n/a	Yes	240
4:1	5	0–21	n/a	Yes	260
5:1–16:1	42–70	≥ 0	n/a	Yes	53
1:1–1:4	4–7	20–40	n/a	Yes	417–730
1:2	21	30–60	n/a	Yes	182
1:2	21	30–60	n/a	Yes	182
1:2	21	30–60	n/a	Yes	218
1:1–1:4	4–7	20–40	n/a	Yes	571–999
1:2	21	30–60	n/a	Yes	255
4:1	5	0–21	n/a	Yes	520
1:2–1:3	28–56	30–60	≥ 2	Yes	1 ton input yields 1 cubic yard output

Volume Ratio of Food Scraps : High-Carbon Feedstocks	System Retention Time (Days)	Additional Time Required to Finish Compost (Days)	# of Units Required	System Hits ≥ 131°F Reliably Year-Round	Estimated Yards Compost Produced (@ Max Capacity)
1:1–1:4	4–7	20–40	n/a	Yes	725–1,268
1:2	21	30–60	n/a	Yes	328
1:1–1:4	4–7	20–40	n/a	Yes	912–1,597
1:2	21	30–60	n/a	Yes	400
1:2	21	30–60	n/a	Yes	437
1:1–1:4	4–7	20–40	n/a	Yes	1,137–1,989
1:1	10	30–42	n/a	Yes	2,350
4:1	5	0–21	n/a	Yes	1,040
1:2	21	30–60	n/a	Yes	582

Table A.9. Compost System Requirements (< 1,000 Pounds per Week)

	Company Name	Model Name/Number	Volume Ratio of Food Scraps : High-Carbon Feedstocks	Minimum External System Requirements
≤ 500 Pounds per Week	Jora Compost Canada	NE 401	7:1–9:1	Stable level surface
	Sustainable Ag Technologies Inc.	Model 5x8	1:4	Electrical hookup, cover (e.g., roof), concrete pad, heated space, water
	Green Mountain Technologies	The Earth Cube	1:1	Stable level surface
	Food Waste Experts	Ridan Medium	1:1–6:1	Stable level surface
500–1,000 Pounds per Week	Food Waste Experts	A500 Rocket Composter	1:1	Electrical hookup, stable level surface, cover (e.g., roof)
	Food Waste Experts	Ridan Large	1:1–6:1	Stable level surface
	Green Mountain Technologies	Earth Tub	1:1	Electrical hookup, stable level surface
	EC ALL Ltd.	Big Hanna T60	5:1–16:1	Electrical hookup, stable level surface, cover (e.g., roof)
	Vertal Inc.	CityPod model S	1:6	Electrical hookup, stable level surface, cover (e.g., roof), water, ventilation piping
	Jora Compost Canada	NE 20T	7:1–9:1	Electrical hook-up, heated space (e.g., heated building)
	Food Waste Experts	A700 Rocket Composter	1:1	Electrical hookup, stable level surface, cover (e.g., roof)

Table A.10. Compost System Requirements (1,000–4,000 Pounds per Week)

	Company Name	Model Name/Number	Volume Ratio of Food Scraps : High-Carbon Feedstocks	Minimum External System Requirements
1,000–2,000 Pounds per Week	Jora Compost Canada	NE 401	7:1–9:1	Electrical hookup, stable level surface, cover (e.g., roof)
	O2Compost	Micro-Bin Compost System	1:3	Electrical hookup, stable level surface, separate mixing pad or area, water
	Vertal Inc.	CityPod model M	1:6	Electrical hookup, stable level surface, cover (e.g., roof), water, ventilation
1–2 Tons per Week	Food Waste Experts	A900 Rocket Composter	1:1	Electrical hookup, stable level surface, cover (e.g., roof)
	FOR Solutions	Model 500	4:1	Electrical hookup, concrete pad, cover (e.g., roof)
	EC ALL Ltd.	Big Hanna T240	5:1–16:1	Electrical hookup, stable level surface, cover (e.g., roof)
	Eco Value Technology Inc.	EVT-412B	1:1–1:4	Electrical hookup, stable level surface
	Eco Value Technology Inc.	EVT-412	1:1–1:4	Electrical hookup, stable level surface
	Global Composting Solutions Ltd.	HotRot 1206	1:1	Electrical hookup, stable level surface, separate mixing pad or area, water, biofilter, bin lifter

Recommended External System Components	Requires Heated Space (Where Freezing Occurs)	System Footprint (Square Feet)	Estimated Labor (Hours per Week)	Human- or Loader-Operated
None	No	15	0.5–1	Human-powered
Insulation (e.g., insulated building)	Yes	60	4	Either
Electrical hookup, plug-in aeration is available	No	20	1–2	Human-powered
Concrete, asphalt, wood, or other flat platform	No	81	1	Human-powered
Concrete pad	No	40	1.5	Human-powered
Concrete, asphalt, wood, or other flat platform	No	81	1	Human-powered
Electrical hookup, concrete pad, cover (e.g., roof), heated space (e.g., heated building), water	No	64	1	Human-powered
Electrical hookup, concrete pad, cover (e.g., roof), walls (e.g., building), heated space (e.g., heated building), water, biofilter	No	45	0.5	Human-powered
Concrete pad, insulation (e.g., insulated building)	No (with additional heating/components)	135	2.5	Either
Electrical hookup, walls (e.g., building), insulation (e.g., insulated building), heated space (e.g., heated building)	Yes	45	0.5	Human-powered
Concrete pad	No	98	1.5	Human-powered

Recommended External System Components	Requires Heated Space (Where Freezing Occurs)	System Footprint (Square Feet)	Estimated Labor (Hours per Week)	Human- or Loader-Operated
Electrical hookup, concrete pad, cover (e.g., roof), walls (e.g., building), heated space (e.g., heated building), water, biofilter	No	70	0.75	Either
Concrete pad, bucket loader, cover (e.g., roof)	No	16 Sq. Ft. per bin	3.5	Either
Concrete pad, insulation (e.g., insulated building)	No (with additional heating/components)	180	3.5	Either
Concrete pad	No	140	2	Human-powered
Walls (e.g., building), insulation (e.g., insulated building), heated space (e.g., heated building), water	No (with additional heating/components)	156	5	Either
Electrical hookup, concrete pad, cover (e.g., roof), walls (e.g. a building), heated space (e.g., heated building), water, biofilter	No	102	0.75	Either
Shredder, mixer	No	84	1–3	Either
Shredder, mixer, conveyor	No	96	1–3	Either (also conveyor)
Concrete pad, bucket loader, cover (e.g., roof), conveyer/s, feed hopper	No	300	11	Either

Table A.11. Compost System Requirements (2–5 Tons per Week)

	Company Name	Model Name/Number	Volume Ratio of Food Scraps : High-Carbon Feedstocks	Minimum External System Requirements
2–5 Tons per Week	Food Waste Experts	A1200 Rocket Composter	1:1	Electrical hookup, stable level surface, cover (e.g., roof)
	FOR Solutions	Model 1000	4:1	Electrical hookup, concrete pad, cover (e.g., roof)
	EC ALL Ltd.	Big Hanna T480	5:1–16:1	Electrical hookup, stable level surface, cover (e.g., roof)
	Eco Value Technology Inc.	EVT-424	1:1–1:4	Electrical hookup, stable level surface
	Green Mountain Technologies	Earth Flow—EF16	1:2	Electrical hookup, stable level surface
	Green Mountain Technologies	Earth Flow—EF20 Intermodel	1:2	Electrical hookup, stable level surface
	Green Mountain Technologies	Earth Flow—EF20	1:2	Electrical hookup, stable level surface
	Eco Value Technology Inc.	EVT-432	1:1–1:4	Electrical hookup, stable level surface
	Green Mountain Technologies	Earth Flow—EF24	1:2	Electrical hookup, stable level surface
	FOR Solutions	Model 2000	4:1	Electrical hookup, concrete pad
	Sustainable Generation	SG Mini System	1:2–1:3	Electrical hookup, stable level surface, separate mixing pad or area

Table A.12. Compost System Requirements (> 5 Tons per Week)

	Company Name	Model Name/Number	Volume Ratio of Food Scraps : High-Carbon Feedstocks	Minimum External System Requirements
> 5 Tons per Week	Eco Value Technology Inc.	EVT-440	1:1–1:4	Electrical hookup, stable level surface
	Green Mountain Technologies	Earth Flow—EF30	1:2	Electrical hookup, stable level surface
	Eco Value Technology Inc.	EVT-532	1:1–1:4	Electrical hookup, stable level surface
	Green Mountain Technologies	Earth Flow—EF40 Intermodel	1:2	Electrical hookup, stable level surface
	Green Mountain Technologies	Earth Flow—EF40	1:2	Electrical hookup, stable level surface
	Eco Value Technology Inc.	EVT-540	1:1–1:4	Electrical hookup, stable level surface
	Global Composting Solutions	HotRot 1811	1:1	Electrical hookup, stable level surface, separate mixing pad or area, water, biofilter, bin lifter
	FOR Solutions	Model 4000	4:1	Electrical hookup, concrete pad
	Green Mountain Technologies	Earth Flow—EF53 Intermodel	1:2	Electrical hookup, stable level surface
	Green Mountain Technologies	Earth Flow—EF5010	1:2	Electrical hookup, stable level surface
	Agrilab Technologies Inc.	Hot Skid 250R	1:2–1:10	Electrical hookup, stable level surface, separate mixing pad or area, bucket loader

Recommended External System Components	Requires Heated Space (Where Freezing Occurs)	System Footprint (Square Feet)	Estimated Labor (Hours per Week)	Human- or Loader-Operated
Concrete pad	No	260	2.5	Human-powered
Walls (e.g., building), insulation (e.g., insulated building), heated space (e.g., heated building), water	No (with additional heating/components)	240	5	Either
Electrical hookup, concrete pad, cover (e.g., roof), walls (e.g., building), heated space (e.g., heated building), water, biofilter	No	176	1	Either
Shredder, mixer, conveyor	No	168	2–3	Either (also conveyor)
Concrete pad, bucket loader, water, biofilter	No	8′4″ × 17′	5	Either
Concrete pad, bucket loader, water, biofilter	No	160	5	Either
Concrete pad, bucket loader, water, biofilter	No	8′4″ × 21′	5	Either
Shredder, mixer, conveyor	No	216	2–4	Either (also conveyor)
Concrete pad, bucket loader, water, biofilter		8′4″ × 25′	5	Either
Cover (e.g., roof), walls (e.g., building), insulation (e.g., insulated building), heated space (e.g., heated building)	No (with additional heating/components)	288	5	Either
Concrete pad, bucket loader, water, mixer	No	≥ 3,000	4	Either

Recommended External System Components	Requires Heated Space (Where Freezing Occurs)	System Footprint (Square Feet)	Estimated Labor (Hours per Week)	Human- or Loader-Operated
Shredder, mixer, conveyor	No	264	3–4	Either (also conveyor)
Concrete pad, bucket loader, water, biofilter	No	8′4″ ×31′	5	Either
Shredder, mixer, conveyor	No	252	3–5	Bucket loader or conveyor
Concrete pad, bucket loader, water, biofilter	No	320	5	Either
Concrete pad, bucket loader, water, biofilter	No	8′4″ ×41′	5	Either
Shredder, mixer, conveyor	No	308	3–5	Bucket loader or conveyor
Concrete pad, bucket loader, cover (e.g., roof), conveyer/s, feed hopper	No	1,600	14	Either
Cover (e.g., roof), walls (e.g., building), insulation (e.g., insulated building), heated space (e.g., heated building), water	No (with additional heating/components)	400	10	Either
Concrete pad, bucket loader, water, biofilter	No	8′6″ ×53′	5	Either
Concrete pad, bucket loader, water, biofilter	No	572	5	Either
Concrete pad, cover (e.g., roof), walls (e.g., building), water, biofilter, conveyer/s, thermal load	No	Variable (3,000′ can aerate 1,000 yards)	7+ batch loading and unloading	Bucket-loader-operated

APPENDIX B

Compost Pile Monitoring Logs

Table B.1. Compost Pile Monitoring Log

PILE IDENTIFICATION: **PILE LOCATION:** **DATE PILE BUILT:**

Date	Temps 1	Temps 2	Temps 3	Temps 4	Temps 5	Air Temp	MC	Odor	Visual	Notes (management, weather, vectors)
	1′/3′	1′/3′	1′/3′	1′/3′	1′/3′					

Source: Adapted from Highfields Center for Composting, courtesy of Vermont Sustainable Jobs Fund.

Table B.2. Compost Pile Monitoring Log: ASP Bays

PILE IDENTIFICATION: **PILE LOCATION:** **DATE PILE BUILT:**

Date	Front		Back		Air Temp	MC	Odor	Visual	Notes (management, weather, vectors)
	Temps 1	Temps 2	Temps 4	Temps 5					
	1′ / 3′	1′ / 3′	1′ / 3′	1′ / 3′					

Source: Adapted from Highfields Center for Composting, courtesy of Vermont Sustainable Jobs Fund.

Table B.3. Compost Pile Monitoring Log: Bins

PILE IDENTIFICATION: **PILE LOCATION:** **DATE PILE BUILT:**

Date	Temperature Monitoring				Air Temp	MC	Odor	Visual	Notes (management, weather, vectors)
	Front		Back						
	1′ Depth	1′ from Base	1′ Depth	1′ from Base					

Source: Adapted from Highfields Center for Composting, courtesy of Vermont Sustainable Jobs Fund.

RESOURCES

Books

Compost Facility Operator Manual: A Compost Facility Operator Training Course Reference and Guide by John Paul and Dieter Geesling (2009).

Compost Utilization in Horticultural Cropping Systems by Peter J. Stoffella and Brian A. Kahn (CRC Press, 2001).

Field Guide to On-Farm Composting (NRAES-114) by NRAES (Natural Resource, Agriculture, and Engineering Service) (1999).

Food Foolish: The Hidden Connection Between Food Waste, Hunger and Climate Change by John M. Mandyck and Eric B. Schultz (Carrier Corporation, 2015).

Holy Shit: Managing Manure to Save Mankind by Gene Logsdon (Chelsea Green Publishing, 2010).

On-Farm Composting Handbook (*NRAES-54*) by NRAES (Natural Resource, Agriculture, and Engineering Service) (1992).

Teaming with Microbes by Jeff Lowenfels and Wayne Lewis (Timber Press, 2006).

The Compost-Powered Water Heater by Gaelan Brown (Countryman Press, 2014).

The Practical Guide to Compost Marketing and Sales, 2nd ed., by Ron Alexander (R. Alexander Associates, 2010).

The Practical Handbook of Compost Engineering by Roger Tim Haug (Lewis Publishers, 1993).

The Science of Composting by Eliot Epstein (CRC Press, 1996).

The Worm Farmer's Handbook by Rhonda Sherman (Chelsea Green Publishing, 2018).

Worms Eat My Garbage: How to Set Up and Maintain a Worm Composting System, 2nd ed., by Mary Appelhof (Flower Press, 1997).

Journals

***BioCycle* magazine**
www.biocycle.net

Compost Science & Utilization
https://www.tandfonline.com/action/journalInformation?journalCode=ucsu20

Organizations & Institutions

Compost Technical Services
www.compostechnicalservices.com
This is the website for my business where I will add new resources over time.

Cornell Waste Management Institute
http://compost.css.cornell.edu

Institute for Local Self-Reliance
www.ILSR.org/composting

ReFED
www.refed.com

United States Compost Council
www.CompostingCouncil.org

US Environmental Protection Agency
www.epa.gov/sustainable-management-food

Woods End Laboratories Inc.
www.woodsend.com

In-Depth Composter Training Programs

Maine Compost School
http://composting.org

US Composting Council—Operator Training Courses
http://compostingcouncil.org/training

Community Composting Resources

BioCycle
http://www.biocycle.net/tag/community-composting

Community Composting Forum
http://www.biocycle.net/communitycomposting

Community Composting NYC
https://sites.google.com/site/communitycompostnyc

The Community Composting Network (UK)
http://www.communitycompost.org

Growing Power
http://www.growingpower.org

Highfields Center for Composting
http://highfieldscomposting.org

Institute for Local Self-Reliance
http://www.ilsr.org/initiatives/composting

Compost & Soil Testing Labs

Earthfort
www.earthfort.com

Pennsylvania State Agricultural Analytical Services Lab
https://agsci.psu.edu/aasl/compost-testing

Soil Food Web New York
www.soilfoodwebnewyork.com

Soil Test Farm Consultants
www.soiltestlab.com

Trace Genomics
https://tracegenomics.com/#

Woods End Laboratories, Inc.
www.woodsend.com

NOTES

Introduction: Recycling Organics at the Community Scale

1. Nate Clark, "The Business of Community Composting," *BioCycle* 56, no. 1 (2015): 32. https://www.biocycle.net/2015/01/14/the-business-of-community-composting.
2. John P. Kretzmann and John L. McKnight, *Building Communities from the Inside Out: A Path Toward Finding and Mobilizing a Community's Assets* (Evanston, IL: Institute for Policy Research, 1993), 1–11.
3. *Municipal Solid Waste Generation, Recycling, and Disposal in the United States: Facts and Figures for 2014* (Washington DC: US Environmental Protection Agency, Office of Resource Conservation and Recovery, 2014), table 1, p. 8.
4. Brenda Platt, James McSweeney, and Jenn Davis, *Growing Local Fertility: A Guide to Community Composting* (Hardwick, VT: Highfields Center for Composting and the Institute for Local Self-Reliance, 2014), 7. http://ilsr.org/wp-content/uploads/2014/07/growing-local-fertility.pdf.
5. Daniel A. Marano, "Nature's Bounty: Soil Salvation," *Psychology Today*, September 1, 2008. https://www.psychologytoday.com/us/articles/200809/natures-bounty-soil-salvation.
6. *Municipal Solid Waste Generation, Recycling, and Disposal in the United States: Facts and Figures for 2014* (Washington DC: US Environmental Protection Agency, Office of Resource Conservation and Recovery, 2014), table 2, p. 8.
7. Emily Broad Leib, Christina Rice, and Jill Mahoney, "Fresh Look At Organics Bans and Waste Recycling Laws," *BioCycle* 57, no. 10 (2016): 16. https://www.biocycle.net/2016/11/10/fresh-look-organics-bans-waste-recycling-laws.
8. Leib, Rice, and Mahoney, "Organics Bans."
9. Ned Beecher and Nora Goldstein, "Biosolids Composting in the United States: *BioCycle* Nationwide Survey," *BioCycle* 51, no. 12 (2010): 35.
10. *Advancing Sustainable Materials Management: 2014 Fact Sheet Assessing Trends in Material Generation, Recycling, Composting, Combustion with Energy Recovery and Landfilling in the United States* (Washington DC: US Environmental Protection Agency, 2016). https://www.epa.gov/sites/production/files/2016-11/documents/2014_smmfactsheet_508.pdf.
11. W. F. Brinton et al., "Occurrence and Levels of Fecal Indicators and Pathogenic Bacteria in Market-Ready Recycled Organic Matter Composts," *Journal of Food Protection* 72, no. 2 (2009): 332–39.
12. Elizabeth Royte, "The Compost King of New York," *New York Times*, February 15, 2017. https://www.nytimes.com/2017/02/15/magazine/the-compost-king-of-new-york.html.
13. Nora Goldstein, editor of *BioCycle* magazine, email to the author, January 5, 2018.

Chapter One: Common Models in Community-Scale Composting

1. Brenda Platt, James McSweeney, and Jenn Davis, *Growing Local Fertility: A Guide to Community Composting* (Hardwick, VT: Highfields Center for Composting and the Institute for Local Self-Reliance, 2014). http://ilsr.org/wp-content/uploads/2014/07/growing-local-fertility.pdf.
2. *Advancing Sustainable Materials Management: 2014 Fact Sheet Assessing Trends in Material Generation, Recycling, Composting, Combustion with Energy Recovery and Landfilling in the United States* (Washington, DC: US Environmental Protection Agency, 2016). https://www.epa.gov/sites/production/files/2016-11/documents/2014_smmfactsheet_508.pdf.
3. James McSweeney and Brenda Platt, *Micro-Composting: A Guide to Small Scale & On-Site Food Scrap Composting Systems* (Washington, DC: Institute for Local Self-Reliance, forthcoming).
4. *2014 NYC Community Composting Report* (New York City: City of New York Department of Sanitation, 2015). https://www1.nyc.gov/assets/dsny/docs/about_2014-community-composting-report-LL77_0815.pdf.

5. Craig Lemoult, "Cambridge's Composting Program Isn't Actually Composting. Is What They're Doing as Good?" WGBH, May 22, 2018. https://www.wgbh.org/news/local-news/2018/05/22/cambridges-composting-program-isnt-actually-composting-is-what-theyre-doing-as-good.
6. Marguerite Manela, email to the author, September 29, 2017.
7. *Composting in the Pacific Northwest* (blog). http://seattletilthcomposting.blogspot.com.

Chapter Two: Composting Methods and Technologies

1. Nora Goldstein, "The State of Organics Recycling in the US," *Biocycle* 58, no. 9 (2017): 22–26.
2. Craig Coker, "Food Safety Rule Impacts on Organics Recyclers," *Biocycle* 57, no. 6 (2016): 22–24.
3. Roger Tim Haug, *The Practical Handbook of Compost Engineering* (Boca Raton, FL: Lewis Publishers, 1993), 28.
4. Mary Applehof, *Worms Eat My Garbage* (Kalamazoo, MI: Flowerfield Enterprises, 1997), 3.
5. Bruce R. Eastman et al., "The Effectiveness of Vermiculture in Human Pathogen Reduction for USEPA Biosolids Stabilization," *Compost Science & Utilization* 9, no. 1 (2001): 42.
6. Geoff B. Hill et al., "The Effectiveness and Safety of Vermi- Versus Conventional Composting of Human Feces with *Ascaris suum* Ova as Model Helminthic Parasites," *Journal of Sustainable Development* 6, no. 4 (2013).
7. Rajiv K. Sinha et al., "The Wonders of Earthworms & Its Vermicompost in Farm Production: Charles Darwin's 'Friends of Farmers,' with Potential to Replace Destructive Chemical Fertilizers from Agriculture," *Agricultural Sciences* 1, no. 2 (2010): 78, 81.
8. *Leftovers for Livestock: A Legal Guide for Using Excess Food as Animal Feed* (Harvard Food Law and Policy Clinic and University of Arkansas School of Law, 2016). http://www.chlpi.org/wp-content/uploads/2013/12/Leftovers-for-Livestock_A-Legal-Guide_August-2016.pdf.
9. *Inventory of US Greenhouse Gas Emissions and Sinks: 1990–2014* (Washington, DC: US Environmental Protection Agency, 2016): ES 5–7. https://www.epa.gov/sites/production/files/2016-04/documents/us-ghg-inventory-2016-main-text.pdf.
10. *Greenhouse Gas Emissions and Sinks: 1990–2014*, 45.
11. Zoë Neale, "Analysis of Biodigesters and Dehydrators to Manage Organics On-Site," *BioCycle* 54, no. 10 (2013): 20. https://www.biocycle.net/2013/10/25/analysis-of-biodigesters-and-dehydrators-to-manage-organics-on-site.
12. Neale, "Analysis of Biodigesters and Dehydrators," 20.
13. Zoë Neale, "Biodigesters and Dehydrators—Operational Experiences," *BioCycle* 55, no.1 (2014): 52. https://www.biocycle.net/2014/01/20/biodigesters-and-dehydrators-operational-experiences.

Chapter Three: The Composting Process

1. William F. Brinton, "The Sustainability of Modern Composting: Intensification Versus Costs and Quality," *Biodynamic Farming and Gardening in the 21st Century* (July–August, 1997).
2. *5.2 Dissolved Oxygen and Biochemical Oxygen Demand* (Washington, DC: US Environmental Protection Agency, 2012). https://archive.epa.gov/water/archive/web/html/vms52.html.
3. *A Guidebook for Local Governments for Developing Regional Watershed Protection Plans* (Athens, GA: Northeast Georgia Regional Development Center, 2001). https://epd.georgia.gov/sites/epd.georgia.gov/files/related_files/site_page/devwtrplan.pdf.
4. Jiwan Singh et al., "Estimation of Compost Stability During Rotary Drum Composting of Municipal Solid Waste," *G- Journal of Environmental Science and Technology* 1, no. 1 (2013): 1–7. http://gjestenv.com/Current_Issue/vol_1/Gjest_01.pdf.
5. Eliot Epstein, *The Science of Composting* (Boca Raton, FL: CRC Press, 1997), 37.
6. Epstein, *The Science of Composting*, 54.
7. Jeff Lowenfels and Wayne Lewis, *Teaming with Microbes: A Gardener's Guide to the Soil Food Web* (Portland, OR: Timber Press, 2006), 121–22.
8. Peter J. Stoffella and Brian A. Kahn, *Compost Utilization in Horticultural Cropping Systems* (Baton Rouge: CRC Press, 2001), 368–73.
9. W. D. Burge, "Monitoring Pathogen Destruction," *BioCycle* 24, no. 2 (1983): 48–50.
10. J. Van Rossum and M. J. Renz, "Composting Reduces Seed Viability of Garlic Mustard (*Alliaria petiolata*) and Common Buckthorn (*Rhamnus cathartica*)," *Invasive Plant Science and Management* 8, no. 3 (2017): 284–91.
11. D. A. Neher et al., "Compost for Management of Weed Seeds, Pathogen, and Early Blight on Brassicas in Organic Farmer Fields," *Agroecology and Sustainable Food Systems* 39, no. 1 (2015): 3.

12. US Government Publishing Office, *Code of Federal Regulations, Title 40, Standards for the Use or Disposal of Sewage Sludge*, Appendix B to Part 503—Pathogen Treatment Processes, 1999. https://www.gpo.gov/fdsys/pkg/CFR-2003-title40-vol27/pdf/CFR-2003-title40-vol27-part503-appB.pdf.
13. National Organic Standards Board (NOSB), *NOSB Recommendation for Guidance: Use of Compost, Vermicompost, Processed Manure and Compost Tea*, November 9, 2006.
14. Nancy M. Trautmann and Elaina Olynciw, "Compost Microorganisms," Cornell Waste Management Institute, 1996. http://compost.css.cornell.edu/microorg.html.
15. Epstein, *The Science of Composting*, 74.
16. Roger Tim Haug, *The Practical Handbook of Compost Engineering* (Boca Raton, FL: Lewis Publishers, 1993), 253–54.
17. Xiying Hao and Monica B. Benke, "Nitrogen Transformation and Losses During Composting and Mitigation Strategies," *Dynamic Soil, Dynamic Plant* 2, no. 1 (2008): 12.
18. Epstein, *The Science of Composting*, 82.
19. Nancy M. Trautmann et al., "Compost Chemistry," Cornell Waste Management Institute, 1996. http://compost.css.cornell.edu/chemistry.html.
20. Marja Tuomela, "Degradation of Lignin and Other ^{14}C-Labelled Compounds in Compost and Soil with an Emphasis on White-Rot Fungi" (academic dissertation in microbiology, University of Helsinki, 2002), 17. http://ethesis.helsinki.fi/julkaisut/maa/skemi/vk/tuomela/degradat.pdf.
21. Epstein, *The Science of Composting*, 82.
22. Haug, *The Practical Handbook of Compost Engineering*, 253–54.
23. M. Tuomela et al., "Biodegradation of Lignin in a Compost Environment: A Review," *Bioresource Technology* 72, no. 2 (2000): 169–83.
24. Yitzhak Hadar and Kalliope K. Papadopoulou, "Suppressive Composts: Microbial Ecology Links Between Abiotic Environments and Healthy Plants," *Annual Review of Phytopathology* 50, no. 1 (2012): 136.
25. K. Maeda, "Microbiology of Nitrogen Cycle in Animal Manure Compost," *Microbial Biotechnology* 4, no. 6 (2011): 700–09.
26. Hao and Benke, "Nitrogen Transformation," 12.
27. Epstein, *The Science of Composting*, 367.
28. Epstein, *The Science of Composting*, 43.
29. Robert Rynk, ed., *On-Farm Composting Handbook* (Ithaca, NY: Natural Resource, Agriculture, and Engineering Service, 1992), 13.
30. Joe Jenkins, "Thermophilic Bacteria, Composting Stages, and the Sanitization of Compost," A Growing Culture, August 18, 2011. http://www.agrowingculture.org/humanure-part-iii-thermophillic-bacteria-composting-stages-the-sanitization-of-compost-joe-jenkin.
31. Lowenfels and Lewis, *Teaming with Microbes*, 117.
32. Epstein, *The Science of Composting*, 91–95.

Chapter Four: Compost Recipe and Feedstocks

1. Nancy M. Trautmann and Marianne E. Krasny, *Composting in the Classroom: Scientific Inquiry for High School Students* (Dubuque, IA: Kendall/Hunt Publishing Company, 1998), 45. https://ecommons.cornell.edu/bitstream/handle/1813/3338/CompostingInTheClassroom.pdf.
2. Michael Day and Kathleen Shaw, "Biological, Chemical, and Physical Processes of Composting," in *Compost Utilization in Horticultural Cropping Systems*, ed. Peter Stoffela and Brian Kahn (Boca Raton, FL: CRC Press, 2001), 27.
3. Fred Magdoff and Harold Van Es, *Building Soils for Better Crops: Sustainable Soil Management* (College Park, MD: Sustainable Agriculture Research and Education, 2010), 95. http://www.sare.org/content/download/841/6675/Building_Soils_For_Better_Crops.pdf.
4. Vicki Bess, "Evaluating Microbiology of Compost," *BioCycle* 40, no. 5 (1999): 84.
5. NPCS Board of Consultants & Engineers, *The Complete Book on Organic Farming and Production of Organic Compost* (Delhi, India: Asia Pacific Business Press, 2008).
6. Day and Shaw, "Processes of Composting," 27.
7. Eric S. Gale et al., "Estimating Plant-Available Nitrogen Release from Manures, Composts, and Specialty Products," *Journal of Environmental Quality* 35, no. 6 (2006): 2330.
8. Dan Sullivan and Robert Miller, "Compost Quality Attributes, Measurements, and Variability," in *Compost Utilization in Horticultural Cropping Systems*, ed. Peter Stoffela and Brian Kahn (Boca Raton, FL: CRC Press, 2001), 111.
9. Harold B. Gotaas, *Composting: Sanitary Disposal and Reclamation of Organic Wastes* (Geneva, Switzerland: World Health Organization, monograph series no. 31, 1956), 92.
10. Lew Naylor, "Fire Prevention at Composting, Mulch Facilities," *BioCycle* 45 (December 2004), 31.

11. Robert Rynk, ed., *On-Farm Composting Handbook* (Ithaca, NY: Natural Resource, Agriculture, and Engineering Service, 1992), 12.
12. Xiying Hao and Monica B. Benke, "Nitrogen Transformation and Losses During Composting and Mitigation Strategies," *Dynamic Soil, Dynamic Plant* 2, no. 1 (2008): 12.
13. Natalie Rector and Al Sutton, *Best Environmental Management Practices: Farm Animal Production, Manure Nutrient Recycling* (East Lansing, MI: Michigan State University Extension). https://www.extension.purdue.edu/extmedia/ID/ID-307.pdf.
14. Ron Hoover and Douglas Beegle, "Increasing Interest in Manure Injection in Pennsylvania?" PennState Extension, March 21, 2018. https://extension.psu.edu/increasing-interest-in-manure-injection-in-pennsylvania.
15. Robert Rynk, email to the author, 2016.
16. William Brinton, "Volatile Organic Acids in Compost: Production and Odorant Aspects," *Compost Science & Utilization* 6, no. 1 (1998): 75–82.
17. Brinton, "Volatile Organic Acids in Compost."
18. X. Hao et al., "Greenhouse Gas Emissions During Cattle Feedlot Manure Composting," *Journal of Environmental Quality* 30, no. 2 (2001): 376–86; Yasuyuki Fukumoto et al., "Patterns and Quantities of NH_3, N_2O, and CH_4 Emissions During Swine Manure Composting Without Forced Aeration—Effect of Compost Pile Scale," *Bioresource Technology* 89, no. 2 (2003): 109–14; J. Lopez-Real and M. Babtista, "A Preliminary Comparative Study of Three Manure Composting Systems and Their Influence on Process Parameters and Methane Emissions," *Compost Science & Utilization* 4, no. 3 (1996): 71–82.
19. Robert Rynk, "Monitoring Moisture in Compost Systems," *BioCycle* 49 (2008): 24–29.
20. Rynk, *On-Farm Composting Handbook*, 15.
21. Rynk, *On-Farm Composting Handbook*, 22; Agricultural Analytical Services Laboratory, *Compost Sample Submission Form*, Pennsylvania State University, 2018.
22. K. L. Schulze, "Continuous Thermophilic Composting," *Applied Microbiology* 10, no. 2 (1962): 108–22; J. S. Annan and R. K. White, "Evaluation of Techniques for Measuring Air Filled Porosity in Composts of Municipal Biosolids and Wood Chips," in *Composting in the Southeast, Proceedings of the 1998 Conference*, ed. K. C. Das (Athens: University of Georgia, 1998), 88–96. http://infohouse.p2ric.org/ref/12/11583.pdf.
23. Qunfeng Wang, "Developing a Standard Method for Measuring Free Air Space in Compost Materials" (master's dissertation, University of Manitoba, 2003).
24. Annan and White, "Evaluation of Techniques."
25. Rynk, *On-Farm Composting Handbook*, 10; Chiu-Chung Young et al., "What Happens During Composting?," Food and Fertilizer Technology Center for the Asian and Pacific Region, December 1, 2005. http://www.fftc.agnet.org/library.php?func=view&id=20110913155219.
26. Roger Tim Haug, *The Practical Handbook of Compost Engineering* (Boca Raton, FL: Lewis Publishers, 1993), 253–54.
27. Rynk, *On-Farm Composting Handbook*, 7.
28. Cecilia Sundberg et al., "Effects of pH and Microbial Composition on Odour in Food Waste Composting," *Waste Management* 33, no. 1 (2012): 204–11.
29. Nancy M. Trautmann et al., "Compost Chemistry," Cornell Waste Management Institute, 1996. http://compost.css.cornell.edu/chemistry.html.
30. Cornelius Tokunbo Bamise and Elizabeth Obhioneh Oziegbe, "Laboratory Analysis of pH and Neutralizable Acidity of Commercial Citrus Fruits in Nigeria," *Advances in Biological Research* 7, no. 2 (2013), doi: 10.5829/idosi.abr.2013.7.2.1106.
31. Hao and Benke, "Nitrogen Transformation," 12.
32. Day and Shaw, "Processes of Composting," 32.
33. Rynk, *On-Farm Composting Handbook*, 23.
34. Eliot Epstein, *The Science of Composting* (Boca Raton, FL: CRC Press, 1997), 366.
35. Namratha Reddy and David M. Crohn, "Compost Induced Soil Salinity: A New Prediction Method and Its Effect on Plant Growth," *Compost Science & Utilization* 20, no. 3 (2012): 134.
36. Agricultural Analytical Services Laboratory, *Compost Analysis Report—Interpretation of % Organic Matter* (State College: Pennsylvania State University, 2013).
37. *Recipe Development Worksheet* (Hardwick, VT: Highfields Center for Composting, last updated 2010).
38. *Feedstock Sampling Protocol* (Hardwick, VT: Highfields Center for Composting, last updated 2010).
39. Fatih Buyuksonmez and Robert Rynk, "Occurrence, Degradation and Fate of Pesticides During Composting," *Compost Science & Utilization* 7, no. 4 (1999): 80; Fatih Buyuksonmez et al., "Occurrence, Degradation and Fate of Pesticides During Composting," *Compost Science & Utilization* 8, no. 1 (2000): 74–77.
40. *Advancing Sustainable Materials Management: 2014 Fact Sheet* (Washington, DC: US Environmental Protection Agency, 2016), 4. https://www.epa.gov/sites/production/files/2016-11/documents/2014_smmfactsheet_508.pdf.
41. *Advancing Sustainable Materials Management: 2014 Fact Sheet*, 4.

42. US Environmental Protection Agency, *Solid Waste Generation, Recycling, and Disposal in the United States: Facts and Figures for 2012* (Washington, DC: US Environmental Protection Agency, 2014), 2. https://www.epa.gov/sites/production/files/2015-09/documents/2012_msw_fs.pdf.
43. *Leaf and Yard Waste Composting Guidance Document* (Massachusetts Department of Environmental Protection, 1992), 8. https://www.mass.gov/files/documents/2016/08/oj/leafguid.pdf.
44. Rynk, *On-Farm Composting*, 112.
45. Rynk, *On-Farm Composting*, 109.
46. Rynk, *On-Farm Composting*, 109.
47. Rynk, *On-Farm Composting*, 112.
48. *Labeling Guidelines for Compostable Plastics Associated with Food Scraps or Yard Trimmings* (Bethesda, MD: United States Composting Council, 2013), 1. https://compostingcouncil.org/wp-content/plugins/wp-pdfupload/pdf/8095/USCC%20Compostable%20Plastic%20Labeling%20Guidelines.pdf.
49. Nora Goldstein, "Unraveling the Maze of Persistent Herbicides in Compost: Part I," *BioCycle* 54, no. 6 (2013): 17–18.
50. Dan Sullivan, "Another Persistent Herbicide: DuPont Says 'Do Not Compost' Grass Clippings," *BioCycle* 52, no. 6 (2011): 23. https://www.biocycle.net/2011/06/16/dupont-label-says-do-not-compost-grass-clippings.

Chapter Five: Processing Capacity and Site Assessment

1. "Reader's Q&A," *Biocycle* 43, no. 11 (2002): 14.
2. Gary A. Breitenbeck and David Schellinger, "Calculating the Reduction in Material Mass and Volume During Composting," *Compost Science & Utilization* 12, no. 4 (2004): 365–71.
3. Eric Walter, "Postmortem of a Food Scraps Composting Facility," *BioCycle* 58, no. 3 (2017): 55–56.
4. "7 CFR 205.203—Soil Fertility and Crop Nutrient Management Practice Standard," Legal Information Institute. https://www.law.cornell.edu/cfr/text/7/205.203.
5. Thea Rittenhouse, *Tipsheet: Manure in Organic Production Systems* (Washington, DC: US Department of Agriculture, ATTRA Project, National Center for Appropriate Technology, 2015), 1. https://www.ams.usda.gov/sites/default/files/media/Manure%20in%20Organic%20Production%20Systems_FINAL.pdf.
6. Craig Coker, "Food Safety Rule Impacts on Organics Recyclers," *Biocycle* 57, no. 6 (2016): 22–24.

Chapter Six: Compost Site Infrastructure and Equipment

1. Robert Rynk, ed., *On-Farm Composting Handbook* (Ithaca, NY: Natural Resource, Agriculture, and Engineering Service, 1992), 52.
2. Rynk, *On-Farm Composting Handbook*, 52.

Chapter Seven: Bin and Bay Composting Systems

1. James McSweeney, *Designing a Bin System for Hot Composting* (Hardwick, VT: Highfields Center for Composting, 2013). Updated version available here: http://dec.vermont.gov/sites/dec/files/wmp/SolidWaste/Documents/ANR%20On-Site%20Composting%20Designing%20a%20Bin%20System%20for%20Hot%20Composting.pdf.

Chapter Eight: Turned Windrow Composting Systems

1. Nora Goldstein, "The State of Organics Recycling in the US," *Biocycle* 58, no. 9 (2017): 22–26.
2. "Standards for the Use or Disposal of Sewage Sludge," US Government Publishing Office, Code of Federal Regulations, Title 40 (2006).

Chapter Nine: Aerated Static Pile Composting Systems

1. Roger Tim Haug, *The Practical Handbook of Compost Engineering* (Boca Raton, FL: Lewis Publishers, 1993), 33.
2. Nora Goldstein, "The State of Organics Recycling in the US," *BioCycle* 58, no. 9 (2017): 22–26.
3. Bruce Fulford, *Composting Greenhouse at New Alchemy Institute: A Report on Two Years of Operation and Monitoring, Research Report No. 3* (Hatchville, MA: New Alchemy Institute, 1986).
4. Gaelan Brown, *The Compost-Powered Water Heater* (Woodstock, VT: Countryman Press, 2014), 46–47.
5. Haug, *Compost Engineering*, 540.
6. Haug, *Compost Engineering*, 280.
7. Craig Coker and Tim O'Neill, "Composting Aeration Floor Functions and Designs, Part II," *Biocycle* 58, no. 6 (2017): 29.
8. Craig Coker and Tim O'Neill, "Aeration Floor Fundamentals: Part I," *Biocycle* 58, no. 5 (2017): 32.
9. G. Hewings et al., "Calculating the Airflow Requirements of Forced Aerated Composting Systems," presented at the 22nd International Conference on Waste Technology & Management, Philadelphia, 2007, 187.

10. Coker and O'Neill, "Aeration Floor Part II," 28; Tom Gibson and Craig Coker, "Pipe and Blower Fan Fundamentals in ASP Design," *Biocycle* 54, no. 2 (2013): 25.
11. Peter Moon, "Aerated Static Pile Composting: Applications & Advancements," presented at the US Compost Council Conference, Orlando, 2013.
12. Coker and O'Neill, "Aeration Floor II," 29.
13. Robert Rynk, ed., *On-Farm Composting Handbook* (Ithaca, NY: Natural Resource, Agriculture, and Engineering Service, 1992), 35.
14. Don Mathson, "Evaluating Compost and Biofilter Aeration Performance," *BioCycle* 45, no. 6 (2004): 22. https://www.biocycle.net/2004/06/15/evaluating-compost-and-biofilter-aeration-performance.
15. Coker and O'Neill, "Aeration Floor II," 28.
16. Rynk, *On-Farm Composting Handbook*, 33.
17. Douglas R. Hartsock, Gerry Croteau, and Jeff Gage, "Uniform Aeration of Compost Media," *National Waste Processing Conference Proceedings ASME*, 1994: 217. http://www.seas.columbia.edu/earth/wtert/sofos/nawtec/1994-National-Waste-Processing-Conference/1994-National-Waste-Processing-Conference-24.pdf.
18. Rynk, *On-Farm Composting Handbook*, 33.
19. Haug, *The Practical Handbook of Compost Engineering*, 539.
20. "Aeration Control Systems," Green Mountain Technologies. http://compostingtechnology.com/aerated-pile-systems/aeration-control-systems/#1447891170753-f37b4154-4d14.
21. Rynk, *On-Farm Composting Handbook*, 32.
22. Richard Nicolai and David Schmidt, "Biofilters, Fact Sheets," Paper 106 (Brookings: South Dakota State University, 2005), 2. http://openprairie.sdstate.edu/extension_fact/106.
23. Nicolai and Schmidt, "Biofilters," 6.
24. Todd O. Williams and Frederick C. Miller, "Biofilters and Facility Operations (Part 2)," *Biocycle* 33, no. 11 (1992): 75–76.
25. Todd O. Williams and Frederick C. Miller, "Odor Control Using Biofilters (Part 1)," *Biocycle* 33, no. 10 (1992): 72.
26. Nicolai and Schmidt, "Biofilters," 2.
27. Williams and Miller, "Biofilters (Part 2)," 75–76.
28. Williams and Miller, "Biofilters (Part 2)," 76.
29. Rynk, *On-Farm Composting*, 34–35.

Chapter Ten: In-Vessel Composting Systems

1. Nora Goldstein, "The State of Organics Recycling in the US," *Biocycle* 58, no. 9 (2017): 22–26.
2. James McSweeney and Brenda Platt, *Micro-Composting: A Guide to On-Site and Small Scale Food Scrap Composting Systems* (Washington, DC: Institute for Local Self-Reliance, forthcoming).
3. McSweeney and Platt, *Micro-Composting*.
4. McSweeney and Platt, *Micro-Composting*.
5. John Culpepper, email message to author, May 8, 2018.
6. McSweeney and Platt, *Micro-Composting*.

Chapter Eleven: Composting with Animals

1. Simon Fairlie, *Meat: A Benign Extravagance* (White River Junction, VT: Chelsea Green Publishing, 2010), 54.
2. *Food Waste: Tier 1 Assessment* (San Francisco: BSR, prepared for the Grocery Manufacturers Association and Food Marketing Institute, 2012), 5. http://www.foodwastealliance.org/wp-content/uploads/2013/06/FWRA_BSR_Tier1_FINAL.pdf.
3. Erasmus K .H. J. zu Ermgassen et al., "Reducing the Land Use of EU Pork Production: Where There's Swill, There's a Way," *Food Policy* 58 (2016): 35–48, doi: https://doi.org/10.1016/j.foodpol.2015.11.001.
4. *Leftovers for Livestock: A Legal Guide for Using Excess Food as Animal Feed* (Harvard Food Law and Policy Clinic and University of Arkansas School of Law, 2016). http://www.chlpi.org/wp-content/uploads/2013/12/Leftovers-for-Livestock_A-Legal-Guide_August-2016.pdf.
5. *Leftovers for Livestock*.
6. M. L. Westendorf and R. O. Myer, *Feeding Food Wastes to Swine* (Gainesville: University of Florida Institute of Food and Agricultural Sciences, 2004), 1. http://edis.ifas.ufl.edu/pdffiles/AN/AN14300.pdf.
7. Westendorf and Myer, *Feeding Food Wastes to Swine*, 1.
8. zu Ermgassen et al., "EU Pork Production: Where There's Swill," 36.
9. Adel El Boushy and Antonius van der Poel, *Handbook of Poultry Feed from Waste: Processing and Waste*, 2nd ed. (Netherlands: Springer Netherlands, 2000), 316.
10. *Egg Safety: What You Need to Know* (Washington, DC: US Food and Drug Administration, 2016). https://www.fda.gov/downloads/Food/FoodborneIllness Contaminants/UCM278445.pdf.
11. *Leftovers for Livestock*, 4.
12. M. MacLeod et al., *Greenhouse Gas Emissions from Pig and Chicken Supply Chains—A Global Life Cycle Assessment* (Rome: Food and Agriculture Organization of the United Nations, 2013).
13. *WARM Composting Overview* (Washington, DC: US Environmental Protection Agency, 2012).

14. "Waste Reduction Model (WARM)," US Environmental Protection Agency. https://www.epa.gov/warm.
15. *WARM Composting Overview*.
16. R. C. Taylor, H. Omed, and G. Edwards-Jones, "The Greenhouse Emissions Footprint of Free-Range Eggs," *Poultry Science* 93 (2014): 231.
17. Boushy and van der Poel, *Handbook of Poultry Feed*, 315.
18. Boushy and van der Poel, *Handbook of Poultry Feed*, 327–31.
19. Michael L. Westendorf, "Food Waste as Swine Feed," in *Food Waste to Animal Feed*, ed. Michael L. Westendorf (Ames: Iowa State University Press, 2000), 74–75.
20. Michael L. Westendorf et al., "Recycled Cafeteria Food Waste as a Feed for Swine: Nutrient Content Digestibility, Growth, and Meat Quality," *Journal of Animal Science* 76 (1998): 2976–83, doi: 10.2527/1998.76122976x.
21. Westendorf and Myer, "Feeding Food Wastes to Swine," 2.
22. Harvey Ussery, *The Small-Scale Poultry Flock*, (White River Junction, VT: Chelsea Green Publishing, 2011), 83.
23. Ting Chen, Yiying Jin, and Dongsheng Shen, "A Safety Analysis of Food Waste-Derived Animal Feeds from Three Typical Conversion Techniques in China," *Waste Management* 45, (2015): 42–46. https://doi.org/10.1016/j.wasman.2015.06.041.
24. X. Liu et al., "Dynamic Changes of Nutrient Composition Throughout the Entire Life Cycle of Black Soldier Fly," *PLOS ONE* 12, no. 8 (2017): 8, doi: https://doi.org/10.1371/journal.pone.0182601.
25. A. J. García et al., "Biodegradable Municipal Solid Waste: Characterization and Potential Use as Animal Feedstuffs," *Waste Management* 25, no. 8 (2005): 780–83, doi: https://doi.org/10.1016/j.wasman.2005.01.006.
26. Boushy and van der Poel, *Handbook of Poultry Feed*, 327–31; García et al., "Biodegradable Municipal Solid Waste," 780–83.
27. *Feeding Community Food Scraps to Laying Hens in an Active Composting System*, SARE Report (Stannard, VT: Black Dirt Farm, 2017), 29. https://projects.sare.org/information-product/feeding-community-food-scraps-to-laying-hens-in-an-active-composting-system-manual/.

Chapter Twelve: Food Scrap Generation and Collection

1. James McSweeney and Brenda Platt, *Micro-Composting: A Guide to On-Site and Small Scale Food Scrap Composting Systems* (Washington, DC: Institute for Local Self-Reliance, forthcoming); Brenda Platt, James McSweeney, and Jenn Davis, *Growing Local Fertility: A Guide to Community Composting* (Hardwick, VT: Highfields Center for Composting and the Institute for Local Self-Reliance, 2014). http://ilsr.org/wp-content/uploads/2014/07/growing-local-fertility.pdf.
2. Eric Walter, "Postmortem of a Food Scraps Composting Facility," *BioCycle* 58, no. 3 (2017): 54.
3. Lisa A. Skumatz et al., *Chittenden Solid Waste District Analysis of Residential Curbside Organics Collection Options: Final Report* (Superior, CO: Skumatz Economic Research Associates, prepared for Chittenden Solid Waste District, 2013), 12, 23.
4. Kat Nigro of Compost Now, email to the author, February 2, 2018; Ivy Young of Santa Cruz Community Compost Co., email to the author, February 6, 2018.
5. *Food Waste: Tier 1 Assessment* (San Francisco: BSR, prepared for the Grocery Manufacturers Association and Food Marketing Institute, 2012), 5. http://www.foodwastealliance.org/wp-content/uploads/2013/06/FWRA_BSR_Tier1_FINAL.pdf.
6. *Estimating Food Scrap Generation by the Commercial/Institutional Sector in Vermont* (Cambridge, MA: Compost Technical Services, 2014).
7. Tom Quested, Robert Ingle, and Andrew Parry, *Household Food and Drink Waste in the United Kingdom 2012* (Banbury, U.K.: Waste and Resource Action Programme, 2013), 41. http://www.wrap.org.uk/sites/files/wrap/hhfdw-2012-main.pdf.pdf.
8. Lisa A. Skumatz et al., *Chittenden Solid Waste District Analysis of Residential Curbside Organics Collection Options: Final Report* (Superior, CO: Skumatz Economic Research Associates, Inc., prepared for Chittenden Solid Waste District, 2013), 11.
9. "Best Practices for Conducting a Waste Assessment" in "Best Practices for WasteWise Participants," US Environmental Protection Agency. https://www.epa.gov/smm/best-practices-wastewise-participants#01.
10. Carol Adaire Jones, "Food Waste Infrastructure in Disposal Ban States," *BioCycle* 58, no. 10 (2017): 23.
11. "How Much Weight Can a Bicycle Carry," *Bikes at Work* (blog), updated February 2, 2012. https://www.bikesatwork.com/blog/how-much-weight-can-a-bicycle-carry.
12. Virginia Streeter and Brenda Platt, "Bike Powered Food Scraps Collection," *Biocycle* 58, no. 1 (2017): 52.
13. Nate Clark, "The Business of Community Composting," *Biocycle* 56, no. 1 (2015): 32–38.
14. "Case Study: Close the Loop! Northeast Kingdom and Lamoille Hauling Routes," Highfields Center for Composting, Hardwick, VT, 2014.

15. *Feeding Community Food Scraps to Laying Hens in an Active Composting System*, SARE Report (Stannard, VT: Black Dirt Farm, 2017). https://static1.squarespace.com/static/5592c1f2e4b0d3a1e1c396d7/t/592ed7cc29687f8352c34feb/1496242176789/Feeding+Community+Food+Scraps+to+Laying+Hens+in+an+Active+Composting+System+manual+%281%29.pdf.
16. Barbara Hesselgrave, "Food Waste Collection Truck Innovations," *BioCycle* 58, no. 2 (2017): 24–25.

Chapter Thirteen: Compost Site Management

1. María Teresa Barral and Remigio Paradelo, "A Review on the Use of Phytotoxicity as a Compost Quality Indicator," *Dynamic Soil, Dynamic Plant* 5, special issue no. 2 (2011): 37.
2. "Test No. 208: Terrestrial Plant Test: Seedling Emergence and Seedling Growth Test," in *OECD Guidelines for the Testing of Chemicals Section 2: Effects on Biotic Systems* (Paris: Organisation for Economic Cooperation and Development, 2006); Will Brinton, "Significance of Stability-Maturity Testing and Plant Bioassays to Assess Composts for Inclusion in Soil Building Projects," Woods End Laboratories, 2012. https://static1.squarespace.com/static/52ec31b2e4b04eb0bbd9c075/t/533b0243e4b00862b216a1f5/1396376131607/Significance+of+Stability-in-Compost.pdf.
3. Everett R. Emino and Phil R. Warman, "Biological Assay for Compost Quality," *Compost Science & Utilization* 12, no. 4 (2004): 342–48; Barral and Paradelo, "A Review on the Use of Phytotoxicity," 38.
4. *Methods Book for the Analysis of Compost* (Stuttgart, Germany: Bundesgütegemeinschaft Kompost e.V., 1994), 64. https://www.kompost.de/fileadmin/docs/shop/Grundlagen_GS/Methods_Book_2002.pdf.
5. Brinton, "Significance of Stability-Maturity Testing," 7.
6. Emino and Warman, "Biological Assay for Compost Quality," 348.
7. Barral and Paradelo, "A Review on the Use of Phytotoxicity," 38.
8. WA-DOE, *Bioassay Test for Herbicide Residues in Compost*.
9. M. Fauci et al., "Development of Plant Bioassay to Detect Herbicide Contamination of Compost at or Below Practical Analytical Detection Limits," *Bulletin of Environmental Contamination and Toxicology* 68 (2002): 79, doi: https://doi-org.ezp-prod1.hul.harvard.edu/10.1007/s00128-001-0222-8.
10. W. F. Brinton, *Compost Quality Standards & Guidelines*, prepared for New York State Association of Recyclers (Mt. Vernon, ME: Woods End Research Laboratory, 2000). http://compost.css.cornell.edu/Brinton.pdf.
11. Jayanta Sinha et al., "Efficacy of Vermicompost Against Fertilizers on Cicer and Pisum and on Population Diversity of N_2 Fixing Bacteria," *Journal of Environmental Biology* 31, no. 3 (2010): 287–92.

Chapter Fourteen: Compost End Uses and Markets

1. Craig G. Cogger, "Potential Compost Benefits for Restoration of Soils Disturbed by Urban Development," *Compost Science & Utilization* 13, no. 4 (2005): 247.
2. Peter Gorrie, "Compost Use Trials on Ontario Farms," *BioCycle* 56, no. 5 (2015): 39.
3. Emmanuel Arthur et al., "Compost Amendment to Sandy Soil Affects Soil Properties and Greenhouse Tomato Productivity," *Compost Science & Utilization* 20, no. 4 (2012): 215–21.
4. Peter Gorrie, "Compost Use on Perennial Fruit Crops," *BioCycle* 57, no. 3 (2016): 66.
5. Abigail A. Maynard, "Low Rates of Compost Increase Vegetable Yields," *Biocycle* 46, no. 11 (2005): 48.
6. Eric S. Gale et al., "Estimating Plant-Available Nitrogen Release from Manures, Composts, and Specialty Products," *Journal of Environmental Quality* 35, no. 6 (2006): 2330.
7. Gale et al., "Estimating Plant-Available Nitrogen," 2328.
8. Gorrie, "Compost Use Trials on Ontario Farms," 39.
9. Abigail A. Maynard, "Applying Leaf Compost to Reduce Fertilizer Use in Tomato Production," *Compost Science & Utilization* 8, no. 3 (2000): 208.
10. Yitzhak Hadar and Kalliope K. Papadopoulou, "Suppressive Composts: Microbial Ecology Links Between Abiotic Environments and Healthy Plants," *Annual Review of Phytopathology* 50, no. 1 (2012): 133–53.
11. Hadar and Papadopoulou, "Suppressive Composts," 134.
12. Hadar and Papadopoulou, "Suppressive Composts," 134; "Suppression of Soilborne Fungal Diseases with Organic Amendments," *Journal of Plant Pathology* 89, no. 3 (2007): 325–40.
13. A. J. Termorshuizen et al., "Suppressiveness of 18 Composts Against 7 Pathosystems: Variability in Pathogen Response," *Soil Biology and Biochemistry* 38 no. 8 (2006): 2461–77.
14. Harry A. J. Hoitink, Matthew S. Krause, and David Y. Han, "Spectrum and Mechanisms of Plant Disease Control with Composts," in *Compost Utilization in Horticultural Cropping Systems*, ed. Peter J. Stoffella and Brian A. Kahn (Baton Rouge: CRC Press, 2001), 263.

15. Gorrie, "Compost Use on Perennial Fruit Crops," 66.
16. R. Khaleel et al., "Changes in Soil Physical Properties Due to Organic Waste Applications: A Review," *Journal of Environmental Quality* 10 (1981): 133.
17. Khaleel et al., "Changes in Soil," 133.
18. M. S. Eden et al., "Organic Waste Recycling in Agriculture and Related Effects on Soil Water Retention and Plant Available Water: A Review," *Agronomy for Sustainable Development* 37 (2017): 11, doi: https://doi-org.ezp-prod1.hul.harvard.edu/10.1007/s13593-017-0419-9.
19. Pavel Zemánek, "Evaluation of Compost Influence on Soil Water Retention," *Acta Universitatis Agriculturae et Silviculturae Mendelianae Brunensis* 59, no. 3 (2011): 227–32.
20. Doug Collins et al., "Commercial Compost Application on Western Washington Farms," *BioCycle* 57, no. 3 (2016): 63.
21. A. M. Thompson et al., "Physical and Hydraulic Properties of Engineered Soil Media for Bioretention Basins," *Transactions of the ASABE* 51, no. 2 (2008): abstract.
22. Nicholas C. Olson et al., "Remediation to Improve Infiltration into COMPACT Soils," *Journal of Environmental Management* 117 (2013): 95.
23. S. Jaiarree et al., "Carbon Budget and Sequestration Potential in a Sandy Soil Treated with Compost," *Land Degradation and Development* 25 (2014): 128, doi: https://doi.org/10.1002/ldr.1152.
24. Mariangela Diacono and Francesco Montemurro, "Long-Term Effects of Organic Amendments on Soil Fertility: A Review," *Agronomy for Sustainable Development* 30, no. 2 (2010): 418, doi: https://doi-org.ezp-prod1.hul.harvard.edu/10.1051/agro/2009040.
25. Hanne Steel et al., "Nematode Communities and Macronutrients in Composts and Compost-Amended Soils as Affected by Feedstock Composition," *Applied Soil Ecology* 61 (2012): 100.
26. Will Brinton, "Significance of Stability-Maturity Testing and Plant Bioassays to Assess Composts for Inclusion in Soil Building Projects," Woods End Laboratories, 2012. https://static1.squarespace.com/static/52ec31b2e4b04eb0bbd9c075/t/533b0243e4b00862b216a1f5/1396376131607/Significance+of+Stability-in-Compost.pdf.
27. Randy Dodson, "Optimum Compost Application for Raised Bed Organic Vegetables" (academic thesis, Tennessee Technological University, 2011), 51.
28. Fred Magdoff and Harold Van Es, *Building Soils for Better Crops: Sustainable Soil Management* (College Park, MD: Sustainable Agriculture Research and Education, 2010), 31. http://www.sare.org/content/download/841/6675/Building_Soils_For_Better_Crops.pdf?inlinedownload=1.
29. Jenny Hill et al., "Comparisons of Extensive Green Roof Media in Southern Ontario," *Ecological Engineering* 94 (2016): 419.
30. Gale et al., "Estimating Plant-Available Nitrogen," 2330.
31. Jeffrey B. Endelman, Jennifer R. Reeve, and Daniel T. Drost, "A New Decay Series for Organic Crop Production," *Agronomy Journal* 102, no. 2 (2010): 457.
32. Gregory Evanylo et al., "Soil and Water Environmental Effects of Fertilizer-, Manure-, and Compost-Based Fertility Practices in an Organic Vegetable Cropping System," *Agriculture, Ecosystems and Environment* 127, no. 1 (2008): 50–58.
33. Stephen R. Carpenter, "Phosphorus Control Is Critical to Mitigating Eutrophication," *Proceedings of the National Academy of Sciences of the United States of America* 105, no. 32 (2008), doi: https://doi.org/10.1073/pnas.0806112105.
34. *WSU Compost Outreach Project: Recommended Best Management Practices for Compost Use (Working Draft)* (Everett, WA: WSU Snohomish County Extension, 2016), 2. http://extension.wsu.edu/snohomish/wp-content/uploads/sites/7/2016/01/2016-WSU-Recommended-Compost-BMP-guide-for-farmers-NEW-_1_22_16-Copy.pdf.
35. Robert Kourik, *Roots Demystified* (Occidental, CA: Metamorphic Press, 2008), 75.
36. Kourik, *Roots Demystified*, 75–77.
37. "Making Your Compost Product Work for You," Northeast Recycling Council, March 2010. https://nerc.org/documents/compost_marketing/making_your_compost_product_work_for_you.pdf.
38. Janet Wallace, "Perfecting the Potting Mix," *The Canadian Organic Grower* (2012): 54. https://cdn.dal.ca/content/dam/dalhousie/pdf/faculty/agriculture/oacc/en/tcog/TCOG_2012_Potting_Mix.pdf; Eliot Coleman, *The New Organic Grower* (White River Junction, VT: Chelsea Green Publishing, 1995), 137–41; Alan Toogood, *The American Horticultural Society: Plant Propagation* (New York: DK Publishing, 1999), 34.
39. Toogood, *The American Horticultural Society: Plant Propagation*, 33.
40. Hudson T. Hartmann et al., *Hartmann and Kester's Plant Propagation: Principles and Practices*, 7th ed. (Upper Saddle River, NJ: Prentice Hall, 2001), 267.
41. Wallace, "Perfecting the Potting Mix," 52.
42. Coleman, *Organic Grower*, 138–41.
43. Charlie Bayrer, personal communication to the author.

44. Jayanta Sinha et al., "Efficacy of Vermicompost Against Fertilizers on Cicer and Pisum and on Population Diversity of N_2 Fixing Bacteria," *Journal of Environmental Biology* 31, no. 3 (2010): 287–92.
45. David McDonald et al., "Building Soils for Storm Water Compliance and Successful Landscapes," *BioCycle* 48, no. 3 (2007): 48. https://www.biocycle.net/2007/03/23/building-soils-for-storm-water-compliance-and-successful-landscapes.
46. Hadar and Papadopoulou, *Suppressive Composts*, 136.
47. Kiyohiko Nakasaki et al., "A New Operation for Producing Disease-Suppressive Compost from Grass Clippings," *Applied and Environmental Microbiology* 64, no. 10 (1998): 4015.

INDEX

Page numbers followed by f refer to figures. Those followed by t refer to tables.

A

C

F

T

ABOUT THE AUTHOR

Crystal Parsons Nunes

James McSweeney is a composting consultant and educator. Through his work at the Highfields Center for Composting and current consultancy, Compost Technical Services, James has worked with hundreds of composters, large and small, on everything from site planning, design, and management to compost heat recovery and livestock feeding systems. He coauthored *Growing Local Fertility: A Guide to Community Composting* and has been an ardent proponent and collaborator in the community composting movement in the United States. With a background in agroecology and permaculture, restoring ecological integrity to our local farm and food systems is at the heart of James's work. He lives in Massachusetts with his wife, Amanda, and three kids.

the politics and practice of sustainable living

CHELSEA GREEN PUBLISHING

Chelsea Green Publishing sees books as tools for effecting cultural change and seeks to empower citizens to participate in reclaiming our global commons and become its impassioned stewards. If you enjoyed *Community-Scale Composting Systems*, please consider these other great books related to food systems and organic recycling.

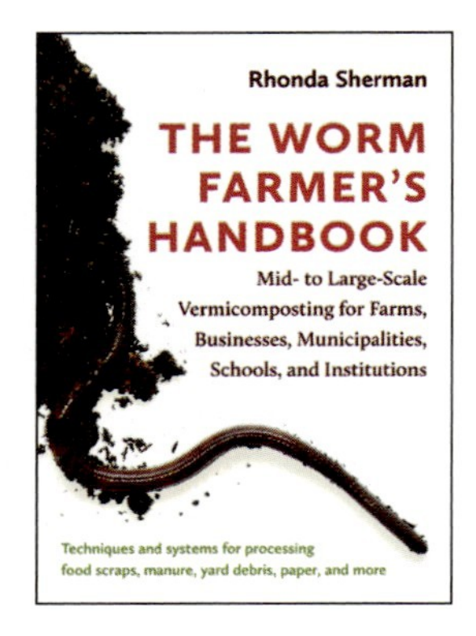

THE WORM FARMER'S HANDBOOK
Mid- to Large-Scale Vermicomposting for Farms, Businesses, Municipalities, Schools, and Institutions
RHONDA SHERMAN
9781603587792
Paperback • $29.95

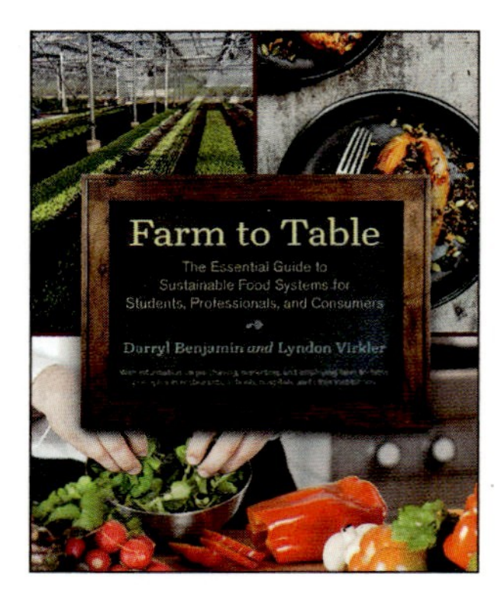

FARM TO TABLE
The Essential Guide to Sustainable Food Systems for Students, Professionals, and Consumers
DARRYL BENJAMIN and LYNDON VIRKLER
9781603586726
Hardcover • $49.95

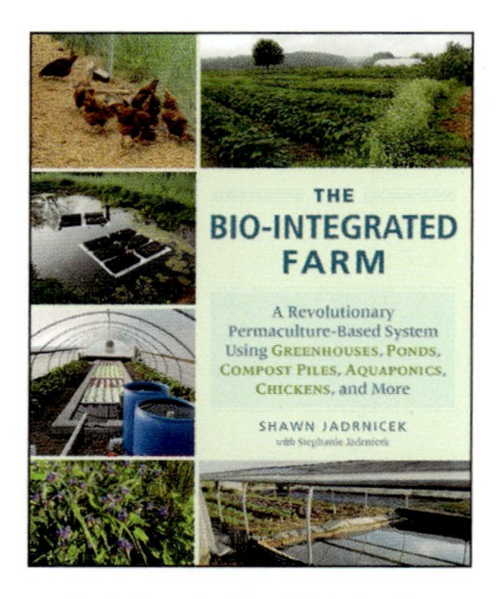

THE BIO-INTEGRATED FARM
A Revolutionary Permaculture-Based System Using Greenhouses, Ponds, Compost Piles, Aquaponics, Chickens, and More
SHAWN JADRNICEK with STEPHANIE JADRNICEK
9781603585880
Paperback • $39.95

ORGANIC MUSHROOM FARMING AND MYCOREMEDIATION
Simple to Advanced and Experimental Techniques for Indoor and Outdoor Cultivation
TRADD COTTER
9781603584555
Paperback • $39.95

For more information or to request a catalog, visit **www.chelseagreen.com** or call toll-free **(800) 639-4099**.

DATE DUE
JUL 09
DISCARDED